Early Main Group Metal Catalysis

Early Main Group Metal Catalysis

Concepts and Reactions

Edited by
Sjoerd Harder

WILEY-VCH

Editor

Prof. Sjoerd Harder
Friedrich-Alexander University
Inorganic and Organometallic Chemistry
Egerlandstr. 1
91058 Erlangen
Germany

Cover

Library of Congress Card No.:
applied for

British Library Cataloguing-in-Publication Data
A catalogue record for this book is available from the British Library.

Bibliographic information published by the Deutsche Nationalbibliothek
The Deutsche Nationalbibliothek lists this publication in the Deutsche Nationalbibliografie; detailed bibliographic data are available on the Internet at <http://dnb.d-nb.de>.

Print ISBN: 978-3-527-34448-2
ePDF ISBN: 978-3-527-81801-3
ePub ISBN: 978-3-527-81803-7
oBook ISBN: 978-3-527-81802-0

Cover Design Tata Consulting Services
Typesetting SPi Global, Chennai, India
Printing and Binding Markono Print Media Pte Ltd, Singapore

Printed on acid-free paper

10 9 8 7 6 5 4 3 2 1

Tá an leabhar seo tiomnaithe do ghrá mo chroí Nollaig agus Darragh James

Contents

Preface

I remember having mixed feelings when Wiley-VCH approached me about editing a book on early main group metal catalysis. On the one hand, I was very excited, because the well-known blue/white Wiley-VCH books stand for quality in scientific publishing. An offer like this meant that the very young and emerging field of s-block metal catalysis has now matured to a discipline of its own. On the other hand, I knew it would be a tough and time-consuming challenge. Currently scientists are rated on successful research grants and the number and impact of scientific papers and therefore it would be hard to find committed authors.

Excitement apparently won from worries about stony roads. The book in your hands is the first monograph fully devoted to catalysis with early main group metals. It is a testimony to the very rapid developments in a field that is normally dominated by precious transition metals. When I started doing chemistry with the more exotic alkaline earth metals Ca, Sr, and Ba in the middle 1990s, I never could have dreamt that these metals catalyze stereoselective alkene polymerization or hydroamination, hydrosilylation, and even alkene hydrogenation. Transition-metal-free hydrogenation breaks the dogma that partially filled d-orbitals are needed for C=C bond activation. Now, more than 20 years later, the urge to replace precious metals by abundant elements is more critical than ever. In view of an exploding world population and increasing demands to share our rising welfare, thoughts about our planet's resources are a moral imperative. Sustainable chemistry not only needs renewable resources but also "green" catalysts based on metals that are not threatened to be depleted soon. Therefore, this publication could not have been launched more timely.

This single book does not have pretensions to be comprehensive – a series would be needed instead. It will, however, cover the most important aspects of early main group metal catalysis. It aims at not only specialists in the field but also beginners. The latter group will profit from the introductory chapter, which discusses the bare fundamentals of s-block chemistry and the working principles of its catalysts. Each of the chapters that follow deals with a specific topic. They aim to be educational rather than encyclopedic, describing major breakthroughs while emphasizing differences between early main group and transition metal catalysis. I am greatly indebted to the many authors who contributed to these specialist chapters. These experts in the field wrote from a personal perspective and their different styles give this book the vibrancy that typifies this field of research.

Although it is clear that s-block metal catalysis still has many shortcomings, there are nevertheless several interesting advantages that make it worthwhile to follow this path. During one of my visits to the TU Berlin, the late chemist Herbert Schumann professed that the ultimate goal should be to do catalysis with those elements you find in your backyard. I hope that this collection of reviews in early main group metal catalysis will inspire a larger group of researchers to join the club and look forward to the many future breakthroughs in this field!

Sjoerd Harder

May 2019
Germany

Friedrich-Alexander-Universität
Erlangen-Nürnberg

1

Introduction to Early Main Group Organometallic Chemistry and Catalysis

Sjoerd Harder

University Erlangen-Nürnberg, Inorganic and Organometallic Chemistry, Egerlandstrasse 1, 91058 Erlangen, Germany

1.1 Introduction

Although organometallic complexes of the early main groups are well known for their very high reactivity and challenging isolation, they are among the first studied during the pioneering beginnings of the field. Their high nucleophilicity and Brønsted basicity have made them to what they are today: strong polar reagents that are indispensible in modern organic synthesis. It is exactly this high reactivity that has made them potent catalysts for organic transformations that are generally catalyzed by transition metal complexes. Despite their lack of partially filled d-orbitals and their inability to switch reversibly between oxidation states, the scope of their application in catalysis is astounding. As the early main group metal catalysis only started to become popular since the beginning of this century, it is still a young field with ample opportunities for further development. This introductory chapter is specifically written for new graduate students in the field. It gives a very compact overview of the history of early main group organometallic chemistry, synthetic methods, bonding and structures, analytical methods, solution dynamics, and some preliminary low-valent chemistry. This forms the basis for understanding their use in catalysis for which the basic steps are described in the second part of this chapter. For further in-depth information, the reader is referred to the individual chapters in this book.

1.2 s-Block Organometallics

1.2.1 Short History

The organometallic chemistry of the highly electropositive early main group metals could not have started without the isolation of these elements in the metallic state. Being only available in nature in the form of their salts, the invention of electricity and electrolysis in the beginning of 1800s has been the key to their isolation. The legendary Humphry Davy (1778–1829) can claim the

Early Main Group Metal Catalysis: Concepts and Reactions,
First Edition. Edited by Sjoerd Harder.

discovery of at least seven s-block metals: the alkali metals Li, Na, and K and the alkaline earth (Ae) metals Mg, Ca, Sr, and Ba.

Most organometallic classes start with Frankland's well-known synthesis of Et_2Zn in 1847 [1] by reaction of Zn and EtI, which was originally an attempt to isolate the Et radical. It is, however, less known that before this experiment, Frankland tried to isolate the Et radical by mixing K and EtI, a reaction that was found to be very violent producing a variety of gaseous products [2]. Wanklyn reacted Frankland's Et_2Zn with Na and isolated the zincate $Na^+ZnEt_3^-$, which, based on its very high reactivity, was described as a solution of EtNa in Et_2Zn. Although not metal pure, this is likely the first preparation of an alkylsodium compound. Numerous early organometallic pioneers attempted to prepare organosodium complexes directly by reaction of the metal with R–X (R = alkyl or aryl; X = Cl, Br, and I) but never isolated the organosodium products that were found to be fleeting intermediates to Wurtz (or Wurtz–Fittig) R–R coupling products.

Switching to the less reactive metal Li, Wilhelm Schlenk was the first to isolate group 1 alkylmetal complexes in a pure form. Reduction of Me_2Hg with Li metal gives metallic mercury and MeLi as a white powder. Schlenk describes the pyrophoric nature of this powder in beautiful words, clearly demonstrating his fascination for these compounds [3]. Although Schlenk is not just the "man behind the flask" [4] but certainly also a major pioneer in organometallic chemistry of the alkali metals, it was Karl Ziegler who developed the simple mercury-free route to organolithium reagents [5], which is the method of choice even today for the synthesis of *n*BuLi: $n\text{BuCl} + 2\text{Li}^0 \rightarrow n\text{BuLi} + \text{LiCl}$. Because alkyllithium reagents are less reactive than Na or K reagents, Wurtz coupling is limited. Also, their much higher solubility in apolar organic solvents added to their successful isolation. The latter strong alkyllithium base is not only the starting point for modern organolithium chemistry but also for the development of superbases based on the heavier alkali metals.

Grignard developed organometallic chemistry of group 2 metals and reported in 1900 a similar protocol to prepare organomagnesium reagents: $\text{R–X} + \text{Mg} \rightarrow \text{RMgX}$ [6]. Foreseeing a great future for these potent reagents, Grignard was awarded the Nobel Prize for this milestone discovery already in 1912. Similar to group 1 metals, the development of the organometallic chemistry of the heavier group 2 metals was found to be more challenging. Beckmann mentioned already in 1905 the first synthetic routes to arylcalcium halides [7] but Gilman had problems reproducing these results [8]. Development of the organometallic chemistry of the heavier Ae metals turned out to be substantially more difficult than simply reproducing the Grignard method with Ca instead of Mg.

1.2.2 Synthesis of Group 1 Organometallics

A short overview of synthetic methods to pure organoalkali metal compounds is shown in Scheme 1.1 (for comprehensive reviews, see [9, 10]). The synthesis of *n*BuLi directly from its metal and *n*BuCl stands central to the further development of lithium chemistry. The highly Brønsted basic $n\text{Bu}^-$ anion (approximate pK_a of CH_3 in butane ≈ 50 [11, 12]) is able to deprotonate a large variety of

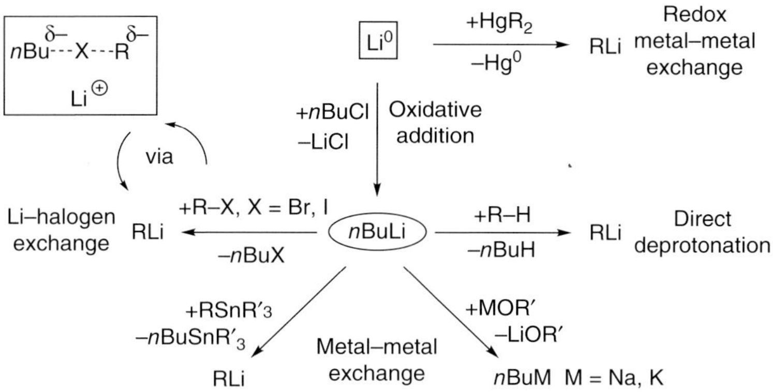

Scheme 1.1 Overview of common synthetic routes to organolithium compounds.

organic substrates with pK_a's up to 45 (e.g. benzene) provided the *n*BuLi reagent is sufficiently activated by polar cosolvents (see Section 1.2.3). For problematic deprotonation reactions, the even stronger base *t*BuLi is routinely used (approximate pK_a of CH in 2-methylpropane ≈ 52 [11, 12]). Direct substrate deprotonation with alkyllithium reagents has the great advantage that the alkane side product is volatile and easily removed. Its disadvantage is that the selectivity of product formation is controlled by the substrate's acidity. In some cases, however, complex formation between substrate and alkyllithium reagent controls the selectivity of deprotonation by the complex-induced-proximity-effect (CIPE) [13] influencing the kinetics and thermodynamics of product formation. The well-known Li–halogen exchange route allows exact regiocontrol of the deprotonation reaction. In polar solvents, this conversion is even extremely fast at −80 °C. The mechanism proceeds through a hypervalent C–X–*n*Bu intermediate because only heavier halogens (X = Br and I) can be used. Also, for these reactions, it holds that the product carbanion should be more stable than the *n*Bu$^-$ anion. For the identity exchange reaction, $C_6F_5I + C_6F_5Li$, the hypervalent intermediate $[Li^+\cdot(TMEDA)_2][C_6F_5{-}I{-}C_6F_5{}^-]$, has been structurally characterized [14]. The Li–halogen exchange is limited to lithiations of sp^2 C atoms because sp^3 C–X groups are prone to S_N2 substitution. The product isolation is sometimes complicated by the subsequent S_N2 reaction of the organolithium product with the side product *n*BuX. Another method for selective lithiation is Li–Sn exchange, which proceeds through a hypervalent intermediate with retention at Sn and C [15].

Reagents based on the heavier alkali metals Na and K cannot be prepared by the direct reaction of an organohalide with the metal because of Wurtz coupling. A simple route for the preparation of the highly reactive superbases *n*BuNa and *n*BuK consists of mixing *n*BuLi with the higher metal alkoxide *t*BuOM (M = Na and K) [16]. This metal exchange reaction is based on the hard-soft-acid-base (HSAB) principle: the smaller (harder) Li^+ prefers to interact with the smaller (harder) O^- of the alkoxide, whereas the larger (softer) Na^+ and K^+ prefer interaction with the larger (softer) C^- in the carbanion. As the early pioneers in this area, Lochmann and Schlosser [17], disputed this invention, these reagents are known as Lochmann or Schlosser bases. However, their roots go back much further when Morton and Chester reported the highly basic properties of RNa/LiOR mixtures [18]. The more correct naming of these mixtures as “superbases” reveals that these reagents are indeed very powerful deprotonation reagents. Their very high reactivity originates from their longer (and weaker) bonds, which facilitate the kinetics of the bond-breaking and bond-making processes. Use of heterobimetallic reagents consisting of an alkalimetal and Mg, Zn, or Al brought the synthetic utility of alkali metal reagents even to a higher level [19]. Template-controlled interaction with the substrate results in deprotonation at unusually mild conditions or with unusual regioselectivities and often polydeprotonations can be achieved.

1.2.3 Synthesis of Group 2 Organometallics

With few exceptions, preparative methods for the organometallic complexes of group 2 were, for a long time, limited to Mg Grignard reagents [20]. Although it is suggested that health risks concerned with Be chemistry can be mastered [21], there are still not many practitioners in this area. The interest in the organometallic chemistry of the heavier metals Ca, Sr, and Ba, however, is rapidly growing because of their application in MOCVD (metal-organic chemical vapor deposition) [22] and homogeneous catalysis [23]. As mentioned previously, organocalcium complexes cannot be prepared by simply replacing Mg for Ca in the Grignard synthesis (Scheme 1.2). Similar to organosodium or potassium compounds, the much more potent “Ca Grignards” may also react with the organohalide to Wurtz-type products. Because nucleophilic aromatic substitution is more difficult, arylcalcium Grignard reagents (ArCaI) could be isolated but only under controlled conditions, which involve low temperature and special activation methods for the Ca metal used [24]. For the magnesium Grignard, ethereal solvents must be used; however, ArCaI is much more reactive

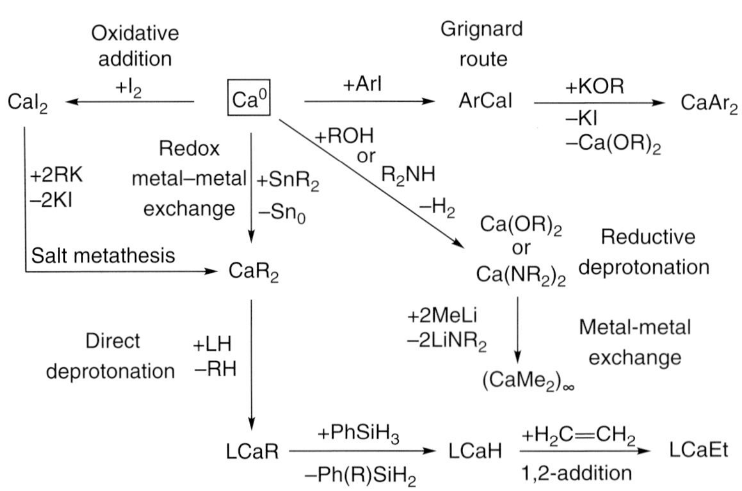

Scheme 1.2 Overview of common synthetic routes to organocalcium compounds.

that it is easily decomposed by ether deprotonation. The temperature during the synthesis is generally maintained below −40 °C, but the Ca Grignard is more stable as thought initially: *para*-tolylcalcium iodide in tetrahydrofuran (THF) has a half-life time of eight days. The arylcalcium complex is more stable in tetrahydropyran (THP) in which the half-life time is increased to 17 days.

Similar to the original magnesium Grignard, Ca Grignard reagents are also susceptible to the Schlenk equilibrium [25]: $2ArCaI \rightleftharpoons Ar_2Ca + CaI_2$. The synthesis of homoleptic organomagnesium reagents can be simply achieved by addition of dioxane to the RMgX solution, which results in immediate precipitation of MgX_2·(dioxane), leaving R_2Mg in solution. Although this procedure does not work for heavier Ca Grignards, Westerhausen and coworkers reported an elegant procedure that involves addition of KO*t*Bu [26]. This results in conversion of ArCaI into insoluble KI and ArCaO*t*Bu, which after ligand exchange leads to precipitation of $[Ca(OtBu)_2]_\infty$, leaving Ar_2Ca in the solution.

The very high reactivity of arylcalcium reagents may be exploited in specialty applications but inherently also makes these complexes highly sensitive toward decomposition by air and/or solvents. Therefore, the most popular Ca reagents are to some extent stabilized by bulky and/or electron-withdrawing groups. Westerhausen and coworkers reported the syntheses of the whole range of bis-trimethylsilylamide complexes $Ae[N(SiMe_3)_2]_2$ (Mg, Ca, Sr, and Ba) by reduction of the Sn(II) complex $Sn[N(SiMe_3)_2]_2$ with Ae^0 [27]. These homoleptic amide complexes, abbreviated by AeN''_2, are likely the most widely used starting reagents in heavier Ae metal chemistry. They can also be prepared by salt metathesis, reacting AeI_2 with 2 equiv of KN″ [28]. Generally, metal iodide salts are chosen as the precursor on account of their better solubility while potassium reagents are favored because of complete insolubility of KI in ethereal solvents. A major disadvantage of the salt metathesis route is the possible formation of "ate" complexes, $[K^+][AeN''^-_3]$, a side product that cannot always be detected by nuclear magnetic resonance (NMR) even in considerable quantities [29]. The other disadvantage of the salt metathesis route is the fact that ethereal solvents have to be used and therefore complexes are always isolated in the form of their etherates. The Sn route can be performed in aromatic solvents giving solvent-free, metal-pure, products AeN''_2, but sometimes, the separation of the very fine Sn particles can be problematic.

The salt metathesis route is probably the most popular pathway to homoleptic complexes and also the procedure of choice for the synthesis of a range of dibenzylcalcium complexes directly from CaI_2 and the benzylpotassium reagent. Complex **1**, which can be obtained in the crystalline pure form in large quantities, is the first example of a heavier Ae metal complex with a benzyl ligand [30]. Although stabilized by an α-$SiMe_3$ substituent and by *ortho*-Me_2N–Ae coordination, it is sufficiently reactive to deprotonate a large variety of substrates while being stable enough for convenient handling and long-term storage. In addition, at higher temperatures, it is also fully stable in THF. Replacing the stabilizing α-$SiMe_3$ substituent with an electron-releasing Me group increases the reactivity significantly (**2**) [31]. At a later stage, it was found that the stabilizing *ortho*-Me_2N-substituent can also be removed (**3**) [32]. In contrast to **1** and **2**, $CaBn_2$·$(THF)_4$ (**3**) can be freed from THF under high vacuum. The more reactive Sr analog of **1** is also accessible by the salt metathesis route [33].

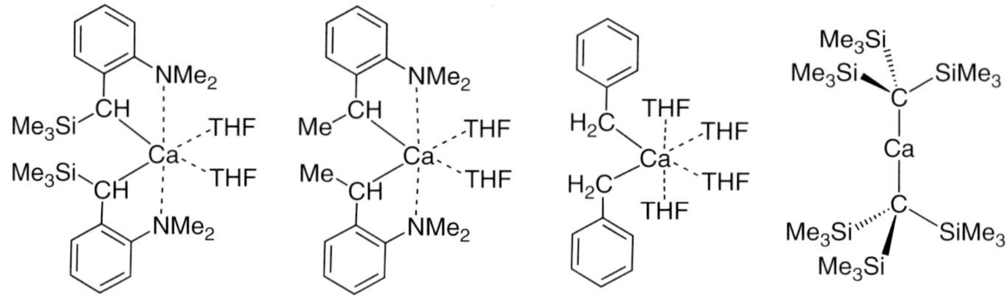

1 **2** **3** **4**

Before this work, dibenzylbarium was prepared in a relatively pure form by a metal–metal exchange method reminiscent of the synthetic protocol for *n*BuNa or *n*BuK: reaction of benzyllithium with BaN″$_2$·(THF)$_2$ resulted in a precipitate of dibenzylbarium and soluble LiN″ [34]. Izod and Waddell reported the syntheses of $[(SrBn_2)_2·(THF)_3]_\infty$ and $[(BaBn_2)_3·(THF)_4]_\infty$ via salt metathesis [35]. While **3** was crystallized as a molecular complex, the Sr and Ba analogs form polymeric ribbon and sheet structures, respectively. The Okuda and coworkers extended the set of reactive benzylcalcium complexes with the simple bis-allyl complex $Ca(C_3H_5)_2$ which was obtained solvent-free via salt metathesis from allyl potassium and CaI_2 [36].

The synthesis of simple, unstabilized, alkylcalcium complexes was found to be a real challenge. Alkylcalcium complex **4** was prepared by salt metathesis but is hardly reactive on account of steric protection and electronic stabilization of the carbanion center by Me_3Si-substituents [37]. The bis-Me_3Si-substituted alkyl complex $Ca[CH(SiMe_3)_2]_2·(dioxane)_2$ was prepared by metal vapor synthesis from Ca^0 and $BrCH(SiMe_3)_2$ [38]. Much later, Hill and coworkers prepared $Ae[CH(SiMe_3)_2]_2·(THF)_3$ (Ae = Ca, Sr, and Ba) by the more convenient salt metathesis route [39]. Since the Ph substituent in benzyl complexes stabilizes a carbanion to a similar extent as a Me_3Si-substituent [40], $Ca[CH(SiMe_3)_2]_2·(THF)_3$ should have a similar reactivity as **1**. Westerhausen and coworkers isolated the Ca Grignard $(Me_3SiCH_2)CaI·(THP)_3$ in which the carbanion is stabilized by only one Me_3Si-substituent [41]. Homoleptic $(Me_3SiCH_2)_2Ca·(THP)_4$ was obtained by subsequent salt metathesis with KCH_2SiMe_3 [42]. The THP solvent was used for increased stability. In THF the complex has a half-life time of c. four hours, clearly demonstrating its increased reactivity.

The synthesis of "true" (unstabilized) alkyl complexes of the heavier Ae metals has only been achieved recently. Because this class of compounds is extremely reactive toward polar solvents such as THF, an ether-free synthetic protocol is crucial. Harder and coworkers attempted to prepare unstabilized alkylcalcium complexes by addition of a THF-free amidinate calcium hydride complex to the highly polarized C=C bond in a N-heterocyclic olefin (NHO, Scheme 1.3a). However, instead of isolating an alkylcalcium complex, the

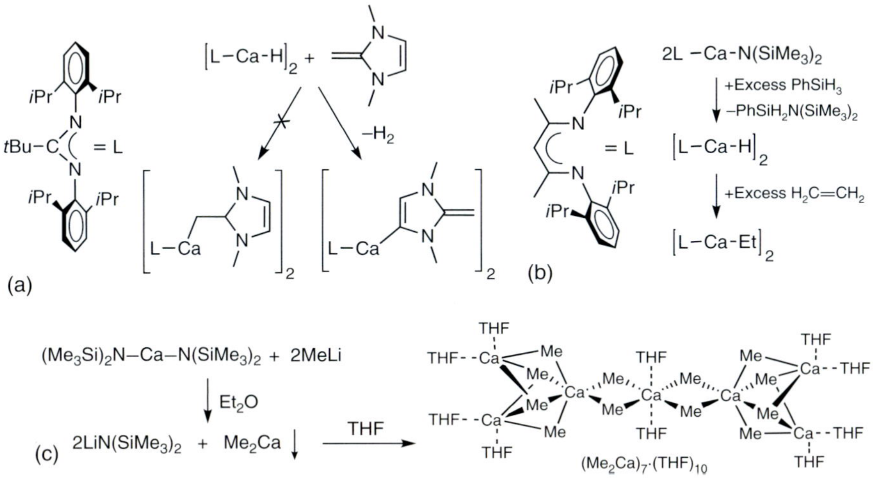

Scheme 1.3 (a) Attempted synthesis of an unstabilized Ca alkyl species. (b) Synthesis of a THF-free Ca hydride complex and further conversion to an ethylcalcium complex. (c) Synthesis of Me_2Ca and its crystal structure from THF.

NHO ligand was deprotonated in the backbone [43]. Attempted synthesis of a THF-free β-diketiminate calcium hydride complex failed because of ligand scrambling and formation of insoluble $(CaH_2)_\infty$ [44], but using excess of $PhSiH_3$, Hill and coworkers successfully isolated the THF-free calcium hydride complex, which formed a highly reactive ethylcalcium complex in reaction with ethylene (Scheme 1.3b) [45]. At the same time, Anwander and coworkers synthesized Me_2Ca by metal exchange between MeLi and $Ca[N(SiMe_3)_2]_2$ in Et_2O (Scheme 1.3c) [46]. The precipitate of $(Me_2Ca)_\infty$ dissolves in THF at low temperatures but also slowly decomposes. Crystals of the larger aggregate $(Me_2Ca)_7{\cdot}(THF)_{10}$ could be isolated.

The Harder group introduced a superbulky β-diketiminate ligand that stabilized a THF-free Sr hydride complex [47]. Reaction with ethylene also formed the first ethylstrontium complex upon polymerization (Scheme 1.4). The very

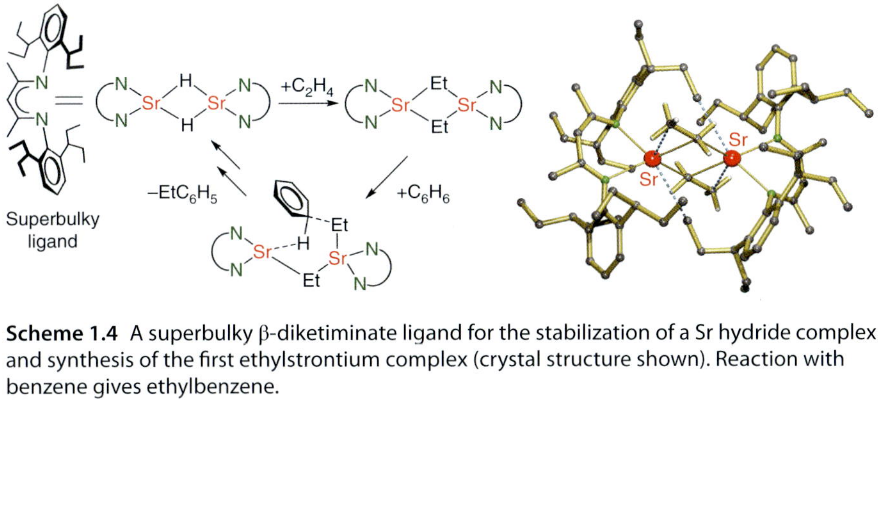

Scheme 1.4 A superbulky β-diketiminate ligand for the stabilization of a Sr hydride complex and synthesis of the first ethylstrontium complex (crystal structure shown). Reaction with benzene gives ethylbenzene.

high reactivity of this complex is demonstrated by its room temperature reaction with benzene, resulting in ethylbenzene and a Sr hydride complex (cf. the ethylcalcium complex shown in Scheme 1.3b reacts with benzene at 60 °C). This formally nucleophilic aromatic substitution likely proceeds through an unstabilized Meisenheimer anion. A similar reactivity observed for the Sr hydride complex is the basis for the Sr-catalyzed deuteration of benzene with D_2.

Apart from these standard methods, procedures that start with Ae metal itself gain popularity. There is some precedence for the direct metal vapor synthesis [38], but there is a lot of room for further development of this technique to activate Ae metals. Unactivated Ae metals may only react with more acidic substrates such as alcohols, but $BaN''_2 \cdot (THF)_2$ could also be obtained by reacting Ba metal and $HN(SiMe_3)_2$ in THF while bubbling through NH_3 gas to partially dissolve the metal [48]. Also, combinations of methods have been practiced. For example, reaction of Ph_2Hg with Ca and subsequent protolysis of Ph_2Ca with cyclopentadiene formed the calcocene [49]. This redox transmetallation protolysis (RTP) approach has the advantage of being a one-pot synthesis starting directly from the Ae metal. Highly poisonous mercury reagents have been replaced with the more benign Ph_3Bi redox reagent [50]. Recently, a simple method was reported in which the Grignard reaction of Ae and PhBr produces highly reactive but undefined "PhAeBr" intermediates (Ae = Mg–Ba), which are *in situ* reacted further with organometallic Ae compounds [51]. It is clear that these methods are limited to deprotonation of substrates with pK_a's that are significantly lower than that of benzene. For more comprehensive reviews on preparative group 2 metal organometallic chemistry, see Ref. [52].

1.2.4 Bonding and Structures of s-Block Organometallics

Some trends for the C–metal bond in early main group organometallics are listed in Scheme 1.5. Metal electronegativities decrease down the groups and the highest and lowest values are found for Be (1.57) and Cs (0.79), respectively. Because

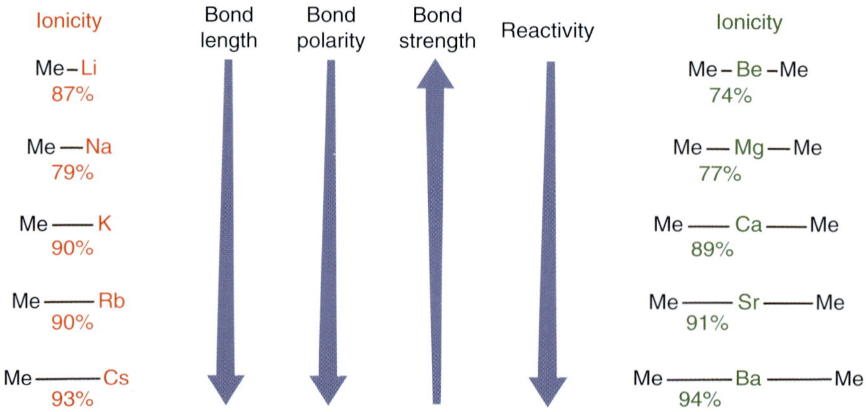

Scheme 1.5 Calculated ionicities of s-block metal–carbon bonds [53] and trends within the groups.

the polarity of the metal–C bond is defined by the electronegativity differences of these bonding partners, the Cs—C bond is among the most ionic one and the Be—C bond has a partial covalent character.

There has been a 25-year long debate on the nature of the C—Li bond. Pauling postulated an empirical formula that expresses its ionicity in terms of electronegativity differences [54], but depending on the scale of electronegativities used, values in the range of 25–50% are calculated. A considerable covalent character was supported by the observation of a substantial ^{13}C–^{7}Li NMR coupling constant of 14.5 Hz for MeLi in diethyl ether [55]. IR studies, however, suggested a considerable ionic character in MeLi based on low frequencies for the H—C—Li bonding [56]. Analyses of NMR chemical shift differences came to the opposite conclusion, estimating only 10% ionicity [55c].

Molecular orbital theory provides precise calculated atomic charges and therefore bond ionicity. Streitwieser described CH_3Li as having essentially no covalent character [57], an extreme viewpoint that was supported by the perfect description of the distorted cubic structure of its tetramer by an electrostatic model using only plus/minus point charges [58]. Lipscomb, however, considered the C—Li bond as being 60% ionic based on high-level calculations using configuration interaction (CI) methods [59]. The problem with atomic charge calculations is that it is not clear where to draw the borderline between C and Li and various methods and/or basis sets give strong contrasting results. The natural population analysis, however, is generally accepted as a method to calculate reliable atomic charges in polar organometallics and is much less sensitive to basis set differences than Mulliken analysis [60]. By the end of 1980s, most key players in the field agreed upon a 80–90% ionicity of the C—Li bond, depending on the organic residue [53].

The very polar nature of early main group metal complexes is the driving force for electrostatic association of monomeric species into larger aggregates. Thus, the simple abbreviation of an organolithium compound by "RLi" is fully inadequate when discussing structures and reactivities. Most student manuals explain the typical tetrameric $(RLi)_4$ or hexameric $(RLi)_6$ with covalent bonding models (Scheme 1.6) using electron-deficient four-center-two-electron models in order to explain bonding. In the light of the predominant ionic nature of the C—Li bond, these representations are far from realistic. These covalent

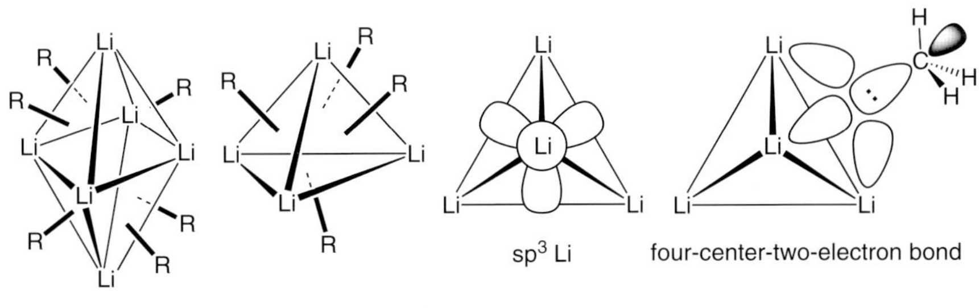

Scheme 1.6 Representations of hexameric and tetrameric organolithium aggregates. Covalent four-center-two-electron bonding model for $(MeLi)_4$ assuming sp^3-hybridized Li and C atoms.

bonding models also falsely suggest the existence of Li—Li bonds, which does not agree with a lack of Li–Li NMR coupling [55b]. From a didactical point of view, ionic bonding models are closer to reality. The polar molecule R^-Li^+ can be seen as a combination of plus and minus charges that interact according to Coulomb's law. Thus, a dimer is the most favorable electrostatic combination of two +/− dipoles for which the total bonding energy can be simply calculated by considering all attractive +/− and repulsive +/+ or −/− Coulomb energies. The distortions from a perfect cube structure for $(MeLi)_4$ have been explained with a fully electrostatic bonding model [58].

Even simpler back-of-a-beer-mat calculations, assuming perfect aggregates with a fixed distance between + and − charges, give a good estimation of the aggregation energy (Scheme 1.7) [61]. These clearly show that a ladder-like trimer is preferred over a cyclic ring and the cube is the most favorable structure for a tetrameric arrangement. The estimated aggregation energies increase with the aggregation number: monomer $-1/r$, dimer $-1.30/r$, trimer $-1.36/r$, tetramer $-1.46/r$, and hexamer $-1.56/r$. This indicates that larger aggregates

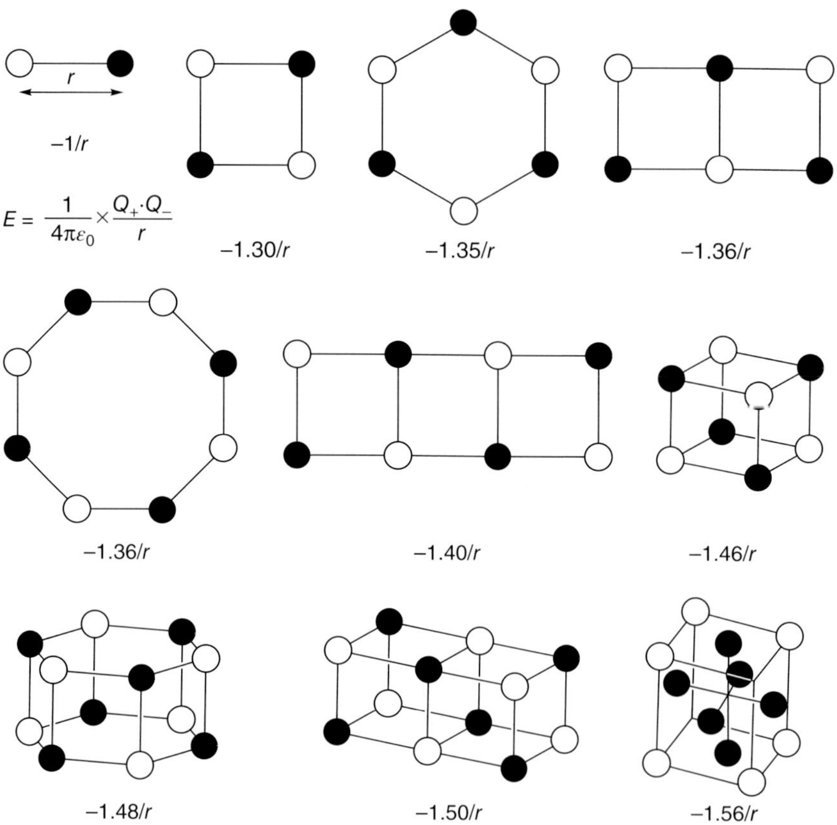

Scheme 1.7 Electrostatic bond energies for combinations of +/− dipoles at a distance of *r*. Following Coulomb's law, the intrinsic energy of one dipole is $-1/r$ au. The electrostatic bond energies for the aggregates (given per dipole) are calculated by considering all attractive (+/−) and repulsive (+/+ and −/−) interactions.

are always more stable than smaller ones. Calculation of an "endless" rock salt structure would converge to an aggregation energy of $-1.748/r$ per dipole, a number which is also known as the Madelung constant. The reason why MeLi is tetrameric and does not form a rock salt structure in the solid state is the nonspherical distribution of negative charge in the Me^- anion. Although the spherically symmetric Cl^- anion can interact with Na^+ in all directions, the hydrogen atoms in CH_3 limit its interaction with the neighboring aggregates, which explains the formation of discrete tetramers. However, in the solid state, there are also important interactions between $(MeLi)_4$ clusters: each Li corner interacts with the Me groups of neighbors with C···Li distances of c. 2.52 Å and vice versa (Scheme 1.8). Although only slightly longer than the C—Li bond of c. 2.28 Å within the tetramer, the interaggregate bonding is considerably weaker than the intra-aggregate bonding because the center of negative charge does not coincide with the C nucleus. The 3D interaggregate linking is still strong enough to make $(MeLi)_4$ fully insoluble in nonpolar solvents. In ethers, however, it dissolves well and maintains its tetrameric structure in which the Li corners are solvated by ether. Increasing the size of the carbanion also increases the interaggregate bonds. Ethyllithium crystallizes as tetrameric aggregates that are only interlinked in two dimensions. The 2D layer structure is shielded on top and bottom by the Et groups and the differences between the intra-aggregate C—Li bonds (2.31 Å) and interaggregate C···Li (2.53 Å) distances is similar to that in MeLi. Because of less extensive interlinking, EtLi is soluble in aromatic solvents but not in hexane. Increasing the size of the carbanion from Et^- to tBu^- fully blocks interaggregate interactions. Consequently, $(tBuLi)_4$ is very well soluble in pentane and even sublimes under vacuum. This clearly shows the enormous influence of nature and size of the carbanion on the structures of organolithium complexes and inherently on their physical properties. Because the cation–anion ratio is decisive, changing Li^+ for the larger cations Na^+ and K^+ will have the same effect as reducing the size of the carbanion. Consequently, Na and K complexes often show pronounced polymeric structures and are less soluble with increasing cation size.

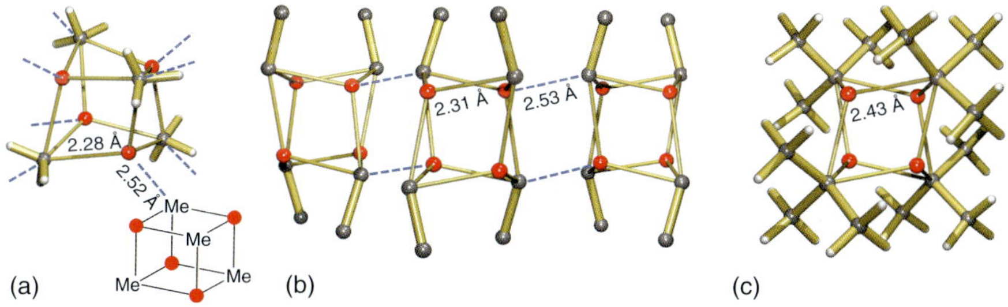

Scheme 1.8 Crystal structures of (a) $(MeLi)_4$ with 3D network, (b) $(EtLi)_4$ with 2D network (H atoms not shown). The 2D plane is shielded on both sides with ethyl groups. (c) $(tBuLi)_4$ is fully shielded.

Group 2 organometallic complexes follow the same bonding principles and their structures can also be thought of as electrostatic combinations of plus and minus charged ions. Bonding to the lightest metal in the group, however, has significant covalent contributions. This can be seen in the structures of Ae metal hydrides AeH_2. The heaviest hydrides (Ca, Sr, and Ba) feature ionic 3D salt structures with a $PbCl_2$ lattice (coordination numbers: Ae = 9, H = 4 and 5), and MgH_2 crystallizes with a rutile lattice (coordination numbers: Mg = 6, H = 3). In contrast, the structure of BeH_2 is often described as the linear polymer $[Be(\mu\text{-}H)_2]_\infty$ (coordination numbers: Be = 4, H = 2) in which Be is sp^3 hybridized and hydrogens are bonded through electron-deficient three-center-two-electron bonds, i.e. a structure with strong similarities to that of B_2H_6. More accurate studies, however, present a 3D network structure of corner sharing BeH_4-tetrahedra (**5**) [62].

Because group 2 organometallics are constituted of two shielding carbanions per metal center, there, is in contrast to group 1 metal chemistry, a much smaller tendency to aggregate to larger complexes. Consequently, the vast majority of all Ae metal complexes is present as a monomer. Dimeric aggregates with μ^2-bridging carbanions are also plentiful, but μ^3-bridging carbanions are rare and usually only found for complexes with smaller anions such as Me^- [46] or with larger Ba^{2+} cations [63].

The tendency of anions to bridge metal centers increases from $R_3C^- < R_2N^- < RO^-$. Amides are either terminally bound or tend to bridge in μ^2-fashion, which is clearly demonstrated by the structures of $(AeN''_2)_2$ that are dimeric over the whole range (Mg–Ba) (**6**) [48, 64]. There are many examples of Ae alkoxide complexes with μ^3-RO^- anions and also spherically symmetric halogenide anions display strong bridging tendencies that increase from I^- to F^-. The highest coordination numbers have been found for O^{2-} and H^- anions, which, as part of a complex, show up to μ^6-bridging [65, 66].

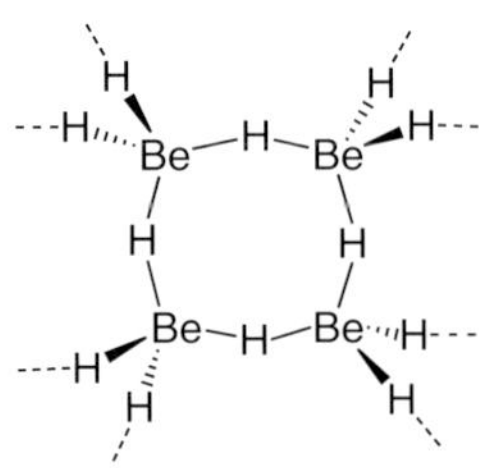

5

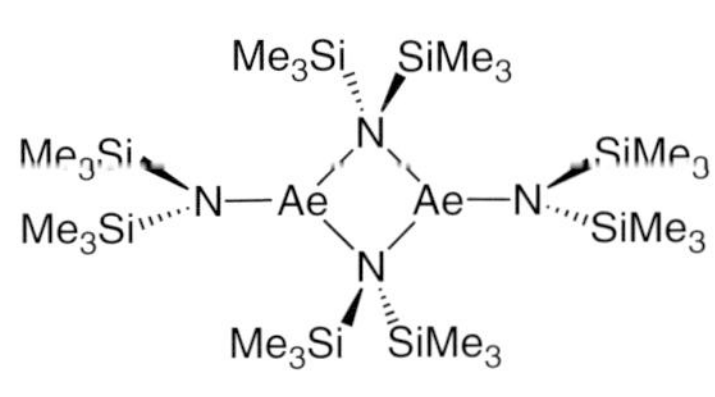

6 Ae = Mg, Ca, Sr, Ba

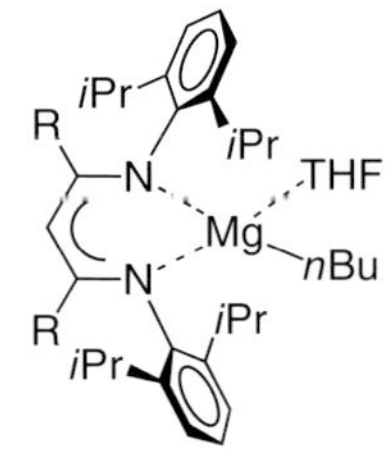

7a R = Me
7b R = *t*Bu

Aromatic groups (or other functionalities rich in π-electrons) also show great potential to involve in metal coordination. These ligands especially bind to the larger softer metal cations. Ba complexes often crystallize with coordinated benzene ligands (even despite the presence of THF) [63]. The importance of weak secondary interactions on structure and stability of Ae complexes are becoming increasingly clearer [67]. These interactions are electrostatic in nature and rely heavily on metal-induced polarization. Consequently, they are more pronounced for complexes with Ae^{2+} cations than for alkali metal complexes with a singly charged cation. Apart from interactions with π-electron density [68], Ae complexes often feature strong agostic metal···H–C interactions. Although the original definition of an agostic interaction is specifically related to intramolecular three-center-two-electron M—H—C bonding with transition metals in which d(M) → σ*(C–H) back-donation is part of the bonding description, the wording for early main group metals is arguably not correctly chosen [69]. There are indeed clear differences in geometry or effects on NMR chemical shifts, and therefore for s-block metals, the new definition "anagostic interaction" would be more appropriate. Because early main group metals do not possess partially filled d-orbitals, the anagostic interaction is of different nature and best described by polarization of a C—H bond (or CH_3 group) by the positively charged Ae^{2+} cation, creating an attractive electrostatic force between the positive point charge and the induced dipole. Especially, $Ae^{2+}\cdots Me^{\delta-}–Si^{\delta+}$ interactions are strong and can often be witnessed in Ae–N$(SiMe_3)_2$ complexes [44, 64, 70]. Because of the low electronegativity of Si, strong $Ae^{2+}\cdots H^{\delta-}—Si^{\delta+}$ bonding contributes to the stability of complexes with the Anwander N$(SiMe_2H)_2$ ligand [71]. Fluorinated ligands or substituents can also interact with metal centers via electrostatic $Ae^{2+}\cdots F^{\delta-}–C^{\delta+}$ contacts, thus filling the metal's coordination sphere [67, 68, 72].

For more detailed information on the structures of group 2 metal complexes, the reader is referred to Ref. [52].

1.2.5 Dynamics of s-Block Organometallics in Solution

The solubility of early main group metal complexes strongly depends on saturation of the metal's coordination sphere. Whether the species is monomer or aggregated, large anionic and/or neutral ligands can block the metal for further intermolecular (or interaggregate) interactions, thus increasing its solubility in apolar solvents. For solubilization of complexes with smaller ligands, often polar ethereal or amine solvents are crucial. Because of the higher number of anionic ligands, group 2 metal complexes are generally better soluble than the alkali metal compounds. Because the coordination sphere of larger metals is more difficult to saturate, solubilities decrease with increasing metal size.

Highly aggregated organolithium compounds display fascinating dynamics in solution. The very rapid metal exchange processes within or between aggregates originate from the ionic nature of the bonds. Weak, mainly ionic, bonds to Li^+ are exploited in the well-known Li batteries for which fast dynamics of the Li^+ cation are essential. The observation of $^{13}C–^{7}Li$ NMR coupling facilitated investigations on the aggregation state and dynamics of organolithium compounds tremendously [73]. Because of the much lower quadrupole moment and spin quantum number of ^{6}Li ($I = 1$), studies of $^{13}C–^{6}Li$ NMR coupling are even more informative [74]. The multiplicity of the ^{13}C signal not only reveals the number of ^{6}Li bonding partners (Scheme 1.9) but information can also be extracted from the NMR chemical shift or the coupling constant, which decreases with increasing number of ^{6}Li contacts [75]. Because of the highly dynamic nature of Li complexes, NMR samples often have to be cooled to observe these coupling patterns. At higher temperature, fast exchange between aggregates gives a singlet signal. Lowering the temperature freezes interaggregate exchange, but fast exchange within the aggregate allows for coupling with all Li nuclei. Further cooling results in static bonding, reducing the multiplicity. The presence of

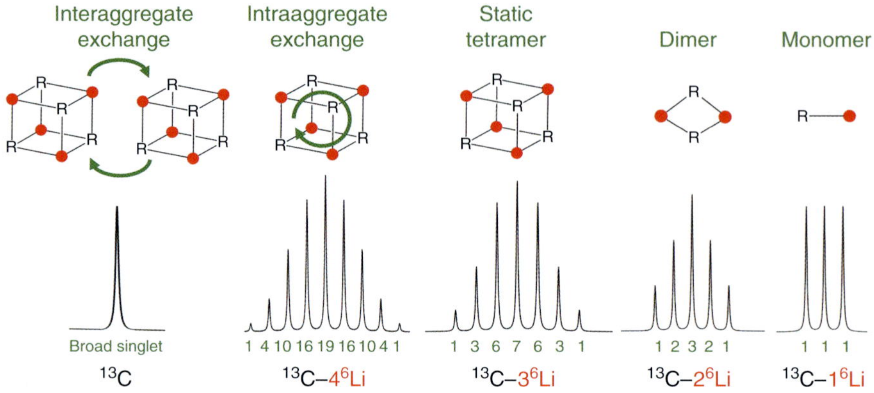

Scheme 1.9 ^{13}C NMR coupling with ^{6}Li (I = 1) gives information on dynamics and the aggregation state.

equilibria between aggregates (tetramer–dimer–monomer) reduces the splitting further but increases the coupling constants from c. 5 to 18 Hz. The rate of these exchange equilibria decides whether coupling constants can be detected ($\Delta G^{\ddagger}$ can vary between c. 5 and 25 kcal/mol) [76]. Polar solvents generally accelerate exchange processes and lower the aggregation number. *n*BuLi, which is a hexamer in hexane, is dissolved in THF in a tetramer–dimer equilibrium. Rapid injection NMR studies have shown that the dimers in this mixture react at least 4 orders of magnitude faster with an aldehyde than the tetramers [77]. The higher reactivity of smaller aggregates partially explains why polar solvents accelerate organolithium reactions.

Unfortunately, the favorable magnetic properties of Li are unique for the s-block metals, and these highly useful NMR methods are limited to studies on organolithium compounds. There are, however, other methods that can be used to gain information on aggregation. Cryoscopy, a technique that can determine the molecular weight of a dissolved molecule by measuring the solvent's melting point lowering, has been used successfully in organolithium chemistry, but in the case of equilibria, it always gives an averaged molecular weight [78]. Recently, diffusion-ordered NMR spectroscopy (DOSY), a technique that determines molecular weight by measuring diffusion rates, has been introduced to s-block metal chemistry [79]. This technique is becoming an increasingly popular method to extract information of s-block organometallics in solution.

As bonds to the twofold positively charged Ae^{2+} cations are stronger than those to group 1 metal cations, the dynamic processes for Ae metal complexes are slower. This is nicely illustrated by water ligand exchange rates for their hydrated cations, which increase with a decrease in the cation's surface–charge area [80]. The fastest exchange rate is found for metal ions of large ionic radius and with a low charge. Exchange rates for the alkali metal cations are among the fastest and vary roughly from 10^{8} s^{-1} (Li^{+}) to 10^{10} s^{-1} (Cs^{+}), whereas exchange at Ae^{2+} cations is much slower varying from 10^{6} s^{-1} (Mg^{2+}) to 10^{9} s^{-1} (Ba^{2+}); the very small Be^{2+} has a very slow exchange rate of 10^{3} s^{-1}. Ligand exchange at s-block metal cations is generally much faster than that at transition metals, a property that is certainly advantageous in catalysis.

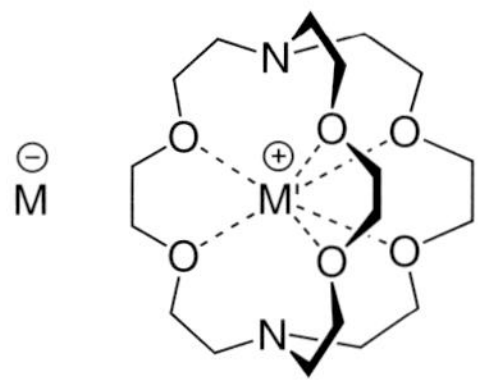

8

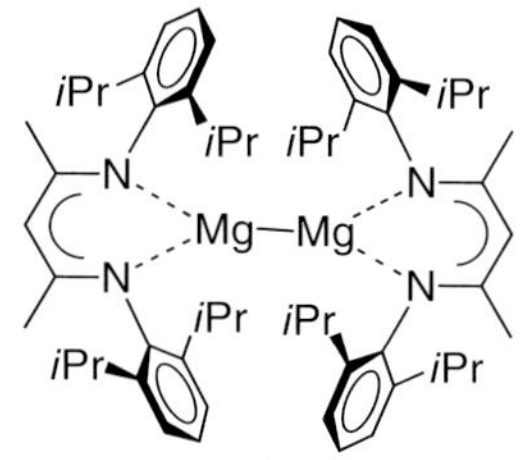

9

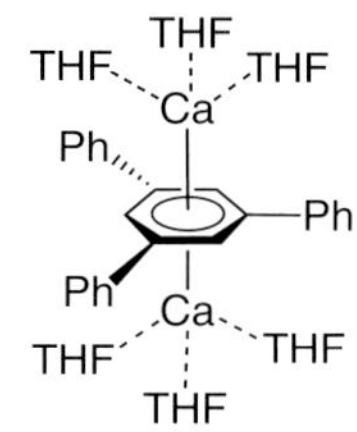

10

Chisholm and coworkers reported an in-depth study on THF exchange rates in two β-diketiminate Mg complexes that vary in bulk (**7**) [81]: in **7b**, the larger *t*Bu backbone substituents create larger metal shielding by forcing the aryl groups to bend toward the metal. Although the less shielded Mg in **7a** can exchange THF in an associative process, solvent exchange in the sterically encumbered **7b** is considerably slower and purely follows the more difficult dissociative route ($\Delta H^{\ddagger} = 7.0(5)$ kcal/mol, $\Delta S^{\ddagger} = 11.6(4)$ cal/mol).

Although ligand exchange processes in Ae metal complexes are slower than for group 1 metal complexes, group 2 metal chemistry is complicated by another serious problem: the Schlenk equilibrium. Originally discovered by Wilhem Schlenk Jr., who wrote his dissertation on this subject under supervision of his father Wilhem Schlenk Sr., the Schlenk equilibrium has been formulated for ligand exchange processes in Grignard reagents: $2RMgX \rightleftharpoons R_2Mg + MgX_2$ [4, 25]. In Et_2O, this equilibrium lies slightly toward heteroleptic RMgX (two different ligands), whereas for the more polar, less bulky, solvent THF, predominantly homoleptic species (equal ligands) are observed. Density functional theory (DFT) calculations are in agreement with this observation and attribute this difference to the fact that the less bulky THF solvent especially stabilizes MgX_2 by coordination of up to four THF ligands [82]. Recent *ab initio* molecular dynamics calculations on MeMgCl in THF give detailed insight into these ligand exchange processes and underscore that the solvent is a direct key player (Scheme 1.10) [83]. The most stable dimer is bridged by two Cl^- anions, but the

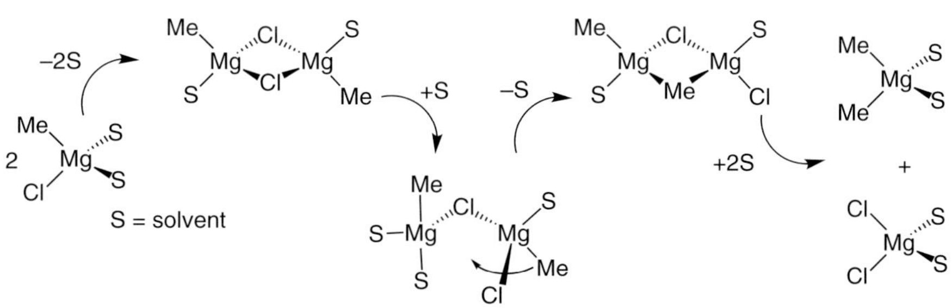

Scheme 1.10 Solvent dynamics play a crucial role in ligand exchange by the Schlenk equilibrium [83].

four-membered Mg_2Cl_2 ring can open up by filling empty coordination sites with a solvent ligand illustrating that polar solvents support these dynamics.

This equilibrium between heteroleptic and homoleptic complexes becomes increasingly faster and more problematic for complexes with larger Ae metals, especially when dissolved in polar solvents. Schlenk equilibria, however, also exist for monomeric heteroleptic complexes in nonpolar solvents. These mechanisms are less well understood but because of the lack of polar solvents to stabilize free coordination sites it is likely that such processes follow an associative pathway. Therefore, open coordination sites should be avoided. For this reason, strongly coordinating multidentate bulky ligands can stabilize heteroleptic complexes of heavier Ae metals [84] but need to be bulkier with increasing metal size [47, 85]. It seems possible to also stabilize heteroleptic complexes with a combination of electron-withdrawing and electron-releasing ligands [72], but clearly, more research is needed to understand and prevent ligand exchange processes.

1.2.6 Low-Valent s-Block Chemistry

In contrast to the late main group metals, the chemistry of the s-block elements has always been characterized by the metal's distinct oxidation state that equals the group number. Apart from the metallic state, there are hardly exceptions to the alkali metal +I and Ae metal +II oxidation states. One of the few exceptions is the −I oxidation states of the alkali metals in an alkalide salt, which is a combination of a metal anion and a metal cation stabilized by a crown ether of cryptand (e.g. **8**) [86]. The energy gained by cation complexation is the driving force for electron transfer from one alkali metal to the other alkali metal. The negatively charged metal can be the same or preferably should be more electronegative than the positively charged metal.

For the Ae metal Mg, there is a rich chemistry in its subvalent form Mg(I) [87, 88]. Although Mg(I) species have been detected in outer space or could be isolated at low temperatures in a matrix, Jones and coworkers reported in 2007 the first examples of a Mg(I) complex stabilized against disproportionation by a bulky bidentate amidinate or β-diketiminate ligand (**9**) [89]. Meanwhile, a large variety of these Mg—Mg bond complexes have been isolated, mainly with the conveniently tunable β-diketiminate ligand. These electron-rich complexes can be used as reducing agents that are soluble, selective, and safe, allowing for facile control of stoichiometry, thus avoiding overreduction. They have been the key to isolation of several novel compound types that could not be prepared using conventional reducing agents (e.g. K mirror or KC_8) [88]. Similar complexes with heavier Ca could hitherto not be obtained, probably on account of the weaker Ca—Ca bond and more facile disproportionation. Surprisingly, the lighter Be–Be complexes have also not been isolated. It is likely that these exist as persisting radicals that decompose via a different pathway.

Westerhausen and coworkers reported a serendipitous example of a Ca(I) complex (**10**) [89]. Several observations support the assumption that the triphenylbenzene moiety is doubly negatively charged, that the formal oxidation state of Ca is +I, and that there are no hidden hydride or other anions. Charge calculation by the natural population analysis (NPA) resulted in +1.03 charges for the Ca atoms and consequently −1.94 for the arene, and analysis of the THF-solvated molecule led to a charge of −3.68 at the arene and +1.84 charges at the Ca atoms. This has been attributed to stabilization of a high positive charge at the metal by the THF ligands and would better fit for Ca in the oxidation +II. The challenge to isolate an unequivocal Ca(I) complex with a Ca—Ca bond is still open.

1.3 s-Block Organometallics in Catalysis

In classical organic syntheses, early main group organometallics have traditionally been the highly reactive nucleophiles or Brønsted bases. It is only since the start of this century that they emerged as catalysts for a growing variety of transformations [23]. This development is strongly motivated by the search for replacement of precious metals by more abundant elements. Although the cost factor is often mentioned as an advantage for doing catalysis with metals that do not belong to the platinum group ("cheap metals for noble tasks" [90]), the generally higher catalyst loadings and also the price of the ligand as well as recyclability should be considered. The worldwide availability of most early main group metals is therefore a much stronger argument for their application. This avoids price instabilities by monopoly situations and is especially for industries an important consideration for long-term investments.

Another major advantage for most s-block metals is their superb biocompatibility. Calcium, the fifth most abundant element in the earth's crust [91], is present in large quantities in the human body. Avoiding the more harmful metals such as Pt is in particular for the pharmaceutical industry, in which catalysis plays a key role, an important issue. These driving forces have motivated chemists from different backgrounds to investigate s-block metal catalysis. There seems to be two schools of investigators: the organic chemists generally focus more on Lewis acid catalysis centered mainly at the metal cation, whereas the inorganic (or organometallic) researchers exploit the strongly basic and nucleophilic properties of the anionic part or the combined organometallic species. Although addition to an unsaturated bond can be described either from the cationic or the anionic side, the majority of catalytic reactions are likely a combination of both, strongly depending on the extent of Lewis acidity and nucleophilicity (Scheme 1.11). Supposing that these areas complement and support each other, the current book describes examples of both schools.

1.3.1 Working Principles in Lewis Acid Catalysis

Chapters 11 and 12 deal with Lewis acidic early main group metal catalysis (some recent reviews are listed in Ref. [92]). For example, Niggemann and coworker focused mainly on the development of the highly Lewis acidic Ca^{2+} cation, working with a system consisting of Ca^{2+} and the weakly coordinating anions Tf_2N^- ($Tf = CF_3S(O)_2O$) and PF_6^- [92a]. The anions are merely present as innocent spectators and the Lewis acidic Ca^{2+} cation is the active part of the catalyst.

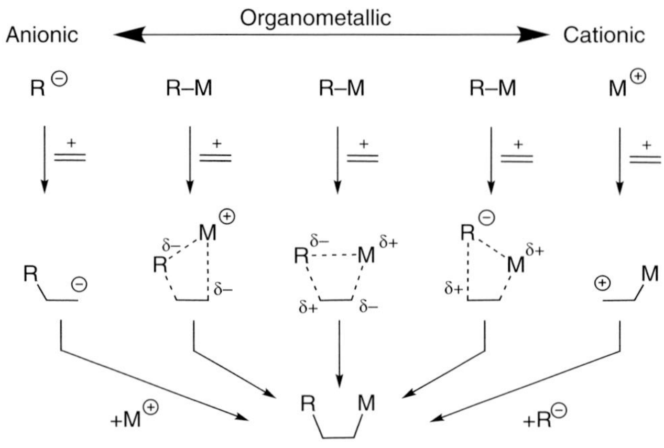

Scheme 1.11 The continuous transition from nucleophilic to electrophilic addition to an unsaturated bond.

Some selected examples of cation–substrate activation by pure electrostatics are shown in Scheme 1.12. Early calculational studies demonstrated that polarized double bonds such as C=O or C=N are strongly activated for nucleophilic attack by cation coordination [93]. This mode of electrostatic activation is general for any C—X bond and also holds for C=C bonds, which will be discussed in detail in Section 3.3. Another mode of substrate activation is demonstrated by ROH coordination. The metal Lewis acid is not only the key to generation of highly reactive carbocations, it could also bind an alcohol leading to polarization and acidification of the O—H group causing facile loss of H^+ which itself could be a catalyst. Cases of assumed "transition-metal-catalyzed" reactions have been shown to be H^+ catalyzed [94]. It is likely that "hidden Brønsted catalysis" also plays a major role in many Lewis acid-catalyzed reactions, especially as it is known that traces of water are often essential for catalytic activity. This topic is discussed in greater detail in Chapter 11. Notwithstanding the fact that the "true" catalyst may be a simple proton, the activating influence of Li^+ or Ca^{2+} cations has been substantiated by calculation and this concept of activation likely plays an important role in biological oxidation [95].

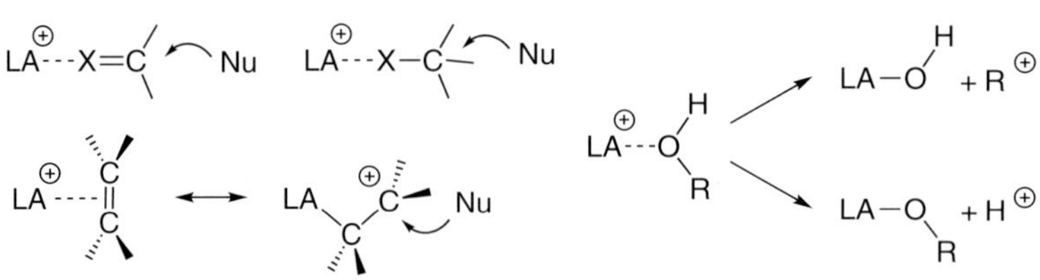

Scheme 1.12 Selected examples of Lewis acid (LA) substrate activation.

Truly cationic Lewis acidic catalysis (i.e. without marked influences of the anion) cannot be controlled by the anion. However, the metal cation can be modified by addition of neutral ligands that can control its sterics and electronics. This is especially exploited in enantioselective Lewis acid catalysis and exemplified by numerous transformations typically catalyzed by acids or metal cations: e.g. cycloadditions, ring-opening reactions, (hetero)-Diels–Alder, Mannich- or Michael-type, Aldol condensations, or Friedel–Craft reactions. In Chapter 12, Kobayashi and coworkers specifically focus on enantioselective Lewis acid catalysis.

1.3.2 Working Principles in s-Block Organometallic Catalysis

The majority of the catalytic reactions described in this book are of organometallic nature, i.e. the anion and metal cation operate in concert (Scheme 1.11). These reactions can be much easier controlled by ligand design. This is especially true for group 2 metal catalysts L–Ae–R that are often build up from a passive spectator ligand (L) and a reactive group R. In contrast to transition metals, which show fast and reversible switching of oxidation states, the early main group metals generally favor only one oxidation state. This excludes catalytic pathways with redox steps such as oxidative addition and reductive elimination. In this respect, s-block metal catalysis has strong similarities to lanthanide metal catalysis for which redox reactions also do not play a role. This simplifies the possible steps enormously, and most catalytic cycles are built around basic dipolar transformations such as deprotonation/protonation, addition/elimination, nucleophilic substitution, or nucleophilic ring opening. Most of the catalytic reactions can be classified as (hetero)functionalization of unsaturated bonds including hydrogenation, hydroboration, hydrosilylation, hydroamination, or hydrophosphination. These all follow the same simplified protocol shown in Scheme 1.13a: catalyst initiation and substrate coordination is followed by addition and nucleophilic substitution. This kind of reactivity is extensively described in Chapters 3–5 (hydroamination and hydrophosphination), Chapter 6 (hydrosilylation), Chapter 7 (hydrogenation), and Chapter 8 (hydroboration). It is important to note that the substrates X–Y react in a dipolar manner. For example, in alkene hydrogenation, the H_2 molecule reacts protic (H^+) and hydridic (H^-). This means that for hydrogenation, hydroboration, and hydrosilylation, the catalytic active species could be a metal hydride species.

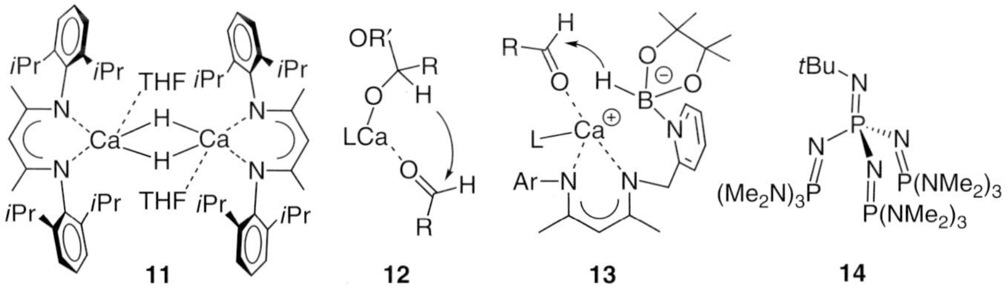

Metal hydride complexes of group 1 and 2 metals have been oddities for a long time. Their intermediacy in catalytic cycles, however, accelerated research in this area tremendously and after isolation of the first calcium hydride complex (**11**) lately several review articles on s-block metal hydrides have appeared [96]. Intermediate metal hydride species also likely play a role in dehydrogenative coupling

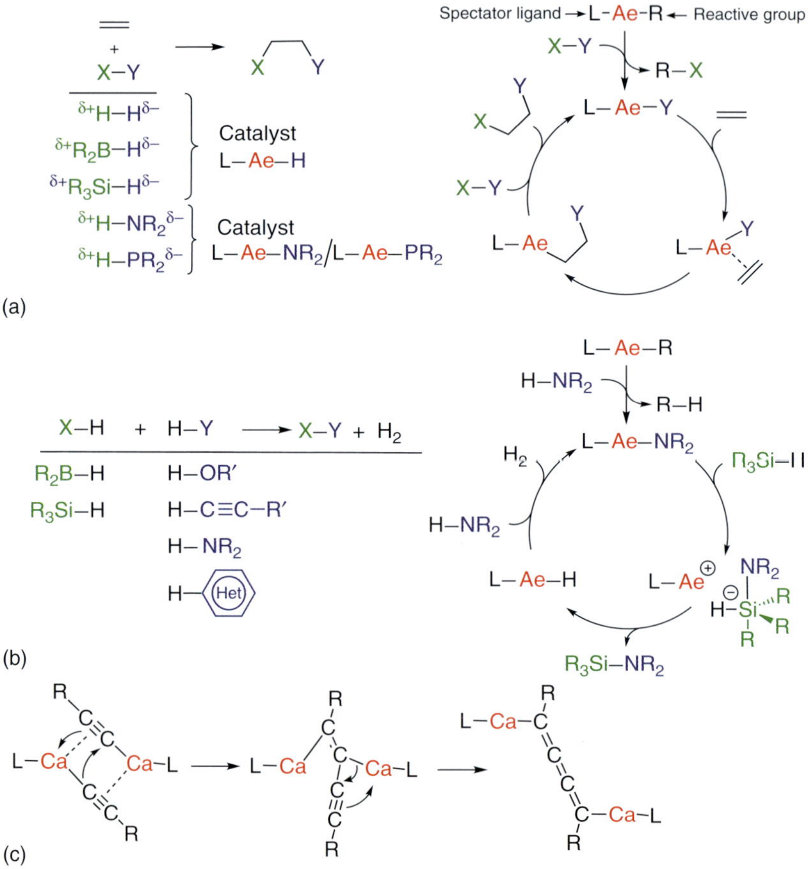

Scheme 1.13 (a) General mechanism for hydrogenation, hydroboration, hydrosilylation, hydroamination, and hydrophosphination of unsaturated bonds using catalyst L–Ae–R. (b) General mechanism for dehydrogenative cross-coupling exemplified by the cycle for amine hydrosilylation. (c) An unusual example of anion–anion coupling.

reactions, which is another group of transformations that follows a collective mechanism (Scheme 1.13b). In this case, the catalyst is formed by deprotonation of the substrate, which, after a substitution reaction with a borane or silane, forms a hydride that in reaction with relatively acidic substrates produces H_2, regenerating the catalyst. Depending on the substrate, these reactions can be much more complicated, and this subject is extensively discussed in detail in Chapter 9. The recent dehydrogenative silylation of C—H bonds in aromatic heterocycles shows that also less acidic substrates can be converted, but the mechanism may be of the radical type [97].

Chapter 2 discusses application of s-block organometallics in polymerization chemistry which involves standard reactivity such as nucleophilic ring opening or successive addition reactions. Finally, Chapter 13 describes miscellaneous reactions in which the metal catalyst shows a different reactivity. For example, in the Ca-catalyzed Tischtschenko aldehyde dimerization, hydride transfer from an alcoholate to an aldehyde proceeds through combined β-hydride elimination/addition reaction (**12**) [98]. Another case of such reactivity was recently postulated for aldehyde hydroboration using a catalyst without a highly reactive group. Instead, the pendant pyridine ring assists in creating a hydridic borate that directly transfers the hydride to the aldehyde (**13**) [99]. This concept differs from hydroboration according to Scheme 1.13a in which metal hydride intermediates are proposed. A recent report on ketone hydroboration with an amidinate Ca iodide catalyst may also follow a hydride-free route, which would explain its unusually highly functional group tolerance [100]. Harder and coworkers suggested similar hydride-free routes for the Mg-catalyzed pyridine hydroboration and proposed direct transfer of the hydride from borate to pyridine [101]. This conclusion was based on differences in regioselectivity between the stoichiometric reaction of an Mg hydride complex with pyridine and its catalytic reactivity.

An odd case of C–C coupling of alkynide carbanions is described by Hill and coworkers (Scheme 1.13c) [102]. This reaction represents the unusual nucleophilic attack of an alkynide anion at another alkynide anion. The driving force is likely the formation of an extended C=C=C=C system. Also, side-on coordination of the alkynide anion to Ca^{2+} may be important to C≡C bond activation.

1.3.3 Substrate Activation by s-Block Metals

Because early main group metals do not possess partially filled d-orbitals for substrate activation by $d \rightarrow \pi^*$ back-bonding, it is questionable to what extent the metal plays a role in the mechanism. Alkene–metal coordination is well established in transition metal chemistry, but for the main group metals, examples are scarce [67, 103]. Metal· · ·C distances are long and bonds should be considered weak [104]. This raises the question: how important is the metal in s-block metal catalysis? The Harder group reported a series of investigations that scrutinize the role of the metal [105]. The metal in an s-block metal amide complex was simply replaced by the Me_4N^+ cation. It was found that the salt $[Me_4N^+][Ph_2N^-]$ catalyzes the hydroamination of carbodiimides equally well as the Ca catalyst $Ca[N(SiMe_3)_2]_2$, thus approaching the anionic limit in Scheme 1.11. DFT calculations, however, show that Me_4N^+ is not truly noncoordinating and also has a minor activating effect on the C=N bond, albeit much less than that of the Li^+ cation in the model catalyst $LiNPh_2$ (Scheme 1.14a). The conclusion is that for highly activated C=N bonds in a carbodiimide, activation by metal–substrate coordination is rather unimportant. On the other hand, the "naked" anion catalyst did not perform in the intramolecular alkene hydroamination, which runs smoothly with $Ca[N(SiMe_3)_2]_2$ (Scheme 1.14b). This was attributed to the low basicity of the Ph_2N^- anion and therefore the same reaction was reinvestigated with a strong neutral metal-free organic base: the Schwesinger base P4 (**14**). Only under harsh conditions (90 °C, five days), substrate conversion was observed, but although the catalyst P4 was able to deprotonate the substrate, the reaction followed a different course. The absence of ring closure suggests that alkene activation by metal coordination plays a crucial role. Indeed, addition of a catalytic

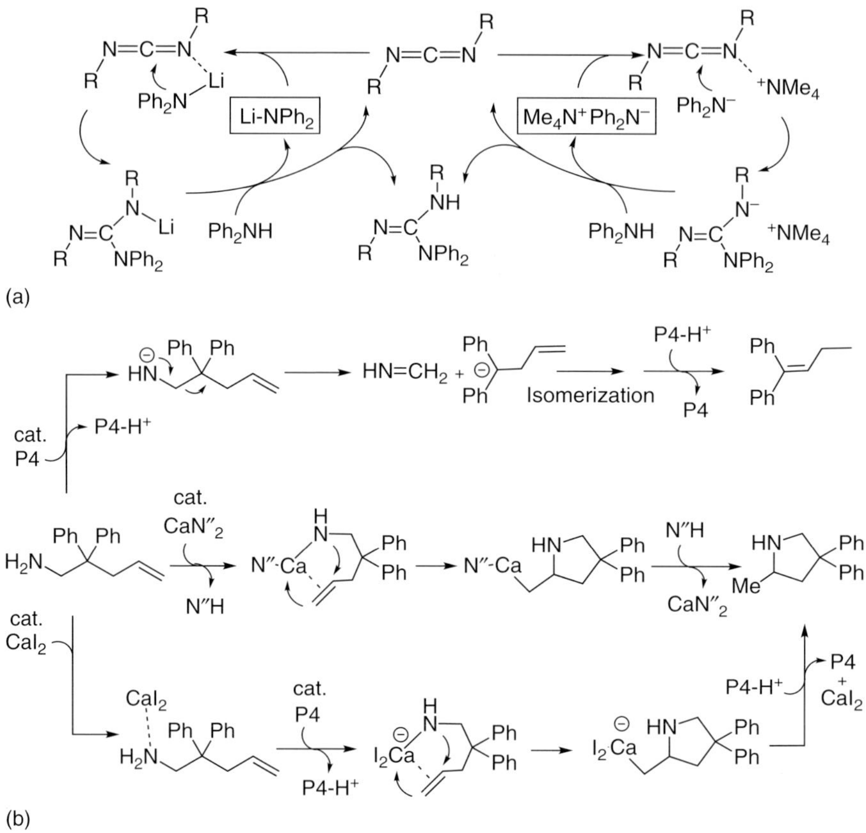

Scheme 1.14 (a) Hydroamination of carbodiimide catalyzed by $LiNPh_2$ (left) or by $[Me_4N^+][Ph_2N^-]$. (b) Intramolecular alkene hydroamination catalyzed by CaN''_2 ($N'' = N(SiMe_3)_2$) or by the P4/CaI_2 combination. Reaction with only P4 gives a different product.

quantity of CaI_2 to a P4/aminoalkene mixture formed the ring closure product under mild conditions (25 °C, two hours). It is suggested that Ca^{2+}-aminoalkene coordination acidifies the NH_2 group, facilitating its deprotonation and that Ca^{2+}-alkene coordination activates the C=C bond enabling smooth ring closure. This hybrid catalyst, consisting of a neutral organic base and a metal salt, offers ample opportunities for variation. Most importantly, it clearly demonstrates the activating influence of the metal cation.

Given the importance of the metal cation in alkene functionalization, Harder and coworkers studied the details of early main group metal coordination to unsaturated substrates [68, 106]. The goal was to isolate a Lewis base-free cationic Mg complex with an open space for substrate coordination. In cationic complex **15**, the Mg metal interacts with the weakly coordinating borate anion (Scheme 1.15). Addition of aromatic solvents partially breaks the cation–anion contacts and a tightly bound Mg–(η^3-benzene) complex (**16**) is formed. In the presence of EtC≡CEt, the first unsupported Mg–alkyne complex (**17**), which also persists in solution, is formed. Compared to the neutral Fe(I)–alkyne

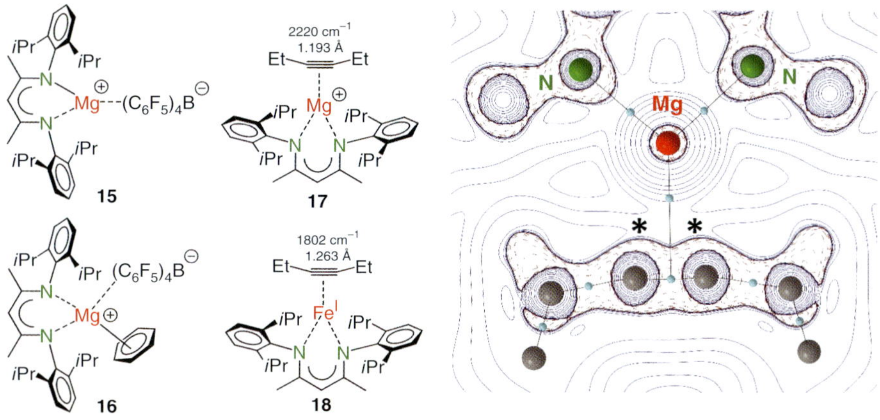

Scheme 1.15 A cationic β-diketiminate complex with a nearly "naked" Mg cation (**15**) binding benzene (**16**) or alkyne (**17**). The inset shows an AIM representation of the Laplacian of the electron density in **17**: π-electron density is polarized toward Mg (see *).

complex (**18**) [107], the triple bond in the Mg complex is hardly stretched, but the lowering of the C≡C stretching frequency by 40 cm^{-1} compared to free EtC≡CEt (2260 cm^{-1}) is significant. The Et substituents are bent by c. 11° away from the C≡C axis, thus shifting electron density toward the metal. This can be clearly seen in an atoms-in-molecules (AIM) representation of the Laplacian of the electron density in **17**, which shows polarization of alkyne π-electron density toward Mg^{2+}. Ion-induced polarization is particularly strong for asymmetrically bound substrates, which polarizes the C≡C bond also along its axis [104], leading to nucleophilic attack. The power of this type of substrate activation has been recently demonstrated by the dearomatization of benzene, which was activated by a similar cationic Ca complex [108].

1.3.4 Future of Early Main Group Metal Catalysis

s-Block metal catalysis experienced a rapid development in the past two decades. This is especially true for the group 2 metals. Transformations typical in transition metal catalysis now routinely can be mediated by early main group metals and it is to be expected that the field will steadily grow further.

Hitherto, enantioselective transformations have been found to be very challenging. In contrast to transition metal catalysts, which have defined coordination geometries that are dictated by orbital interactions, ionically bound s-block metal catalysts are highly dynamic and exercise much less control over diastereoselective transition states. Because both enantioselective catalysis and replacement of benign and expensive platinum group metals are especially important for the pharmaceutical industry, further development of this field is highly desirable. Development of ligand systems that do not allow for much fluxionality may be the key to improve ee values.

At the same time, functional group tolerance is crucial for a wider application of s-block metal catalysts. The recently presented unstabilized Ae metal alkyl species of the heavier metals Ca and Sr [45–47] may display unique reactivity but are too reactive to cope with the many functional groups in complicated pharmaceutical building blocks. Broader functional group tolerance may be achieved with more covalently bound Mg complexes and less reactive alkoxide or amide ligands.

Secondary interactions between ligand and metal become increasingly more important to stabilize complexes [67, 71]. Ligand–substrate secondary interactions could, similar to that in organocatalysis, also play a role in catalysis. There are cases of “organocatalysis” in which Ca^{2+} that leached from the silica column used in catalyst “purification” has been shown to be essential [109]. The possibilities of such inorganic/organic hybrid catalysts are hitherto largely unexplored and certainly deserve attention.

This topic is also related to design of catalysts with noninnocent ligands. The recognition that “spectator” ligands can sometimes be active in catalysis [110] revealed interesting possibilities for future catalyst design. Catalysts in which an unreactive ligand is coupled to a reactive group would be much less susceptible to Schlenk equilibria and could especially in enantioselective catalysis be important.

In contrast to Al catalysis [111], the potential of redox-active noninnocent ligands has so far not been exploited in early main group metal catalysis. Ligands that can act as reversible electron reservoirs could significantly expand the toolbox of s-block metal catalysis bringing them further at par with transition metals. First, reversible redox reactions with Mg complexes have been reported [112], and there is also increasing interest in complexes with highly charged π-systems, which can act as electron supply [113].

Finally, given the importance of heterogeneous catalysis in the industry, support of catalysts on surfaces or synthesis of defined insoluble catalysts from molecular precursors, possibly with control over morphology and/or porosity, can expand the possibilities even further, thus creating continuously moving horizons.

List of Abbreviations

Ae	alkaline earth
AIM	atoms-in-molecules
N″	$N(SiMe_3)_2$
NPA	natural population analysis
Tf	triflate = $CF_3S(O)_2O$
THF	tetrahydrofuran
THP	tetrahydropyran

References

1 Frankland, E. (1849). *Ann. Chem. Pharm.* 71: 213–215.
2 For historical reviews see: (a)Seyferth, D. (2001). *Organometallics* 20: 2940–2955. (b) Seyferth, D. (2006). *Organometallics* 25: 2–24. (c) Seyferth, D. (2009). *Organometallics* 28: 2–33.

3 "In trockenem Zustand stellt das Lithiummethyl ein vollkommen farbloses mikrokrystallinisches Pulver dar. An die Luft gebracht, verbrennt es mit explosionsartiger Heftigkeit. Die auftretende Flamme besitzt die schöne Rotfärbung, welche fur die Lithiumflamme charakteristisch ist; da zahlreiche aus der roten Flamme fortgeschleuderte Funken gelbglühend sind, gewährt das Abbrennen von Lithiummethyl einen prächtigen Anblick." Schlenk, W. and Holtz, J. (1917). *Ber. Dtsch. Chem. Ger.* 50: 271.
4 Tidwell, T.T. (2001). *Angew. Chem. Int. Ed.* 40: 331–337.
5 Ziegler, K. and Colonius, H. (1930). *Liebigs Ann. Chem.* 479: 135–149.
6 Grignard, V. (1900). *C.R. Acad. Sci.* 130: 1322–1324.
7 Beckmann, E. (1905). *Ber. Dtsch. Chem. Ges.* 38: 904–906.
8 Gilman, H. and Schulze, F. (1926). *J. Am. Chem. Soc.* 48: 2463–2467.
9 Schlosser, M. (2002). *Organometallics in Synthesis: A Manual*, 2e. New York: Wiley.
10 (a) Brandsma, L. and Verkruijsse, H. (1987). *Preparative Polar Organometallic Chemistry 1*. Berlin: Springer-Verlag. (b) Brandsma, L. and Verkruijsse, H. (1990). *Preparative Polar Organometallic Chemistry 2*. Berlin: Springer-Verlag.
11 Smith, M.B. (2013). *March's Advanced Organic Chemistry*, 7e, 314. Hoboken, NJ: Wiley.
12 pK_a values of alkanes are difficult to estimate and different scales have been reported: (a)Cram, D.J. (1965). *Fundamentals of Carbanion Chemistry*. New York: W. A. Benjamin, Inc. (b) Dessy, R.E., Kitching, W., Psarras, T. et al. (1966). *J. Am. Chem. Soc.* 88: 460–467. (c) Robertson, P.J., Scurrell, M.S., and Kemball, C. (1975). *J. Chem. Soc., Faraday Trans. 1* 71: 903–912.
13 (a) Whisler, M.C., MacNeil, S., Snieckus, V., and Beak, P. (2004). *Angew. Chem. Int. Ed.* 43: 2206–2225. (b) Tilly, D., Magolan, J., and Mortier, J. (2012). *Chem. Eur. J.* 18: 3804–3820.
14 Farnham, W.B. and Calabrese, J.C. (1986). *J. Am. Chem. Soc.* 108: 2449–2451.
15 Fu, P.-F., Khan, M.A., and Nicholas, M. (1992). *J. Am. Chem. Soc.* 114: 6579–6580.
16 Lochmann, L. and Janata, M. (2014). *Cent. Eur. J. Chem.* 12: 537–548.
17 (a) Lochmann, L., Pospíšil, J., and Lim, D. (1966). *Tetrahedron Lett.*: 257–260. (b) Schlosser, M. (1967). *J. Organomet. Chem.* 8: 9–16.
18 Morton, A.A. and Chester, E.C. (1954). *J. Am. Chem. Soc.* 76: 4935–4938.
19 Mulvey, R.E., Mongin, F., Uchiyama, M., and Kondo, Y. (2007). *Angew. Chem. Int. Ed.* 46: 3802–3824.
20 Bickelhaupt, F. (1994). *J. Organomet. Chem.* 475: 1–14.
21 Naglav, D., Buchner, M.R., Bendt, G. et al. (2016). *Angew. Chem. Int. Ed.* 55: 10562–10576.
22 (a) Condorelli, G.G., Malandrino, G., and Fragalà, I.L. (2007). *Coord. Chem. Rev.* 251: 1931–1950. (b) Kim, S.B., Yang, C., Powers, T. et al. (2016). *Angew. Chem. Int. Ed.* 55: 10228–10233.
23 (a) Harder, S. (2010). *Chem. Rev.* 110: 3852–3876. (b) Barrett, A.G.M., Crimmin, M.R., Hill, M.S., and Procopiou, P.A. (2010). *Proc. R. Soc. Math. Phys. Eng. Sci.* 466: 927–963. (c) Westerhausen, M. (2009). *Z. Allg. Anorg. Chem.* 635: 13–32. (d) Hill, M.S., Liptrot, D.J., and Weetman, C. (2016).

Chem. Soc. Rev. 45: 972–988. (e) Sarazin, Y. and Carpentier, J.-F. (2016). *Chem. Rec.* 16: 2482–2505. (f) Crimmin, M.R. and Hill, M.S. (2013). *Alkaline-Earth Metal Compounds: Oddities and Applications*, vol. 45 (ed. S. Harder), 191–241. Heidelberg: Springer. (g) Rochat, R., Lopez, M.J., Tsurugi, H., and Mashima, K. (2016). *ChemCatChem* 8: 10–20.

24 (a) Westerhausen, M., Koch, A., Görls, H., and Krieck, S. (2017). *Chem. Eur. J.* 23: 1456–1483. (b) Westerhausen, M., Langer, J., Krieck, S. et al. (2013). *Alkaline-Earth Metal Compounds: Oddities and Applications*, vol. 45 (ed. S. Harder), 29–72. Heidelberg: Springer.

25 Schlenk, W. and Schlenk, W. Jr., (1929). *Ber. Dtsch. Chem. Ges.* 62B: 920–924.

26 Langer, J., Gärtner, M., Fischer, R. et al. (2007). *Inorg. Chem. Commun.* 10: 1001–1004.

27 Westerhausen, M. (1991). *Inorg. Chem.* 30: 96–101.

28 Brady, E.D., Hanusa, T.P., Pink, M., and Young, V.G. Jr., (2000). *Inorg. Chem.* 39: 6028–6037.

29 Johns, A.M., Chmely, S.C., and Hanusa, T.P. (2009). *Inorg. Chem.* 48: 1380–1384.

30 Harder, S., Feil, F., and Weeber, A. (2001). *Organometallics* 20: 1044–1046.

31 Feil, F., Müller, C., and Harder, S. (2003). *J. Organomet. Chem.* 683: 56–63.

32 Harder, S., Müller, S., and Hübner, E. (2004). *Organometallics* 23: 178–183.

33 Feil, F. and Harder, S. (2001). *Organometallics* 20: 4616–4622.

34 Weeber, A., Harder, S., and Brintzinger, H.-H. (2000). *Organometallics* 19: 1325–1332.

35 Izod, K. and Waddell, P.G. (2015). *Organometallics* 34: 2726–2730.

36 Jochmann, P., Dols, T.S., Spaniol, T.P. et al. (2009). *Angew. Chem. Int. Ed.* 48: 5715–5719.

37 Eaborn, C., Hawkes, S.A., Hitchcock, P.B., and Smith, J.D. (1997). *Chem. Commun.*: 1961–1962.

38 Cloke, F.G.N., Hitchcock, P.B., Lappert, M.F. et al. (1991). *J. Chem. Soc., Chem. Commun.*: 724–725.

39 Crimmin, M.R., Barrett, A.G.M., Hill, M.S. et al. (2008). *Chem. Eur. J.* 14: 11292–11295.

40 Feil, F. and Harder, S. (2000). *Organometallics* 19: 5010–5015.

41 Köhler, M., Koch, A., Görls, H., and Westerhausen, M. (2016). *Organometallics* 35: 242–248.

42 Koch, A., Wirgenings, M., Krieck, S. et al. (2017). *Organometallics* 36: 3981–3986.

43 Causero, A., Elsen, H., Pahl, J., and Harder, S. (2017). *Angew. Chem. Int. Ed.* 56: 6906–6910.

44 Causero, A., Ballmann, G., Pahl, J. et al. (2017). *Dalton Trans.* 46: 1822–1831.

45 Wilson, A.S.S., Hill, M.S., Mahon, M.F. et al. (2017). *Science* 358: 1168–1171.

46 Wolf, B.M., Stuhl, C., Maichle-Mössmer, C., and Anwander, R. (2018). *J. Am. Chem. Soc.* 140: 2373–2383.

47 Rösch, B., Gentner, T.X., Elsen, H. et al. (2019). *Angew. Chem. Int. Ed.* 58: 5396–5401.

48 Vaartstra, B.A., Huffmann, J.C., Streib, W.E., and Caulton, K.G. (1991). *Inorg. Chem.* 30: 121–125.

49 Deacon, G.B., Jaroschik, F., Junk, P.C., and Kelly, R.P. (2015). *Organometallics* 34: 2369–2377.

50 Takahashi, Y., O'Brien, A., Deacon, G.B. et al. (2017). *Inorg. Chem.* 56: 11480–11489.

51 de Bruin-Dickason, C.N., Deacon, G.B., Jones, C. et al. (2019). *Eur. J. Inorg. Chem.*: 1030–1038.

52 (a) Buchanan, W.D., Allis, D.G., and Ruhlandt-Senge, K. (2010). *Chem. Commun.* 46: 4449–4465. (b) Torvisco, A. and Ruhlandt-Senge, K. (2013). *Alkaline-Earth Metal Compounds: Oddities and Applications*, vol. 45 (ed. S. Harder), 1–28. Heidelberg: Springer.

53 Schleyer, P.v.R. and Lambert, C. (1994). *Angew. Chem. Int. Ed. Engl.* 33: 1129–1140.

54 Pauling, L. (1960). *The Nature of the Chemical Bond*. Ithaca, NY: Cornell University Press.

55 (a) McKeever, L.D., Waack, R., Doran, M.A., and Baker, E.B. (1968). *J. Am. Chem. Soc.* 90: 3244. (b) Brown, T.L., Seitz, L.M., and Kimura, B.Y. (1968). *J. Am. Chem. Soc.* 90: 3245. (c) McKeever, L.D., Waack, R., Doran, M.A., and Baker, E.B. (1969). *J. Am. Chem. Soc.* 91: 1057–1061.

56 Andrews, L. and Carver, T.G. (1968). *J. Chem. Phys.* 72: 1743–1747.

57 Streitwieser, A. Jr.,, Williams, J.E. Jr.,, Alexandratos, S., and McKelvey, J.M. (1976). *J. Am. Chem. Soc.* 98: 4778–4784.

58 Streitwieser, A. Jr., (1978). *J. Organomet. Chem.* 156: 1–3.

59 Graham, G.D., Marynick, D.S., and Lipscomb, W.N. (1980). *J. Am. Chem. Soc.* 102: 4572–4578.

60 Reed, A.E., Weinstock, R.B., and Weinhold, F. (1985). *J. Chem. Phys.* 83: 735–746.

61 Harder, S. (1990). A study on the structure and reactivity of aryllithium compounds with an α- or β-heteroatom. PhD thesis. University of Utrecht.

62 Smith, G.S., Johnson, Q.C., Smith, D.K. et al. (1988). *Solid State Commun.* 67: 491–494.

63 Wiesinger, M., Maitland, B., Färber, C. et al. (2017). *Angew. Chem. Int. Ed.* 56: 16654–16659.

64 (a) Westerhausen, M. and Schwarz, W. (1992). *Z. Anorg. Allg. Chem.* 609: 39–44. (b) Westerhausen, M. and Schwarz, W. (1991). *Z. Anorg. Allg. Chem.* 604: 127–140. (c) Westerhausen, M. and Schwarz, W. (1991). *Z. Anorg. Allg. Chem.* 606: 177–190.

65 Bock, H., Hauck, T., Näther, C. et al. (1995). *Angew. Chem. Int. Ed. Engl.* 34: 1353–1355.

66 Maitland, B., Wiesinger, M., Langer, J. et al. (2017). *Angew. Chem. Int. Ed.* 56: 11880–11884.

67 Rosca, S.-C., Dinoi, C., Caytan, E. et al. (2016). *Chem. Eur. J.* 22: 6505–6509.

68 Pahl, J., Brand, S., Elsen, H., and Harder, S. (2018). *Chem. Commun.* 54: 8685–8688.

69 Brookhart, M., Green, M.L.H., and Parkin, G. (2007). *Proc. Natl. Acad. Sci. U.S.A.* 104: 6908–6914.

70 Scherer, W., Sirsch, P., Grosche, M. et al. (2001). *Chem. Commun.*: 2072–2073.
71 Liu, B., Roisnel, T., Carpentier, J.-F., and Sarazin, Y. (2013). *Chem. Eur. J.* 19: 13445–13462.
72 Fischer, C.A., Rösch, A., Elsen, H. et al. (2019). *Dalton Trans.* 48: 6757–6766.
73 Günther, H. (1999). *J. Braz. Chem. Soc.* 10: 241–262.
74 Seebach, D., Hässig, R., and Gabriel, J. (1983). *Helv. Chim. Acta* 66: 308–337.
75 Günther, H., Moskau, D., Bast, P., and Schmalz, D. (1987). *Angew. Chem. Int. Ed. Engl.* 26: 1212–1220.
76 Reich, H.J. (2013). *Chem. Rev.* 113: 7130–7178.
77 (a) McGarrity, J.F., Ogle, C.A., Brich, Z., and Loosli, H.-R. (1985). *J. Am. Chem. Soc.* 107: 1810–1815. (b) Jones, A.C., Sanders, A.W., Bevan, M.J., and Reich, H.J. (2007). *J. Am. Chem. Soc.* 129: 3492–3493.
78 Bauer, W. and Seebach, D. (1984). *Helv. Chim. Acta* 67: 1972–1988.
79 (a) Neufeld, R. and Stalke, D. (2015). *Chem. Sci.* 6: 3354–3364. (b) Neufeld, R., Teuteberg, T.L., Herbst-Irmer, R. et al. (2016). *J. Am. Chem. Soc.* 138: 4796–4806.
80 Lincoln, S.F. (2005). *Helv. Chim. Acta* 88: 523–545.
81 Balasanthiran, V., Chisholm, M.H., Choojun, K. et al. (2016). *Polyhedron* 103: 235–240.
82 Tammiku-Taul, J., Burk, P., and Tuulmets, A. (2004). *J. Phys. Chem. A* 108: 133–139.
83 Peltzer, R.M., Eisenstein, O., Nova, A., and Cascella, M. (2017). *J. Phys. Chem. B* 121: 4226–4237.
84 Ruspic, C. and Harder, S. (2007). *Inorg. Chem.* 46: 10426–10433.
85 Gentner, T.X., Rösch, B., Thum, K. et al. (2019). *Organometallics* 38: 2485–2493.
86 Dye, J.L. (1984). *J. Phys. Chem.* 88: 3842–3846.
87 Stasch, A. and Jones, C. (2011). *Dalton Trans.* 40: 5659–5672.
88 (a) Jones, C. (2017). *Nat. Chem. Rev.* 1: 0059. (b) Green, S.P., Jones, C., and Stasch, A. (2007). *Science* 318: 1754–1757.
89 Krieck, S., Görls, H., Yu, L. et al. (2009). *J. Am. Chem. Soc.* 131: 2977–2985.
90 Bullock, R.M. (2004). *Chem. Eur. J.* 10: 2366.
91 Wiberg, N. (2007). *Lehrbuch der Anorganischen Chemie*, 102e. Berlin: de Gruyter.
92 (a) Begouin, J.-M. and Niggemann, M. (2013). *Chem. Eur. J.* 19: 8030–8041. (b) Hong, L., Sun, W., and Wang, R. (2016). *RSC Green Chem. Ser.* 38 (Sustainable Catalysis, Part 1): 49–88. (c) Tsubugo, T., Yamashita, Y., and Kobayashi, S. (2013). *Alkaline-Earth Metal Compounds: Oddities and Applications*, vol. 45 (ed. S. Harder), 243–270. Heidelberg: Springer. (d) Yamashita, Y., Tsubogo, T., and Kobayashi, S. (2012). *Chem. Sci.* 3: 967–975. (e) Pelissier, H. (2017). *Org. Biomol. Chem.* 15: 4750–4782.
93 Kaufmann, E., Schleyer, P.v.R., Houk, K.N., and Wu, Y.-D. (1985). *J. Am. Chem. Soc.* 107: 5560–5562.
94 Dang, T.T., Boeck, F., and Hintermann, L. (2011). *J. Org. Chem.* 76: 9353–9361.

95 (a) Horn, A.H.C. and Clark, T. (2003). *J. Am. Chem. Soc.* 125: 2809–2816. (b) Schürer, G. and Clark, T. (1998). *Chem. Commun.*: 257–258.
96 (a) Harder, S. (2012). *Chem. Commun.* 48: 11165–11177. (b) Fohlmeister, L. and Stasch, A. (2015). *Aust. J. Chem.* 68: 1190–1201. (c) Mukherjee, D., Schuhknecht, D., and Okuda, J. (2018). *Angew. Chem. Int. Ed.* 57: 9590–9602. (d) Mukherjee, D. and Okuda, J. (2018). *Angew. Chem. Int. Ed.* 57: 1458–1473.
97 (a) Banerjee, S., Yang, Y.-F., Jenkins, I.D. et al. (2017). *J. Am. Chem. Soc.* 139: 6880–6887. (b) Liu, W.-B., Schuman, D.P., Yang, Y.-F. et al. (2017). *J. Am. Chem. Soc.* 139: 6867–6879.
98 Crimmin, M.R., Barrett, A.G.M., Hill, M.S., and Procopiou, P.A. (2007). *Org. Lett.* 9: 331–333.
99 Yadav, S., Dixit, R., Bisai, M.K. et al. (2018). *Organometallics* 37: 4576–4584.
100 Yadav, S., Pahar, S., and Sen, S.S. (2017). *Chem. Commun.* 53: 4562–4564.
101 Intemann, J., Lutz, M., and Harder, S. (2014). *Organometallics* 33: 5722–5729.
102 Barrett, A.G.M., Crimmin, M.R., Hill, M.S. et al. (2009). *Chem. Commun.*: 2299–2301.
103 Schumann, H., Schutte, S., Kroth, H.-J., and Lentz, D. (2004). *Angew. Chem. Int. Ed.* 43: 6208–6211.
104 Freitag, B., Elsen, H., Pahl, J. et al. (2017). *Organometallics* 36: 1860–1866.
105 Penafiel, J., Maron, L., and Harder, S. (2015). *Angew. Chem. Int. Ed.* 54: 201–206.
106 (a) Pahl, J., Stennet, T., Volland, M. et al. (2019). *Chem. Eur. J.* 25: 2025–2034.
107 Yu, Y., Smith, J.M., Flaschenriem, C.J., and Holland, P.L. (2006). *Inorg. Chem.* 45: 5742–5751.
108 Brand, S., Elsen, H., Langer, J. et al. (2018). *Angew. Chem. Int. Ed.* 57: 14169–14173.
109 Hatano, M., Moriyama, K., Maki, T., and Ishihara, K. (2010). *Angew. Chem. Int. Ed.* 49: 3823–3826.
110 Freitag, B., Fischer, C.A., Penafiel, J. et al. (2017). *Dalton Trans.* 46: 11192–11200.
111 Berben, L.A. (2015). *Chem. Eur. J.* 21: 2734–2742.
112 Dange, D., Gair, A.R., Jones, D.D.L. et al. (2019). *Chem. Sci.* 10: 3208–3216.
113 Causero, A., Elsen, H., Ballmann, G. et al. (2017). *Chem. Commun.* 53: 10386–10389.

2

Polymerization of Alkenes and Polar Monomers by Early Main Group Metal Complexes

Sjoerd Harder

University Erlangen-Nürnberg, Inorganic and Organometallic Chemistry, Egerlandstrasse 1, 91058 Erlangen, Germany

2.1 Introduction

The discovery of the first synthetic polymer *Bakelite* by the Belgian chemist Baekeland [1] is a chemical milestone that is arguably of equal importance as the invention of the Haber–Bosch ammonia synthesis. Large-scale synthesis and application of this phenol formaldehyde resin started the "age of plastic." Since this pioneering work, the further development of polymers came a long way and it is fair to say that today's plastics are a main chemical industry market. Although Baekeland's phenol formaldehyde copolymerization is largely uncontrolled, over the years, the precision of polymerization increased continuously. This led to an explosion in possible applications, spurring on further research and resulting in an exponential increase of plastic sales. Polymers are ideal materials for consumer products; however, the enormous growth of oil-based, single-use, plastics also generated a major environmental disaster that is especially noticeable in world's largest rubbish bins, the oceans [2]. Currently, a strong movement promotes the development of biodegradable polymers made from renewable monomers.

Group 1 and 2 metals certainly played a major role in these developments and they will also likely be of interest in future for application in polymer science. Historically, the invention of living anionic polymerization is one of the first technologies that enabled precise control over polymer architectures. Anionic polymerization comprises per definition an anionic chain end from which it follows that the most electropositive s-block metals are the counter cations. Most anionic polymerizations are living, meaning that there is no termination reaction and propagation keeps going until all monomer has been consumed. Because for living polymerizations the organometallic complex that starts the polymerization reaction is never formed back, it is called *initiator* rather than *catalyst*, but the principles of catalysis can certainly be applied to each insertion step. Szwarc observed that intensely green Na-naphthalide in tetrahydrofuran (THF) initiates styrene polymerization [3]. The resulting intensely red solution initiates polymerization of other activated alkenes (isoprene and butadiene), thus allowing for the

Early Main Group Metal Catalysis: Concepts and Reactions,
First Edition. Edited by Sjoerd Harder.

synthesis of block copolymers in which block lengths can be easily varied [4]. Ever since, the alkali metals played a major role in the rapidly growing field of anionic polymerization. The simplest, and most frequently used, initiator is the commercially available bulk product *n*BuLi, but there is a wealth of research on its modification or on mixed-metal alternatives. One of the latest developments is retarded anionic polymerization (RAP) in which the new low-cost initiator NaH/Et_3Al polymerizes styrene with essentially 100% conversion under similar conditions as radical polymerization [5].

A review of all research activities on the living alkene polymerization with s-block initiators would fill a complete book series. This is due to the wealth of research on alkali metal initiators. In contrast, very little is known about group 2 metal initiators, but despite scanty information, it was already clear at an early stage that these doubly charged cations have a noticeably different effect on polymerization activities and especially on selectivities [6]. Therefore, we will limit this chapter to the more exotic field of group 2 initiators that have seen major impetus only since the last decades.

The second part of the chapter will cover the rapidly growing field of polyoxygenates, i.e. biodegradable polymers partially made from renewable monomers. Although most catalysts are based on Zn, Al, the lanthanides, or the group 4 metals Ti and Zr, there are some detailed studies on group 2 metal initiators which will be discussed.

2.2 Alkene Polymerization

Although there are early patents on ethylene polymerization by *n*BuLi·(*N*, *N*, *N*′, *N*′-tetramethylethylenediamine [TMEDA]) [7], it is clear that this initiator is far from competitive to highly active Ziegler–Natta catalysts based on Ti or Zr. At 120 °C and 3.5 bar ethylene pressure, only a waxy product with molecular weight (MW) of 1000–3000 was obtained. Polymerization of unactivated (isolated) alkenes such as ethylene and propene, or generally 1-alkenes, apparently needs C=C double bond activation by substantially more Lewis acidic Ti^{4+} or Zr^{4+} cations. Although to some extent lanthanide (Ln) catalysts with Ln^{3+} may also be used for ethylene polymerization, Ae^{2+} metal cations are already too weak in Lewis acidity for efficient polymerization (Ae = alkaline earth). Indeed, MgR_2 has been added to Nd catalysts as a chain transfer reagent that stores growing polymer chains in a dormant state [8]. This polymerization technique, known as catalyzed chain growth (CCG), has the advantage that the living polymerization by the Nd species delivers more than one polymer chain per Nd center, thereby being more economical (Scheme 2.1). This is especially true when larger quantities of MgR_2 are used. The dormant state clearly shows that ethylene polymerization by Mg reagents is not feasible. Very recently, Harder and coworkers isolated a stable but highly reactive, Sr hydride complex and showed it to be active in homologation of ethylene to higher MW species but comprehensive details are hitherto unknown [9]. However, this remains an isolated example of group 2-initiated ethylene polymerization. The s-block metal complexes are much more

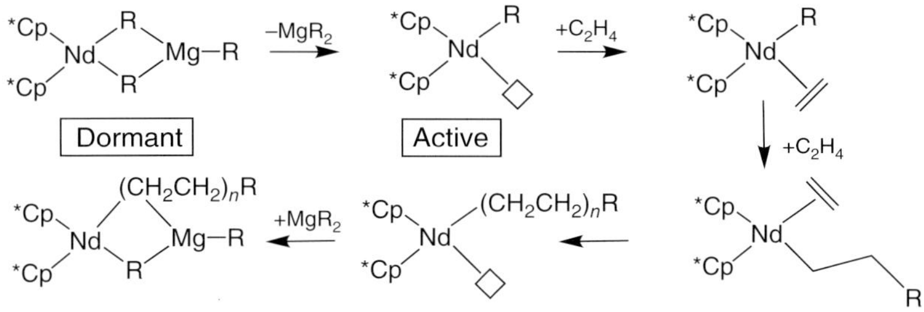

Scheme 2.1 Living ethylene polymerization by a Nd(III) catalyst and MgR_2 as a chain transfer reagent.

popular as initiators for polymerization of activated (conjugated) alkenes such as styrene, butadiene, isoprene, or methylmethacrylate.

2.2.1 Styrene Polymerization

Although pure styrene itself is very sensitive to spontaneous oligomerization, controlled living polymerization to give long-chain polystyrene (PS) of defined molecular weight is generally achieved by alkyllithium initiators. For group 2, at least initiators with metals as heavy as Ca are needed. This is clearly demonstrated by the preparation of the styryl Mg Grignard *p*-CH_2=$CH(C_6H_4)MgCl$, a complex that is fully stable at room temperature [10]. Organocalcium complexes, however, are substantially more nucleophilic and, depending on the organic rest, allow for alkene addition. Eaborn's highly stabilized alkylcalcium complex $Ca[C(SiMe_3)_3]_2$ (**1**) [11] is not able to polymerize styrene. Likewise, complex **2** was found to be inactive [12]. In order to boost the reactivity of Ca species, less stabilized organocalcium complexes have been targeted. It was found that 2-dimethylamino-α-trimethylsilyl-benzyl $(DMAT)_2Ca{\cdot}(THF)_2$ (**3**) is an excellent initiator for styrene polymerization, allowing for a living polymerization giving polystyrene (PS) with a relatively narrow dispersion index ($M_n = 1.1 \times 10^5$, polydispersity index [PDI] = 1.369) [13]. Because the reactivity of the Ae—C bond increases going down group 2 (because of increasing polarity and decreasing bond strength), heavier Sr and Ba organyls will be increasingly more reactive in styrene polymerization.

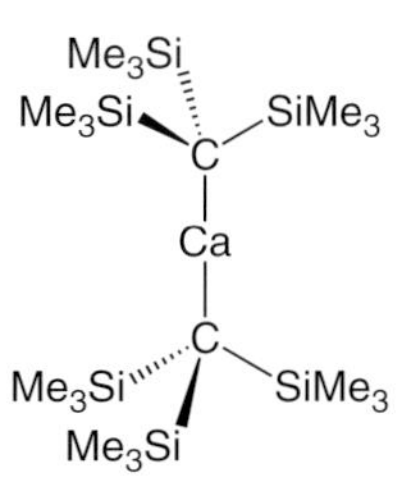

1

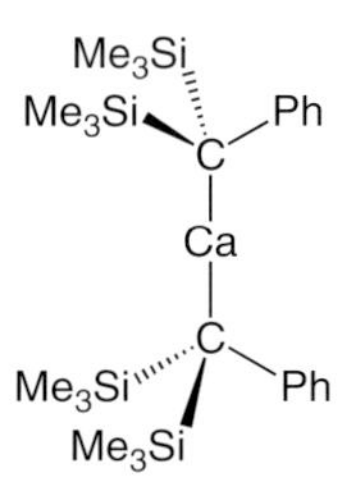

2

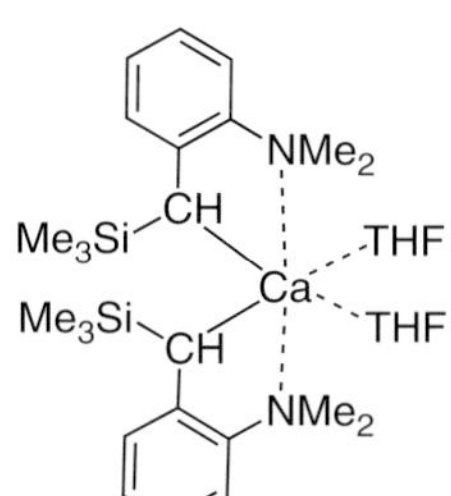

3

The main objective to use exotic organometallic Ca, Sr, or Ba initiators has been the idea to combine the advantages of a living polymerization with those of a stereocontrolled polymerization. Polystyrene is a polymer in which each second C atom is substituted with a Ph group and therefore atactic (*a*-PS, random stereocenters), isotactic (*iso*-PS, all stereocenters are the same), or syndiotactic (*syn*-PS, the stereocenters show alternating chirality) are possible. Although both *iso*-PS and *syn*-PS are regular crystalline polymers with high melting points, *syn*-PS (mp = 273 °C) is clearly the most desired material because of its very fast crystallization rate [14]. It has been discovered by the Idemitsu company in 1985 [15] and can be obtained with a very high regularity ($r > 99\%$) using titanocene half-sandwich complexes activated by methylaluminoxane (MAO) generating the cation $Cp^{*}TiMe^{+}$ (Scheme 2.2). This highly Lewis acidic species activates styrene by coordination after which migratory insertion in the Ti—Me bond follows. This coordination–insertion mechanism strongly contrasts with that for anionic styrene polymerization by a *n*BuLi initiator. In this case, the anion is responsible for polymerization activity and the loosely bound, less Lewis-acidic Li^{+} is less important for alkene activation. Anionic styrene polymerization is living and therefore has attractive advantages such as MW control, block copolymer design, and choice of end group. Because anionic polymerization is not metal-controlled, the polystyrene obtained is fully atactic and has a low glass

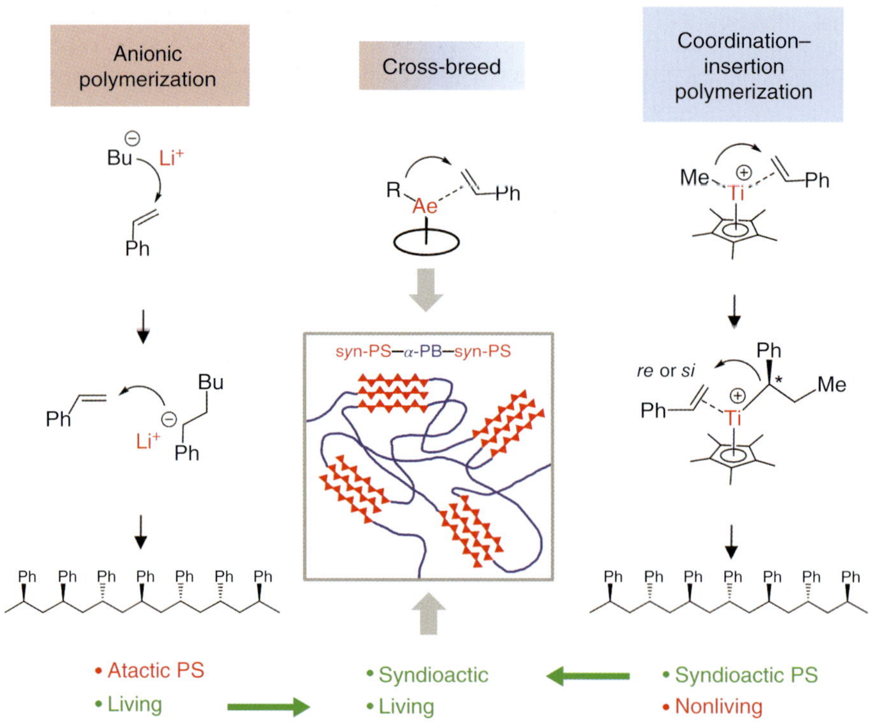

Scheme 2.2 Left: Classical anionic styrene polymerization. Right: Ti-catalyzed styrene polymerization by a coordination–insertion mechanism. Middle: A crossbreed Ae-based catalyst that is living (anionic polymerization) as well as syndioselective (coordination–insertion polymerization). These combined advantages allow for synthesis of a (*syn*-PS)–(*a*-PB)–(*syn*-PS) block copolymer.

temperature and melting point, clearly limiting its applications. Because the Ti-mediated polymerization is not living, the idea was to construct a crossbreed catalyst that is ionic in nature and therefore gives living polymerization, while it also contains a more Lewis acidic Ae^{2+} metal cation (Ca^{2+}, Sr^{2+}, or Ba^{2+}) that controls the stereoselectivity by a coordination–insertion mechanism. Styrene polymerization with a heteroleptic neutral LAeR complex (L = spectator ligand and R = reactive initiator) could combine the advantages of a living polymerization with those of stereoregular chain growth. Such catalysts would be able to produce a styrene–butadiene block copolymer PS–PB–PS in which the polystyrene is preferably syndiotactic. In that case, the polymer consists of highly crystalline, strongly bound *syn*-PS blocks attached to each other by multiple flexible atactic *a*-PB wires (Scheme 2.2). Such a designer plastic would be a high-impact elastomer with thermoplastic properties, i.e. it could simply be processed by melting the *syn*-PS junctions that, after cooling, crystallize again, cross-linking the polymer chains.

Research on "crossbreed" LAeR catalysts, similar to that shown in Scheme 2.2, started with the synthesis of a potential Ba initiator. The large size of this metal was thought to be advantageous for stereocontrol as the polystyryl chain will likely show secondary Ba···Ph interactions, effectively restricting the degrees of freedom of the growing chain. Two main problems were identified: (i) simple synthetic access to highly reactive organobarium complexes was needed, and (ii) the highly ionic rather than long Ba–ligand bonds are weak and ligand exchange is facile. Ligand scrambling by the Schlenk equilibrium would result in the formation of two polymerization active species: the heteroleptic complex LBaR and the homoleptic complex BaR_2. The latter would initiate the growth of two polymer chains and therefore generate two chiral chain ends, increasing the number of stereochemically different structures and therefore decreasing stereoselectivity.

The first problem was solved by development of a straightforward synthesis of $Ba(CH_2Ph)_2$ [16]. Similar to the synthesis of heavier group 1 metal reagents such as superbasic BuK, this was achieved by a metal exchange reaction. Reaction of 2 equiv of benzyllithium with either $Ba[N(SiMe_3)_2]_2$ or $Ba(OAr)_2$ (Ar = 2,4,6-*t*Bu-phenyl) produced dibenzylbarium as a highly reactive, dark-red powder. Previously, this compound was obtained by reaction of $Hg(CH_2Ph)_2$ with a Ba metal mirror [17], a route that is unfavorable because of the use of a highly poisonous mercury reagent. In order to stabilize and solubilize $BaBn_2$, it was reacted with 1,1-diphenylethylene (DPE, Scheme 2.3) giving a complex in which the large Ba^{2+} ion is better shielded by secondary Ba···Ph interactions (**4**). This complex dissolves well in aromatic solvents and polymerizes styrene in a living manner ($M_n = 9 \times 10^4$, PDI = 1.20) [16]. For conversion of **4** into a heteroleptic complex of type LBaR, the large multidentate ligand $Me_2Si(Ph)(C_5Me_4)$ was chosen (Scheme 2.3). Although this ligand can chelate the Ba^{2+} ion by interaction with the Cp and Ph groups, the heteroleptic complex **5** was not stable and is in equilibrium with the homoleptic species **4** and **6**. This clearly shows the pitfalls of heavier group 2 metal chemistry: the long and weak bonds to Ba are prone to ligand exchange. Homoleptic complex **6** could be isolated and was structurally characterized [16], Scheme 2.3. The Schlenk equilibrium could be steered to the heteroleptic side by addition of **6**, and this

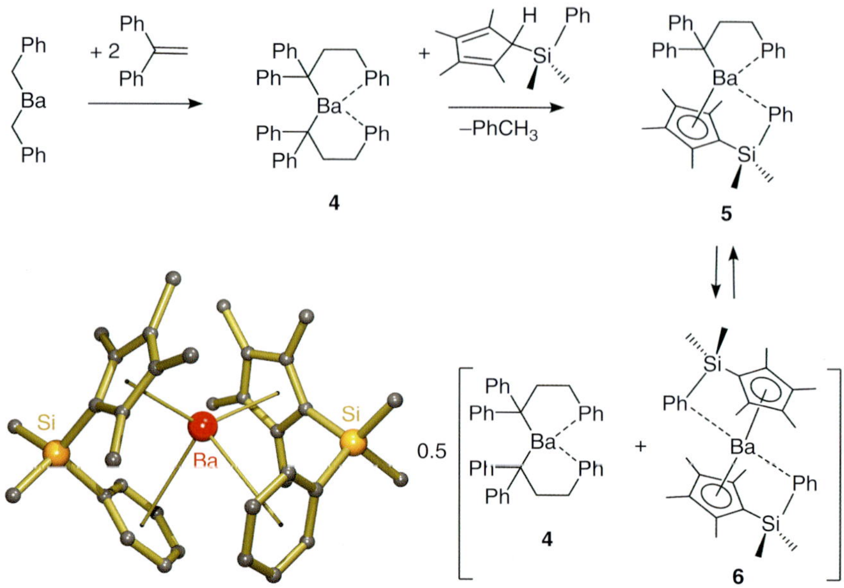

Scheme 2.3 Reaction of poorly soluble dibenzylbarium ($BaBn_2$) with 1,1,-diphenylethylene to give a soluble, stabilized initiator for anionic styrene polymerization (**4**). Further conversion into a heteroleptic initiator (**5**) is hindered by the Schlenk equilibrium. The crystal structure of the homoleptic complex **6** is shown.

mixture of mainly complex **5** and **6** has been used as a catalyst for styrene polymerization. Although long-chain polystyrene with $M_n = 1.2 \times 10^5$ and a relatively narrow MW distribution (PDI = 1.22) was obtained, it was found to be essentially atactic [16].

Reasoning that the larger Ba metal may have ample space for secondary interactions but that inherently these are also very weak, research was redirected to initiators based on Ca, a metal that is still to be considered large but is significantly more Lewis acidic than Ba, thus forming stronger secondary interactions. Heteroleptic complexes of type LCaR (L = inactive spectator ligand, R = active initiator) could be synthesized by different routes. (i) Salt metathesis of CaI_2 with two different potassium reagents LK and RK, (ii) deprotonation of a spectator ligand LH by reactive CaR_2, and (iii) ligand exchange between the homoleptic species L_2Ca and CaR_2. The synthesis of well-defined heteroleptic Ca complexes was not as straightforward as anticipated. Cyclopentadienide, amidinate, or β-diketiminate spectator ligands produced homoleptic complexes (or at its best homoleptic–heteroleptic mixtures) [18]. One ligand class, however, was found to be privileged to stabilize heteroleptic complexes. The ligand 9-Me_3Si-fluorenyl produced heteroleptic complexes that are even at higher temperatures in toluene-d_8 fully stable toward ligand exchange (**7**–**9**, Scheme 2.4). Also, the active initiating group varied. The DMAT ligand is an anion in which the negative charge is stabilized by delocalization into the aromatic ring and by negative hyperconjugation of the benzylic Me_3Si substituent. In complex **8**, the Me_3Si substituent is substituted by an electron-donating Me group making

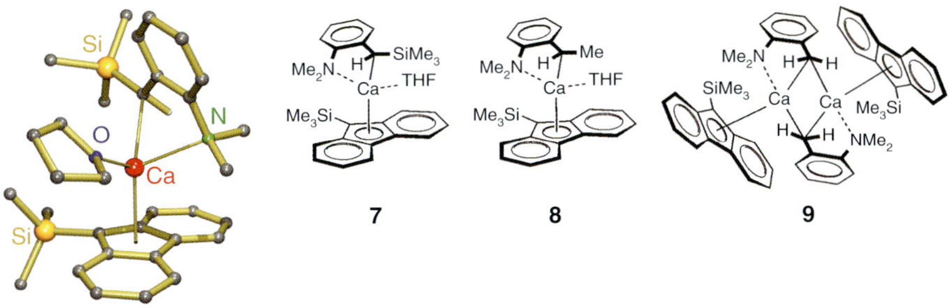

Scheme 2.4 Heteroleptic complexes stabilized by the 9-Me_3Si-fluorenyl ligand.

the initiating group considerably more reactive and faster in initiation styrene polymerization [19]. Complex **9** contains a benzylic CH_2 group that is small enough to bridge two Ca^{2+} ions, thus resulting in a dimeric structure [20]. Because of the bridging nature of the benzylic group, the complex could be obtained free of THF.

Styrene polymerization with these well-defined heteroleptic Ca initiators in cyclohexane formed polymers of $M_n = 10^5$ but with PDIs that are clearly higher (2.2–2.3), which is due to slow initiation rather than chain termination as could be shown by stepwise polymerization [19–21]. The polystyrene obtained, however, showed that there is hardly a preference for syndiotactic insertion. In respect to polystyrene prepared with an *n*BuLi initiator, the polymer was only slightly enriched in syndiotactic diads (*r*). It was suggested that poor stereocontrol may be due to configurational instability of the chiral chain end [21]. This would mean that any stereocontrol is racemized before the next insertion. Racemization of the chiral benzyl C atom in **7** already hinted to this problem. At room temperature, the fluorenyl ligand in **7** gives eight different aromatic ^{1}H NMR (nuclear magnetic resonance) signals as both sides of the ligand are diastereotopic. At 90 °C, coalescence of the two diastereotopic sides is achieved ($\Delta G = +18.8$ kcal/mol). Chirality change at the benzylic C center proceeds by a Ca—C bond breaking process and was found to be accelerated by addition of THF.

Considering that stereoerrors originate from facile inversion of the chiral chain end, temperature and monomer concentration should have an effect on the stereoselectivity. Although lower temperatures showed a slightly higher syndioselectivity, styrene concentration changes had the largest influence (Scheme 2.5) [21]. Polymerization in pure styrene was quite controlled and produced polystyrene with at least 88% *r*-diads as determined by ^{13}C NMR analysis. It was found that the previous assignments of ^{13}C signals based on a pentad/hexad level of signal resolution (which means that the ^{13}C signal is influenced by two/three units further down the chain) is wrong. According to Bernoullian statistics, a hexad/heptad assignment was proposed [22], which fits the 1/1/1 ratio for signals assigned to *mrrrrr*, *rmrrrr*, and *rrmrrr* (see Scheme 2.5b). This means that the ^{13}C chemical shift is influenced by at least three units further down the chain. The syndioselectivity of styrene polymerization with **7** could be slightly improved from $r = 88\%$ to 92% by lowering the temperature to −20 °C.

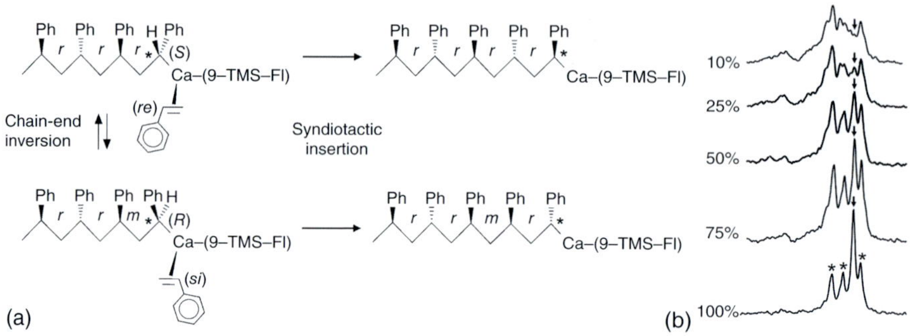

Scheme 2.5 (a) Chain end inversion results in a single error (*rrrr* → *rrmr*); 9-TMS-Fl = 9-Me_3Si-fluorenyl. (b) ^{13}C NMR ($C_2Cl_2D_2$) signal for the C_{ipso} in the Ph group of polystyrene obtained with initiator **7** at different styrene concentrations (wt% in cyclohexane). The signal for the syndiotactic heptade *rrrrrr* is marked by an arrow. The error heptades *mrrrrr*, *rmrrrr*, and *rrmrrr* are marked by *.

The strong dependency of polystyrene tacticity on the concentration of styrene monomer can be explained as follows. Inversion of the chiral chain end is a process that is independent of styrene concentration; however, styrene insertion accelerates with increasing styrene concentration. The ratio (chain end inversion) : (styrene insertion), which corresponds to the percentage of stereoerrors, is therefore drastically decreased by increasing styrene concentration.

As the aforementioned chirality change at the benzylic C center in **7** was found to be accelerated by addition of THF, the influence of polar solvents on the syndioselectivity of styrene polymerization was investigated. Styrene polymerization with the THF-free initiator **9** was expected to show better selectivities, but it was found that catalysts **7** and **9** produced polystyrene of essentially similar syndiotacticity [20]. Addition of 5 equiv of THF, however, resulted in decreased stereoselectivity. This demonstrates that polar solvents should be avoided.

It has been attempted to further improve syndiotacticity by steric control: increasing the bulk of the fluorenyl ligand in **7** by replacement of the Me_3Si substituent for a hypersilyl $(Me_3Si)_3$Si substituent (complex **10** in Scheme 2.6) could lead to shorter distances between the chiral chain end and coordinated styrene, improving communication and stereoselectivity. Although the hypersilyl

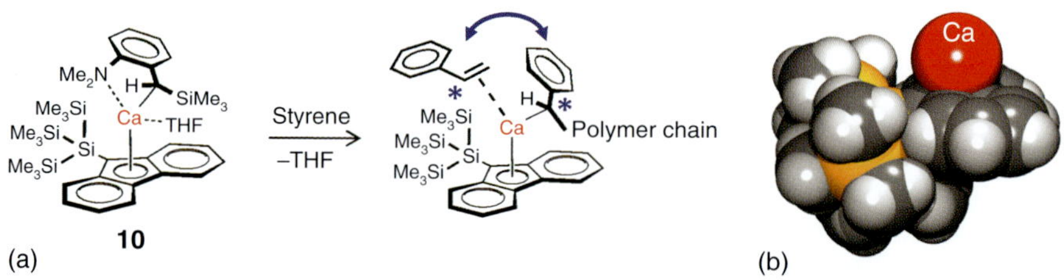

Scheme 2.6 (a) Using a benzylcalcium initiator with a superbulky hypersilyl substituent increases communication between chiral chain end and styrene monomer. (b) Space-filling model of the 9-$(Me_3Si)_3$Si-fluorenyl calcium unit.

substituent led to a slight increase of syndioselectivity, it drastically reduced the activity of the initiator giving chain lengths that are only 10% that of those obtained with initiator **7** (under same conditions) [23]. This is likely due to blocking of free coordination sites at the Ca center by the very large hypersilyl group (Scheme 2.6b). Indeed, a benzylcalcium initiator with an intramolecular coordinating $Me_2NCH_2CH_2$ substituent in the 9-position of the fluorenyl ligand completely inactivated the initiator [24]. This observation demonstrates that free coordination sites at Ca are essential and supports the proposed coordination–insertion mechanism. Large groups in close vicinity of the metal center will always block free coordination space at the Ca metal, thus retarding insertion. This increases the ratio (chain end inversion) : (styrene insertion) and decreases the syndioselectivity. For that reason, it was investigated whether large substituents in remote 2- and 7-positions of the fluorenyl ligand have a positive effect on stereocontrol. It was found that increasing the bulk of these substituents has a substantial effect on the syndioselectivity [24]. Styrene polymerization in dilute cyclohexane solution (10 wt%) could indeed be controlled by ligand modification. Although the spectator ligand 9-Me_3Si-fluorenyl showed hardly any syndioselectivity ($r = 60\%$), introduction of *t*Bu groups in 2- and 7-positions already showed a significant increase in selectivity ($r = 75\%$). Further increase of the substituent size gradually improved the syndioselectivity up to $r = 95\%$ for (p-*t*Bu-phenyl)$_2$MeC substituents (Scheme 2.7). The crystal structure analysis of

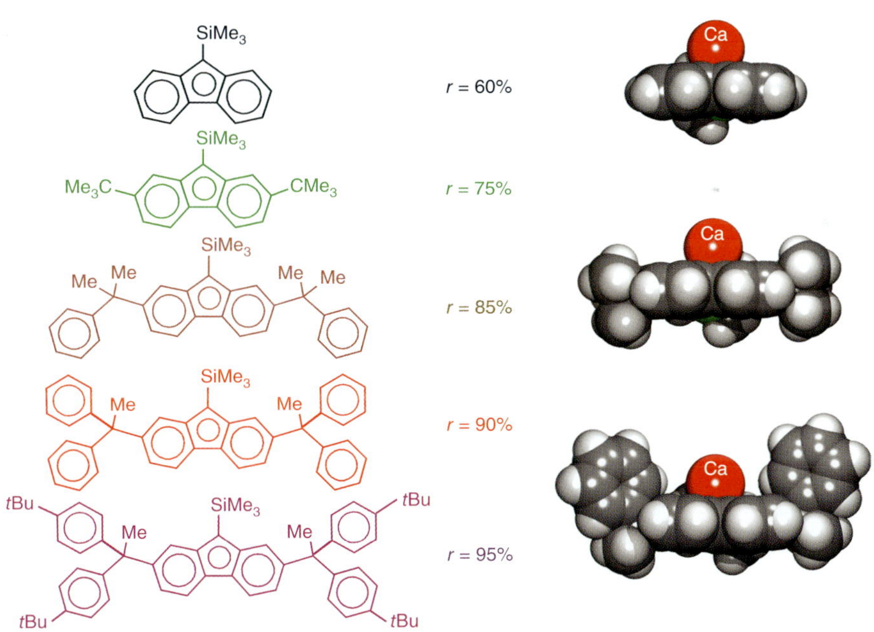

Scheme 2.7 Influence of the remote substituents in 2- and 7-positions of the fluorenyl ligand in initiator **7** on the syndioselectivity in styrene polymerization (10 wt% styrene in cyclohexane). Increasing the substituent size along the row H < *t*Bu < $C(Me)_2Ph$ < $C(Me)Ph_2$ < $C(Me)(C_6H_4\text{-}t\text{Bu})_2$ showed an increase of *r*-diads from 60% to 95%. Shown are space-filling models for Ca bound to the fluorenyl ligands with 2,7-substituents H, *t*Bu, and $C(Me)_2Ph$.

some of the catalysts shows that large substituents in 2- and 7-positions generate a cage around the Ca center, thus restricting the dynamics of the growing polymer chain. It is believed that large substituents in remote positions create an enzyme-like pocket around the metal that controls the selectivity and rate of chain end inversion by secondary interactions between the polystyrene chain and the substituents.

These results show that the activity and stereoselectivity of benzylcalcium initiators in styrene polymerization can be controlled by the spectator ligand. This means that the general principles of organometallic catalysis can be applied to Ca catalysis. Organocalcium-initiated alkene polymerization is not simply anionic polymerization in which the metal is less important. Instead, it follows a coordination–insertion mechanism in which the metal has, apart from alkene activation, also a templating role that allows for stereocontrol by the spectator ligands. The syndioselectivities reached by organocalcium initiators (up to 95% *r*) are not as good as those obtained by Ti(III) [15] or Ln(III) [25] catalysts (99% *r*); Ln = lanthanide. This is most likely due to the much faster inversion of chain end chirality which is due to the weaker Ca^{2+}—C bond vs. the Ti^{3+}—C and Ln^{3+}—C bonds (Scheme 2.8). The Li^{+}—C bond is even weaker and chiral carbanions bound to Li racemize fast (the $\Delta G^{\ddagger}$ for inversion of benzylic carbon in chiral benzyllithium complexes is c. 9 kcal/mol) [26]. It is therefore unlikely that organolithium initiators may be used for highly stereoselective styrene polymerization. In this respect, organocalcium initiators fill the gap between anionic living polymerization and stereoselective polymerization. Noteworthy is the fact that the highly syndioselective Ti(III) styrene polymerization catalysts can now also be used in a (semi)living manner by addition of alkylaluminum transfer reagents [27]. This shows that the main group metal and transition metal chemists work hand in hand to achieve efficient living and stereoselective styrene polymerization.

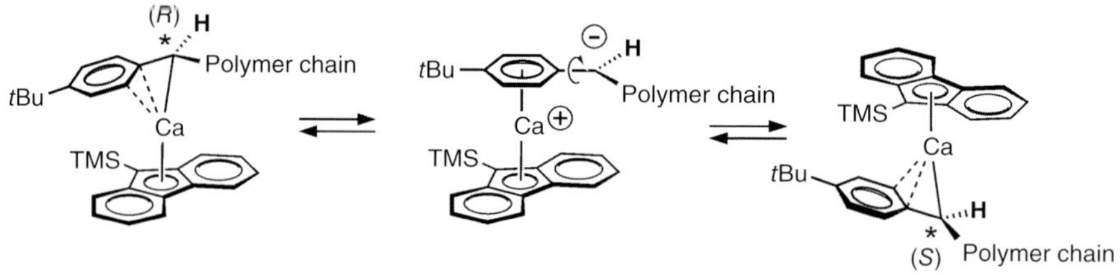

Scheme 2.8 The proposed mechanism for chain end inversion includes a Ca—C bond breaking step. Introduction of a *t*Bu group in the para-position of styrene diminishes delocalization of the negative charge impeding Ca—C bond cleavage.

2.2.2 Polymerization of Modified Styrene

As modified styrene monomers are costly compared to the bulk chemical styrene itself, (co)polymers thereof would only be attractive for specialty applications. It is known that a *t*Bu group in the para-position of styrene increases polymer solubility [28]. Also, statistical styrene/*p*-*t*Bu–styrene copolymerization has been reported [29]. Variation of substituents in the para-position of styrene influences the inclusion of statistical copolymer: the insertion rates clearly follow Hammett's law [30]. Slower insertion of *p*-*t*Bu–styrene can be explained by the inductive +I effect of the *t*Bu group, raising the contribution of a resonance structure with a positive charge at the *para*-C and a formal negative charge at the terminal vinylic C, compromising a nucleophilic attack at this C. The effect of para-substitution on the stereoselectivity of styrene polymerization has been less studied [31]. It is hard to predict the influence of styrene substituents on stereoselectivity. For the Ca-initiated *p*-*t*Bu–styrene polymerization, a slower insertion of *p*-*t*Bu–styrene suggests that chain end inversion becomes relatively fast, thus resulting in less stereocontrol. On the other hand, the inductive effect of the *para*-*t*Bu group also results in localization of the negative charge at the chiral chain end (Scheme 2.8). This makes the Ca—C bond to the chain end stronger, inversion of chirality slower, and therefore the tacticity higher.

The polymerization of *p*-*t*Bu–styrene with the benzylcalcium initiators **7** and **9** has been investigated (cyclohexane, 50 °C) (S. Harder and S. Müller, unpublished results). For an initiator/*p*-*t*Bu–styrene ratio of c. 100, polymers of c. 500 styrene units were isolated. This demonstrates that insertion is faster than initiation. The following molecular weights and PDIs were determined: $M_n = 9.5 \times 10^4$ and PDI = 2.79 for initiator **7** and $M_n = 8.7 \times 10^4$ and PDI = 1.67 for initiator **9**. The tacticity of the polymers was determined by high-temperature ^{13}C NMR in tetrachloroethane-d_2. The diagnostic signal in the ^{13}C NMR spectrum is the C_{ipso} signal in the aryl ring (142.5 ppm), see Scheme 2.9. Assuming a resolution at the hexad/heptad level, three minor signals due to a single error insertion were detected: *mrrrrr*, *rmrrrr*, and *rrmrrr*. The probability for a syndiotactic insertion is r = 94% for polymerization in 10 wt% solution (or 97% for polymerization in bulk *p*-*t*Bu–styrene). The syndioselectivity for polymerization of normal styrene with complex **7** at 20 °C in 10 wt% solution was r = 60% (or r = 88% for polymerization in pure styrene). These results clearly show that a *t*Bu substituent in a remote position of the styrene monomer substantially increases the selectivity of Ca-initiated polymerization. As of yet, it is not clear whether this originates from slower chain end inversion (because of the inductive effect of the *p*-*t*Bu substituent) or from a decrease in the flexibility of the polymer chain resulting in higher enantioselectivities for the insertion step.

Polyvinylpyridine is another polymer closely related to polystyrene. It has interesting applications such as a carrier for reagents or a support for acids or organometallic catalysts, in polymer engineering, as an electrolyte in batteries, or as a resin for ion exchange [32]. Polymerization is generally initiated by transition metal catalysts, radicals, or Lewis acids. Anionic polymerization of 2-vinyl- or 4-vinyl-pyridine with organolithium initiators is difficult because of possible side reactions with the pyridine ring (deprotonation or addition). Limited research on group 2 metal initiators for 2-vinyl-pyridine polymerization has been published: initiators Ph_3CAe (Ae = Ca, Sr, and Ba) produced poly-2-vinyl-pyridine that is slightly isotactically enriched with stereoselectivities decreasing the row Ca > Sr > Ba [33]. Recently, calculations for the Ca-initiated polymerization of 2-vinyl-pyridine have been reported

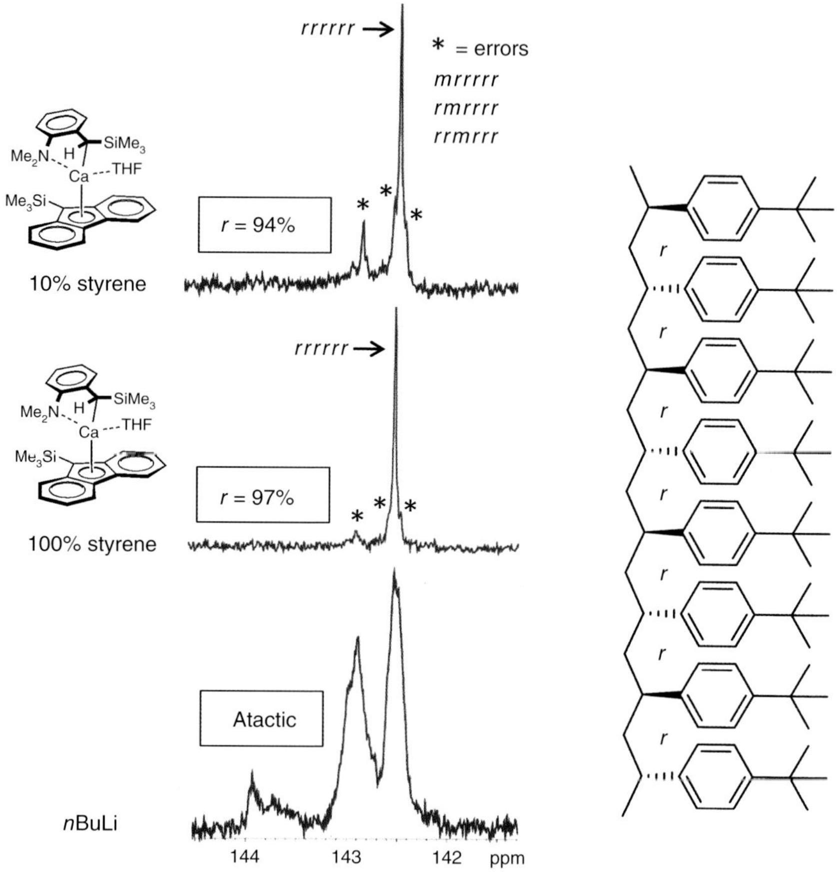

Scheme 2.9 Syndioselective polymerization of *p*-*t*Bu–styrene by benzylcalcium complex **7** compared to atactic polymerization with *n*BuLi. Solution polymerization (10 wt%, 50 °C) already gives high selectivities with *r* = 94%. Shown is the ^{13}C NMR signal for C_{ipso} in the Ph group of polystyrene.

(Scheme 2.10) [34]. This calculational study is related to catalytic alkene hydrophosphination with the β-diketiminate catalyst (DIPP-nacnac)$CaPPh_2$ [35]. Although hydrophosphination of styrene is an efficient transformation, attempts to hydrophosphinate 2-vinyl-pyridine led to exclusive polymerization. Calculations show that the C,N-chelating carbanion Ph_2PCH_2CH(pyridine)$^-$ has mesomeric forms in which formal charges on C or N can be formulated (Scheme 2.10). Insertion of 2-vinyl-pyridine proceeds through an N-bound amide and needs an activation energy of less than 10 kcal/mol. The product, an N-bound amide, contains a planar benzylic C atom and it is currently unclear which insertion product is favored (*r*- or *m*-insertion). The facile anionic polymerization of vinylpyridine and the general possibility to control stereoselectivities by ligand design may stimulate further research in Ca-initiated vinyl-pyridine polymerization.

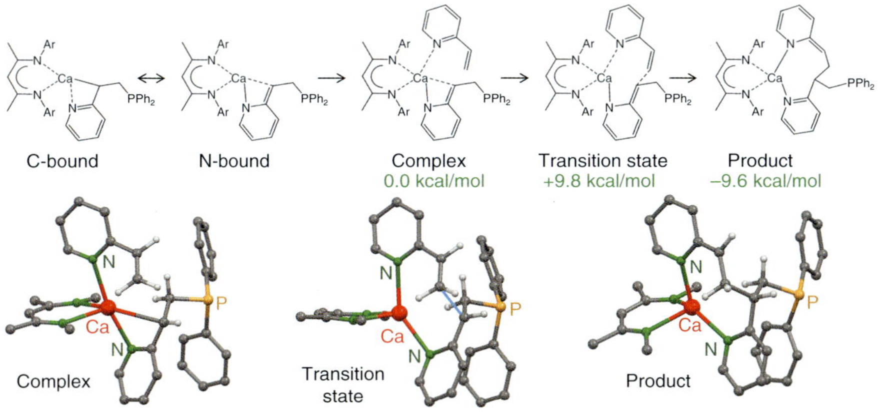

Scheme 2.10 Density functional theory (DFT) calculated pathway for 2-vinyl-pyridine polymerization (ΔG values) [34].

2.2.3 Polymerization of Butadiene or Isoprene

Block copolymers containing polybutadiene (PB) or polyisoprene (PiP) segments are highly important synthetic materials on account of their rubber-like properties that, especially for the car tire market, are of enormous commercial value. Controlled catalytic polymerization of these dienes is of high interest, not only from an industrial point of view but also for its scientific challenge [36].

Compared to alkene polymerization, the polymerization of dienes is inherently more complicated. Not only the stereochemistry but also the regiochemistry of insertion has to be taken into account. Polymerization of a simple diene like butadiene can give three different products (Scheme 2.11). 1,4-Polymerization can give cis- or trans-isomers and 1,2-polymerization affords a polymer chain with a chiral center giving rise to various polymer microstructures (atactic, isotactic, and syndiotactic). Polymerization of isoprene, a substituted butadiene, is even more complicated and at least four different types of insertion should be considered (Scheme 2.11).

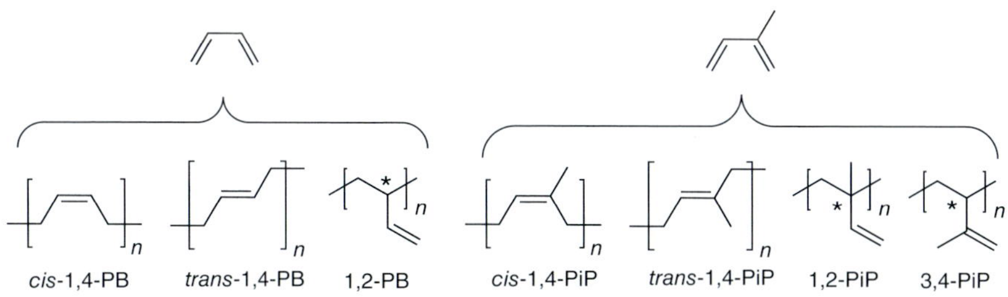

Scheme 2.11 Stereoregular different polybutadienes (PBs) and polyisoprenes (PiPs).

Most important for car tire manufacturing is the 1,4-*cis*-PB structure which, when incorporated in a block copolymer such as PS–(1,4-*cis*-PB)–PS, is essential to the production of SBS rubber. Control over the regio- and stereoselectivity in diene polymerization is therefore an important scientific goal. The key to controlled polymerization is the organometallic catalyst. It has been suggested that catalysts with large vacant sites could bind dienes in a η^4-manner giving rise to 1,4-insertion, whereas η^2-diene coordination preferentially gives 1,2-insertion (Scheme 2.12) [37]. Unfortunately, a full mechanistic consideration is somewhat more complicated [36] and also the initial 1,2-insertion process may lead to 1,4-insertion through a range of isomerization processes in the growing polymer chain. From this complexity, it is clear that selectivity can be steered to a great extent by choice of metal, ligand, and reaction conditions.

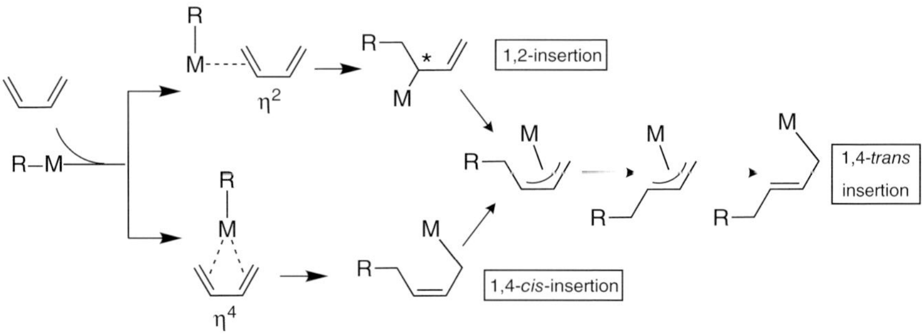

Scheme 2.12 Different modes of diene insertion for a general organometallic catalyst R–M.

The industrial catalyst for PB with a high 1,4-*cis*-PB content is a rather undefined, empirically designed, multicomponent catalytic system composed of Nd(III) carboxylates (or alkoxides or halides) and aluminum cocatalysts containing alkyl, halide, or hydride ligands. Also, Ni, Co, or Ti catalysts have been intensively investigated. The s-block metal catalysts are much less researched, but it was found that the larger Lewis acidic group 2 metal cations in general control the tacticity of butadiene polymerization favorably. The Firestone Tire and Rubber company filed a patent in which *n*BuLi is used as an initiator giving PB with 40% 1,4-*cis*, 50% 1,4-*trans*, and 10% 1,2-insertion [38]. The very high dynamics of organolithium compounds is reflected in the low selectivity. Because lithium reagents can be controlled substantially by solvation, it is no surprise that the selectivity of butadiene polymerization can be influenced by polar solvents. Indeed, addition of small quantities of ethers showed an increase to 90% 1,2-addition. Group 2 metal initiators or additives have been used to control selectivity. The Michelin company filed a patent in which Ba-containing anionic initiators formed a PB with 5% 1,2-insertion and 95% 1,4-insertion with a very high trans-content [39]. An initiator obtained by the reaction of Ba metal with DPE formed a complex identified as $Ba^{2+}(Ph_2C{-}CH_2{-}CH_2{-}CPh_2)^{2-}$, which is a highly active initiator for butadiene polymerization in benzene to give 11% 1,2-insertion, 76% 1,4-*cis*, and 13% 1,4-*trans*-insertion [6]. In polar solvents, the 1,4-*trans*-content is significantly higher. Unfortunately, the Ba initiators in these studies have not been fully characterized. Hitherto very little is known on butadiene polymerization with well-defined group 2 metal initiators. Okuda and coworkers polymerized butadiene with the bis-allylcalcium initiator $Ca(C_3H_5)_2$ in THF [40]. An initiator/butadiene ratio of 1/100 resulted in polymers with $M_n = 6.6 \times 10^3$ and a narrow PDI of 1.15. Then, selectivity was found: 48% 1,2-insertion, 23% 1,4-*cis*-insertion, and 29% 1,4-*trans*-insertion. The well-defined Ca complex **9** (Scheme 2.4) and $(DMAT)_2Ca{\cdot}(THF)_2$ (**3**) have been used as the initiator for isoprene polymerization in cyclohexane [18]. Complex **9**: 1,2-/3,4-/1,4-insertion = 11/37/52. $(DMAT)_2Ca{\cdot}(THF)_2$ (**3**): 1,2-/3,4-/1,4-insertion = 22/26/52. The sum of 1,2- and 3,4-addition is for both systems, 48%, which equals the percentage of 1,2-insertion observed for butadiene insertion with $Ca(C_3H_5)_2$. Although these experiments demonstrate that the microstructure of diene polymers can be tuned by choice of metal and ligands, the underlying mechanism that would allow fine-tuning of stereoselectivities is hitherto not well understood and there is ample opportunity for research activities in group 2 metal-controlled diene polymerization.

2.3 Polymerization of Polar Monomers

The vast majority of industrial bulk polymers is based on alkenes and consequently finds their origin in natural oil and gas. The threatening depletion of fossil fuels stimulates investigations in polymers based on renewable sources. The difficult transition from oil to plants is a slow process that is already ongoing for a few decades. Since the last years, however, this urge is reinforced by the growing awareness that the combination of highly resistant oil-based polymers coupled to a single-use-throwaway mentality has major negative impacts on our environment. It is estimated that annually between 5 and 13 million tons of polymer waste ends up in the oceans [2]. If the current rate of pollution is maintained, by 2050, the total weight of all plastic will equal that of all fish swimming in the ocean [41]. Polymer debris from large to small nanoparticles seriously harms sea life. Animals may be entangled, poisoned by plasticizers and other polymer additives, or choked by consuming plastic. Even the detrimental effects of nanoplastic are slowly starting to emerge and it is hitherto unclear how this will affect the seafood eating population. Clear, however, is the fact that polymer waste has to be reduced drastically either by recycling and/or by developing biodegradable plastics. This is an enormous challenge with many pitfalls. Most biodegradable products lack the characteristic flexibility, strength, or toughness of oil-based polymers. Also, the timing of self-degradation needs to be adjusted, which is considering the enormous differences in the conditions of natural environments (desert, ice, or ocean) likely an impossible task [42]. Biodegradable plastic and polymers from renewable resources have, however, one thing in common: they often contain C—O linkages that are weaker than C—C bonds and generally easily broken. This part of the chapter deals with polymerization of some selected polar monomers for which s-block metal catalysts play, or may play, an important role.

2.3.1 Polymerization of Lactides

Polylactide (PLA), the polymer from lactide or alternatively dilactide (the condensation product of two lactic acid molecules), is worldwide attracting considerable interest as a biodegradable thermoplastic that can be produced from renewable feedstocks such as corn or sugarcane. With an annual production capacity of >10^5 tons, it is not only used as a major bioplastic finding applications as a specialty polymer in the biomedical field (e.g. slow drug release or biodegradable sutures) but also becomes increasingly popular as a commodity polymer that may one day replace the polyolefins. Indeed, PLA in which chain lengths and stereoselectivity have been properly tuned can have properties resembling those of the polyolefins. Especially, the stereoregularity of the polymer may have an enormous influence on the rate of biodegradation and its mechanical properties [43].

Scheme 2.13 outlines the full pathway for PLA production. Starting form CO_2 and H_2O, which are converted to biomass by plants using solar energy,

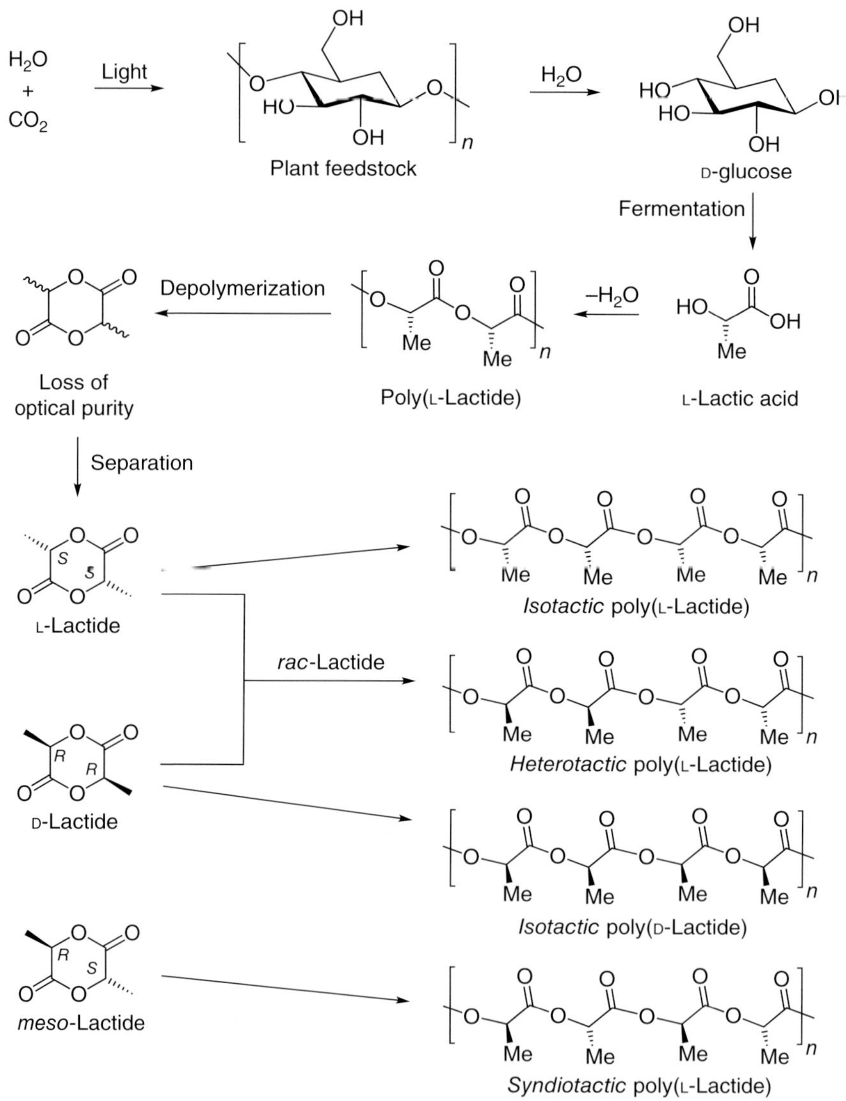

Scheme 2.13 Stereoselective production of PLA starting from CO_2 and H_2O.

starch is depolymerized to D-glucose, which is fermented by bacteria to optically active L-lactic acid. This is polymerized by simple condensation to PLA of molecular weight 1000–5000. A depolymerization process transforms these low-molecular-weight oligomers/polymers into lactide, a process in which the stereocenters are partially racemized. Separation of the stereoisomers gives three different lactides: L-lactide, D-lactide, and *meso*-lactide (the 1/1 mixture of D- and L-lactide is known as *rac*-lactide), the purity of which may crucially affect the properties of the PLA obtained [44]. Ring-opening polymerization generally does not create new chiral centers and also does not change existing stereocenters (racemization may only occur when strong bases are used). A general ring-opening mechanism for lactide polymerization including backbiting, which results in broad molecular weight distributions, is shown in Scheme 2.14. Isotactic poly(L-lactide) is simply obtained by homopolymerization of L-lactide, whereas its antipode is formed by D-lactide polymerization. Polymerization of *rac*-lactide may, depending on stereoselectively control, generate different PLAs: isotactic, heterotactic, atactic, or stereoblock PLA such as $\{[\text{D-PLA}]-[\text{L-PLA}]\}_n$, which arises from an occasional error in the isotactic propagation. Similarly, *meso*-lactide polymerization may produce atactic, syndiotactic, or heterotactic PLA. Stereocontrol can either be achieved by chain end control, in which case the chiral chain end communicates with the next monomer to be inserted, or by site control that makes use of communication between a chiral ligand and the next monomer to be inserted.

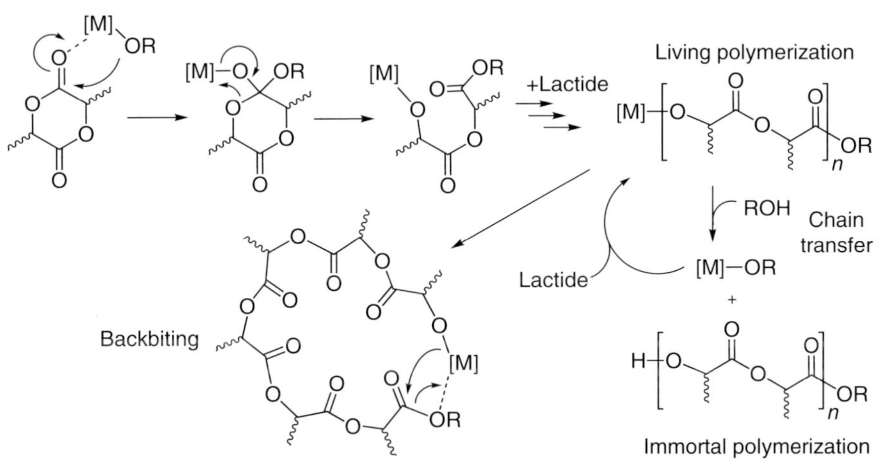

Scheme 2.14 General mechanism for the ring-opening polymerization of lactide by a metal alkoxide initiator showing chain transfer (with excess ROH) and backbiting giving rise to chain shortening and a broad molecular weight distribution.

The main players in the field are catalysts based on metals from group 3 (including the lanthanides) [45], group 4 (Ti, Zr, and Hf) [46], and group 13 (mainly Al) [47], i.e. typical polar organometallics with Lewis acidic metal sites. There is, however, an increasing interest in using group 2 and group 12 metal catalysts (Mg, Ca, and Zn) [48]. As PLAs find biomedical applications in the human body, especially the use of catalysts with metals such as Mg and Ca is highly attractive.

In addition, these very abundant environmentally friendly metals would also be a good choice when producing biodegradable throwaway polymers.

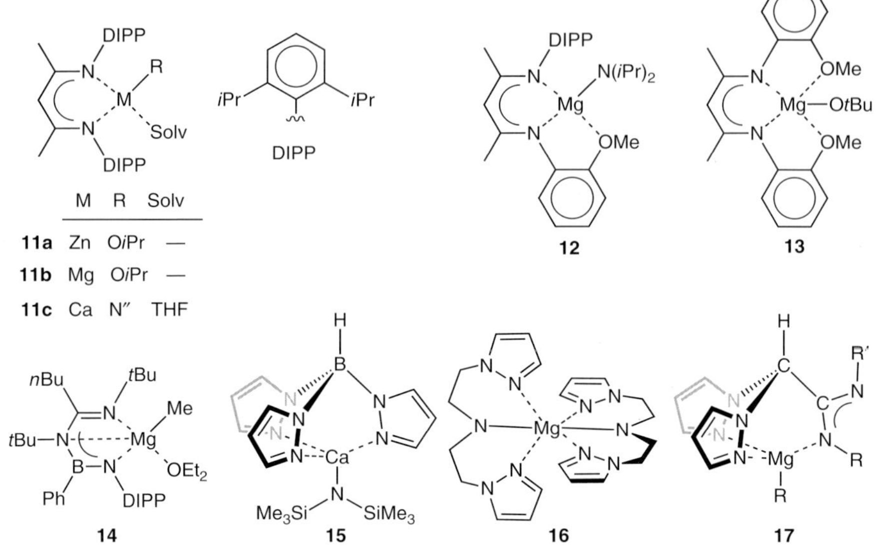

Westerhausen, Feijen, and coworkers found the simple Ca amide complex $CaN''_2\cdot(THF)_2$ ($N'' = N(SiMe_3)_2$) to be a slow initiator for L-lactide polymerization, but increased reaction rates were obtained by addition of small amounts of alcohol giving PLA with an extremely narrow PDI of 1.03 and an isotactic microstructure, demonstrating that there is no base-catalyzed epimerization [49]. Chain end stereocontrol needs a bulkier ligand: the most well-known Zn initiators are based on complexes with a β-diketiminate ligand. The Coates complex **11a** polymerizes *rac*-lactide in a highly controlled way (PDI = 1.10) to give a heterotactic polymer with a probability for a racemic insertion of $P_r = 90\%$ [50]. The analog Mg complex **11b**, however, showed extremely fast *rac*-lactide polymerization but an essentially atactic PLA [51]. Chisholm et al. demonstrated the important role of solvent: although benzene or CH_2Cl_2 resulted in atactic PLA, polymerization in THF afforded a fair heterotacticity with $P_r = 90\%$ [52]. A similar Ca complex (**11c**), however, showed very slow lactide polymerization, which was explained with an ill-defined or polymeric (insoluble) Ca catalyst formed by Schlenk equilibria that, especially for the larger metals, pose a problem [53]. This was verified by reacting **11c** with *i*PrOH, which resulted already at −78 °C in the formation of homoleptic species including polymeric $[Ca(iOPr)_2]_n$.

Similar to enantioselective homogeneous catalysis, there is a delicate balance between catalyst activity and stereoselectivity: fast catalysts are usually not very selective and vice versa. It was therefore probed whether restriction of the metal's open coordination sphere would give a slower but more controlled reactivity. The Mg complexes **12** and **13** were investigated in *rac*-lactide polymerization and found to be still very fast showing poor stereocontrol [54]. Chivers et al. introduced Mg complexes with the *bamam* ligand (**14**) [55] that is closely related to the β-diketiminate ligand but features a slightly different coordination mode in which the backbone N also weakly interacts with the Mg metal. Polymerization of *rac*-lactide, however, was slower and less selective with a slight bias toward heterotactic insertion.

For the larger metal Ca, Chisholm introduced complexes with bulkier scorpionate ligands (**15**) [53]. These led to extremely fast *rac*-lactide insertion (>200 insertions/min), which was relatively well controlled: PDI = 1.74 and P_r = 90% (heterotactic PLA). The pioneering work by Chisholm and coworkers stimulated other research groups to develop tripod ligands. The group of Carpentier introduced amide ligands but could only obtain the homoleptic bis-amide Mg complex **16** in a pure form [56]. Complex **16** is highly active in *rac*-lactide polymerization (PDI = 1.33) giving an atactic polymer. A second modification of the scorpionate ligand has been introduced by Sánchez-Barba et al. [57]. Various alkylmagnesium complexes of type **17** initiated lactide polymerization but, depending on bulk, were found to be slow. Despite their poor performance, the polymers show a narrow molecular weight distribution (PDI = 1.09) but are essentially atactic. Similar, but chiral, heteroscorpionates showed a slow *rac*-lactide polymerization but with an interesting bias to isotactic PLA (P_m = 77%) [58]. Mountford and coworkers reported closely related heteroscorpionate ligands but with moderate activities and no stereocontrol [59].

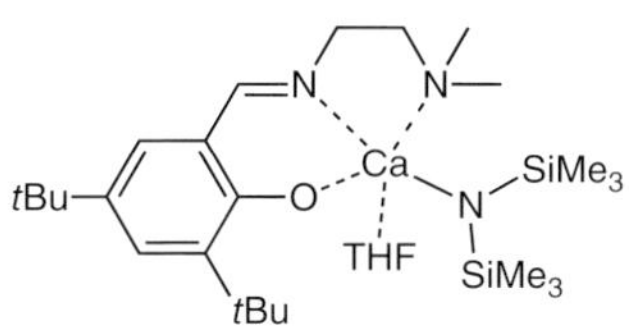

18

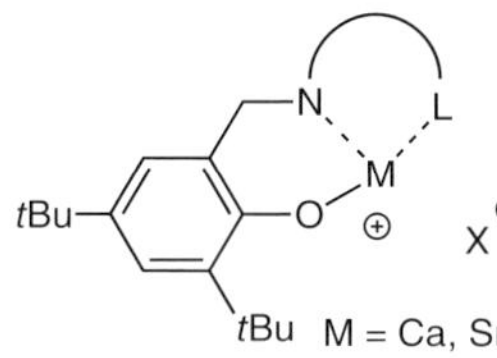

19 M = Ca, Sr, Ba
X = Weakly coordinating
L = Strongly coordinating

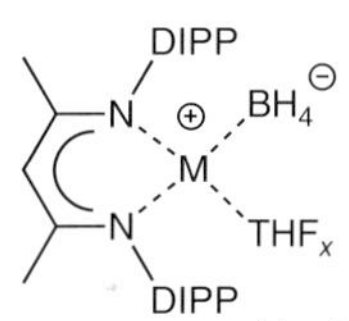

20 M = Mg, Ca
x = 1, 2

Many examples of catalysts based on phenolate or Schiff base ligands have been reported [48], most of these featuring modest to high activities but no stereocontrol. Exemplary is the Schiff base catalyst **18**, which polymerizes *rac*-lactide to essentially atactic PLA with very high activities and a PDI of 1.04 [60]. An interesting group of catalysts are the cationic phenolate (or alcoholate) complexes (e.g. **19**) by the Carpentier and Sarazin groups [61]. Their activity is based on the high Lewis acidity of these positively charged initiators that work in concert with an excess of BnOH or *i*PrOH giving immortal lactide polymerization. The alcohol not only functions as an initiating group but also as a transfer reagent. Thus, the initiator is a true catalyst and can grow more than one chain per metal (see Scheme 2.14). For L-lactide polymerization, the activity Ba ≈ Sr > Ca was found, and even the most reactive cationic Ba complexes showed no epimerization resulting in isotactic PLA. The Mountford group introduced cationic complexes with a weakly bound BH_4^- counter anion (**20**) [61]. Chain end group analysis showed hydroxyl terminus, indicating that the polymerization is initiated by hydride transfer from B–H to lactide. Although the *rac*-lactide polymerization was very fast, essentially, atactic PLA was isolated. An increase of PDI during the chain growth indicates that backbiting is a problem (Scheme 2.14).

Discussed in here are selected examples of group 2 metal catalysts for lactide polymerization that illustrate principles and difficulties. Over the past decade, the field has grown enormously, especially because nearly any group 2 metal compound is able to polymerize lactides or similar cyclic esters such as ε-caprolactone or trimethylene carbonate that are reviewed elsewhere [62]. The challenge to grow chains in a controlled way has only partially been met. Backbiting and high PDIs can be a problem, especially for the larger metals. Although this problem has been partially solved, stereocontrol to give PLA of very high tacticity is still poor, and only in some case, reasonable stereocontrol up to 90% has been achieved. This may be partially inherent to the very dynamic nature of highly ionic group 2 metal complexes. Future challenges include a considerable improvement in stereocontrol to obtain highly syndiotactic or heterotactic PLA from *rac*- or *meso*-lactide. Also, a catalyst system that is able to produce highly isotactic PLA from *rac*-lactide would be a rewarding objective. This challenge has already been met by Al and Y catalyst in which the more Lewis acidic metals give a higher degree of control [63] but is still open for s-block metal catalysts. This process is not only of interest for kinetic resolution of D- and L-lactide mixtures but also for improvement of the PLA plastics. Mixtures of poly-L-lactide and poly-D-lactide, which form chains of opposite helical chirality, cocrystallize to stereocomplex PLA. Such stereocomplexes consist of cocrystallized helical chains of opposite chirality, featuring considerably enhanced mechanical strength and a higher melting point than the enantiopure isotactic PLA [64]. Given the enormous research activity, direct fabrication of highly desired stereocomplex PLA with abundant and biocompatible Mg or Ca initiators is certainly a realistic prospect.

2.3.2 Copolymerization of Epoxides and CO_2

Although the discovery of alternating ring-opening copolymerization of epoxides and CO_2 (ROCOP) dates back to 1969 [65], it is especially since the past decades that this method to produce aliphatic polycarbonates gained enormous interest. This is partially due to the global CO_2 problem that makes any industrial use of this greenhouse gas attractive. Also, the potential biodegradability of polycarbonates/polyethers gave this class of polymers increased attention. There are, however, a number of odds and ends. It is clearly evident that polycarbonate production from CO_2 is not going to compensate for the worldwide extremely high CO_2 emissions. Another major issue is that most catalytic procedures need a reasonably pure CO_2 gas stream, which increases the cost of this ubiquitous C1-feedstock. A dream would be to simply extract the CO_2 straight from air. Using CO_2 directly from power plants is a more practical option that recently has been realized for epoxide/CO_2 copolymerization [66]. This requires catalysts that show good tolerance toward CO_2 impurities such as water, CO, amines, or thiols.

Industrial ROCOP generally makes use of heterogeneous double-metal cyanide (DMC) catalysts of the general formula $[M]_m[M'(CN)_6]_n$ in which M is an oxophilic metal and M′ is a transition metal [67], a system for which homogeneous examples have also been isolated [68]. Another heterogeneous catalyst finding industrial application is the zinc glutarate system introduced by Soga et al. [69]. The historical development of ROCOP catalysis has been summarized by a series of excellent reviews [70]. The aim of this contribution is to give an overview of s-block metals in ROCOP, which is currently limited to very few metals. Because the principles of these catalytic processes are based on Coates landmark discovery of highly active discrete homogeneous Zn catalysts [70e], a short overview of these closely related Zn systems will be given first.

The Zn catalysts are composed of a bulky bidentate β-diketiminate ligand that chelates the Zn center and an initiating group (**21**) [71]. The latter can be an amide, alkoxide, sulfinate, or acetate (but not an alkyl). The reason for this is the fact that in Zn chemistry, alkyls are very poor bridging ligands. A successful catalyst needs ligands that are able to bind two Zn centers forming a dinuclear species. Alternating epoxide/CO_2 copolymerization proceeds through a bimetallic mechanism in which a polymer chain with a carboxylate end group is the resting state and subsequent epoxide insertion is the rate-determining step (Scheme 2.15). The zinc carboxylate intermediate is in a monomer–dimer equilibrium that is dependent on the aryl substituents in the β-diketiminate ligand. Because the nucleophilic ring opening of epoxides is a process with rear-side attack [72], the dimer in which the larger ring allows for an unstrained transition state is considerably

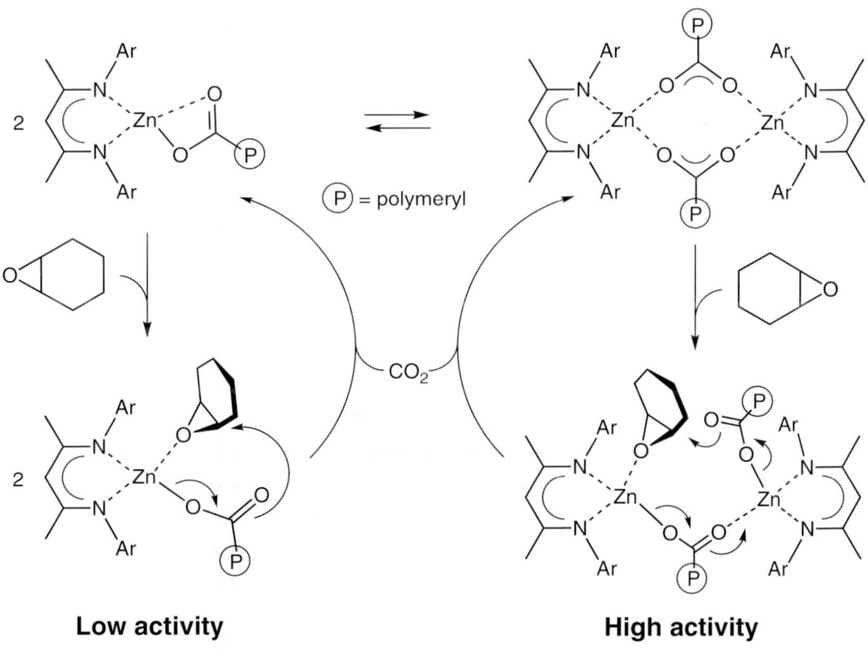

Scheme 2.15 Mechanism of the alternating cyclohexene oxide/CO_2 ring-opening polymerization with a β-diketiminate Zn catalyst.

more active than the monomer. Therefore, the activities of the polymerization catalysts can be controlled by small variations in the β-diketiminate ligand and by catalyst concentration. At very low catalyst concentrations, the monomer–dimer equilibrium shifts to the monomeric side resulting in poor catalytic activity. For this reason, concentrations should be high and ROCOP with these Zn catalysts is carried out without the use of a solvent: the catalyst is dissolved in the pure epoxide and this solution is pressurized with CO_2 (generally >6 bar). Because conversion is limited because of viscosity problems, full epoxide conversion giving long-chain polymers can never be obtained. Dilution of the system with an inert solvent would not only reduce the viscosity but also shift the monomer–dimer equilibrium to the monomeric side, thus reducing the catalyst activity.

This problem has been tackled by introduction of preorganized dinuclear catalysts for which monomer–dimer equilibria are irrelevant. The dinuclear Zn catalyst **22** is even in high dilution highly active for the copolymerization of cyclohexene oxide (CHO) and CO_2 giving high-molecular-weight polymers (M_n up to 284 000) and high yields [73]. Also, catalyst **23** with a much simpler dinucleating ligand led to nearly full monomer conversion [74]. The bimetallic Zn complex based on the Trost ligand (**24-Zn**) was found to be active in diluted solutions and under the remarkably low CO_2 pressure of 1 bar [75]. Interestingly, the analog dinuclear Mg complex (**24-Mg**) was also found to be active in CHO/CO_2 copolymerization under the same conditions, thereby representing the first s-block metal catalyst for this particular conversion [76].

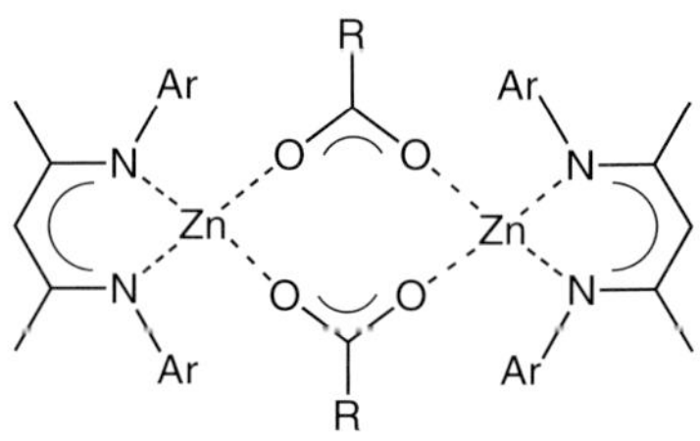

21

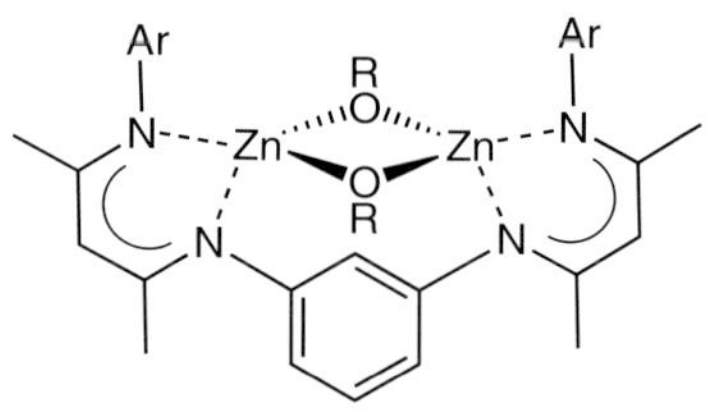

23

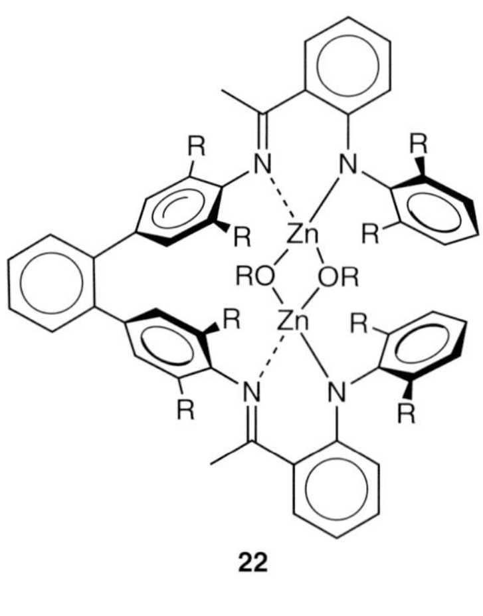

22

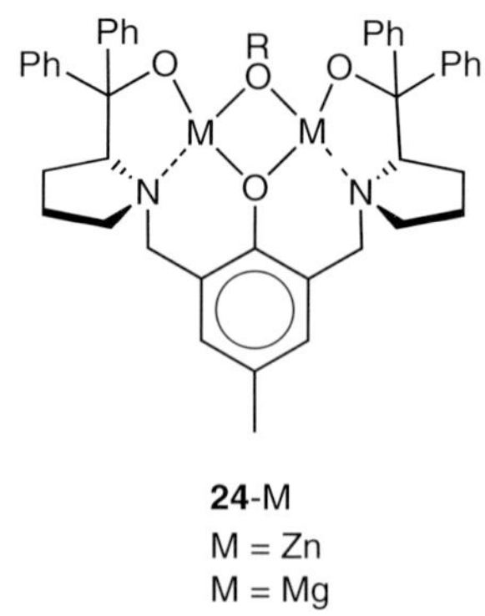

24-M
M = Zn
M = Mg

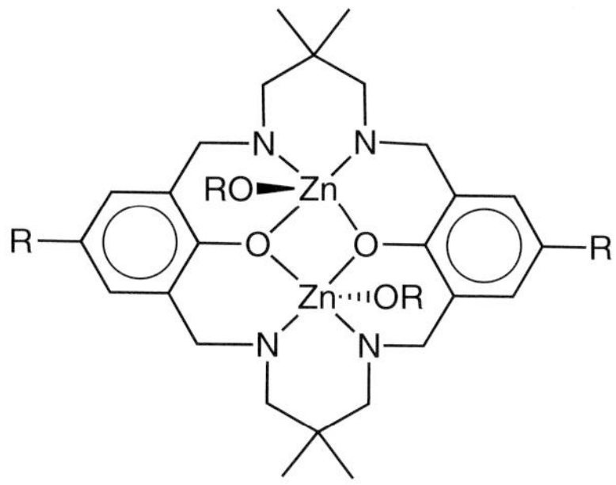

25

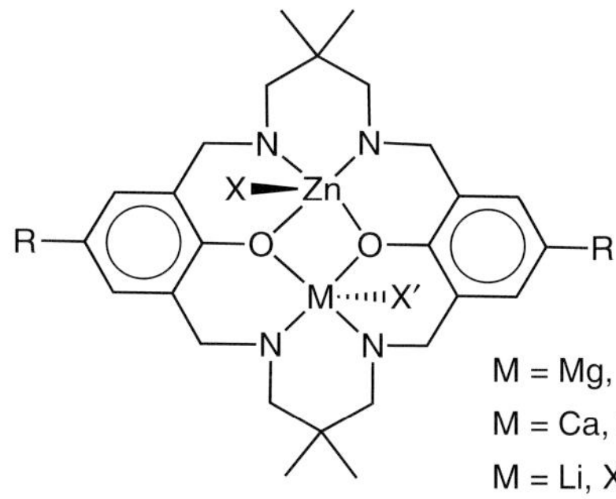

26

M = Mg, X′ = I
M = Ca, X′ = I
M = Li, X′ = –
M = K, X′ = –

Williams and coworkers designed a macrocyclic diphenolate ligand and showed that its dinuclear Zn complex (**25**) is an excellent CHO/CO_2 ROCOP catalyst [77]. The same group also reported mixed metal catalysts (**26**). Mixed Zn–Li and Zn–K catalysts showed low activity, but the mixed Zn–Mg complex clearly outperformed the homonuclear Mg–Mg and Zn–Zn catalysts [78]. Both metals in this heterobimetallic catalyst fulfill different roles: the very Lewis acidic Zn enhances reactivity by epoxide coordination, and Mg, featuring more ionic and dynamic bonds, supplies the growing chain for rear-side nucleophilic attack. Because a similar mixed Zn/Ca catalyst is fully inactive, it may be more likely that the alkaline earth metal binds the epoxide and the growing chain is delivered by Zn [78b]. This is in line with all previous failures to introduce Ca in CHO/CO_2 ROCOP catalysis [74, 76, 79].

2.4 Conclusions

The application of abundant early main group metals in polymerization catalysis is a growing field that is attractive especially for the polyoxygenates. This group of polymers is likely an important future bulk commodity on account of its biodegradable properties and medicinal applications. The biocompatibility of the catalyst is therefore of special importance, giving prominence to metals such as Mg and Ca. Apart from that, group 2 metals have also shown considerable control in the polymerization of activated alkenes such as styrene and butadiene. In contrast to group 1 metal initiators, which are known to polymerize alkenes in an anionic manner in which the metal only has a limited influence, the higher Lewis acidity of the group 2 metal cations plays an important role in monomer coordination, activation, and insertion. Understanding the polymerization mechanisms and metal–ligand interactions is therefore the key to control of polymer microstructures. In this respect, there is certainly strongly parallel with controlled polymerization using transition metal catalysts. Despite pioneering studies in early main group metal polymerization catalysis, there is currently only limited knowledge and experience. It is, however, anticipated that the rapidly growing field of group 2 metal chemistry will especially boost their application in polymerization catalysis with a special emphasis on further control of the regio- and stereoselectivities of the chain growth process. Recent developments in the syntheses of "true" group 2 metal alkyl complexes, i.e. complexes with unstabilized alkyl ligands such as Et, also promises further advancement in polymerization of unactivated alkenes such as ethylene [9, 80]. The observation that a strontium alkyl complex is able to produce polyethylene [9] hints that activities in this direction may unfold.

List of Abbreviations

Ae	alkaline earth
CCG	catalyzed chain growth
CHO	cyclohexene oxide
DMC	double-metal cyanide
DPE	1,1-diphenylethylene
Ln	lanthanide
MAO	methylalumoxane
MW	molecular weight
N″	$N(SiMe_3)_2$
PB	polybutadiene
PDI	polydispersity index
PiP	polyisoprene
PLA	polylactide
PS	polystyrene
ROCOP	ring-opening copolymerization

References

1 Crespy, D., Bozonnet, M., and Meier, M. (2008). *Angew. Chem. Int. Ed.* 47: 3322–3328.
2 Jambeck, J.R., Geyer, R., Wilcox, C. et al. (2015). *Science* 347: 768–771.
3 Szwarc, M. (1956). *Nature* 178: 1168–1169.
4 Szwarc, M., Levy, M., and Milkovich, R. (1956). *J. Am. Chem. Soc.* 78: 2656–2657.
5 Desbois, P., Warzelhan, V., Niessner, N. et al. (2006). *Macromol. Symp.* 240: 194–205.
6 Nakhmanovich, B.I., Basova, R.V., and Arest-Yakubovich, A.A. (1975). *J. Macromol. Sci. Part A Pure Appl. Chem.* A9: 575–596.
7 (a)Eberhardt, G.G. (1967). Preparation of organolithium amine complexes. US Patent 3,321,479. (b) (1971) Methods of producing straight chain waxes by catalytic polymerization of ethylene. US Patent 3,567,703.
8 Ribeiro, R., Ruivo, R., Nsiri, H. et al. (2016). *ACS Catal.* 6: 851–860.
9 Rösch, B., Gentner, T.X., Elsen, H. et al. (2019). *Angew. Chem. Int. Ed.* 58, early view, https://doi.org/10.1002/anie.201901548.
10 Wang, M., Nudelman, F., Matthes, R.R., and Shaver, M.P. (2017). *J. Am. Chem. Soc.* 139: 14232–14236.

11 Eaborn, C., Hawkes, S.A., Hitchcock, P.B., and Smith, J.D. (1997). *Chem. Commun.*: 1961–1962.

12 Feil, F. and Harder, S. (2000). *Organometallics* 19: 5010–5015.

13 Harder, S., Feil, F., and Weeber, A. (2001). *Organometallics* 20: 1044–1046.

14 Malanga, M. (2000). *Adv. Mater.* 12: 1869–1872.

15 Ishihara, N., Seimiya, T., Kamoto, M., and Uoi, M. (1986). *Macromolecules* 19: 2464–2465.

16 Weeber, A., Harder, S., and Brintzinger, H.-H. (2000). *Organometallics* 19: 1325–1332.

17 De Groof, B., Van Beylen, M., and Szwarc, M. (1975). *Macromolecules* 8: 396–400.

18 Feil, F. (2002). Stereoselektive anionische Polymerisation von Styrol mit heteroleptischen Erdalkaliinitiatoren. PhD thesis. Universität Konstanz.

19 Feil, F., Müller, C., and Harder, S. (2003). *J. Organomet. Chem.* 683: 56–63.

20 Harder, S. and Feil, F. (2002). *Organometallics* 21: 2268–2274.

21 Harder, S., Feil, F., and Knoll, K. (2001). *Angew. Chem. Int. Ed.* 40: 4261–4264.

22 Feil, F. and Harder, S. (2003). *Macromolecules* 36: 3446–3448.

23 Feil, F. and Harder, S. (2003). *Eur. J. Inorg. Chem.*: 3401–3408.

24 Piesik, D.F.-J., Häbe, K., and Harder, S. (2007). *Eur. J. Inorg. Chem.*: 5652–5661.

25 Kirillov, E., Lehmann, C.W., Razavi, A., and Carpentier, J.-F. (2004). *J. Am. Chem. Soc.* 126: 12240–12241.

26 Hoffmann, R.W., Rühl, T., Chemla, F., and Zahneisen, T. (1992). *Liebigs Ann. Chem.*: 719–724.

27 Annunziata, L., Sarazin, Y., Duc, M., and Carpentier, J.-F. (2011). *Macromol. Rapid Commun.* 32: 751–757.

28 Oliver, A.M., Cwyther, J., Boott, C.E. et al. (2018). *J. Am. Chem. Soc.* 140: 18104–18114.

29 Chen, J. and Fetters, L.J. (1981). *Polym. Bull.* 4: 275–280.

30 Shima, M., Bhattacharyya, D.N., Smid, J., and Szwarc, M. (1963). *J. Am. Chem. Soc.* 85: 1306–1310.

31 Bella, A.F., Ruiz, A., Claver, C. et al. (2008). *J. Organomet. Chem.* 693: 1269–1275.

32 Borah, K.J., Dutta, P., and Borah, R. (2011). *Bull. Korean Chem. Soc.* 32: 225–228.

33 Lindsell, W.E., Robertson, F.C., and Souter, I. (1983). *Eur. Polym. J.* 19: 115–120.

34 Ward, B.J. and Hunt, P.A. (2017). *ACS Catal.* 7: 459–468.

35 Crimmin, M.R., Barrett, A.G.M., Hill, M.S. et al. (2007). *Organometallics* 26: 2953–2956.

36 Taube, R. and Sylvester, G. (1996). Stereospecific polymerization of butadiene or isoprene. In: *Applied Homogeneous Catalysis with Organometallic Compounds*, vol. 1 (eds. B. Cornils and W.A. Herrmann), 280–318. Weinheim: VCH.

37 Boisson, C., Barbotin, F., and Spitz, R. (1999). *Macromol. Chem. Phys.* 200: 1163–1166.

38 Firestone Tire and Rubber Co. (F. C. Foster) (1967). Butadiene polymers prepared with a lithium based catalyst chracterized by having at least 29 percent of the butadiene as cis-1,4. US 3.317.918.
39 Michelin et Cie. (Comp. General. Etabl. Michelin) (1972). Solution polymerization of conjugated dienes. DE 2.115.462.
40 Jochmann, P., Dols, T.S., Spaniol, T.P. et al. (2009). *Angew. Chem. Int. Ed.* 48: 5715–5719.
41 Schneiderman, D.K. and Hillmyer, M.A. (2017). *Macromolecules* 50: 3733–3749.
42 Narancic, T., Verstichel, S., Reddy Chaganti, S. et al. (2018). *Sci. Technol.* 52: 10441–10452.
43 Chiellini, E. and Solaro, R. (1996). *Adv. Mater.* 8: 305–313.
44 Inkinen, S., Hakkarainen, M., Albertsson, A.-C., and Södergard, A. (2011). *Biomacromolecules* 12: 523–532.
45 Ma, H., Spaniol, T.P., and Okuda, J. (2006). *Angew. Chem. Int. Ed.* 45: 7818–7821.
46 Sauer, A., Kapelski, A., Fliedel, C. et al. (2013). *Dalton Trans.* 42: 9007–9023.
47 Ovitt, T.M. and Coates, G.W. (1999). *J. Am. Chem. Soc.* 121: 4072–4073.
48 Wheaton, C.A., Hayes, P.G., and Ireland, B.J. (2009). *Dalton Trans.*: 4832–4846.
49 Zhong, Z., Schneiderbauer, S., Dijkstra, P.J. et al. (2001). *Macromolecules* 34: 3863–3868.
50 Cheng, M., Attygalle, A.B., Lobkovsky, E.B., and Coates, G.W. (1999). *J. Am. Chem. Soc.* 121: 11583–11584.
51 Chamberlain, B.M., Cheng, M., Moore, D.R. et al. (2001). *J. Am. Chem. Soc.* 123: 3229–3238.
52 Chisholm, M.H., Gallucci, J., and Phomphrai, K. (2002). *Inorg. Chem.* 41: 2785–2794.
53 Chisholm, M.H., Gallucci, J., and Phomphrai, K. (2004). *Inorg. Chem.* 43: 6717–6725.
54 (a)Dove, A.P., Gibson, V.C., Marshall, E.L. et al. (2004). *J. Chem. Soc., Dalton Trans.*: 570–578. (b) Chisholm, M.H., Gallucci, J.C., and Phomphrai, K. (2005). *Inorg. Chem.* 44: 8004–8010.
55 Chivers, T., Fedorchuk, C., and Parvez, M. (2005). *Organometallics* 24: 580–586.
56 Lian, B., Thomas, C.M., Casagrande, C.W. Jr., et al. (2007). *Polyhedron* 26: 3817–3824.
57 Sánchez-Barba, L.F., Garcés, A., Fajardo, M. et al. (2007). *Organometallics* 26: 6403–6411.
58 Honrado, M., Otero, A., Fernández-Baeza, J. et al. (2015). *Organometallics* 34: 3196–3208.
59 Barros, M.L., Cushion, M.G., Schwarz, A.D. et al. *Dalton Trans.*, early view, DOI: https://doi.org/10.1039/C8DT04985H.
60 Darensbourg, D.J., Choi, W., Karroonnirun, O., and Bhuvanesh, N. (2008). *Macromolecules* 41: 3493–3502.
61 Sarazin, Y., Liu, B., Roisnel, T. et al. (2011). *J. Am. Chem. Soc.* 133: 9069–9087.

62 (a) Noureddine, A., Carpentier, J.-F., Guillaume, C. et al. (2010). *Dalton Trans.* 39: 8363–8376. (b) Carpentier, J.-F. and Sarazin, Y. (2013). *Alkaline-Earth Metal Compounds – Oddities and Applications*, Topics in Organometallic Chemistry, vol. 45 (ed. S. Harder), 141–189. Heidelberg: Springer.

63 Ovitt, T.M. and Coates, G.W. (2002). *J. Am. Chem. Soc.* 124: 1316–1326.

64 Bai, H., Deng, S., Bai, D. et al. (2017). *Macromol. Rapid Commun.* 38: 1700454.

65 Inoue, S., Koinuma, H., and Tsuruta, T. (1969). *J. Polym. Sci., Part B: Polym. Lett.* 7: 287–292.

66 Langanke, J., Wolf, A., Hofmann, J. et al. (2014). *Green Chem.* 16: 1865–1870.

67 (a) Kruper, W.J., Jr., and Swart, D.J. (1985). Carbon dioxide oxirane copolymers prepared using double metal cyanide complexes. US Patent 4,500,704. (b) Chen, L.-B. (1992). *Makromol. Chem. Macromol. Symp.* 59: 75–82.

68 Darensbourg, D.J., Adams, M.J., Yarbrough, J.C., and Phelps, A.L. (2003). *Inorg. Chem.* 42: 7809–7818.

69 Soga, K., Imai, E., and Hattori, I. (1981). *Polym. J.* 13: 407–410.

70 (a) Darensbourg, D.J. (2007). *Chem. Rev.* 107: 2388–2410. (b) Darensbourg, D.J. and Wilson, S.J. (2012). *Green Chem.* 14: 2665–2671. (c) Kember, M.R., Buchard, A., and Williams, C.K. (2011). *Chem. Commun.* 47: 141–163. (d) Klaus, S., Lehenmeier, M.W., Anderson, C.E., and Rieger, B. (2011). *Coord. Chem. Rev.* 255: 1460–1479. (e) Coates, G.W. and Moore, D.R. (2004). *Angew. Chem. Int. Ed.* 43: 6618–6639. (f) Trott, G., Saini, P.K., and Williams, C.K. (2016). *Philos. Trans. R. Soc. London, Ser. A* 374: 20150085.

71 (a) Cheng, M., Lobkovsky, E.B., and Coates, G.W. (1998). *J. Am. Chem. Soc.* 120: 11018–11019. (b) Cheng, M., Moore, D.R., Reczek, J.J. et al. (2001). *J. Am. Chem. Soc.* 123: 8738–8749.

72 Harder, S., van Lenthe, J.H., van Eikema Hommes, N.J.R., and Schleyer, P.V.R. (1994). *J. Am. Chem. Soc.* 116: 2508–2514.

73 Lee, B.Y., Kwon, H.Y., Lee, S.Y. et al. (2005). *J. Am. Chem. Soc.* 127: 3031–3037.

74 Piesik, D.F.J., Range, S., and Harder, S. (2008). *Organometallics* 27: 6178–6187.

75 Xiao, Y., Wang, Z., and Ding, K. (2005). *Chem. Eur. J.* 11: 3668–3678.

76 Xiao, Y., Wang, Z., and Ding, K. (2006). *Macromolecules* 39: 128–137.

77 Kember, M.R., Knight, P.D., Reung, P.T.R., and Williams, C.K. (2009). *Angew. Chem. Int. Ed.* 48: 931–933.

78 (a) Garden, J.A., Saini, P.K., and Williams, C.K. (2015). *J. Am. Chem. Soc.* 137: 15078–15081. (b) Deacy, A.C., Durr, C.B., Garden, J.A. et al. (2018). *Inorg. Chem.* 57: 15575–15583.

79 Range, S., Piesik, D.F.-J., and Harder, S. (2008). *Eur. J. Inorg. Chem.*: 3442–3451.

80 Wilson, A.S.S., Hill, M.S., Mahon, M.F. et al. (2017). *Science* 358: 1168–1171.

3

Intramolecular Hydroamination of Alkenes

Sebastian Bestgen[1] and Peter W. Roesky[2]

[1] *University of Oxford, Chemistry Research Laboratory, Department of Chemistry, 12 Mansfield Road, Oxford OX1 3TA, UK*
[2] *Karlsruhe Institute of Technology (KIT), Institute of Inorganic Chemistry, Engesserstraße 15, 76131 Karlsruhe, Germany*

3.1 Introduction

A vast majority of naturally occurring and scientifically or industrially used chemicals contain nitrogen. Thus, compounds such as amines, imines, enamines, amides, nitrates, or azides are widespread, commonly used as bulk and fine chemicals or pharmaceuticals, and show tremendous importance in the value chain. Among all nitrogen-containing substances, N-heterocyclic compounds form a considerable subsection and are found in, e.g. many alkaloids, vitamins, nucleobases, and drugs (Scheme 3.1).

Accordingly, the synthesis of N-heterocyclic compounds is a very relevant process and has therefore been investigated for decades. Famous and established synthetic approaches are, e.g. the Paal–Knorr pyrrole synthesis, the Hantzsch pyridine synthesis, the Bönnemann cyclization, or the Staudinger cycloaddition to form β-lactams. In the light of scarcity of energy and natural resources, demands of an expanding population and environmental protection, synthetic chemistry is enormously challenged to address these issues. Therefore, a paradigm shift is taking place toward the development of a green and sustainable chemistry. Therefore, one major contribution comprises the preparation of bulk and fine chemicals using energy-saving methods and renewable feedstocks, which is reflected in intense research on new catalysts using cheap, nontoxic, and abundant substances. As these requirements are largely fulfilled by most s-block metals, novel catalytic procedures using alkali- or alkaline earth catalysts are a particularly interesting research field. In this chapter, we wish to present an overview on intramolecular hydroaminations (HAs) of alkenes using s-block metal catalysts. This allows access toward N-heterocyclic substances using abundant starting materials and ecologically compatible metals.

Early Main Group Metal Catalysis: Concepts and Reactions,
First Edition. Edited by Sjoerd Harder.

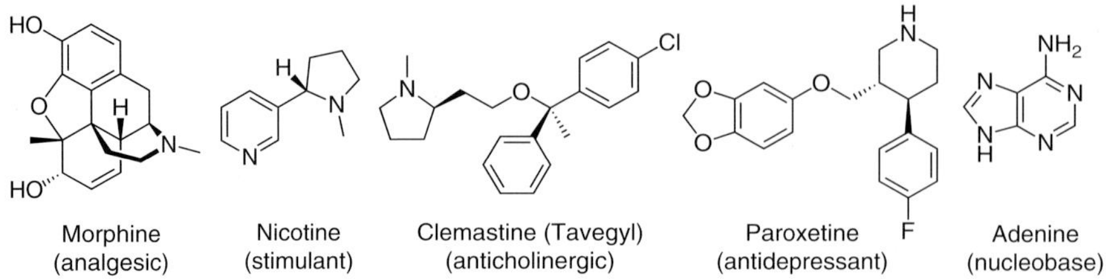

Scheme 3.1 Nitrogen-containing heterocycles in natural products and drugs.

3.2 Hydroamination

Besides many established synthetic procedures toward nitrogen-containing heterocycles, hydroamination reactions are particularly momentous as they are formally atom economical and green and have therefore been intensely studied [1–19]. In hydroaminations (HAs), an amine adds to a multiple bond, which is most often a CC multiple bond, but not limited to it as hydroaminations of C=O and C=N functionalities have also been reported [20]. In the case of unsaturated C—C bonds, the addition of an amine leads to the formation of a new N—C bond and thus offers easy access to nitrogen-containing molecules. The reaction can take place within one accordingly difunctionalized molecule such as aminoalkenes or -alkynes (intramolecular hydroamination) or between a primary/secondary amine and an alkene or alkyne (intermolecular HA) (Scheme 3.2) [21].

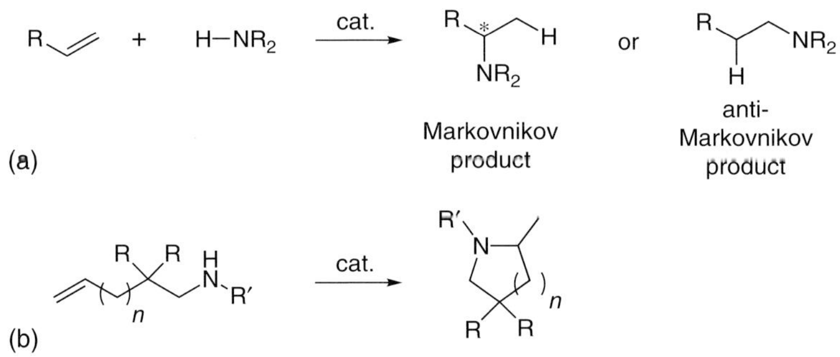

Scheme 3.2 Intermolecular (a) and intramolecular (b) hydroamination of alkenes [21].

From a thermodynamic point of view, hydroaminations are slightly exothermic and viable under ambient conditions, but in total, approximately thermochemically neutral as they are accompanied by a reduction in entropy especially in the intermolecular case. Also, a nucleophilic attack of the amine to the electron-rich alkene causes electrostatic repulsion between the nitrogen lone pair and the C=C π-system. An alternative 2+2 cycloaddition of N—H bonds to unsaturated C—C bonds is orbital symmetry-forbidden and can usually not be accomplished because of the high energy difference of the involved orbitals. Additionally, the highly negative reaction entropy shifts the equilibrium toward the starting materials when the temperature is increased. Thus, the direct addition of amines to alkenes/alkynes comes along with a high reaction barrier, which can be bypassed by using catalysts allowing for alternative reaction pathways. The intermolecular hydroamination of alkenes and alkynes can in principal lead to two different regioisomers as products, which are called the Markovnikov and the anti-Markovnikov product [4, 17]. However, in intermolecular HAs of unactivated alkenes/alkynes, the Markovnikov product is frequently observed and many efforts are taken to, e.g. obtain reversal of the regiochemistry. Also, hydroaminations of alkynes proceed easier compared to those of alkenes because of their higher reactivity and electron density at the C≡C bond, thus making alkene hydroamination more challenging. In general, intramolecular hydroaminations of aminoalkenes and -alkynes proceed more readily than intermolecular reactions, and as secondary amines are more nucleophilic than primary ones, they undergo intramolecular HA more facilely. In intramolecular cyclizations, the variation of substituents in β-position to the amine additionally affects the reaction rate as, e.g. bulky groups lead to bending of the structure and facilitate a conformation that is beneficial for ring closure. This effect is commonly known as Thorpe–Ingold effect and often utilized in catalytic processes [22–25]. Although hydroaminations are thermodynamically feasible, the use of a catalyst fosters or even enables the process via different reaction pathways and overall lowering of the energetic barrier. The development of catalytically active compounds basically follows three strategies to activate the substrates and facilitate hydroamination (Scheme 3.3a–c) [26]. First, N—H bonds can oxidatively add to transition metals (TMs) to generate a complex that allows an alkene to either insert into the M—N or the M—H bond. However, this naturally requires a redox-active metal (Scheme 3.3a). Second, the alkene can transitionally be coordinated to a metal via π-bonding, which enhances its affinity toward nucleophilic attacks by the nitrogen lone pair (Scheme 3.3b). Lastly, deprotonation of the primary or secondary amine leads to a metal amide

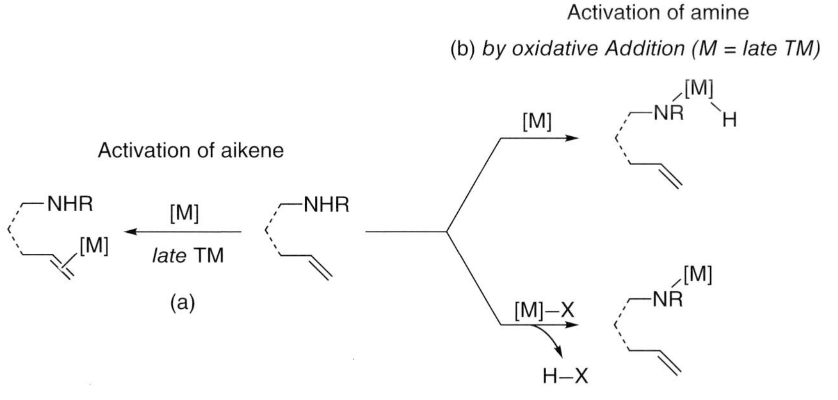

Scheme 3.3 Three different approaches toward substrate activation in catalytic hydroamination [26].

M–N, which is more nucleophilic than the amine and thus more prone to attack the olefin (Scheme 3.3c). As oxidative additions and metal π-complexes in hydroaminations are typically observed for low-coordinate and late transition metals, a detailed overview on both will not be given as this chapter focuses on intramolecular hydroaminations using s-block metal catalysts. Instead, a short summary of the enormous scope of hydroamination is presented to illustrate the basic approaches and differences throughout the periodic table.

3.2.1 Scope

Hydroaminations can not only be achieved by the use of organometallic catalysts but also by Brønsted acids, bases, or amino- and amidomercuration–demercuration of alkenes and alkynes [17]. Brønsted acids have been used both as heterogeneous and as homogeneous catalysts [3, 17, 27]. However, the preferred formation of ammonium salts, rather than protonation of the π-system as the initial catalytic step, somehow limits their applicability to amines with low basicity [17]. For organometallic catalysis, a huge amount of catalysts has been developed and investigated for their utilization in hydroamination. Besides alkali and alkaline earth metals, which will be discussed thoroughly later, early transition metals [3, 9, 21], late transition metals [3, 7, 9, 12, 14, 15, 21], lanthanides, and actinides [3, 7–9, 11–13, 15–17] were utilized as catalytically active metals. As the inherent properties of these groups of metals differ, their mechanism in action varies (Scheme 3.3):

Late transition metals: In the case of late transition metals (see Scheme 3.4 for examples), the olefinic functionality of the substrate is usually activated first by coordination of the metal to the π-system, as late TMs such as palladium [7, 9, 12, 15, 21], platinum [9, 21], copper [9], or gold [14] show a higher affinity toward π-coordination compared to hard nitrogen donors. Upon activation of the multiple bond via, e.g. d → π* orbital interactions, nucleophilic attack of the amine can occur at the coordinated π-system. Furthermore, the insertion of an alkene/alkyne into a metal–hydride bond has been observed for hydridic complexes. On the other hand, a second reaction mechanism is possible if the late transition metals are in low oxidation state and thus allow for oxidative addition of the amine. The amine N–H moiety can be oxidatively added to, e.g. coordinatively unsaturated Pd(0) or Pt(0) complexes to form a M(H)(N) species. Subsequently, an alkene is then able to be added into the M—N or M—H bond to finally form a hydroamination product after reductive elimination and cleavage of the substrate. In the case of oxidative additions, the amine

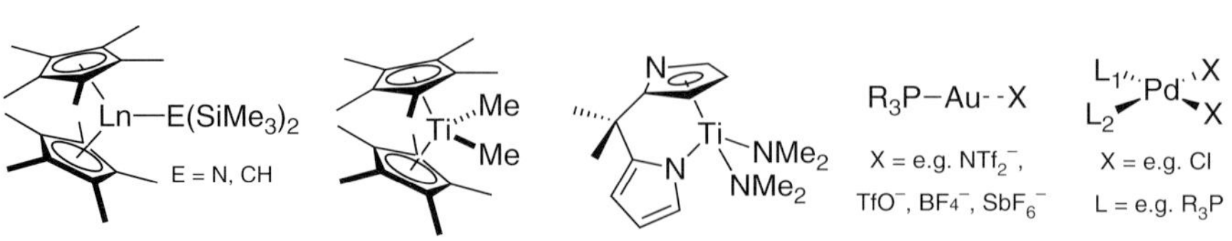

Scheme 3.4 Rare earth, early, and late transition metal (pre-)catalysts used in hydroamination reactions.

is activated first instead of the CC multiple bond and thus alters the reactivity. Generally, late transition metal catalysts are more tolerant toward substrates with polar functional groups and also less sensitive toward air and moisture compared with, e.g. s-block or lanthanide catalysts. However, they are also less reactive and require higher reaction temperatures and catalyst loadings, which is especially problematic because most of the late transition metals are rare, expensive, and partly even toxic. Thus, the ecological friendliness and compatibility of a catalytic hydroamination is somewhat flawed when a rare and toxic metal catalyst has to be utilized.

Early transition metals: An improvement in reactivity was achieved by using catalysts composed of group 4 and 5 metals (see Scheme 3.4 for examples), which in principal catalyze hydroaminations at lower temperature and catalyst loading [4, 6, 7, 12, 13, 18]. Therefore, the catalytically active species is believed to be a metal imido species M=N, which is formed upon double deprotonation of an amine. This is commonly achieved by using a (pre-)catalyst with two basic ligands (e.g. Me and NR_2), which reacts with amines via α-elimination. The metal–imido complex then undergoes a reversible and rate-determining [2+2]-cycloaddition with an alkyne or alkene to form an azametallacyclobutene or -butane, which is cleaved from the metal center as hydroamination product upon protonation by the succeeding protic amine substrate. Despite their higher reactivity, early transition metal catalysts are less tolerant toward polar functional groups because of their high oxophilicity and Lewis acidity.

Rare earth metals: Although rare earth metal catalysts suffer from similar drawbacks (see Scheme 3.4 (left) for examples), they have shown to be very active catalysts for hydroaminations providing high turnover frequencies and excellent stereoselectivities [3, 7–9, 11–13, 15–17]. The high reactivity applies especially to intramolecular hydroaminations of alkenes, alkynes, allenes, and dienes but is decreased in the intermolecular case [28]. This can be rationalized by the strong coordinative interaction between very Lewis acidic trivalent rare earth metals and basic amines or amides, which, in competition, severely hampers binding of weakly coordinating alkenes at vacant sites of the catalytically active metal center. Rare earth metal (pre-)catalysts show very similar functionality to alkaline earth catalysts. The amine is usually deprotonated only once by a rare earth metal amido or alkyl species via σ-bond metathesis to form a Ln–N metal amide. In a second and commonly rate-determining step, the olefin inserts into the metal–amido bond by forming a transient metallacycle as the transition state. Subsequent protonolysis and product cleavage occurs rapidly by the succeeding amine substrate. Rare earth metal catalysts are usually trivalent, although examples with divalent Sm(II) precatalysts have been reported [29]. At least one labile, basic, and typically σ-bonded ligand, such as a hydride, alkyl, or amide, is attached to the metal center, acting as a leaving group upon protonolysis by a substrate amine. For charge balance as well as control of steric bulk and regio- or enantioselectivity, the coordination sphere is completed by spectator ligands, which stay attached to the metal center throughout the catalytic cycle and were traditionally based on cyclopentadienyl-type systems. Since then, many efforts were undertaken to develop and improve novel ligand systems, which are nowadays known as post-metallocene catalysts and exhibit significantly more diverse properties than cyclopentadienyl ligands.

3.3 s-Block Metal Catalysis

3.3.1 General Remarks

This chapter discusses intramolecular hydroamination reactions using s-block metal catalysts. Although hydroamination has been achieved by using both alkali and alkaline earth metal catalysts, distinct differences apply for group 1 and group 2 metals regarding their organometallic chemistry, mechanism of action, element range, and most importantly, their scope. In general, group 2 metal catalysts and their applicability outnumber group 1 catalysts and examples for well-defined alkali metal catalysts are scarce. As group 1 metals feature the lowest electronegativities throughout the periodic table, they form predominantly ionic compounds with alkali metal ions in the oxidation state +1, and with the exception of Li, almost no covalent bonding interactions. Considering the high reactivity of elemental K, Rb, and Cs and their compounds, inherently less research progress has been made on the application of their complexes in hydroamination catalysis. Elements of the first group show a strikingly homologous behavior, with Li being different from Na, K, Rb, and Cs. In contrast to the heavier alkali metals, the coordination and organometallic chemistry of lithium is rich, more covalent, and a variety of defined compounds is known, including complexes with chiral ligands. However, almost every alkali metal compound studied in catalysis so far is also equally a base, which is why examples for group 1 catalysis are most frequently described as base-catalyzed reactions in the literature. Abundant bases such as KOH, $NaNH_2$, bulk alkali metals, or *n*-butyllithium (*n*-BuLi) have been frequently used, but often high reaction temperatures/pressures are necessary to obtain reasonable yields. In contrast, the coordination chemistry of alkaline earth metals is more comprehensive and subject of research for over a century, with a particular focus on the well-known Grignard reagents. However, the development and investigation of Ae complexes for catalysis is a rather recent research field, which has attracted increasing attention over the last years. Except for Be, the lighter Ae metals are abundant, nontoxic, cheap, ecologically unproblematic, and therefore meet today's requirements for a green and sustainable chemistry [30]. Group 2 metals ions exhibit many similarities to alkali metals ions and the prevalently redox inactive d^0 lanthanide ions, as they are also highly electropositive, and show mostly nondirectional ionic bonding interactions with increasing polarizability descending the group. The only exception is beryllium, which has a somewhat different chemistry due to covalent bonding interactions and moreover, it is toxic. Because of its toxicity, Be has not been used in hydroamination catalysis and thus is not further discussed herein. With some exceptions (Mg(I) being the most famous example), alkaline earth metals generally feature only the very stable oxidation state +II. Their redox inactivity therefore excludes oxidative addition or reductive elimination reactivity. However, their charge of +2 enhances ion–ligand binding and together with their wide range of ionic radii, varying from 0.78 Å (Mg^{2+}) to 1.43 Å (Ba^{2+}), which allows for a rich coordination chemistry. One important characteristic of alkaline earth metal complexes, which has often hampered their applicability in catalysis, is their tendency to undergo ligand rearrangements in solution. This phenomenon, which has become known as the Schlenk equilibrium for magnesium compounds, is also observed for the heavier homologs (Scheme 3.5) [31]. Thus, the synthesis of well-defined and coordinatively stable heteroleptic Ae complexes is one major challenge in organometallic chemistry.

$$2\ \mathrm{LAeX} \rightleftharpoons \mathrm{L_2Ae} + \mathrm{AeX_2}$$

Scheme 3.5 A Schlenk-type ligand redistribution typical for alkaline earth metal complexes [31].

In order to overcome ligand redistribution and retain defined heteroleptic catalysts in solution, a number of ligand frameworks have been developed, which suppress these exchange reactions. These spectator ligands often comprise polydentate monoanionic scaffolds with hard donor sites. Sterically demanding chelating ligands were found to successfully suppress Schlenk-like redistributions and provide kinetic stability, thus making catalytic studies feasible. A second approach to circumvent problematic ligand exchange reactions comprises the use of homoleptic Ae pre-catalysts featuring two reactive ligands. Commonly, highly basic amido- or alkyl ligands are used. For instance, bis-(trimethylsilyl)amide $[Ae(N\{SiMe_3\})_2(THF)_x]_y$ ($x = 2$, $y = 1$ or $x = 0$, $y = 2$) or bis-(trimethylsilyl)methanide complexes $[Ae(CH\{SiMe_3\}_2)_2(THF)_x]$ (Mg, Ca: $x = 2$; Sr, Ba: $x = 3$) of all Ae metals have been reported and were proven to be catalytically active [32, 33].

3.3.2 Mechanistic Aspects

For early main group metal (EMGM) catalysis dealing with heterofunctionalizations of π-systems, a universally valid reaction mechanism covering all alkali and alkaline earth metals cannot be given and mechanistic investigations are still in progress. However, after the discovery of the first truly group 2 metal-mediated intramolecular hydroamination of aminoalkenes and -alkynes, many efforts have been made to gain deeper insight and understanding into their reaction mechanisms [34–40]. The catalytic cycles may depend on many factors such as the nature of the substrate and catalyst, polarizability and Lewis acidity of the metal center, or the polarity of the solvent. Harder and coworkers also raised the question about the importance of the metal in EMGM catalysis itself. They found that some hydroamination reactions, in which an s-block metal catalyst was used, could also be accomplished by a metal-free base catalyst [41]. Given the weakness of EMGM–alkene interactions, some reactions are better defined as organocatalysis. Reactions, in which the metal cation plays a crucial role, however, should be defined as organometallic catalysis.

So far, a clear differentiation for EMGM catalytic processes cannot generally be made, and a greater unification of reactivity models is a desirable goal, as it would probably allow for the targeted design of main group catalysts to address synthetic challenges. However, general trends in reactivity and key steps in main group metal catalysis are well understood, and also for alkaline earth metals compared to catalytic systems based on lanthanides. The two dominantly observed and redox-independent mechanistic steps are σ-bond metathesis and insertion reactions of polarized systems (Scheme 3.6) [30, 42].

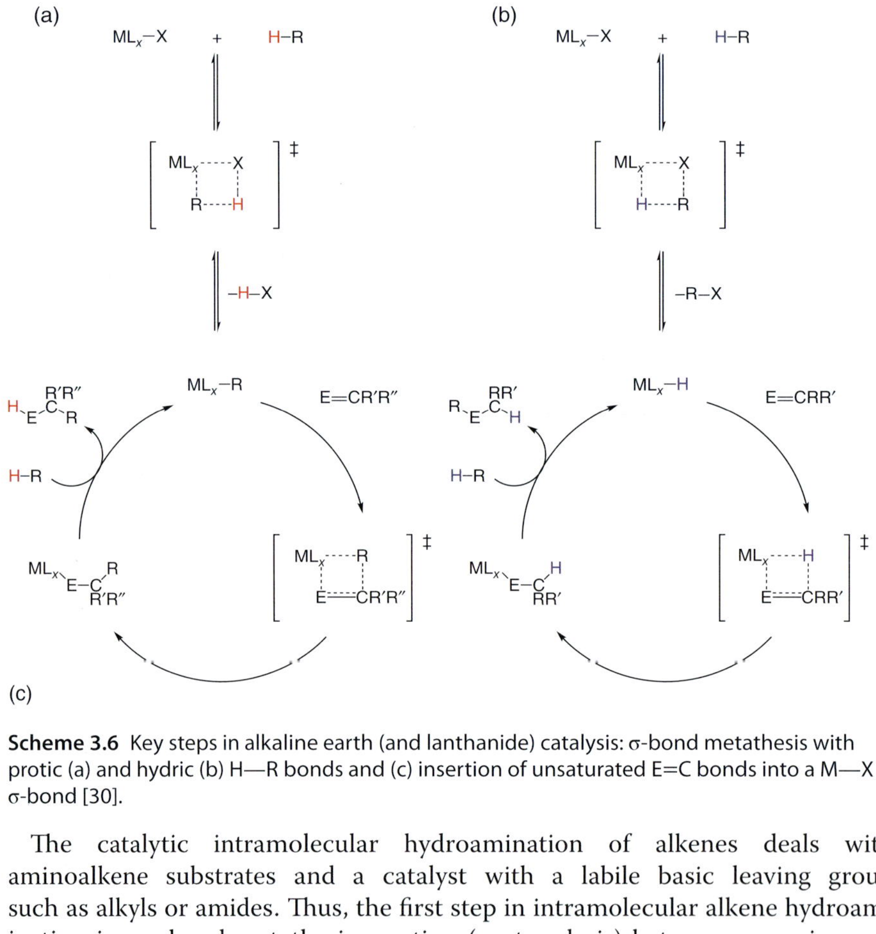

Scheme 3.6 Key steps in alkaline earth (and lanthanide) catalysis: σ-bond metathesis with protic (a) and hydric (b) H—R bonds and (c) insertion of unsaturated E=C bonds into a M—X σ-bond [30].

The catalytic intramolecular hydroamination of alkenes deals with aminoalkene substrates and a catalyst with a labile basic leaving group such as alkyls or amides. Thus, the first step in intramolecular alkene hydroamination is a σ-bond metathesis reaction (protonolysis) between an amine and the organometallic complex L_xM–R. This is a concerted process with a highly ordered four-membered transition state, which is generally fast. Depending on the nature of the metal and the basicity of the leaving group R^-, this step can be an equilibrium or an irreversible reaction. It results in a metal–amido complex (Scheme 3.7, **A**).

The second step involves the insertion of the olefinic bond into the metal–nitrogen M—N bond and proceeds, in the case of n = 1, via a

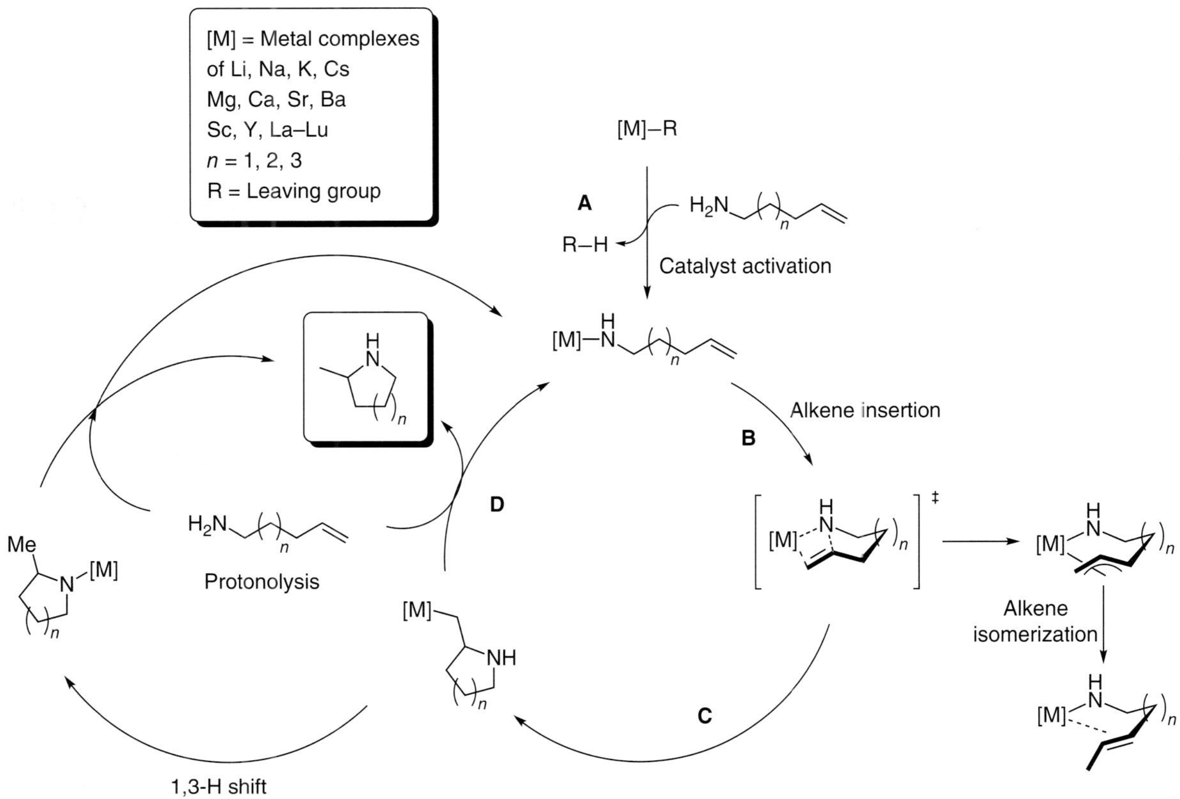

Scheme 3.7 Simplified reaction mechanism for the catalytic intramolecular hydroamination–cyclization of aminoalkenes using early main group or lanthanide metals.

chair-like transition state (Scheme 3.7, **B**). Noteworthy, this step is only enabled by the electrostatic activation of the double bond through the Lewis acidic cationic metal center and its intramolecular nature, resulting in a minimization of entropy loss. Commonly, the insertion step is also rate determining and results in a very reactive and unstabilized metal–alkyl complex (Scheme 3.7, **C**). Although shown here for the formation of a five-membered ring, six- and seven-membered rings are also possible (with the latter requiring special catalysts). It is in all cases regioselective, as only the more stable primary alkyl–metal intermediates are formed. This alkyl–metal intermediate is then rapidly protonated by a succeeding substrate amine resulting in the cleavage of the cyclization product and the beginning of the next catalytic cycle (Scheme 3.7, **D**). Although the formation of the highly unstabilized metal–alkyl intermediate RCH_2–M seems counterintuitive, it has been proven by labeling experiments with deuterated amines RND_2 that evidenced RCH_2D as products [37]. Two competing side reactions are 1,3-H shifts and alkene isomerization, with the latter being only occasionally observed. In the case of 1,3-H shifts, the N–H proton shifts to the alkyl group to form a more stabilized secondary metal–amide complex, which can then be protonated by the following substrate molecule [37, 43]. Alkene isomerization occurs via intramolecular deprotonation in the allylic position followed by protonation of the terminal carbon atom.

3.3.3 Group 1-Based Catalysis

3.3.3.1 Concerted Reaction

Group 1-based hydroamination catalysis is known for decades and because of the nature of alkali metal catalysts, most often described as base-mediated catalysis. Frequently used catalytic systems involve either the metals itself or alkali metal bases such as *n*-BuLi, NaOEt, or KO^{t}Bu [5, 44]. Alkali metal salts such as LiCl can also play a role as additives or in heterometallic catalysts, but these cooperative effects are beyond the scope of this chapter [45–47].

Recent developments in s-block catalysis comprise the intramolecular hydroamination of aminoalkenes using catalytic amounts of *n*-BuLi or other basic lithium complexes, after the usability of *n*-BuLi for intramolecular hydroaminations was reported in 2003 [48]. Therefore, primary or secondary aminoalkenes are exposed to catalytic amounts of *n*-BuLi (16 mol%) to deliver the corresponding heterocyclic products in high yields (Scheme 3.8) [49]. The hydroamination of vinyl sulfides proceeded more rapidly, under much

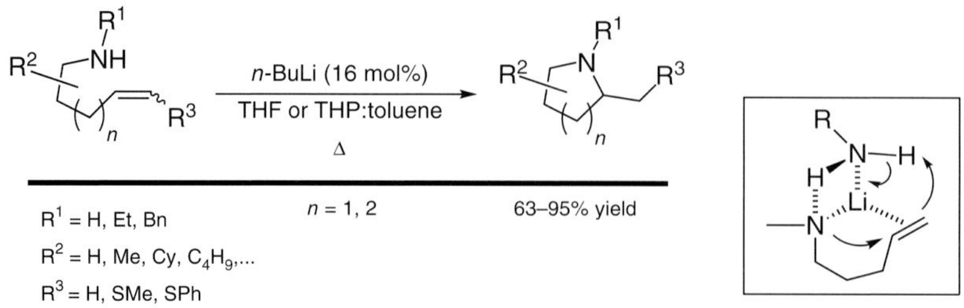

Scheme 3.8 *n*-BuLi-catalyzed cyclohydroamination.

milder conditions (25–50 °C in tetrahydrofuran [THF]) and enabled five- and six-membered ring formation as well as tandem hydroamination [50].

In contrast, cyclization of unactivated terminal olefins required harsher reaction conditions (up to 110 °C) and a solvent mixture (THP : toluene 1 : 1, THP = tetrahydropyran) was applied in order to suppress alkene isomerization [51]. Quantitative deprotonation (by using stoichiometric amounts of base) and subsequent heating of the fully deprotonated amide did not lead to cyclization while addition of small amounts of amine (diisopropylamine) to lithiated amidoalkenes again induced hydroamination. Thus, a coordinating and proton-donating amine is likely important in the transition state. Despite the unfavorable reaction conditions, the large scope of the reaction and its convenient access toward various N-heterocycles are synthetically beneficial.

Given that many natural and pharmaceutical products are chiral, the development of diastereo- and enantioselective catalysts for asymmetric hydroaminations is a worthwhile challenge and many efforts have been made to develop catalytic systems with metals throughout the periodic table [26, 49].

In group 1 chemistry, using *n*-BuLi and N-substituted aminocyclohexa-2,5-dienes allowed for a diastereoselective version of lithium base-catalyzed cyclohydroamination. By attaching chiral glycinol ether auxiliaries on the amino group, intramolecular hydroamination took place with complete diastereocontrol but was limited to aforementioned special substrates [52]. The first chiral main group metal-based lithium catalyst for hydroamination/cyclization was reported by Hultzsch and coworkers in 2006 [53]. They employed a symmetrically *N*-methylpyrrolidine-substituted diamidobinaphthyl dilithium salt for cyclizations of aminopentene derivatives and satisfactory ee's up to 75% and nearly quantitative yields were achieved (Scheme 3.9). Within the dimeric, tetranuclear lithium complex, all lithium atoms feature different coordination environments and the (*S*,*S*,*S*) configured species was used in catalytic studies. Although its

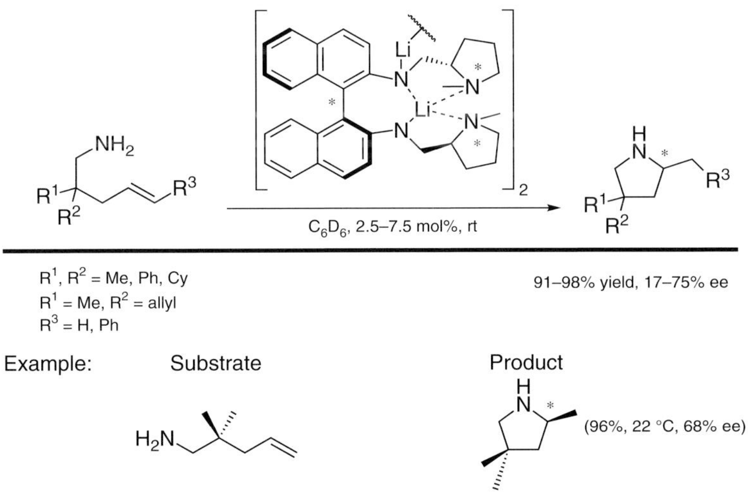

Scheme 3.9 Asymmetric cyclohydroamination using a dimeric diamidobinaphthyl dilithium complex.

exact molecular structure under catalytic conditions remained unknown, the dimeric nature was found to be responsible for enantioselectivity. Noteworthy, a catalytic system composed of (−)-sparteine and $LiN(SiMe_3)_2$ did not achieve enantiomeric excess, and in comparison, the corresponding Mg(*n*-Bu) and Zn(Et) complexes of the (*S*,*S*,*S*)-ligand showed only very poor ee's but good conversion [54].

The work on asymmetric intramolecular hydroamination was extended by Tomioka and coworkers, who used chiral bisoxazoline ligands in combination with *n*-BuLi and diisopropylamine (Scheme 3.10) [55]. At low temperature (−60 °C), excellent yields and ee's up to 91% were realized in a kinetically controlled reaction simply using an *in situ* prepared chiral lithium catalyst. Therefore, diisopropylamine acts as a coordinating and proton-donating reagent.

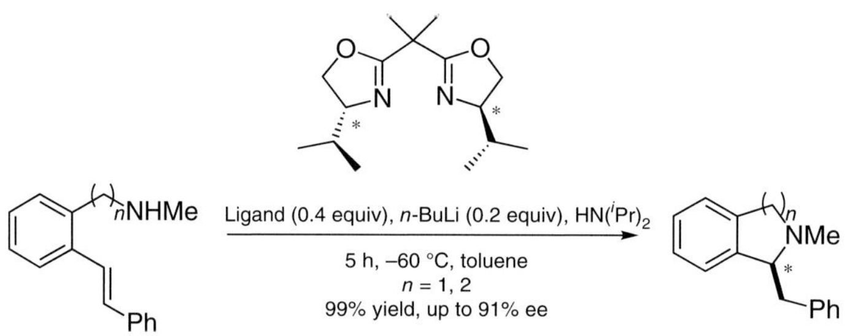

Scheme 3.10 Asymmetric cyclohydroamination using bisoxazoline ligands and *n*-BuLi.

The *in situ* combination of N-substituted (*R*)-(+)-1,1′-binaphthyl-2,2′-diamine and commercially available methyllithium was reported as an efficient diastereo- and enantioselective catalytic system for the cyclohydroamination of conjugated primary amino-1,3-dienes (Scheme 3.11) [56]. Various side chains were introduced to the amines, which show a significant impact on conversion and selectivity. Although the crystal structure of a dilithium diamido species could be established, the best catalyst was found to be a mixture of 4 equiv of MeLi and 1 equiv of the ligand and an excess of base was indeed necessary to facilitate catalysis. It was suggested that the chirally ligated Li center activates the C=C double bond in a stereodefined environment, whereas a slight excess of base

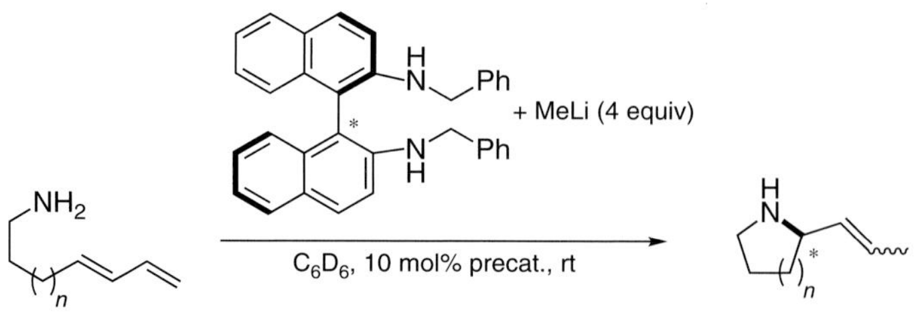

Scheme 3.11 Asymmetric hydroamination of amino-1,3-dienes.

was believed to induce substrate activation by deprotonation and concomitantly enhanced nucleophilicity of the nitrogen atom.

While maintaining high substrate conversion, a diastereoselectivity up to $(E)/(Z) = 92:8$ and enantioselectivity up to 72% ee were observed for a five- and six-membered ring formation.

In subsequent studies with a broader substrate and catalyst scope, hydroamination of amino-1,3-dienes and aminoalkenes was investigated using again a variety of binaphthyldiamine ligands and MeLi or $LiCH_2SiMe_3$ as the base [57]. The cyclization of primary and secondary aminoalkenes occurred efficiently, but with low to moderate enantioselectivities reaching an ee of 58% for pyrrolidine derivatives.

3.3.3.2 Radical-Mediated Intramolecular Hydroamination

Inspired by radical-mediated catalytic cross-coupling reactions, a mixture of KO^tBu and dimethylformamide (DMF) was successfully applied as a catalytic system in the intramolecular hydroamination of *ortho*-vinyl benzamides [58]. With a catalyst loading of 0.3 equiv KO^tBu in DMF, dihydroisoquinolinones and 3-benzylisoindolinones were obtained in moderate to excellent yields (Scheme 3.12).

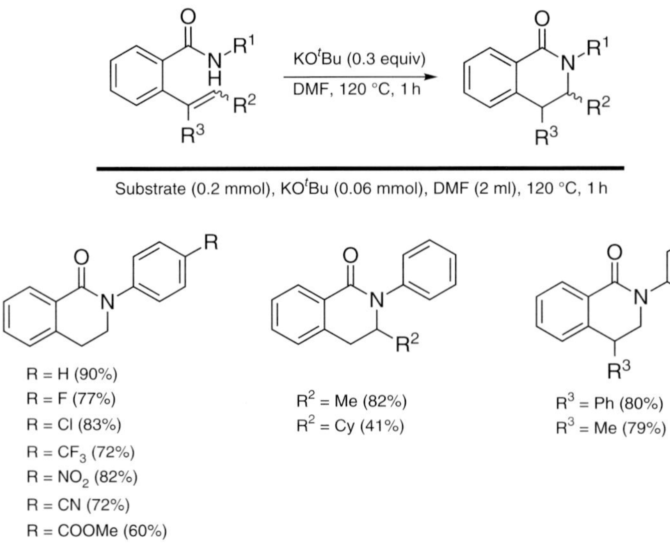

Scheme 3.12 Intramolecular hydroamination of *ortho*-vinyl benzamides catalyzed by KO^tBu and DMF.

While the reaction does not require N-prefunctionalization or strong oxidants, elevated reaction temperatures of 120 °C had to be applied. Other alkali metal bases such as NaO^tBu, KOMe, and NaOEt were also tested and found to be applicable but significantly lower yields were obtained. The utilization of K_2CO_3 or *n*-BuLi as well as different solvents (THF or dimethylacetamide [DMA]) was inefficient and also catalyst loadings below 0.3 equiv led to lower yields. Different substitutions on *N*-phenyl groups were well tolerated, but electron-withdrawing groups showed slightly lower yields. The reaction is believed to take place via a radical mechanism that was supported by radical-trapping and inhibition experiments (Scheme 3.13) [58]. Upon reaction with KO^{t}Bu, DMF is deprotonated and forms a carbamoyl radical **A** (Scheme 3.13) via single-electron transfer (SET). This reactive species abstracts a hydrogen atom from benzamide to form an amidyl radical **B**, which leads to an intramolecular radical cyclization and the intermediate radical **C** (Scheme 3.13). Subsequent hydrogen abstraction from another DMF molecule generates the product and regenerates the carbamoyl radical **A**. Thus, the solvent and its reactivity play a rather crucial role here, whereas there is only a minor contribution from the alkali metal cation.

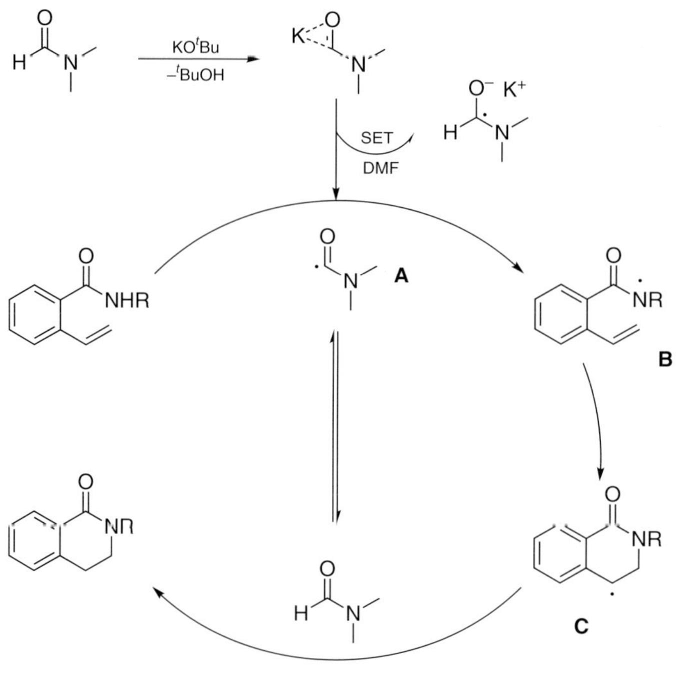

Scheme 3.13 Proposed reaction mechanism for the radical-mediated intramolecular hydroamination using KO^{t}Bu and DMF [58].

3.3.3.3 Reactions of *N*-Arylhydrazones and Ketoximes

Potassium *tert*-butoxide was also found to mediate the intramolecular hydroamination of γ,δ-alkenyl *N*-arylhydrazones with the outer nitrogen to afford tetrahydropyridazine derivatives (Scheme 3.14) [59]. Reactions were carried out in *o*-xylene at 135 °C and produced the cyclic hydroamination products in moderate to good yields. However, catalytic amounts of KO^{t}Bu were not sufficient and the additive Et_3COH was added to improve the yields. Other alkali metal bases such as MO^{t}Bu (M = Li, Na) or K_2CO_3/K_3PO_4 did not produce hydroamination products. Therefore, the additive presumably acts as a proton source donating a

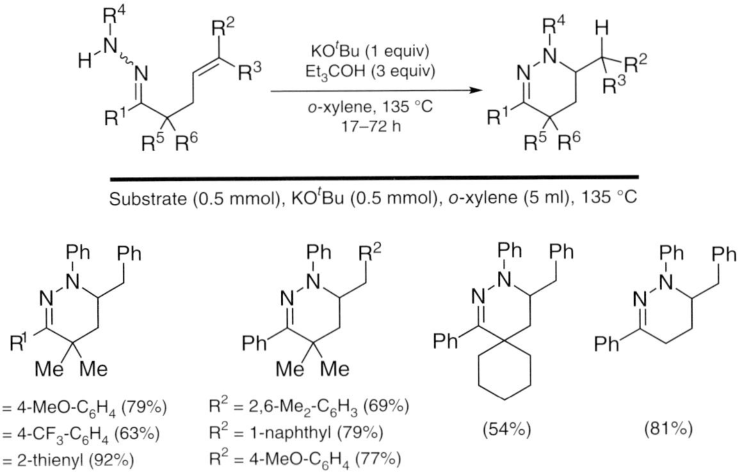

Scheme 3.14 Intramolecular hydroamination of γ,δ-alkenyl *N*-arylhydrazones mediated by KO^{t}Bu and Et$_3$COH.

proton to an intermediate carbanion, which facilitates the reaction toward product formation. However, the prior deprotonation of the hydrazone is essential for bypassing the otherwise very high reaction barrier for the cyclization of a neutral hydrazone. Noteworthy, steric effects of the alkene substituents did not influence the reaction outcome and the synthesis of tetrahydropyridazines with a simple reagent is straightforward. The concept of potassium base-mediated cyclohydroamination under otherwise very similar reaction conditions can also be transferred to different substrates providing access to other heterocycles. When γ,δ-alkenyl *N*-benzyl-hydrazones (Note: NCH_2R) are subjected to the KO^{t}Bu/Et$_3$COH system, a diastereo-divergent synthesis of saturated azaheterocycles is possible [60]. Therefore, hydrazone–hydrazone isomerization occurs upon deprotonation followed by cyclization and subsequent protonation. Modifications of the hydrazone substituents switches the diastereoselectivity and density functional theory (DFT) studies suggest that certain selectivities are caused by cation–π interactions between K^+ and aryl substituents on the hydrazone moiety.

When alkenyl oximes are used as starting materials, potassium bases initiate hydroamination and give rise to cyclic nitrones in moderate to excellent yields (Scheme 3.15) [61]. γ,δ-Unsaturated ketoximes do not spontaneously undergo cyclization to form a five-membered cyclic nitrone even after prolonged heating in different solvents. However, the addition of alkali metal bases facilitates hydroamination as the energy barrier for the reaction of the deprotonated anionic oxime is significantly lower compared with the neutral species. The K^+ cation (or Na^+) also stabilizes accumulating negative charges during the cyclization process as evidenced by DFT calculations. In contrast to the hydroamination of γ,δ-alkenyl *N*-arylhydrazones, the reaction of alkenyl oximes is catalytic with regard to the metal base and an essential role of the alkali metal cation is suggested.

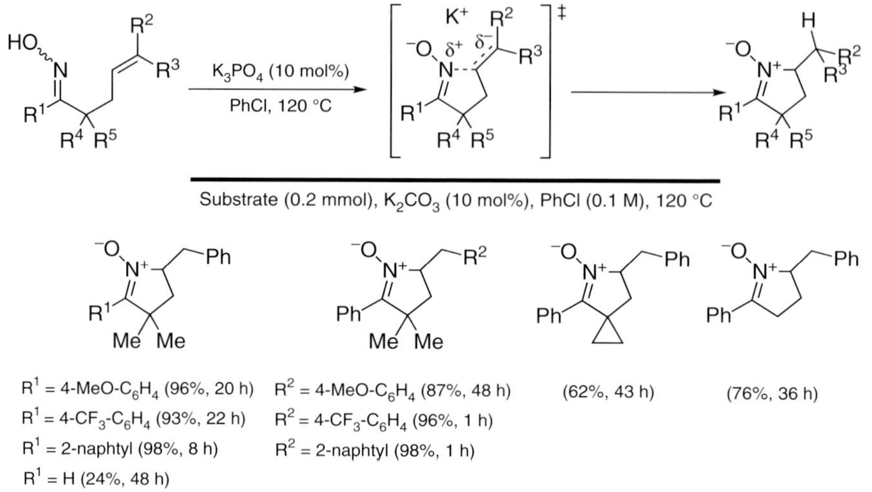

Scheme 3.15 Intramolecular hydroamination of γ,δ-unsaturated ketoximes catalyzed by K_2CO_3.

3.3.4 Group 2 Metal-Mediated Catalysis

The first alkaline earth-mediated hydroamination catalysis was achieved by Hill and coworkers in 2005 [34]. After the synthetically convenient heteroleptic β-diketiminato complex $[\{HC(C(Me)N\text{-}2,6\text{-}^iPr_2C_6H_3)_2\}Ca\{N(SiMe_3)_2\}(THF)]$ ($\{HC(C(Me)N\text{-}2,6\text{-}^iPr_2C_6H_3)_2\}$ = DippNacNac) had been found to smoothly react with primary amines [62–64], the complex was reacted with aminoalkenes in stoichiometric and catalytic ratios giving the desired cyclic products (Scheme 3.16). First studies revealed that the reaction can be accompanied by incomplete production of $HN(SiMe_3)_2$ and the formation of primary calcium–amide intermediates or calcium-adducted heterocylic products. As the β-diketiminato ligand provides only moderate kinetic stability, the catalytically inactive homoleptic complex $[(DippNacNac)_2Ca]$ also formed via solution exchange equilibria. This underpins the relevance of Schlenk-type equilibria and/or irreversible ligand redistribution in organometallic alkaline earth chemistry. However, hydroamination of geminally substituted 1-aminopent-4-enes proceeded rapidly at room temperature and quantitative conversion was achieved. Additionally, the 6-*exo*-trig cyclization of 1-amino-2,2-dimethylhex-1-ene to form 2-methyl-5,5-dimethylpiperidine was possible at increased temperature and reaction time. Noteworthy, no sign of alkene hydroamination was observed when the simple homoleptic complex $[Ca\{N(SiMe_3)_2\}_2(THF)_2]$ was utilized, despite complete hydrolysis of the bis(trimethylsilyl)amide groups and elevated reaction temperatures. Inspired by these first results and the concomitant widespread search for new homogeneous hydroamination catalysts for hydroaminations, a more detailed investigation was undertaken on alkaline earth β-diketiminato complexes [37]. The synthesis of these precatalysts is facile and applicable to every group 2 metal; however, complexes with the heavier metals Sr and Ba were found to be very unstable

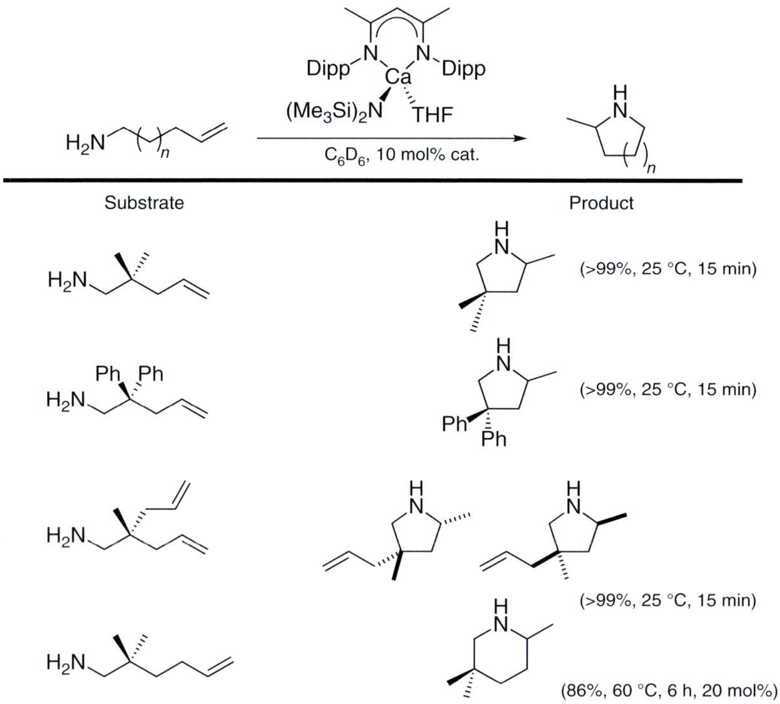

Scheme 3.16 Catalytic reactivity of [(DippNacNac)Ca{N(SiMe$_3$)$_2$}(THF)] and aminoalkenes.

toward ligand rearrangement and thus contaminated with the homoleptic bis(trimethylsilyl) amide by-products (Scheme 3.17) [63]. The initially futile attempts of using a (pre)catalyst without spectator ligands underscore the importance of the ligands in group 2 chemistry. However, the synthesis of desired heteroleptic compounds LMX (L = spectator ligand, X = leaving group) is challenging. Only certain combinations of L, M, and X prevent ligand redistribution and the formation of homoleptic compounds L_2M/MX_2, and

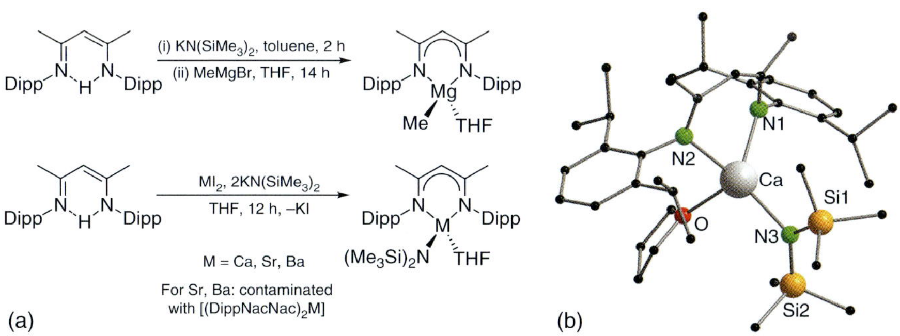

Scheme 3.17 Synthesis of group 2 β-diketiminato precatalysts (a) and (b) molecular structure of [(DippNacNac)Ca{N(SiMe$_3$)$_2$}(THF)] in solid state. Hydrogen atoms are not shown for clarity [63].

for L_2M, the metal is often completely shielded and thus catalytically inactive. Even incomplete ligand exchange processes lead to a mixture of organometallic species, which exacerbates precise studies on the catalytic system and its optimization. Based on previous studies dealing with low-coordinate and subvalent main group metal complexes [65], the DippNacNac ligand turned out to enable the synthesis of heteroleptic Ae complexes, thereby providing sufficient kinetic stabilization. However, coincident formation of the undesired homoleptic compounds is observed for the heavier Ae's Sr and Ba, which constrained the use of these complexes for catalytic application. The $LCaN(SiMe_3)_2$ and LMgMe precatalysts mediated numerous cyclizations toward pyrrolidine, piperidine, and hexahydroazepine derivatives under mild conditions and low catalyst loading. Detailed investigations revealed several commonalities and trends. These could partly be explained by the different ionic radii [66] of Mg and Ca, their different polarities/electronegativities, and the various nature of substrates. In contrast to Mg, Ca-mediated catalysis tends toward irreversible ligand redistribution to form the homoleptic L_2Ca complex. This is especially observed for runs at higher temperatures and extended reaction times. On the other hand, the Ca amide affected the cyclization of aminoalkenes at higher reaction rates and is more efficient than the Mg catalyst. This can be explained by the larger ionic radius for Ca^{2+}, which allows for more efficient alkene activation. Earlier studies on lanthanide-mediated hydroaminations revealed that the turnover frequency (TOF) increases with increasing coordinative unsaturation at the metal center, which frequently implies a higher reactivity for the larger cations [67, 68]. Some coordinative unsaturation is necessary in order to enable the insertion of the alkene into the M—N bond, which is generally proposed to be the rate-determining step in hydroamination catalysis.

With regard to the substrate scope, the geminal disubstitution of aminoalkenes beneficially influenced cyclization and the shortest reaction times with low catalyst loading were observed for these substrates. With the increasing steric demand of the substituents (H,H < Me,Me < Cy < Ph,Ph), both reaction times and catalyst loadings decreased. The geminal substituents lower the conformational freedom of the aminoalkene substrate and facilitate a reactive conformation (Thorpe–Ingold effect). In contrast, substitution on the terminal C of the C=C double bond hampered or even completely inhibited the cyclization. This is presumably because of the steric factors and unfavorable transition-state geometries. Depending on the substrate, different ring sizes (5, 6, 7) were obtained, and being consistent with Baldwin's discourse on ring formation, five-membered rings are formed more easily than seven-membered rings ($5 > 6 > 7$) [69]. Noteworthy, the formation of a seven-membered ring from 1-amino-2,2-diphenyl-6-heptene could not be facilitated by the more reactive Ca complex, but was achieved by the hitherto less reactive Mg catalyst, although very long reaction times and elevated temperature were needed. This is likely explained by the enhanced stability of the Mg diketiminate under the harsh reaction conditions. The reaction mechanism was further examined and found to be similar to lanthanide-based systems (Scheme 3.18). The first step of alkaline earth-mediated hydroamination involves the σ-bond metathesis (protonolysis) between the NacNacM-precatalyst and the amine to form a metal–amide

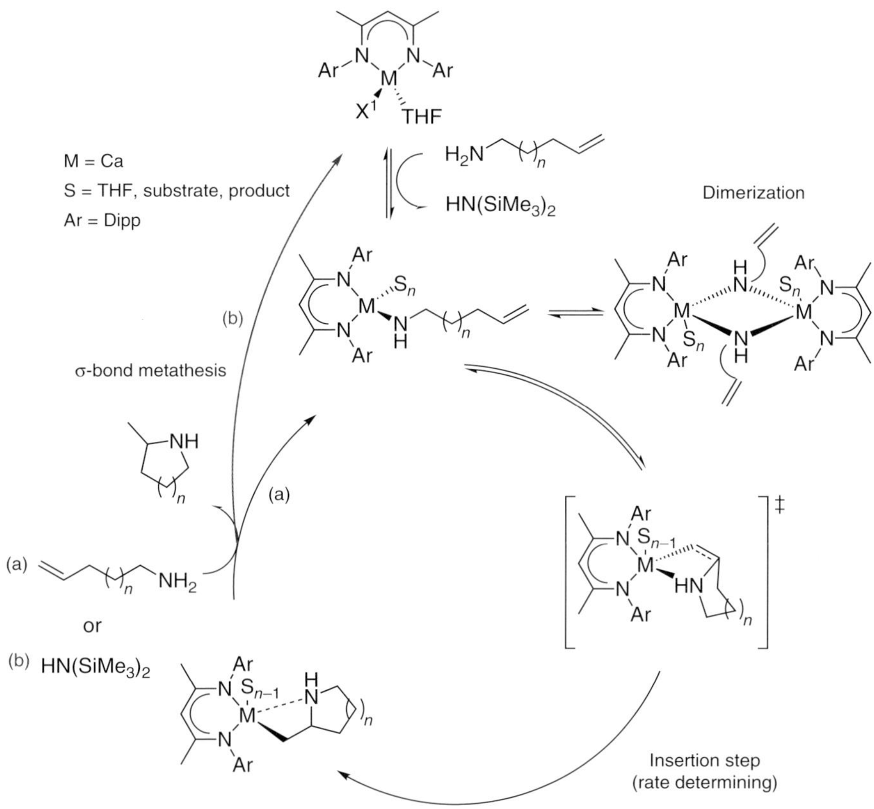

Scheme 3.18 Proposed reaction mechanism for the intramolecular hydroamination of aminoalkenes using [(DippNacNac)Ca{N(SiMe$_3$)$_2$}(THF)] [37].

complex. This step is usually fast, and depending on the basicity of the leaving group, protonolysis can be an irreversible or an equilibrium process. Both lead to monomeric primary metal amide species or dimeric metal complexes with bridging amide ligands. Subsequently, the rate-determining insertion of the C=C π-system into the M—N bond takes place via an ordered annular transition state. Thereby, a transient and highly reactive metal alkyl intermediate is formed. This could be proven by deuterium labeling experiments. This intermediate then reacts toward the more stable alkaline earth metal amide by either inter- or intramolecular protonolysis (see also Scheme 3.7) [37]. The cyclization product is finally cleaved from the metal center by reaction of the complex with a succeeding substrate molecule. Besides alkene isomerization processes, the catalytic turnover can be inhibited by coordinative saturation of the intermediate species, dimerization of the metal amides, or Schlenk-like and/or irreversible ligand redistribution resulting in catalytically inactive complexes.

Succeeding the initial report on calcium-mediated hydroamination, P. Roesky and coworkers utilized a different ligand framework for the synthesis of heteroleptic alkaline earth complexes and investigated their applicability in catalysis. Based on a system that earlier had been successfully employed in rare earth and zinc chemistry [70, 71], 2-(isopropylamino)tropone, (^{i}PrAT)H, and *N*-isopropyl-2-(isopropylamino)troponimine, {(^{i}Pr)$_2$ATI}H were established in group 2 and divalent lanthanide chemistry [72, 73]. The dimeric Ca complex [(^{i}PrAT)Ca{N(SiMe$_3$)$_2$}(THF)]$_2$ as well as the monomeric complex [{(^{i}Pr)$_2$ATI}M{N(SiMe$_3$)$_2$}(THF)$_2$] (M = Ca, Sr) were obtained via amide elimination or salt metathesis reactions, respectively (Scheme 3.19). A similar approach starting from BaI$_2$ or [Ba{N(SiMe$_3$)$_2$}$_2$] did not result in the formation of the expected heteroleptic barium complex. Instead, the homoleptic [{(^{i}Pr)$_2$ATI}$_2$Ba(THF)$_2$] was obtained, which is not suitable for catalytic studies because of the absence of a basic leaving group. All aminotroponiminate complexes were found to be active in the challenging intramolecular hydroamination of unactivated aminoalkenes. The mono- and dimeric Ca complexes facilitate smooth and almost quantitative conversion of terminal aminoalkenes to five-membered heterocycles. The lower reactivity of the dimeric Ca complex is presumably caused by incomplete dissociation in solution and thus enhanced coordinative saturation. The monomeric Ca complex, however, displays similar turnovers and reactivity compared with the DippNacNac system. Although it is slightly less reactive in the cyclization of some five-membered rings, it showed superior reactivity in converting the substrate 2,2-dimethyl-1-aminohex-5-ene into a product with a six-membered ring. Most remarkably, the strontium complex was found to be significantly less reactive than the isostructural calcium complex. The same observation was made in the follow-up studies using 2,5-bis{*N*-(2,6-diisopropylphenyl)iminomethyl}pyrrolyl complexes of calcium and strontium (Scheme 3.20) [74]. This observation is counterintuitive as for the lanthanides, an increase in reactivity was observed with increasing ionic radii: The bigger the metal cation, the more accessible is the metal center for alkene coordination in the rate-limiting insertion step. In view of an increase in reactivity from Mg to Ca followed by a decrease in reactivity from Ca to Sr, a clear trend with regard to cation size and reactivity cannot be established. The postulation of a metal–reactivity relationship is additionally hampered by the lack of a homologous series of kinetically stable group 2 precatalysts as well as contrary reports on the reactivity of comparable Ca and Sr catalysts. For

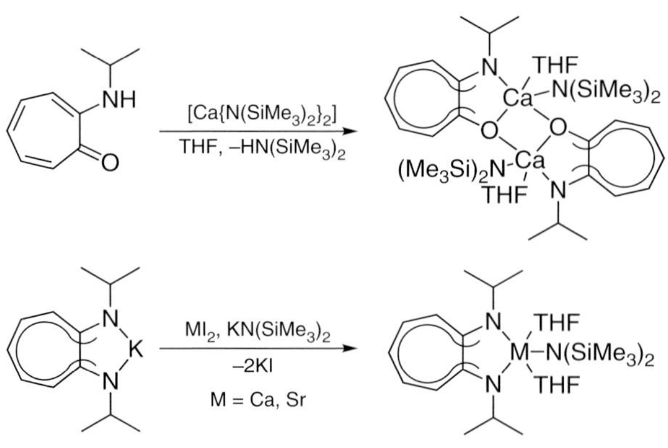

Scheme 3.19 Synthesis of mono- and dimeric aminotroponiminate complexes of Ca and Sr.

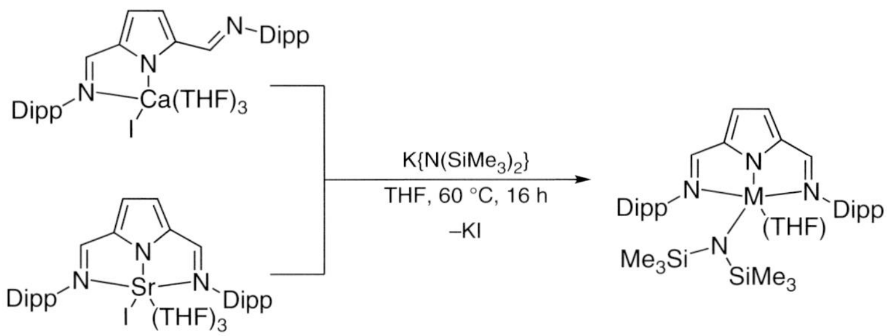

Scheme 3.20 2,5-Bis{*N*-(2,6-diisopropylphenyl)iminomethyl}pyrrolyl-based catalysts.

bis(imidazolin-2-ylidene-1-yl)borate and triazenide complexes of the heavier alkaline earths, Sr was found to be catalytically superior [75, 76]. Thus, the reactivity of alkaline earth metal catalysts cannot be broken down to the intrinsic properties of one metal. More likely, kinetic protection, the propensity to redistribution of the spectator ligands, and the peculiarities of the metal have to be mutually taken into account.

A full series of heteroleptic alkyl and amide iminoanilide, β-diketiminato, and tetradentate amino-ether phenolate alkaline earth complexes was prepared by Carpentier and Sarazin (Scheme 3.21) [77, 78]. All complexes feature either bis(trimethylsilyl)amido, bis(dimethylsilyl)amido, or bis(trimethylsilyl)methanide as leaving groups and are accessible in good yields and fully soluble and stable in solution (even the Ba complex). These catalysts have been evaluated in the inter- and intramolecular alkene hydroamination or intermolecular hydrophosphination [31, 79].

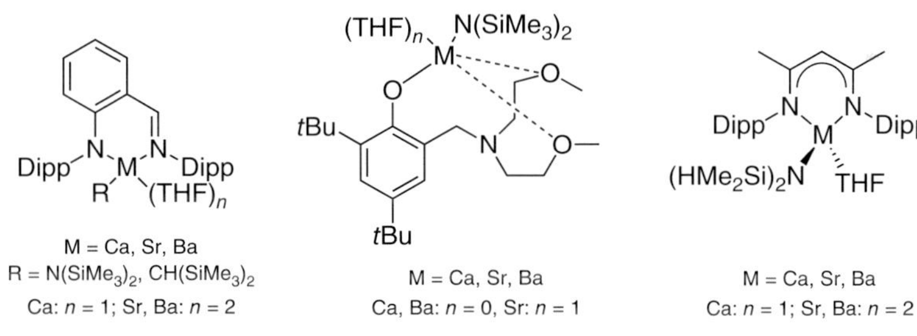

Scheme 3.21 Alkaline earth metal complexes that are stable to ligand redistribution by Schlenk-type equilibria.

Overall, the catalytic performance in intramolecular alkene cyclohydroaminations decreased in the order Ca > Sr ≫ Ba independent of the ancillary ligand. This is exactly the opposite of what had been observed for intermolecular alkene hydroaminations between styrene/isoprene and amines (Ba > Sr > Ca (>Mg)). Sadow and coworkers utilized oxazolinylborate (and cyclopentadiene–bis(oxazoline)) complexes of magnesium and investigated their mechanism in action for hydroamination catalysis (Scheme 3.22) [80–82]. The scorpionate ligand features an interesting and hemilabile coordination environment with two or three oxazolinyl rings bonded to the metal. The

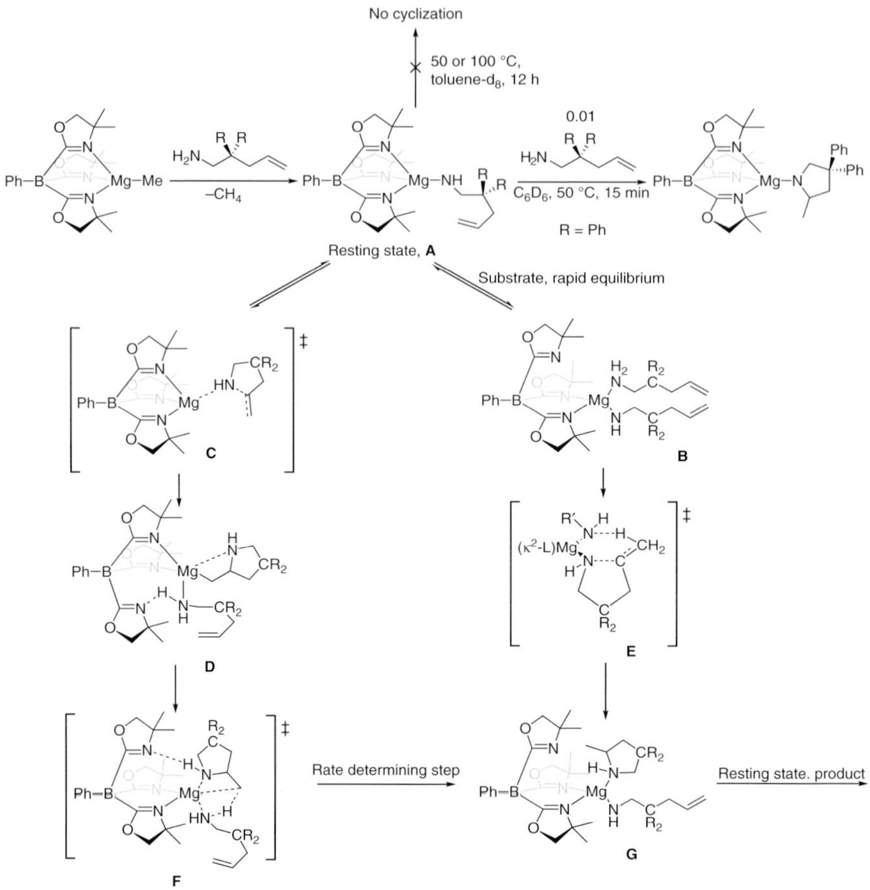

Scheme 3.22 Two possible mechanistic pathways for the intramolecular hydroamination with Sadow's magnesium oxazolinylborate catalyst [83].

LMgMe species smoothly reacts with aminoalkenes to form the corresponding magnesium–amido complex, which can be isolated. Unexpectedly, heating of these compounds does not lead to cyclization and product formation, which was unexpected, as no alkene insertion into the Mg—N bond occurs. However, in the presence of even very small amounts of additional substrate (0.01 equiv), cyclohydroamination rapidly takes place. Thus, the authors suggested a substrate-assisted, concerted, and noninsertive reaction mechanism with a transient, six-membered annular transition state (Scheme 3.22, **A** → **B** → **E** → **G**). However, in upcoming theoretical investigations by Tobisch, a σ-insertive pathway was proposed: **A** → **C** → **D** → **F** → **G** [84]. Therefore, the magnesium–amido complex **A** is in rapid equilibrium with its substrate-adduct **B** (Scheme 3.22). A cyclometallated complex **C** is reversibly formed from **A**, which can add an additional substrate amine to form **D**. This species can undergo intramolecular aminolysis via transition state **F** to finally regenerate the resting state and cyclic product, with the latter being the rate-determining step.

Comparable Ca and Sr silylamido complexes of a highly fluorinated 3-phenyl hydrotris(indazolyl)borate ligand were later reported by Etienne, Dinoi, and coworkers [85]. They were found to be inert toward ligand redistribution processes and the Ca complex was highly active in the intramolecular hydroamination of 2,2-dimethylpent-4-en-1-amine at very low catalyst loading, converting 200 equiv of aminoalkene within 16 minutes at 25 °C (0.5 mol%). Again, much lower activity was observed for the Sr analog and catalyst decomposition at large substrate loadings occurred likely because of the high sensitivity of the Ae species. Hill and coworkers synthesized dearomatized bis(imino)-acenaphthene (Dipp–BIAN) alkaline earth alkyl complexes of Mg, Ca, and Sr and utilized them for hydroamination of 15 different substrates comprising sterically hindered aminoalkenes and also substrates giving challenging seven-membered ring formation (Scheme 3.23) [86, 87]. The complexes are easily accessible from the Dipp-BIAN precursor and the alkaline earth dialkyl compounds [Ae{CH($SiMe_3$)$_2$}$_2$(THF)$_2$], which led to dearomatization of the BIAN backbone (Scheme 3.23). Enhanced stability toward Schlenk-type ligand redistributions was noticed only for complexes with bulky {CH($SiMe_3$)$_2$}-substituents, but an isostructural barium complex could not be obtained. Instead, sterically induced reductive C–C coupling of the ligand took place. The calcium catalyst was even able to cyclize substrates bearing alkyl substituents at either of the alkenyl positions. Its overall reactivity was found to be superior to that of its magnesium and strontium analogs (Ca > Mg ≫ Sr).

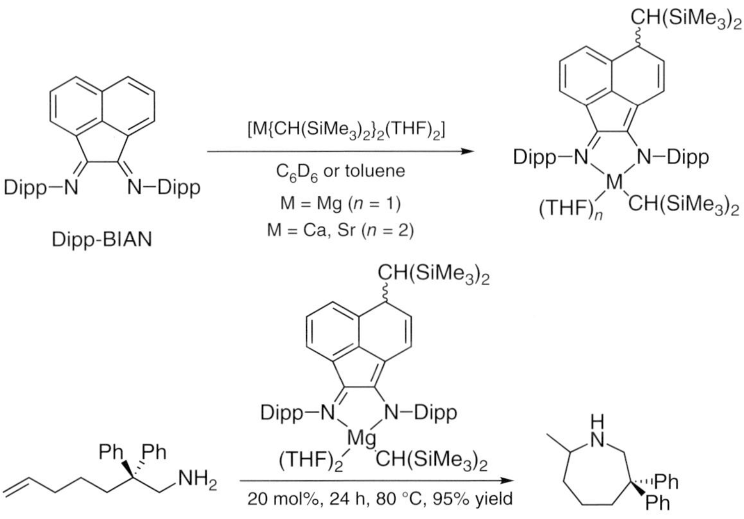

Scheme 3.23 Synthesis of alkaline earth Dipp–BIAN metal–alkyl complexes and a selected alkene hydroamination reaction.

In contrast to catalytically active heteroleptic complexes bearing a basic leaving group such as N{$SiMe_3$}$_2$, Harder and coworkers recently utilized a dianionic bora-amidinate (bam) ligand ^{Dipp}NBN to synthesize the alkaline earth metal complexes [(^{Dipp}NBN)Ae(THF)$_x$] (Mg: $x = 3$; Ca, Sr: $x = 4$), ^{Dipp}NBN = HB[N(2,6-^{i}Pr$_2$-C$_6$H$_3$)]$_2$ (Scheme 3.24 and Figure 3.1) [88, 89]. Except

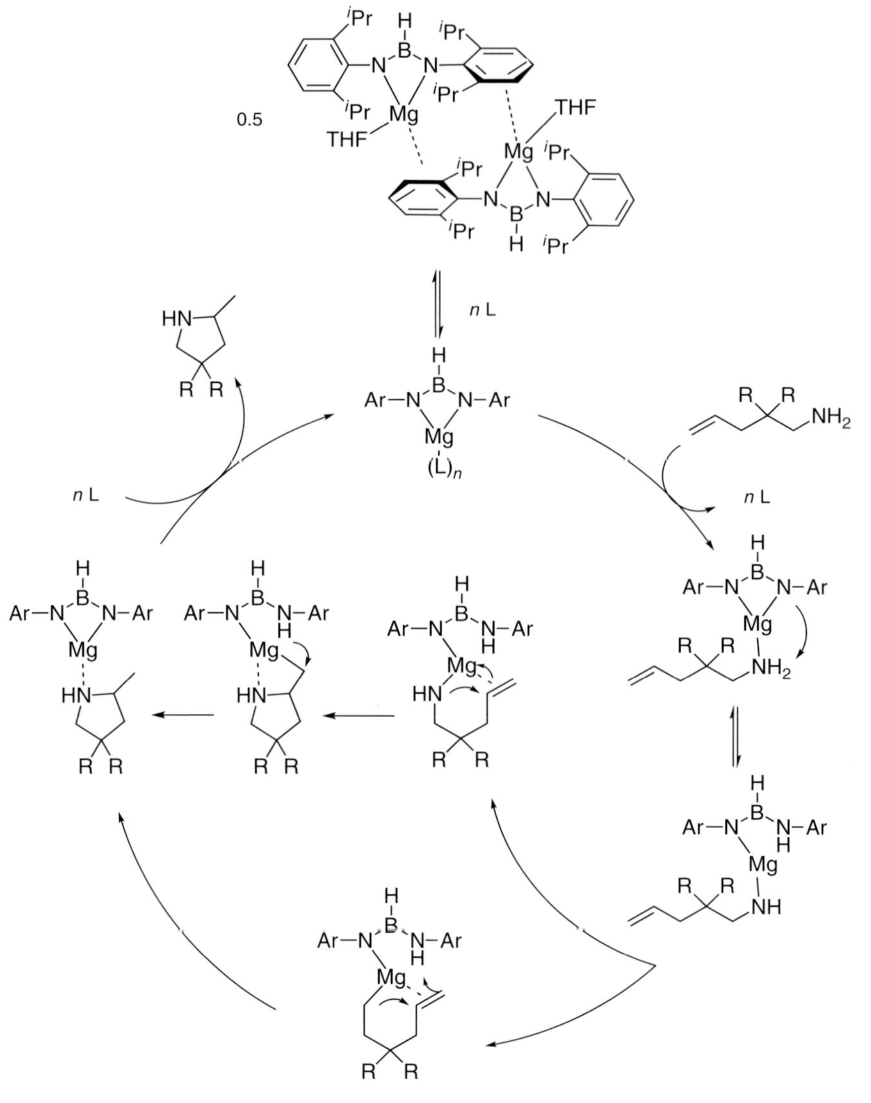

Scheme 3.24 Proposed catalytic cycle for the intramolecular alkene hydroamination with a bora-amidinate magnesium catalyst (nL = n equiv of a donor ligand, e.g. THF, RNH_2, or R_2NH) [89].

for Mg, the compounds are believed to exist as monomeric species in solution. Although the widely used amidinate ligands bear a single negative charge, the isoelectronic bora-amidinates are dianionic and thus fewer ligands are required for charge balance in case of Ae^{2+} cations, which implies a more open coordination site suitable for substrate coordination. Indeed, all compounds were found to smoothly catalyze the formation of pyrrolidine derivatives from different aminoalkenes under mild conditions with Ca showing the highest activity (Mg < Ca > Sr). Although the reactivity of the M—N bond increases from Mg

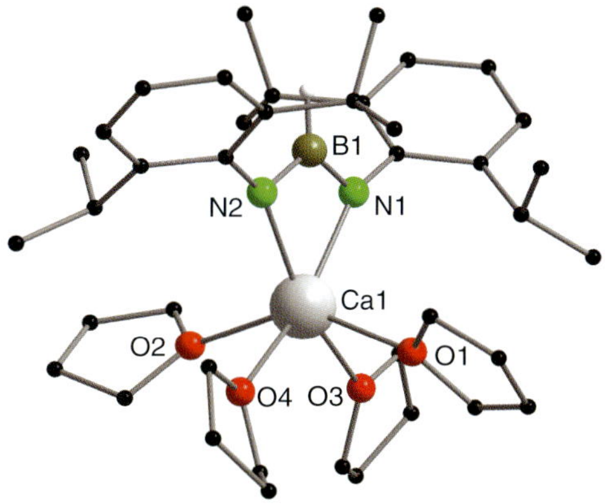

Figure 3.1 Molecular structure of $[(^{Dipp}NBN)Ca(THF)_4]$ in solid state. Hydrogen atoms are omitted for clarity.

to Sr, the Lewis acidity, which is crucial for alkene activation, decreases. These antiparallel trends are best balanced for Ca and explain its superior reactivity. The reaction mechanism was suggested to actively involve the strongly basic bam-ligand, which as a noninnocent ligand plays an important role in substrate deprotonation (Scheme 3.24). Therefore, the noninnocent bam-ligand shows flexible coordination behavior and also acts as an intermediate proton transfer source.

3.3.5 Group 2-Mediated Asymmetric Cyclohydroamination

As Schlenk-type equilibria hampered enantioselective alkene hydroaminations with diamidobinaphthyl-based catalysts (Mg and Zn), more rigid polydentate phenoxyamine ligands were developed by Hultzsch and coworkers [38]. Their corresponding isopropylmagnesium complexes were accessible in good yields and coordinatively stable in solution, with Mg^{2+} adopting a tetrahedral geometry. For various aminoalkenes, full conversion to their cyclic amination products was achieved (3–20 mol% cat. loading) but rather long reaction times and harsh conditions had to be applied. Importantly, spectroscopic investigations did not reveal any signs of Schlenk-type ligand redistribution processes, which is crucial for potential chiral catalysis.

Consequently, a chiral version of a phenoxyamine magnesium complex with bulky triphenylsilyl substituents was developed by the same group and obtained after recrystallization as pure $(R,R)Mg^R$ complex (Scheme 3.25, **A**) [90].

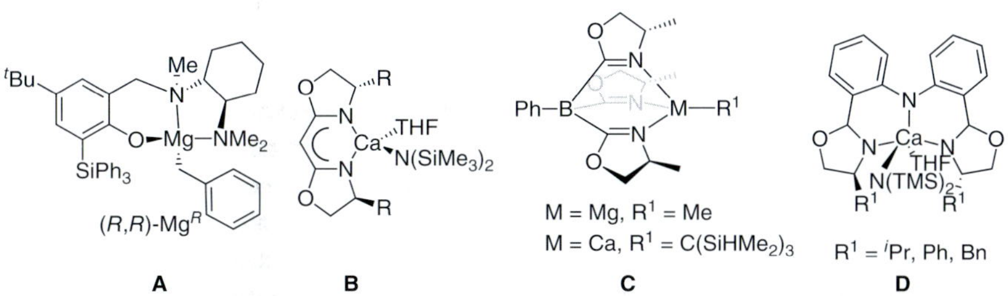

Scheme 3.25 Chiral complexes (**A–D**) utilized in asymmetric alkaline earth-mediated cyclohydroamination.

For aminoalkenes profiting from the Thorpe–Ingold effect, high activity and outstanding enantioselectivities (ee's up to 92%) were observed. This demonstrates that Schlenk equilibria can be suppressed, thus enabling enantioselective catalysis. Further studies on the mechanism revealed that at least two substrate molecules are required for substrate cyclization. Two rivaling mechanistic pathways, a σ-insertive and a concerted noninsertive route, were examined. It was suggested that the σ-insertive pathway is the more likely option [38].

First attempts on asymmetric calcium-mediated hydroamination were performed by Harder and coworker in 2008 [91]. A chiral enantiopure β-diketiminato ligand, synthesized from acetylacetone and (*S*)-α-Me-benzylamine, and the commercially available bisoxazoline (*S*)-Ph-BOX ligand were reacted with $[(Me_2N\text{-}\alpha\text{-}Me_3Si\text{-benzyl})_2Ca(THF)_2]$ or $[Ca\{N(SiMe_3)_2\}_2(THF)_2]$ (Scheme 3.25, **B**). Unfortunately, their corresponding $LCaN(SiMe_3)_2$ complexes underwent partial ligand redistribution to form the homoleptic L_2Ca compounds, and despite good conversion of aminoalkenes, poor ee's (<10%) were achieved. A slight improvement in enantioselectivity was made by Ward using $LCaN(SiMe_3)_2$ complexes of chiral diamine ligands, and albeit an ee of 26% was observed for the cyclization of $H_2NCH_2CPh_2CH_2CHCH_2$. The reaction proceeded very slowly and Schlenk equilibria were not optimally suppressed [92]. A chiral and pseudo C_3-symmetric ligand tris(4*S*-*tert*-butyl-2-oxazolinyl)phenylborate was used for the preparation of magnesium and calcium complexes by Sadow and coworkers (Scheme 3.25, **C**). Its resistance to ligand redistribution allowed for ee's ranging between 18% (Ca) and 36% (Mg) [81]. Subsequently, Ca–bismidazoline complexes were also found to undergo ligand rearrangements and thus provided only poor ee's [93]. However, a significant improvement was made by the same group introducing chiral bis(oxazolinylphenyl)amide ligands (BOPA) into the coordination sphere of calcium (Scheme 3.25, **D**). Because of very low ligand redistribution, various pyrrolidine derivatives were catalytically obtained with excellent conversion rates and ee's ranging between 14% and 50%. For piperidines, however, drastically deteriorated activities and selectivities were observed [94].

3.3.6 Lewis Acidic Metal Cation Catalysis

Because of the weak interaction between EMGM ions and unactivated alkenes, the question arises whether the metal actually plays a crucial role in hydroamination catalysis of aminoalkenes and if the reaction should be considered organometallic catalysis or organocatalysis [41]. Recently, Harder and coworkers found that the intramolecular hydroamination of aminoalkenes with unactivated C=C bonds requires a Lewis acidic metal in order to activate the substrate. However, a simple hybrid catalyst consisting of CaI_2 and the strong organic Schwesinger base showed smooth and quantitative conversion of $H_2C{=}CHCH_2CPh_2CH_2NH_2$ to the cyclic five-membered amine under mild and fast conditions. This indicates that very simple catalytic systems instead of sophisticated organometallic species may provide alternative access to hydroamination catalysts. Although calcium-based Lewis acidic catalysis is known to enable various transformations [95], a highly efficient promoter system for inter- and intramolecular hydroaminations was discovered only very recently. Using a mixture of calcium triflimide $Ca(NTf)_2$, $[nBu_4N^+][PF_6^-]$, and hexafluoroisopropanol (HFIP), N-heterocycles were obtained in good to excellent yields with good diastereoselectivities from *N*-tosyl- and other N-protected aminoalkenes (Scheme 3.26) [96]. The addition of $[nBu_4N^+][PF_6^-]$ is believed to promote the formation of a $[Ca(NTf)^+][PF_6^-]$ species with enhanced Lewis acidity. Subsequent ligation of HFIP to the "naked" Ca^{2+} ion further increases its high acidity and results in protonation of the alkene functionality. Thus, $[Ca(NTf)(HFIP)_n]^+$ can be seen as a metal-templated organocatalyst in which the meta-coordinated HFIP functions as a proton source.

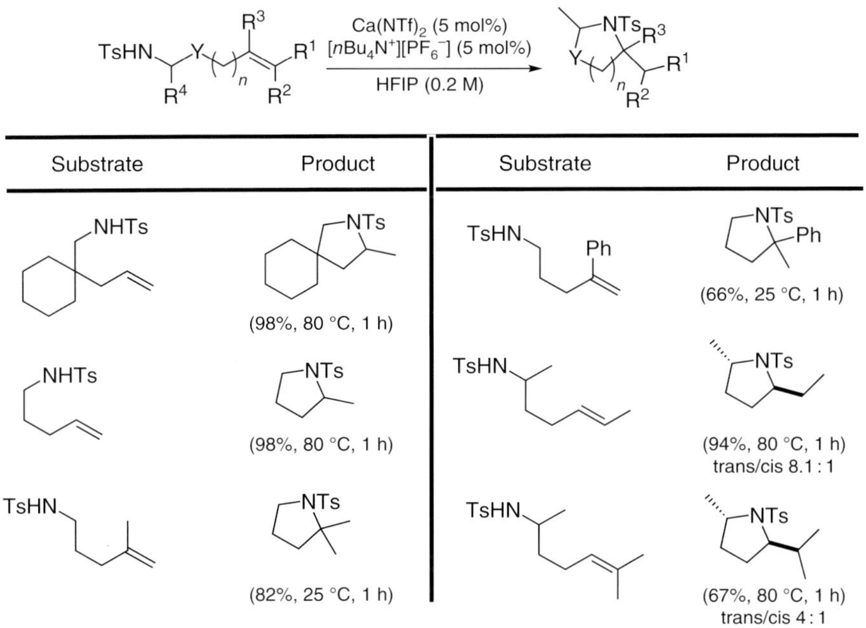

Scheme 3.26 Intramolecular hydroamination of *N*-tosylated aminoalkenes using a $Ca(NTf)_2/[nBu_4N^+][PF_6^-]$/HFIP system.

3.3.7 Miscellaneous

Although it is known that group 2 and group 4 metals are catalytically active in hydroamination, heterometallic systems that potentially show cooperative effects were not investigated until 2011. H. Roesky and coworkers utilized a well-defined organometallic hydroxide precursor $[Cp^*_2Zr(Me)(OH)]$ and reacted it with $[Ca\{N(SiMe_3)_2\}_2(THF)_2]$ to obtain an oxygen-bridged bimetallic species (Scheme 3.27) [97]. The heterometallic compound mediated cyclohydroaminations of unactivated aminoalkenes, providing excellent conversion and good reactivity. In a second step, hydroamination of secondary aminoalkenes was investigated, as methyl zirconocene was known to activate secondary amines in the presence of an activator $[PhNMe_2H^+][B(C_6F_5)_4^-]$. The activator is believed to abstract the zirconium-bound methyl group, thereby generating a

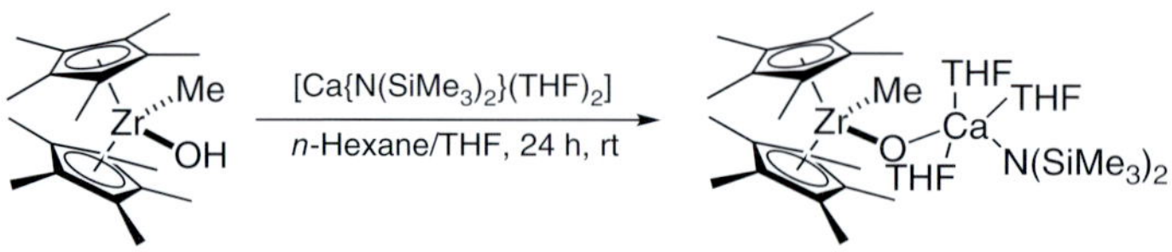

Scheme 3.27 Synthesis of a heterobimetallic Zr–Ca hydroamination catalyst.

cationic zirconium species. Indeed, a variety of different secondary amines were in most cases quantitatively converted to their cyclic derivatives, but reaction temperatures of 110 °C were needed to assure full conversion. Preliminary investigations suggest that the calcium center activates the primary amines, whereas zirconium activates the secondary amines. Later, a similar concept of heterometallic catalysis comprising group 2 and 4 elements was established. Although there are many reports on heterogeneous hydroamination catalysts and immobilized metal complexes [17], a metal–organic framework (MOF) with single-site magnesium alkyl catalysts for hydroamination reactions was reported only recently [98]. The secondary building units of a Zr-TPHN-MOF (TPHN = 4,4′-bis-(carboxyphenyl)-2-nitro-1,1′-biphenyl) were metalated with $MgMe_2$. This showed a highly active and reusable solid-state catalyst featuring a MgMe functionality that is active in the hydroamination of aminopentenes (Scheme 3.28). Despite the rather long reaction times, full conversion was achieved. Following ICP-MS analyses of the product, only very little metal leaching was observed.

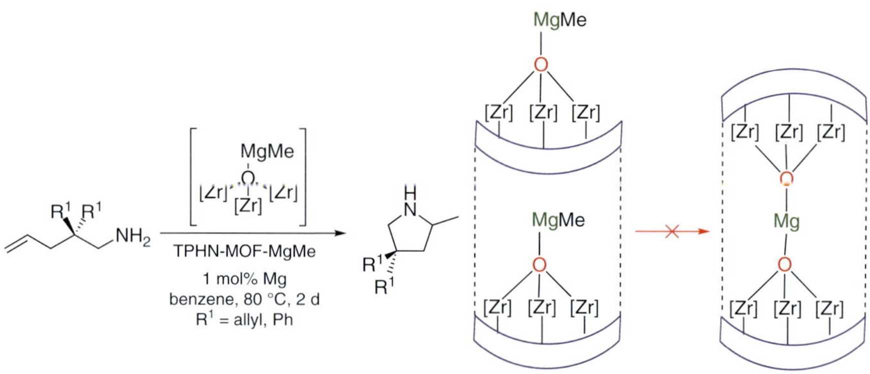

Scheme 3.28 MOF-MgMe catalysis and an illustration of the beneficial effect of active site isolation of heteroleptic Mg species, thus preventing ligand redistribution processes.

3.4 Outlook

In view of the need for sustainable and environmentally compatible chemistry, atom-efficient and energy-saving catalytic reactions such as hydroamination reactions will become more and more important. The discovery of catalytically active alkali and alkaline earth metal complexes has made an important contribution to the field of sustainable catalysis; however, many challenges such as improvements in enantioselective catalysis still have to be addressed. Furthermore, a greater unification of reactivity models, especially for group 2 metals, is needed. This would be very beneficial for the design of highly active catalysts. Although important contributions have been made in s-block-catalyzed alkene hydroamination, there is room for further improvements. Especially, in view of the scarcity of our resources, intense research efforts should be directed to this important field.

Acknowledgments

SB gratefully acknowledges the Deutsche Forschungsgemeinschaft (DFG, BE 6401/1-1) (project number 380155090) for supporting this work. Support from the DFG-funded transregional collaborative research center SFB/TRR 88 "Cooperative Effects inHomo and Heterometallic Complexes (3MET)", projects B3 is acknowledged.

List of Abbreviations

Ae	alkaline earth
Am	alkali metal
bam	bora-amidinate
Bn	benzyl
BOPA	bis(oxazolinylphenyl)amide
BTSA	bis(trimethylsilyl)amide
cat.	catalyst
Cp	cyclopentadienyl
Cp*	pentamethylcyclopentadienyl
Cy	cyclohexyl
DFT	density functional theory
Dipp	2,6-diisopropylphenyl
DMA	dimethylacetamide
DMBA	α-deprotonated dimethylaminobenzyl
DME	1,2-dimethoxyethane
DMF	dimethylformamide
ee	enantiomeric excess
EMGM	early main group metal
Et	ethyl
*i*Bu	isobutyl
*i*Pr	isopropyl
Ln	Sc, Y, Lanthanide
Me	methyl
MeLi	methyllithium
MOF	metal–organic framework
NacNac	1,3-diketimine
*n*Bu	*n*-butyl

Ph	phenyl
PhMe	toluene
rt	room temperature
*t*Bu	*t*-butyl
THF	tetrahydrofuran
TM	transition metal
TOF	turnover frequency
σ-BM	sigma-bond metathesis

References

1 Roundhill, D.M. (1992). *Chem. Rev.* 92: 1–27.
2 Johannsen, M. and Jorgensen, K.A. (1998). *Chem. Rev.* 98: 1689–1708.
3 Brunet, D.J.-J. and Neibecker, D.D. (2001). Catalytic hydroamination of unsaturated carbon–carbon bonds. In: *Catalytic Heterofunctionalization* (ed. Prof. Dr. A. Togni and Prof. Dr. H. Grützmacher), 91–141. Wiley.
4 Nobis, M. and Drießen-Hölscher, B. (2001). *Angew. Chem.* 113: 4105–4108.
5 Seayad, J., Tillack, A., Hartung, C.G., and Beller, M. (2002). *Adv. Synth. Catal.* 344: 795–813.
6 Bytschkov, I. and Doye, S. (2003). *Eur. J. Org. Chem.*: 935–946.
7 Pohlki, F. and Doye, S. (2003). *Chem. Soc. Rev.* 32: 104–114.
8 Roesky, P.W. and Müller, T.E. (2003). *Angew. Chem.* 115: 2812–2814.
9 Alonso, F., Beletskaya, I.P., and Yus, M. (2004). *Chem. Rev.* 104: 3079–3160.
10 Hartwig, J.F. (2004). *Pure Appl. Chem.* 76: 507–516.
11 Hong, S. and Marks, T.J. (2004). *Acc. Chem. Res.* 37: 673–686.
12 Hultzsch, K.C. (2005). *Adv. Synth. Catal.* 347: 367–391.
13 Hultzsch, K.C. (2005). *Org. Biomol. Chem.* 3: 1819–1824.
14 Widenhoefer, R.A. and Han, X. (2006). *Eur. J. Org. Chem.* 2006: 4555–4563.
15 Aillaud, I., Collin, J., Hannedouche, J., and Schulz, E. (2007). *Dalton Trans.*: 5105–5118.
16 Hunt, P.A. (2007). *Dalton Trans.*: 1743–1754.
17 Müller, T.E., Hultzsch, K.C., Yus, M. et al. (2008). *Chem. Rev.* 108: 3795–3892.
18 Roesky, P.W. (2009). *Angew. Chem.* 121: 4988–4991.
19 Patton, D.A. and Cremeens, M.E. (2014). *Rev. J. Chem.* 4: 1–20.
20 Harder, S. (2010). *Chem. Rev.* 110: 3852–3876.
21 Müller, T.E. and Beller, M. (1998). *Chem. Rev.* 98: 675–704.
22 Beesley, R.M., Ingold, C.K., and Thorpe, J.F. (1915). *J. Chem. Soc. Trans.* 107: 1080–1106.
23 Ingold, C.K. (1921). *J. Chem. Soc. Trans.* 119: 305.
24 Jung, M.E. and Gervay, J. (1991). *J. Am. Chem. Soc.* 113: 224–232.
25 Jung, M.E. and Piizzi, G. (2005). *Chem. Rev.* 105: 1735–1766.
26 Hannedouche, J. and Schulz, E. (2013). *Chem. Eur. J.* 19: 4972–4985.
27 Schlummer, B. and Hartwig, J.F. (2002). *Org. Lett.* 4: 1471–1474.
28 Ryu, J.-S., Li, G.Y., and Marks, T.J. (2003). *J. Am. Chem. Soc.* 125: 12584–12605.

29 Gagne, M.R., Nolan, S.P., and Marks, T.J. (1990). *Organometallics* 9: 1716–1718.

30 Hill, M.S., Liptrot, D.J., and Weetman, C. (2016). *Chem. Soc. Rev.* 45: 972–988.

31 Sarazin, Y. and Carpentier, J.F. (2016). *Chem. Rec.* 16: 2482–2505.

32 Westerhausen, M. (1991). *Inorg. Chem.* 30: 96–101.

33 Crimmin, M.R., Barrett, A.G.M., Hill, M.S. et al. (2008). *Chem. Eur. J.* 14: 11292–11295.

34 Crimmin, M.R., Casely, I.J., and Hill, M.S. (2005). *J. Am. Chem. Soc.* 127: 2042–2043.

35 Arrowsmith, M., Crimmin, M.R., Barrett, A.G.M. et al. (2011). *Organometallics* 30: 1493–1506.

36 Barrett, A.G.M., Crimmin, M.R., Hill, M.S., and Procopiou, P.A. (2010). *Proc. R. Soc. London, Ser. A* 466: 927–963.

37 Crimmin, M.R., Arrowsmith, M., Barrett, A.G.M. et al. (2009). *J. Am. Chem. Soc.* 131: 9670–9685.

38 Zhang, X., Tobisch, S., and Hultzsch, K.C. (2015). *Chem. Eur. J.* 21: 7841–7857.

39 Tobisch, S. (2015). *Chem. Eur. J.* 21: 6765–6779.

40 Tobisch, S. (2014). *Chem. Eur. J.* 20: 8988–9001.

41 Penafiel, J., Maron, L., and Harder, S. (2015). *Angew. Chem. Int. Ed.* 54: 201–206.

42 Kelly, R.P. and Roesky, P.W. (2016). Catalytic sigma-bond metathesis and the polymerization of 1,3-dienes by rare-earth metal complexes. In: *50 Years of Structure and Bonding – The Anniversary Volume* (ed. D.M.P. Mingos), 85–117. Cham: Springer International Publishing.

43 Lauterwasser, F., Hayes, P.G., Bräse, S. et al. (2004). *Organometallics* 23: 2234–2237.

44 Patel, M., Saunthwal, R.K., and Verma, A.K. (2017). *Acc. Chem. Res.* 50: 240–254.

45 Chapurina, Y., Guillot, R., Lyubov, D. et al. (2013). *Dalton Trans.* 42: 507–520.

46 Michon, C., Abadie, M.-A., Medina, F., and Agbossou-Niedercorn, F. (2017). *J. Organomet. Chem.* 847: 13–27.

47 Spallek, T. and Anwander, R. (2016). *Dalton Trans.* 45: 16393–16403.

48 Ates, A. and Quinet, C. (2003). *Eur. J. Org. Chem.* 2003: 1623–1626.

49 Rodriguez-Ruiz, V., Carlino, R., Bezzenine-Lafollee, S. et al. (2015). *Dalton Trans.* 44: 12029–12059.

50 Quinet, C., Sampoux, L., and Markó, I.E. (2009). *Eur. J. Org. Chem.* 2009: 1806–1811.

51 Quinet, C., Jourdain, P., Hermans, C. et al. (2008). *Tetrahedron* 64: 1077–1087.

52 Rousseau, G., Lebeuf, R., Schenk, K. et al. (2014). *Chem. Eur. J.* 20: 14771–14782.

53 Martinez, P.H., Hultzsch, K.C., and Hampel, F. (2006). *Chem. Commun.*: 2221–2223.

54 Horrillo-Martínez, P. and Hultzsch, K.C. (2009). *Tetrahedron Lett.* 50: 2054–2056.

55 Ogata, T., Ujihara, A., Tsuchida, S. et al. (2007). *Tetrahedron Lett.* 48: 6648–6650.
56 Deschamp, J., Olier, C., Schulz, E. et al. (2010). *Adv. Synth. Catal.* 352: 2171–2176.
57 Deschamp, J., Collin, J., Hannedouche, J., and Schulz, E. (2011). *Eur. J. Org. Chem.* 2011: 3329–3338.
58 Chen, Z.Y., Wu, L.Y., Fang, H.S. et al. (2017). *Adv. Synth. Catal.* 359: 3894–3899.
59 Peng, X., Kaga, A., Hirao, H., and Chiba, S. (2016). *Org. Chem. Front.* 3: 609–613.
60 Kaga, A., Peng, X., Hirao, H., and Chiba, S. (2015). *Chem. Eur. J.* 21: 19112–19118.
61 Peng, X., Tong, B.M.K., Hirao, H., and Chiba, S. (2014). *Angew. Chem. Int. Ed.* 53: 1959–1962.
62 Avent, A.G., Crimmin, M.R., Hill, M.S., and Hitchcock, P.B. (2004). *Dalton Trans.*: 3166–3168.
63 Chisholm, M.H., Gallucci, J., and Phomphrai, K. (2003). *Chem. Commun.*: 48–49.
64 Avent, A.G., Crimmin, M.R., Hill, M.S., and Hitchcock, P.B. (2005). *Dalton Trans.*: 278–284.
65 Bourget-Merle, L., Lappert, M.F., and Severn, J.R. (2002). *Chem. Rev.* 102: 3031–3066.
66 Shannon, R. (1976). *Acta Crystallogr., Sect. A* 32: 751–767.
67 Gagne, M.R., Stern, C.L., and Marks, T.J. (1992). *J. Am. Chem. Soc.* 114: 275–294.
68 Gagne, M.R. and Marks, T.J. (1989). *J. Am. Chem. Soc.* 111: 4108–4109.
69 Baldwin, J.E. (1976). *J. Chem. Soc., Chem. Commun.*: 734–736.
70 Bürgstein, M.R., Berberich, H., and Roesky, P.W. (1998). *Organometallics* 17: 1452–1454.
71 Zulys, A., Dochnahl, M., Hollmann, D. et al. (2005). *Angew. Chem. Int. Ed.* 44: 7794–7798.
72 Datta, S., Roesky, P.W., and Blechert, S. (2007). *Organometallics* 26: 4392–4394.
73 Datta, S., Gamer, M.T., and Roesky, P.W. (2008). *Organometallics* 27: 1207–1213.
74 Jenter, J., Köppe, R., and Roesky, P.W. (2011). *Organometallics* 30: 1404–1413.
75 Barrett, A.G.M., Crimmin, M.R., Hill, M.S. et al. (2008). *Inorg. Chem.* 47: 7366–7376.
76 Arrowsmith, M., Hill, M.S., and Kociok-Köhn, G. (2009). *Organometallics* 28: 1730–1738.
77 Liu, B., Roisnel, T., Carpentier, J.F., and Sarazin, Y. (2012). *Angew. Chem. Int. Ed.* 51: 4943–4946.
78 Liu, B., Roisnel, T., Carpentier, J.F., and Sarazin, Y. (2013). *Chem. Eur. J.* 19: 13445–13462.
79 Liu, B., Roisnel, T., Carpentier, J.F., and Sarazin, Y. (2013). *Chem. Eur. J.* 19: 2784–2802.

80 Dunne, J.F., Fulton, D.B., Ellern, A., and Sadow, A.D. (2010). *J. Am. Chem. Soc.* 132: 17680–17683.
81 Neal, S.R., Ellern, A., and Sadow, A.D. (2011). *J. Organomet. Chem.* 696: 228–234.
82 Eedugurala, N., Hovey, M., Ho, H.-A. et al. (2015). *Organometallics* 34: 5566–5575.
83 Rochat, R., Lopez, M.J., Tsurugi, H., and Mashima, K. (2015). *ChemCatChem* 8: 10–20.
84 Sven, T. (2011). *Chem. Eur. J.* 17: 14974–14986.
85 Romero, N., Roşca, S.C., Sarazin, Y. et al. (2015). *Chem. Eur. J.* 21: 4115–4125.
86 Arrowsmith, M., Hill, M.S., and Kociok-Köhn, G. (2014). *Organometallics* 33: 206–216.
87 Arrowsmith, M., Hill, M.S., and Kociok-Köhn, G. (2011). *Organometallics* 30: 1291–1294.
88 Harder, S., Freitag, B., Stegner, P. et al. (2015). *Z. Anorg. Allg. Chem.* 641: 2129–2134.
89 Freitag, B., Fischer, C.A., Penafiel, J. et al. (2017). *Dalton Trans.* 46: 11192–11200.
90 Zhang, X., Emge, T.J., and Hultzsch, K.C. (2012). *Angew. Chem. Int. Ed.* 51: 394–398.
91 Buch, F. and Harder, S. (2008). *Z. Naturforsch., B: Chem. Sci.* 63: 169–177.
92 Wixey, J.S. and Ward, B.D. (2011). *Chem. Commun.* 47: 5449–5451.
93 Wixey, J.S. and Ward, B.D. (2011). *Dalton Trans.* 40: 7693–7696.
94 Nixon, T.D. and Ward, B.D. (2012). *Chem. Commun.* 48: 11790–11792.
95 Begouin, J.M. and Niggemann, M. (2013). *Chem. Eur. J.* 19: 8030–8041.
96 Qi, C., Hasenmaile, F., Gandon, V., and Lebœuf, D. (2018). *ACS Catal.* 8: 1734–1739.
97 Mukherjee, A., Nembenna, S., Sen, T.K. et al. (2011). *Angew. Chem. Int. Ed.* 50: 3968–3972.
98 Manna, K., Ji, P., Greene, F.X., and Lin, W. (2016). *J. Am. Chem. Soc.* 138: 7488–7491.

4

Molecular s-Block Catalysts for Alkene Hydrophosphination and Related Reactions

Yann Sarazin and Jean-François Carpentier

Univ Rennes, CNRS, ISCR (Institut des Sciences Chimiques de Rennes) – UMR 6226, F-35000 Rennes, France

4.1 Introduction

Organophosphorus compounds constitute a vast and important class of compounds that find widespread applications, in particular as valuable ligands in coordination chemistry and for metal-catalyzed organic transformations [1], in industry as pest control and pesticides, and also owing to their biological activity, such as antibiotics and antitumor agents [2–5]. As a result, the creation of C—P bonds is attracting persistent interest from synthetic organic and organometallic chemists alike. The addition of H—P bonds across unsaturated bonds represents an attractive method to generate a plethora of different C—P bonds. Typical electrophilic carbon atoms are hence those found in alkynes, alkenes, aldehydes, and imines. These hydroelementation reactions offer the main advantage of 100% atom efficiency, without formation of any by-product.

The metal-mediated formation of C—P bonds can be divided into different categories (Scheme 4.1). Phosphination, which can be regarded as a cross-coupling reaction, involves substitution at a C_{sp2}—X or C_{sp3}—X carbon atom, typically in the presence of a base and upon release of HX. This reaction, although of obvious importance, does not enter the category of hydroelementation reactions, and is beyond the scope of this review; it has been reviewed elsewhere on several occasions [6–8]. Depending on the nature of the substrate added across the C≡C, C=C, C=O, or C=N unsaturation, P-hydroelementation reactions are subdivided into three main categories, according to the number of oxygen atoms contained in the substrate [8]. Hydrophosphinations involve a phosphine HPR_2, no oxygen atom; this reaction is also sometimes referred to as hydrophosphanylation. Hydrophosphinylation (for $HP(O)R_2$ and $HP(O)(OR)R'$, one and two oxygen atoms, respectively; for $HP(O)R_2$, the term hydrophosphorylation has also been used at times) and hydrophosphonylation (for $HP(O)(OR)_2$, three oxygen atoms) are the two other families of reactions (Scheme 4.1).

Of these three families, hydrophosphination arguably constitutes the most important class of reactions. This net addition of H—P-containing moieties to unsaturated substrates has been shown to occur in the absence of metal catalysts [9, 10], with a free radical mechanism ensuing from homolytic cleavage of the

Early Main Group Metal Catalysis: Concepts and Reactions,
First Edition. Edited by Sjoerd Harder.

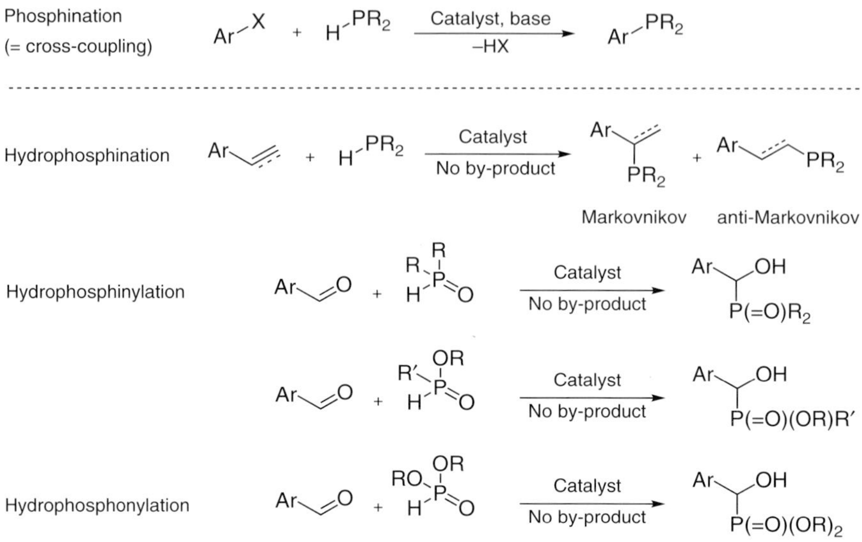

Scheme 4.1 Examples of common metal-mediated reactions leading to the creation of C—P bonds.

P—H bond under the action of temperature, light or radical initiators [11, 12], or upon treatment with simple acid or base catalysts [13, 14]. On the other hand, the utilization of well-defined metal catalysts has yielded key breakthroughs by enabling reactions with increased reaction rates, regioselectivity, and enantioselectivity. Catalysts built on metals spanning across the periodic table are known to mediate the hydrophosphination of unsaturated substrates. In the wake of Pringle's seminal report of the hydrophosphination of acrylonitrile with PH_3 or $HP(CH_2CH_2CN)_2$ catalyzed by the Pt(0) complex $[Pt\{P(CH_2CH_2CN)_3\}_3]$ [15], late transition metals, especially Pt, Pd, and Ni, have been at the forefront of catalyst development in this area for nearly three decades. Regarding C=C bonds, these systems are effective chiefly (although not exclusively) for electron poor substrates, such as acrylonitrile, enones, and other Michael acceptors; they can also catalyze alkyne hydrophosphination reasonably well. On the other hand, they tend to be less efficacious for styrene derivatives and nonactivated olefins, although a few late transition metal systems can efficiently hydrophosphinate styrene and alkynes. Another potential drawback is that the addition of a base, e.g. NEt_3, can be required as a cocatalyst for these systems to be operative. Following Marks' footsteps in the 1990s, lanthanide(III) complexes have also emerged as potent base-free and relatively versatile catalysts for the hydrophosphination of alkenes and alkynes; group 4 catalysts typically based on titanium and zirconium appeared shortly thereafter. For more information, the interested reader can refer the several available reviews that survey the development of these catalytic systems since the 1990s, including enantioselective processes [8, 16–24].

The first report of a discrete catalyst based on an alkaline earth metal (calcium) for the hydrophosphination of alkenes is comparatively recent, dating back only to 2007 [25]. This initial report was followed by a series of contributions from the group of Westerhausen describing the alkaline earth promoted hydrophosphination of alkynes and diynes [26–29], as well as by several other studies dealing with the hydrophosphination of activated alkenes (styrene derivatives and conjugated dienes) and heterocumulenes catalyzed by complexes of calcium and of the larger alkaline earths, strontium and barium. In addition, a few contributions also detailed the use of Ca, Sr, and Ba precatalysts to promote the hydrophosphonylation of nonactivated alkenes and of aldehydes and ketones. By contrast, the alkali metals lithium, sodium, and potassium have seldom been involved in this area, and there is to date no report of *molecular* catalysts based on these s-block elements for the controlled creation of C—P bonds.

The present chapter will focus mostly on the developments of alkaline earth precatalysts for alkene hydrophosphination and will also include the few examples known for alkaline earth-mediated hydrophosphonylation and hydrophosphinylation. Precatalysts for alkyne and diyne hydrophosphinations will be discussed elsewhere in Chapter 5.

4.2 General Considerations

The recent impetus in the area of hydrophosphination catalysis, and more generally of homogeneous catalysis, mediated by complexes of the large alkaline earth metals (=Ae) calcium, strontium, and barium are closely associated with the development of sufficiently stable and soluble homoleptic and heteroleptic complexes of the general type $[AeX_2{\cdot}(D)_p]$ and $[\{L_n\}AeX{\cdot}(D)_p]$, respectively, where Ae is chosen among Ca, Sr, and Ba, X is a reactive group that can be either nucleophilic or basic, $\{L_n\}^-$ is a monoanionic, multidentate ancillary ligand, and D is a Lewis base including tetrahydrofuran (THF), pyridine (py), or diethyl ether. Ca^{2+} (effective ionic radius 1.00 Å for a coordination number (CN) of 6), Sr^{2+} (1.18 Å) and Ba^{2+} (1.35 Å) are three very large cationic centers; magnesium is comparatively much smaller (0.82 Å). They are highly electropositive and polarizable and generate d^0 complexes where the highly ionic bonding is governed by electrostatic and steric factors, with little or no ability for electron sharing; these features increase upon going down in group 2 (Figure 4.1). As a result, the problem of kinetic stability in the solution of heteroleptic complexes has long plagued the solution organometallic and inorganic chemistry of these elements.

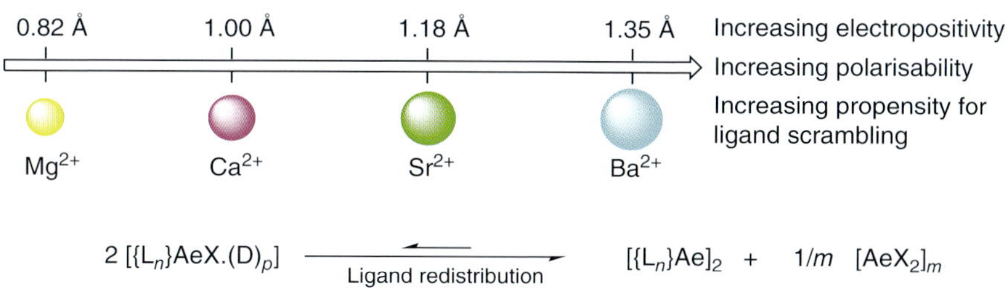

Figure 4.1 Figure 4.1 Properties of the alkaline earths and ligand redistribution (aka Schlenk equilibrium).

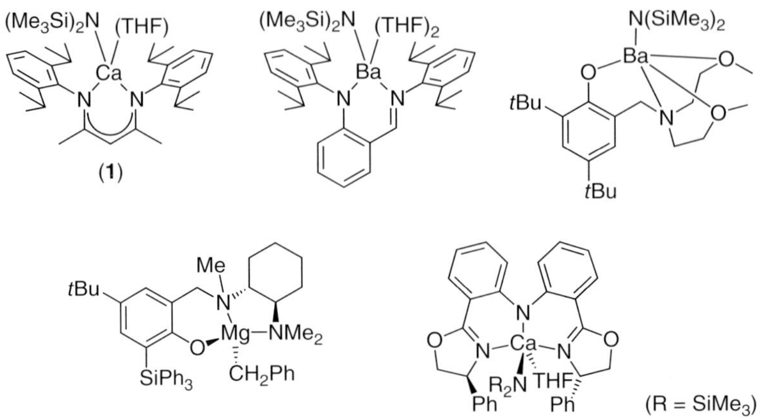

Figure 4.2 Examples of heteroleptic alkaline earth complexes for hydroelementation catalysis.

The presence of a bulky, stabilizing ligand $\{L_n\}^-$ is not mandatory to observe catalytic turnovers, e.g. in alkene hydroamination [30, 31] or the hydrophosphonylation of aldehydes and ketones [32] catalyzed by $[Ae\{N(SiMe_3)_2\}_2 \cdot (THF)_p]$ ($p = 0$ or 2), or in the dehydrocoupling of amines and hydrosilanes mediated by $[Ae\{CH(SiMe_3)_2\}_2 \cdot (THF)_2]$ [33, 34]. Yet, by far and large, the majority of homogeneous Ae catalysts obey the composition $[\{L_n\}AeX \cdot (D)_p]$. One of the main roles of the ancillary ligand $\{L_n\}^-$ consists in conferring sufficient kinetic stability to the resulting heteroleptic complex and in preventing its decomposition caused by ligand redistribution that occurs during Schlenk equilibria, that is, ligand scrambling, in solution. It also provides steric shielding against substrates that usually contain N, O, or P heteroatoms, which could lead to catalyst decomposition under the conditions required for catalysis. Ultimately, it may also enable enantioselective catalysis, but such examples remain scarce to this day and have met limited success [35, 36] except for one remarkable case of a magnesium hydroamination catalyst [37]. Examples of heteroleptic complexes that have proved efficient in group 2 metal catalysis are displayed in Figure 4.2, the β-diketiminato calcium complex **1** first synthesized by Chisholm [38] and used to great effect by Hill and others [39–43], certainly proving the most versatile among these.

4.3 Hydrophosphination of Alkenes

Alkene hydrophosphination catalyzed by alkaline earths has solely been concerned by *intermolecular* reactions, typically between an activated alkene, such as vinylarenes and conjugated dienes, and phenylphosphine ($PhPH_2$) and diphenylphosphine ($HPPh_2$); the range of substrates has been otherwise little explored, and to date, the assembly of phosphines with a regular, nonactivated alkenes (e.g. 1-hexene, cyclopentene) has seldom been successful. The focus on intermolecular Ae-mediated alkene hydrophosphination catalysis contrasts with the situation for alkaline earth hydroamination catalysis, which, in the first years, was mostly dealing with the *intramolecular* cyclohydroamination of aminoalkenes [40–42]. It also diverges from lanthanide(III)-promoted hydrophosphination reactions, which for the most involved the cyclohydrophosphination of phosphinoalkenes [44].

Alkaline earth precatalysts for alkene hydrophosphination are for the largest part heteroleptic complexes bearing nitrogen-based ancillary ligands. Only few precatalysts bear other types of ligands, such as phenolates and alkoxides.

4.3.1 Precatalysts with Nitrogen-Based Ligands

The seminal report by the Hill group of the calcium-catalyzed intermolecular hydrophosphination of activated alkenes – styrene derivatives and conjugated dienes – [25] triggered the interest in the utilization of alkaline earth catalysts for the creation of C—P bonds, which had until then mostly been confined to the hydroamination (both intra- and intermolecular) of alkenes.

The heteroleptic, four-coordinate β-diketiminato complex [{BDI^{Dipp}}CaN(SiMe$_3$)$_2$·(THF)] (**1**, Scheme 4.2; {BDI^{Dipp}}$^-$ = {HC(C(Me)N-2,6-*i*Pr$_2$-C$_6$H$_3$)$_2$}) was shown to catalyze the addition of HPPh$_2$ across the vinyl bond in styrene under relatively mild conditions (temperature range 25–75 °C, reaction time 13–24 hours, solvent benzene-d$_6$), affording neatly the anti-Markovnikov

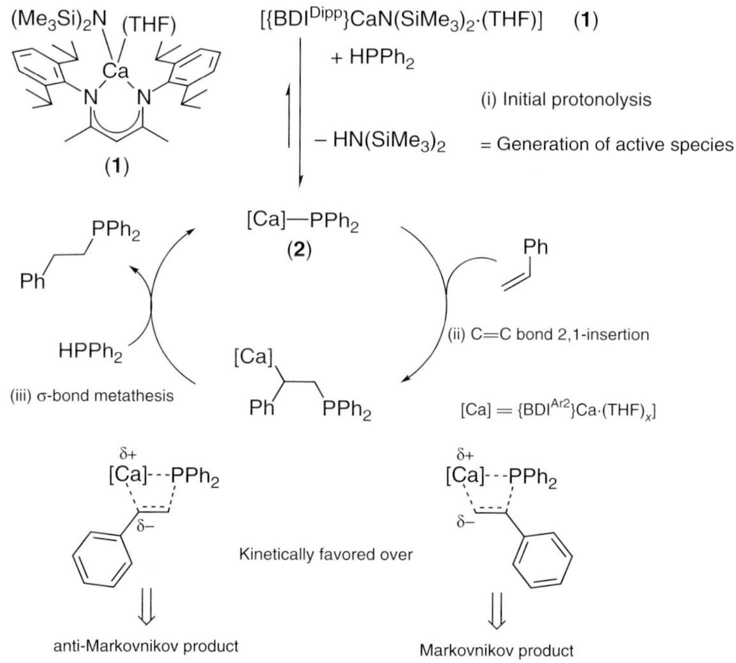

Scheme 4.2 Proposed mechanism for the hydrophosphination of styrene with HPPh$_2$ catalyzed by [{BDI^{Dipp}}CaN(SiMe$_3$)$_2$·(THF)] (**1**) and [{BDI^{Dipp}}CaPPh$_2$·(THF)] (**2**) [25].

addition product $PhCH_2CH_2PPh_2$ in near-quantitative yields [25]. Based on the close analogy between alkaline earth and trivalent lanthanide complexes, the mechanism was speculated to occur via insertion of the C=C double bond into the Ca—P bond (step ii, Scheme 4.2), followed by the slow σ-bond metathesis (protonolysis) of the newly created Ca—C bond upon action of an incoming $HPPh_2$ molecule (step iii). The 2,1-regioselective insertion of the C=C unsaturation into the Ca—P bond resulting in the exclusive formation of the anti-Markovnikov product is consistent with the proposed mechanism. The presence of the electron-withdrawing phenyl group on the carbon atom directly linked to the calcium ion and that will undergo the accumulation of a partial negative charge is indeed likely to stabilize the transition state leading to C—P bond formation, hence resulting in selective formation of $PhCH_2CH_2PPh_2$. Such stabilization cannot occur in the case of the 1,2-insertion event, which could therefore be ruled out on the basis of kinetic considerations.

The slow rate of σ-bond metathesis in step (iii) was attributed to a coordination effect. Coordination of the soft phosphine $HPPh_2$ onto the hard calcium center is likely to be hampered by the presence of the harder $HN(SiMe_3)_2$ in the reaction medium. Consistent with this hypothesis, the reaction of **1** with $HPPh_2$, resulting in the formation of $[\{BDI^{Dipp}\}CaPPh_2{\cdot}(THF)]$ (**2**), was shown to be extremely slow and highly dependent on the concentration in $HPPh_2$, while it could be inhibited upon addition of excess $HN(SiMe_3)_2$. The relative basicity and steric features of the reagents were of little importance in this case. In contrast with the more acidic and less hindered $HPPh_2$, $HNPh_2$ demonstrated high reactivity toward **1** and returned $[\{BDI^{Dipp}\}CaNPh_2{\cdot}(THF)]$ in quantitative yield. Besides, a competitive experiment indicated that in the presence of a 1 : 1 mixture of $HNPh_2$ and $HPPh_2$, **1** reacted rapidly and selectively with the amine to generate $[\{BDI^{Dipp}\}CaNPh_2{\cdot}(THF)]$, hence clearly demonstrating the importance of precoordination.

The bis-amido precatalyst $[Ca\{N(SiMe_3)_2\}_2{\cdot}(THF)_2]$ proved less effective than **1**, requiring 36 hours at 75 °C with a 10 mol% catalyst loading to convert the substrates. The reaction in this case was plagued by the formation/precipitation of $[Ca\{PPh_2\}_2{\cdot}(THF)_2]$, a yellow compound with very limited solubility in hydrocarbons, including benzene-d_6, or of its higher aggregates $[Ca\{PPh_2\}_2{\cdot}(THF)_x]_y$. The role of the ancillary ligand $\{BDI^{Dipp}\}^-$ thus proved crucial in order to generate a stable, soluble catalytically active species thought to be **2** or a congeneric $[\{BDI^{Dipp}\}CaPPh_2{\cdot}(THF)_x]$. Complex **2**, synthesized independently by the authors on a preparative scale, efficiently generated $PhCH_2CH_2PPh_2$ under experimental conditions identical to those used for **1**. Yet, it proved prone to ligand redistribution upon prolonged exposure to elevated temperatures, eventually yielding mixtures of $[\{BDI^{Dipp}\}_2Ca]$ and species postulated to be $[Ca\{PPh_2\}_2{\cdot}(THF)_2]$ or $[Ca\{PPh_2\}_2{\cdot}(HPPh_2)_x]$.

Precatalyst **1** also competently promoted the addition of $HPPh_2$ to isoprene, 1,3-cyclohexadiene, and 1,1′-diphenylacetylene (Table 4.1). On the other hand, catalytic turnover did not occur for more hindered substrates, e.g. 1,1-diphenylethylene, α-methylstyrene, or *trans*-stilbene. Also, attempts to utilize **1** or **2** to catalyze the addition of $HPPh_2$ onto the more electrophilic 2-vinylpyridine did not result in C—P bond formation, but instead produced

Table 4.1 Hydrophosphination of activated alkenes with $HPPh_2$ catalyzed by **1** and **2** [25].

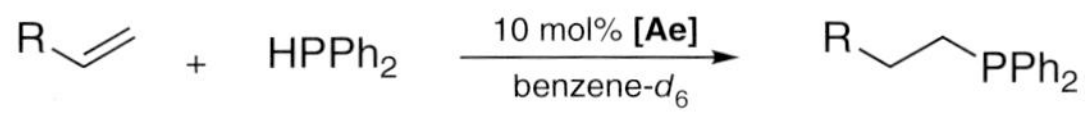

Alkene	Product	(Pre)Catalyst	T^{re} (°C)	Time (h)	Yield (%)
Styrene		**1**	75	20	95
Styrene	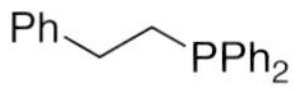	$[Ca\{N(SiMe_3)_2\}_2\cdot(THF)_2]$	75	36	92
Styrene	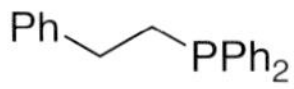	**2**	60	24	89
Isoprene	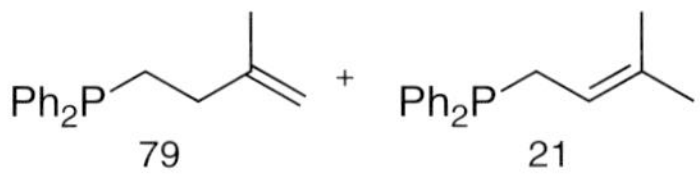	**1**	25	24	95
1,3-Cyclohexadiene	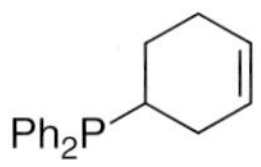	**1**	75	24	78

phosphine-capped poly(2-vinylpyridine) (Scheme 4.3). This was assumed to result from a more favorable rate of insertion of C=C bond into the Ca—C bond, in comparison with the slow rate of the σ-bond metathesis in the presence of $HPPh_2$.

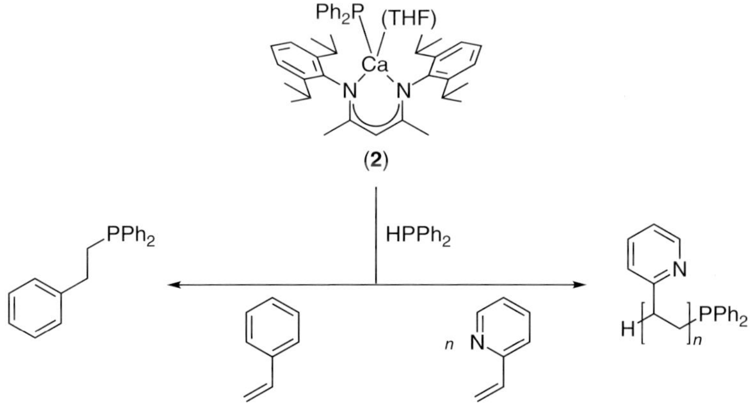

Scheme 4.3 Styrene hydrophosphination vs. 2-vinylpyridine polymerization catalyzed by **2** [25].

Following the initial report by Hill, a thorough computational study by Ward and Hunt examined the intimate mechanisms leading to these divergences observed between styrene and 2-vinylpyridine toward complex **2** [45]. Their detailed density functional theory (DFT) calculations showed that insertion of the C=C bond did not follow the traditional route. Their findings hence contradicted the proposition of a classical insertive mechanistic pathway initially suggested by Hill [25] and later by Sarazin and Carpentier [46] on the basis of experimental data and seemingly supported by the analogy with the operative mechanism established by DFT computations for the intermolecular hydroamination of alkenes [47]. By contrast with the hydroamination mechanism where the alkylamine product is stabilized exclusively through a dative Ca ← N dative interaction, it was discovered that for the hydrophosphination of vinylarenes, insertion proceeds via an unusual outer sphere, conjugative addition. The computed Gibbs free energy profile for the hydrophosphination of styrene catalyzed by [{HC(C(Me)NPh)$_2$}CaPPh$_2$], i.e. a simpler version of **2**, is depicted in Figure 4.3. The rate-limiting step, consisting of the insertion of the alkene, was found to be kinetically affordable, with $\Delta G_1^{\ddagger} = 76$ kJ/mol.

In this mechanistic scenario, there is no direct interaction of the metal center with the C=C double bond. Alkene insertion constitutes the initial turnover-limiting step, following which two distinct, competitive mechanisms

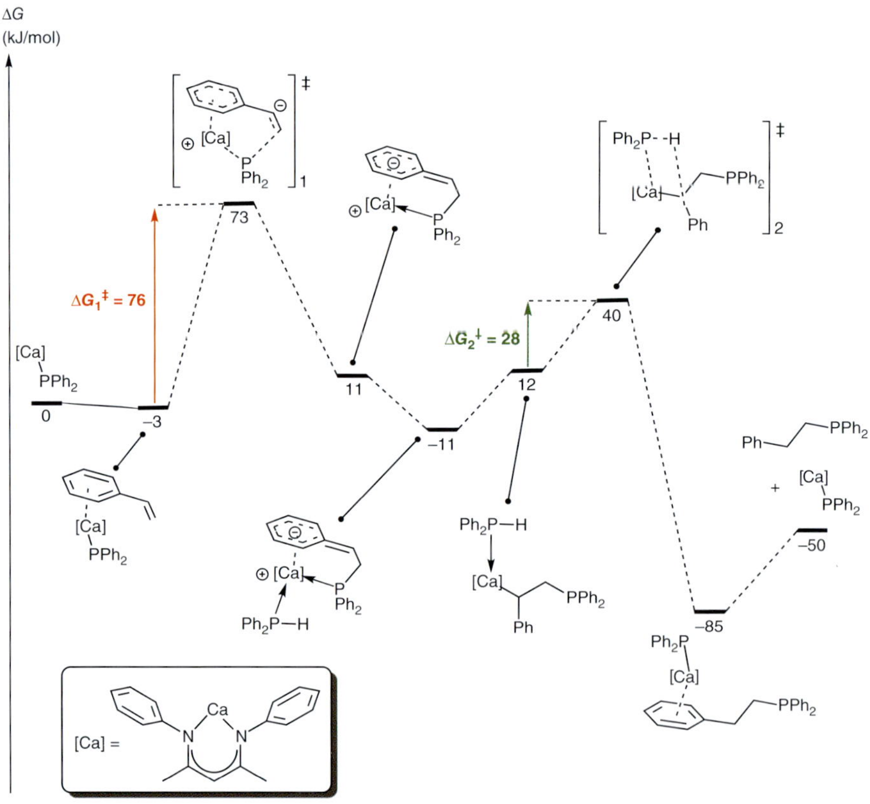

Figure 4.3 Gibbs energy profile (all energies in kJ/mol) for the hydrophosphination of styrene catalyzed by [{HC(C(Me)NPh)$_2$}CaPPh$_2$] [45].

can be envisaged: protonolysis or polymerization. In agreement with Hill's experimental data, protonolysis was shown to be energetically favored in the case of styrene (which binds to the metal mostly through Ca···C_π interactions), while for 2-vinylpyridine (coordinated to calcium through Ca ← N dative interaction), polymerization resulting from multiple C=C insertion was found to be more facile. The aromatic ring of the vinylarene moiety therefore plays a key role in both hydrophosphination and polymerization mechanisms. Not only do the multiple Ca···C_π, CH···C_π, and Ca ← E (E = P or N) noncovalent interactions dictate the modes of coordination of the reaction substrates and products onto the metal but also the vinylarene aromatic ring counterbalances unfavorable charge localization in the transition state.

Cui reported in 2012 the high yield synthesis of a five-coordinate calcium precatalyst of composition [{2-NC(Ph)-NArC_6H_4CHNAr}CaN$(SiMe_3)_2$·(THF)] (**3**) supported by an imino-amidinate [48]. Its ytterbium(II) congener [{2-NC(Ph)-NArC_6H_4CHNAr}YbN$(SiMe_3)_2$·(THF)] (**3′**) was also synthesized. Complex **3** catalyzes the fully anti-Markovnikov addition of diphenylphosphine across the vinylic bond in styrene and *p*-Me-styrene, using a 2–5 mol% catalyst loading at room temperature (Table 4.2). It is also efficient for the regioselective 1,4-addition of $HPPh_2$ to isoprene, 2,3-Me_2-1,3-butadiene, and myrcene. Notably, the 1,4-regioselectivity of this addition differs from that observed with **1**, which preferentially generated the products of 1,2-addition to conjugated dienes (see Table 4.1). Besides, unlike **1**, catalyst **3** competently converts bulky substrates such as α-methylstyrene and *cis*-stilbene; the reactions, however, require in these cases more forcing conditions than for styrene and *p*-Me-styrene. Beyond alkenes, **3** also catalyzes the hydrophosphination of alkynes. Interestingly, the Yb(II) derivative **3′** turned out to be a more efficient hydrophosphination catalyst than **3**. The structural and reactivity analogies between calcium and ytterbium(II) complexes are well documented [49–52], in particular because both dications present the same electronic configurations and ionic radii (Ca^{2+}, 1.00 Å; Yb^{2+}, 1.02 Å); complete filling of the low-lying f orbitals, with their limited radial extension, in Yb(II) complexes is not known to induce fundamentally different reactivity compared to the calcium congeners.

Sarazin and Carpentier disclosed in a communication released in 2012 that the series of β-diketiminato precatalysts [{BDI^{Dipp}}AeN$(SiMe_2H)_2$·$(THF)_p$] (Ae = Ca, p = 1, **4**; Ae = Sr, p = 2, **5**; Ae = Ba, p = 2, **6**) and their related imino-anilido derivatives [{N^N}AeN$(SiMe_3)_2$·$(THF)_p$] (Ae = Ca, p = 1, **7**; Ae = Sr, p = 2, **8**; Ae = Ba, p = 2, **9**; Figure 4.4) enabled the hydrophosphination of styrene with $HPPh_2$ or $HPCy_2$ (Cy = cyclohexyl) [53]. The Sr and Ba complexes **5** and **6** proved stable under the conditions required for catalysis owing to the presence of stabilizing intramolecular Ae···β-Si–H anagostic interactions; their parents [{BDI^{Dipp}}AeN$(SiMe_3)_2$·$(THF)_2$], which do not exhibit these profitable interactions, had been shown earlier to be unstable [54].

For both series of complexes, the catalytic activity in the hydrophosphination of styrene (catalyst loading 2 mol%, 60 °C, neat substrates) increased upon descending group 2: Ca < Sr < Ba. The imino-anilido precatalysts **7–9** demonstrated greater efficiency than the β-diketiminates **4–6**. The reactions proceeded

Table 4.2 Hydrophosphination of activated alkenes with $HPPh_2$ catalyzed by **3** and **3′** [48].

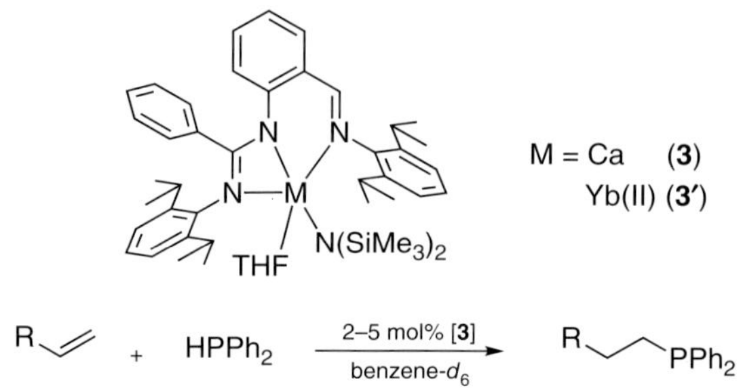

Alkene	Product	Precatalyst	T^{re} (°C)	Time (h)	Yield (%)
Styrene	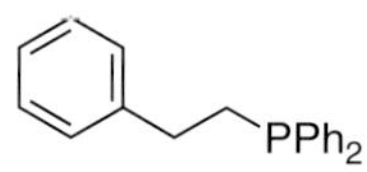	**3** **3′**	25 25	3 2	100 100
p-Me-styrene	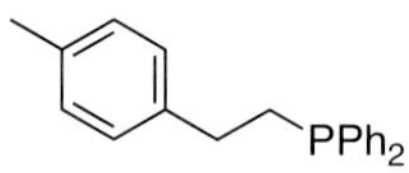	**3**	25	3	99
Isoprene	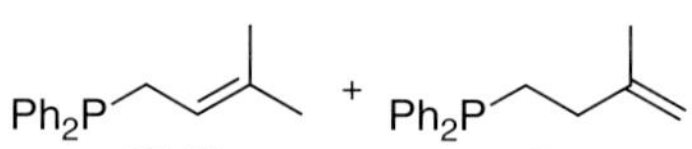 95 (**3**) / 5; 100 (**3′**) / 0	**3**	25 25	2 1.5	100 100
2,3-Me_2-1,3-butadiene	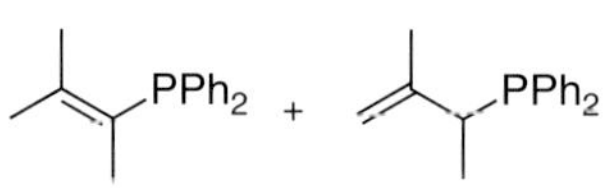 100 (**3**) / 0; 100 (**3′**) / 0	**3** **3′**	60 25	24 24	86 90
Myrcene	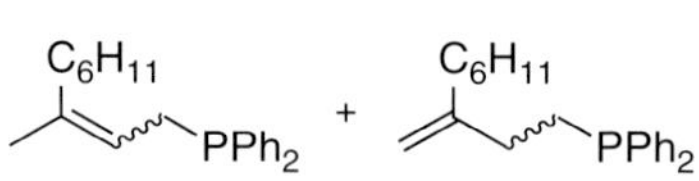 94 (**3**) / 6; 97 (**3′**) / 3	**3** **3′**	25 25	8 1.5	100 100
Cis-Stilbene	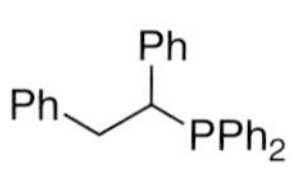	**3**	60	18	72
α-Me-styrene	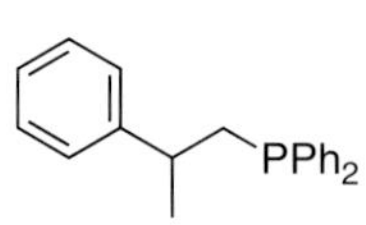	**3**	60	24	85

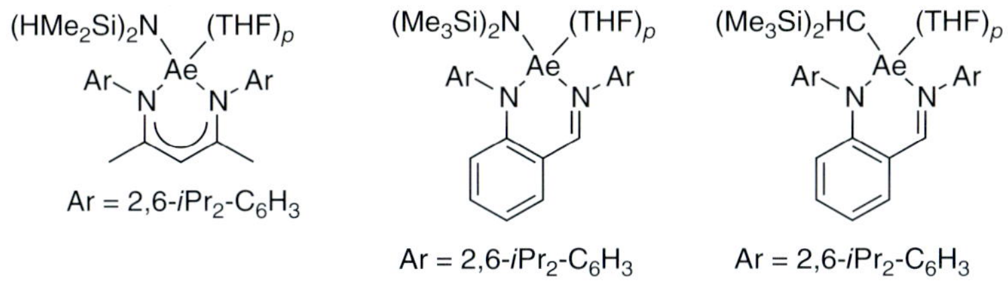

Ae = Ca	$p = 1$	**4**	Ae = Ca	$p = 1$	**7**	Ae = Ca	$p = 1$	**10**
Ae = Sr	$p = 2$	**5**	Ae = Sr	$p = 2$	**8**	Ae = Sr	$p = 2$	**11**
Ae = Ba	$p = 2$	**6**	Ae = Ba	$p = 2$	**9**	Ae = Ba	$p = 2$	**12**

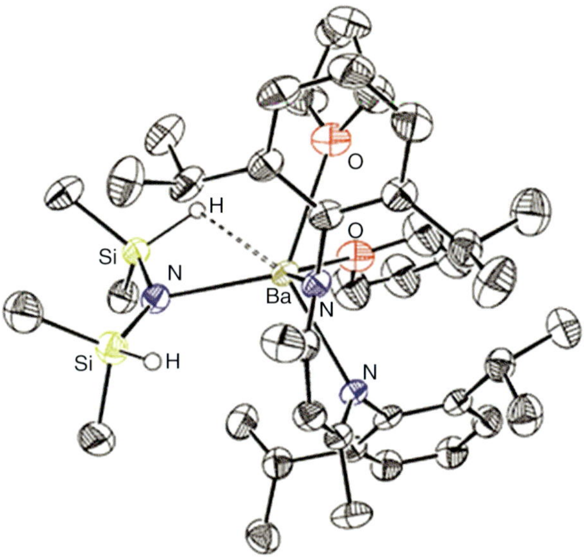

Figure 4.4 β-Diketiminato and imino-anilido Ca–Ba hydrophosphination precatalysts [46, 53], showing the ORTEP representation (ellipsoid at 50% probability level; H atoms except Si*H* omitted for clarity) of the molecular solid-state structure of [{BDI^{Dipp}}BaN(SiMe$_2$H)$_2$·(THF)$_2$] (**6**). Ba1···H1–Si1 β-anagostic stabilization depicted by a dashed bond (Ba1–H1 = 3.005(34) Å).

with perfect anti-Markovnikov regioselectivity. With $HPCy_2$, only partial conversion to $PhCH_2CH_2PCy_2$ was achieved after 18.5 hours, even with **4–6**, the most active complexes for this transformation (c. 42% with **9**). In contrast, the reactions with the more acidic $HPPh_2$ were faster, yielding $PhCH_2CH_2PPh_2$ with 96% conversion in as little as 15 minutes using precatalyst 9 (turnover frequency, TOF = 192 $mol_{subst}/(mol_{Ba}\ h)$).

In a subsequent study [46], the authors extended their investigations to the heteroleptic alkyl imino-anilido complexes [{N^N}AeCH(SiMe$_3$)$_2$·(THF)$_p$] (Ae = Ca, $p = 1$, **10**; Ae = Sr, $p = 2$, **11**; Ae = Ba, $p = 2$, **12**; Figure 4.4), and the scope was enlarged to include electronic effects in the vinylarenes. The addition of $HPCy_2$ and $HPPh_2$ across the vinylic bond in various para-substituted styrene derivatives was examined with a catalyst loading of 2.0 mol% (Table 4.3).

Table 4.3 Hydrophosphination of vinylarenes catalyzed by Ae amido (**7–9**) and alkyl (**10–12**) precatalysts [46].[a)]

Precatalyst		HPR_2	X	*t* (min)	Conversion(%)[b)]
[{N^N}CaN(SiMe$_3$)$_2$·(THF)]	7	$HPCy_2$	H	1080	31
[{N^N}SrN(SiMe$_3$)$_2$·(THF)$_2$]	8	$HPCy_2$	H	1080	41
[{N^N}BaN(SiMe$_3$)$_2$·(THF)$_2$]	9	$HPCy_2$	H	1080	50
[{N^N}CaN(SiMe$_3$)$_2$·(THF)]	7	$HPPh_2$	H	15	42
[{N^N}SrN(SiMe$_3$)$_2$·(THF)$_2$]	8	$HPPh_2$	H	15	92
[{N^N}BaN(SiMe$_3$)$_2$·(THF)$_2$]	9	$HPPh_2$	H	15	> 96
[{N^N}CaCH(SiMe$_3$)$_2$·(THF)]	10	$HPPh_2$	H	3	20
[{N^N}CaCH(SiMe$_3$)$_2$·(THF)]	10	$HPPh_2$	H	15	76
[{N^N}SrCH(SiMe$_3$)$_2$·(THF)$_2$]	11	$HPPh_2$	H	3	55
[{N^N}BaCH(SiMe$_3$)$_2$·(THF)$_2$]	12	$HPPh_2$	H	3	95
[{N^N}CaN(SiMe$_3$)$_2$·(THF)]	7	$HPPh_2$	CF_3	10	92
[{N^N}CaN(SiMe$_3$)$_2$·(THF)]	7	$HPPh_2$	Cl	10	80
[{N^N}BaN(SiMe$_3$)$_2$·(THF)$_2$]	9	$HPPh_2$	CF_3	5	80
[{N^N}BaN(SiMe$_3$)$_2$·(THF)$_2$]	9	$HPPh_2$	Cl	<10	95
[{N^N}BaN(SiMe$_3$)$_2$·(THF)$_2$]	9	$HPPh_2$	Me	15	79
[{N^N}BaN(SiMe$_3$)$_2$·(THF)$_2$]	9	$HPPh_2$	*t*Bu	30	44
[{N^N}BaN(SiMe$_3$)$_2$·(THF)$_2$]	9	$HPPh_2$	OMe	30	27

a) $[\text{precatalyst}]_0/[\text{styrene}]_0/[\text{phosphine}]_0 = 1:50:50$, no solvent, precatalyst: 10.0 μmol, 60 °C.
b) Conversion determined by ^{1}H NMR spectroscopy.

The reactions afforded exclusively the anti-Markovnikov product. As seen previously, reactions involving styrene and the basic $HPCy_2$ were slow, with substrate conversion in the range 31–50% after 18 hours. Reaction rates were greater with $HPPh_2$, with full conversion of the substrate within 3–15 minutes with the Ba derivatives **9** and **12**. The increasing efficiency Ca < Sr < Ba was confirmed during this study. The Ae-alkyl precatalysts **10–12** were more productive than their Ae-amide congeners 7–**9**, most likely because they generate a larger amount of the Ae-phosphide catalytically active species through irreversible release of $CH_2(SiMe_3)_2$. Overall, these intermolecular hydrophosphination precatalysts, and in particular the barium ones, range among the best ones known to date. Ba precatalysts afforded the highest activity by some margin. The Ba-amido precatalyst **9** was used to qualitatively assess the influence of substitution at the para position of the aromatic ring in the styrenic substrates.

The electron-withdrawing inductive effect of chlorine led to increased activity, as full conversion of *p*-Cl-styrene was achieved within less than 10 minutes. With the even more electron-withdrawing CF_3, high to complete conversions were achieved with **10** or **12** within minutes. By contrast, with electron-donating para substituents (Me, *t*Bu, or OMe) reaction rates decreased with increasing electron-donating strength according to (CF_3 > Cl >) H > Me > *t*Bu > OMe. This recalled observations made for the intermolecular hydroamination of vinylarenes [53, 55]. Kinetic studies performed with the calcium precatalyst **7** demonstrated the empirical rate law for the hydrophosphination of styrene with $HPPh_2$ expressed by Eq. (4.1):

$$R_{HP} = k \cdot [HPPh_2]^0 \cdot [\text{styrene}]^1 \cdot [\mathbf{7}]^1 \quad (4.1)$$

There was therefore no dependence on phosphine concentration. Investigations performed for the Sr and Ba congeners **8** and **9** provided the same outcome. The activation energy values, E_a = 70.5(5.4) (**7**), 43.3(6.0) (**8**), and 34.2(2.4) (**9**) kJ/mol, were determined by Arrhenius analysis. They were in good agreement with the qualitative trend disclosed in Table 4.3, corroborating the activity trend Ba > Sr > Ca.

Based on the experimental data, the mechanism represented in Scheme 4.4 was postulated, where two parameters affect the overall kinetics: the amount of catalytically active species existing in the medium, and the nature of the rate-determining step. On account of the zero dependence in [$HPPh_2$], the rate-limiting transition state was suggested to consist of insertion of the polarized alkene into the metal–phosphide bond. Yet, the computational investigation reported by Ward and Hunt ([45], see discussion above) were not

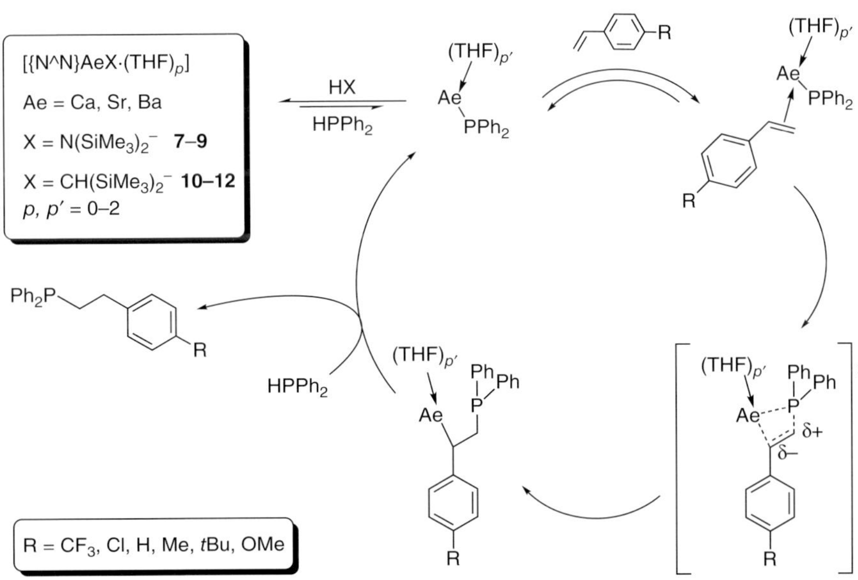

Scheme 4.4 Catalytic manifold initially proposed for the hydrophosphination of vinylarenes with $HPPh_2$ catalyzed by **7–12** [46]. Later studies by Ward and Hunt indicated that a different mechanism was likely operating [45], see Figure 4.3 and pertaining discussion.

in favor of this hypothesis. It is indeed likely that the mechanisms at work for the β-diketiminato (**4–6**) and imino-anilido (**7–12**) precatalysts are identical, and a mechanism involving outer sphere, conjugative addition highlighting the important role of Ae···C_π interactions, is thus probably favored. The higher acidity of $HPPh_2$ with respect to $HPCy_2$ agrees with the greater reaction rates observed with the former. The extent of deprotonation/reprotonation equilibria also explains the somewhat better efficiency seen for Ae-alkyl precatalysts **10–12** in comparison with the Ae-amides **7–9**.

At 60 °C, precatalysts **7–9** and **10–12** also proceeded to the rapid hydrophosphination of conjugated dienes (isoprene and myrcene) using 2.0 mol% in neat substrates, but with limited regioselectivity. For instance, with isoprene, the calcium and strontium precatalysts slightly favored the product of 3,4-addition, whereas barium systems preferably returned the products of 1,4-addition with a 3 : 1 preference. Interestingly, the barium precatalyst **9** mediated very efficiently (90% conversion of 50 equiv of each substrates within four hours at 60 °C) the anti-Markovnikov addition of $HPPh_2$ to the more reluctant α-methylstyrene, whereas under otherwise identical experimental conditions, the calcium analog proved essentially ineffective.

The attempt to improve on imino-anilido (**4–12**) and imino-amidinato (**3**) heteroleptic precatalysts by introducing chalcogen-containing ligands proved an interesting synthetic challenge, albeit not rewarding in terms of catalytic efficiency [56]. The calcium complexes [{$Ph_2P(E)$-N-C_6H_4-CH=N(2,6-$i$$Pr_2$-$C_6H_3$)}$CaN(SiMe_3)_2$·(THF)] (E = S, **13**; Se, **14**) were synthesized in high yields. The addition of Ph_2PH to styrene was examined with a precatalyst loading of 2.0 mol%, with $[\text{styrene}]_0/[HPPh_2]_0/[Ca]_0$ = 50 : 50 : 1. It was found that the reaction was fully regioselective and afforded exclusively the anti-Markovnikov product (Table 4.4). The reactions were reasonably fast in neat substrates, achieving near-complete conversion within 12 hours. Activity (TOF) values in the region of 18 $mol_{subst}/(mol_{Ca}\ h)$ were recorded for both **13** and **14**, which showed generally comparable efficiency regardless of the chalcogen bound to calcium. These values were at the time only outclassed by those attributed to **1** and **7–12** [25, 46]. Precatalysts **13–14** displayed low reaction rates in benzene-d_6; poor conversion was observed after 24 hours, which was thought to reflect catalyst decomposition upon prolonged exposure to elevated temperature.

The group of P. Roesky devised homoleptic six-coordinate chiral benzamidinato complexes [{(S)-PEBA}$_2$Ae·(THF)$_2$] (Ae = Ca, **15**; Sr, **16**) and [{(S)-PEBA}$_2$Ba]$_2$ (**17**), where (S)-{PEBA}$^-$ is the enantiomerically pure N,N'-bis(1-phenylethyl)-benzamidinato ligand (Figure 4.5) and used them as Lewis acid catalysts to promote the addition of $HPPh_2$ onto vinylarenes [57]. Low catalytic activity was recorded. The products of the catalytic hydrophosphination, performed with a precatalyst loading of 5 mol% at 60 °C in benzene-d_6, were obtained in moderate to high yields (conversion ranged from 40% to 97%), but over extended periods of time, spreading from 48 to 120 hours. The anti-Markovnikov products were always obtained; hence, potential enantioselective control was irrelevant. The catalytic activity increased very substantially, upon going down in group 2: Ca < Sr < Ba. On the other hand, there was no obvious trend regarding the presence of groups in para position of the aromatic

Table 4.4 Hydrophosphination of styrene with $HPPh_2$ catalyzed by **13** and **14** [56].[a)]

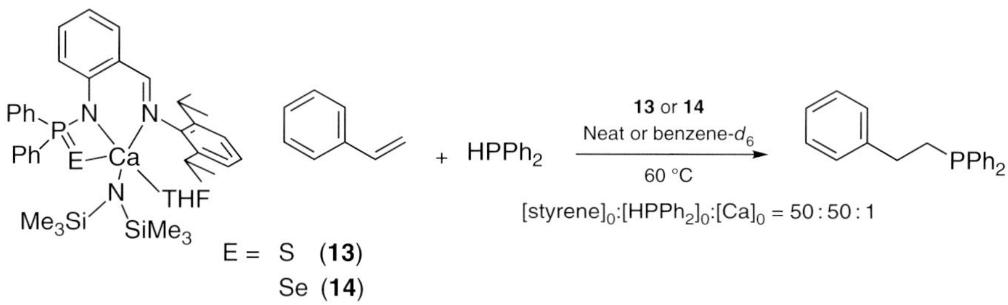

Precatalyst	Time (h)	Conversion (%)	Solvent
14	1	16	Neat
14	2	23	Neat
14	12	93	Neat
13	12	94	Neat
13[b)]	2	8	Benzene-d_6
13[b)]	24	22	Benzene-d_6

a) Reactions performed at 60 °C, using 2.0 mol% of precatalyst.
b) Reactions in 0.3 ml of benzene-d_6 at 0.51 mM in substrates.

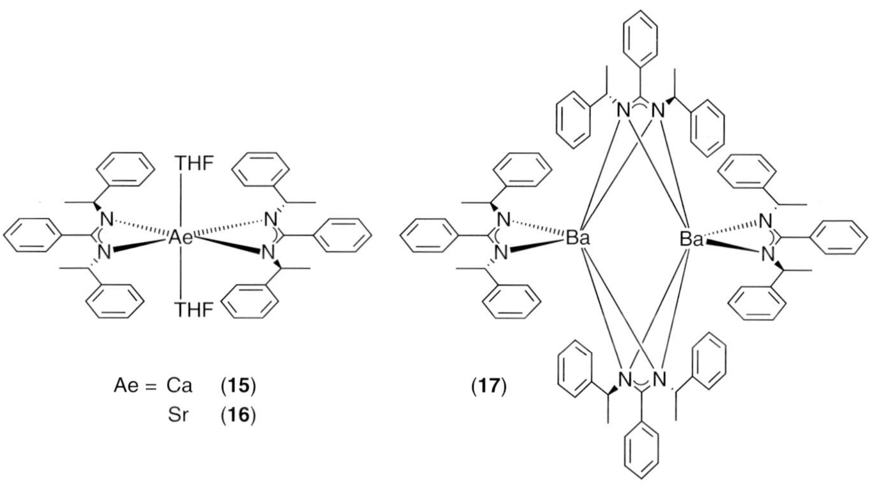

Figure 4.5 [{(*S*)-PEBA}$_2$Ae·(THF)$_2$] alkaline earth complexes **15–17** bearing the enantiomerically pure *N*,*N*′-bis(1-phenylethyl)-benzamidinato ligand [57].

ring in *p*-X-C_6H_4-CH=CH_2, for X = H, Me, OMe, and Cl. The mechanisms involved in this catalyzed reaction were not discussed.

As part of a study mostly focused on the divalent lanthanides Yb(II) and Sm(II) [58], Trifonov and coworkers synthesized the four-coordinate amidinato calcium complex [{*t*BuC(NC_6H_3-*i*Pr_2-2,6)$_2$}CaN(SiMe_3)$_2$·(THF)] (**18**, Scheme 4.5). This complex catalyzed the regiospecific anti-Markovnikov addition of $HPPh_2$ onto styrene carried out in benzene-d_6. Full conversion was measured in two hours at 60 °C for a precatalyst loading of 2 mol%; the calcium compound in particular proved more effective than its direct Yb(II) and Sm(II) derivatives (Ca > Sm > Yb). More unusually, it also allowed for the addition of $PhPH_2$ with excellent selectivity, affording the secondary phosphine $PhCH_2CH_2PHPh$ with a 96 : 4 preference over the formation of the tertiary phosphine $(PhCH_2CH_2)_2PPh$. At 60 °C, the reaction required a long time (70 hours, TOF < $\mathbf{1\ mol_{subst}\ mol_{Ca}/h}$) but afforded a complete conversion of the neat substrates ($[PhPH_2]_0/[styrene]_0/[\mathbf{18}]_0 = 50:50:1$).

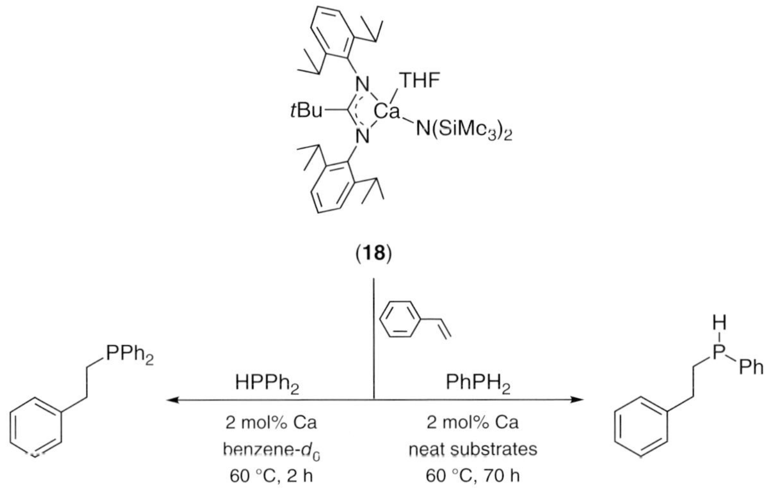

Scheme 4.5 Hydrophosphination of styrene with $HPPh_2$ and $PhPH_2$ catalyzed by **18** [58].

The Trifonov group also reported on the hydrophosphination of vinylarenes with a range of primary and secondary phosphines catalyzed by five-coordinate amido Ca and Yb(II) complexes [59] incorporating bulky, tridentate amidine-amidopyridinato ligands initially designed and developed by Kirillov and coworkers [60, 61]. The authors probed the ability of the calcium compounds **19–22** of general formulae $[\{N^\wedge N^\wedge N^\wedge N\}CaN(SiMe_3)_2\cdot(THF)]$ (Figure 4.6) to catalyze the hydrophosphination of vinylarenes variously substituted in **para** position (H, Me, OMe, Cl, and F), α-Me-styrene, and 2,3-dimethyl-1,3-butadiene with a variety of phosphines ($PhPH_2$, 2,4,6-Me_3-$C_6H_2PH_2$, 2-pyridyl-PH_2, $HPPh_2$ and $HPCy_2$).

The catalytic activities of compounds **19–22** in styrene hydrophosphination with $PhPH_2$ and Ph_2PH were globally high. The reactions were fully regiospecific, only affording the anti-Markovnikov addition products. For the reactions with the primary phosphine $PhPH_2$ and 1 equiv of styrene, **19** and **20** showed high chemoselectivity, as they afforded solely the product of mono-addition, i.e. the secondary phosphine $(PhCH_2CH_2)PHPh$; no trace of the tertiary phosphine $(PhCH_2CH_2)_2PPh$ could be detected. On the other

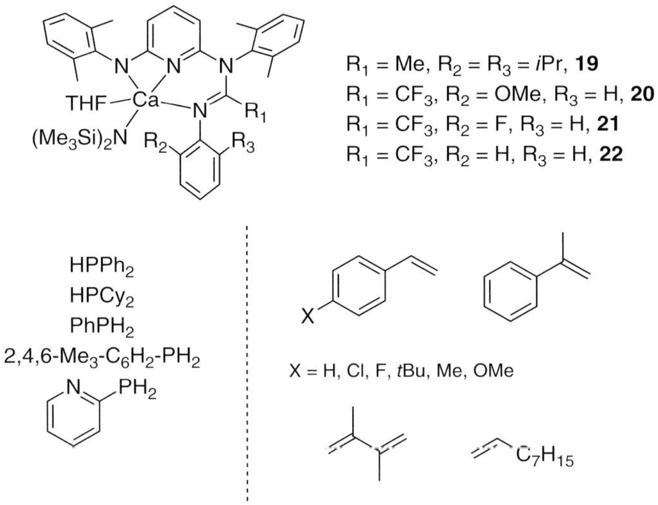

Figure 4.6 Complexes **19–22** used for alkene hydrophosphination reactions [59].

hand, **21** and **22** exhibited lower chemoselectivity, returning mixtures of secondary and tertiary phosphines in 80 : 20 and 86 : 14 ratios, respectively. When performed with 2 equiv of styrene vs. $PhPH_2$, compound **20** selectively produced the tertiary phosphine as a single product, while the other precatalysts generated a mixture of secondary and tertiary phosphines. In comparison with $PhPH_2$, the hydrophosphinations of styrene with the bulky 2,4,6-Me_3-$C_6H_2PH_2$ and the less nucleophilic 2-pyridylphosphine catalyzed by **19–22** proceeded with considerably lower reaction rates, albeit the reactions maintained their anti-Markovnikov regioselectivity. Compared to styrene, the kinetics slowed down upon introduction of electron-donating groups (Me, *t*Bu, and OMe) in para position to the aromatic ring of the vinylarenes; there was no obvious influence with electron-withdrawing groups. Also, complexes **19–22** promoted the regiospecific and quantitative addition of $PhPH_2$ onto α-Me-styrene, albeit at a rate significantly lower than for styrene (full conversion of 50 equiv. of each substrate required 40 hours). Compounds **19** and **20** also enabled the 1,2-addition of $PhPH_2$ onto 2,3-Me_2-1,3-butadiene with excellent regio- and chemoselectivity, yielding exclusively the linear secondary phosphines; by contrast, **21** and **22** returned mixtures of mono- and double-addition products.

Remarkably, the hydrophosphination of the nonactivated 1-nonene with Ph_2PH was achieved with precatalyst **20**; 40% conversion of the substrates ($[\text{1-nonene}]_0/[HPPh_2]_2/[\mathbf{20}]_0 = 50 : 50 : 1$) was recorded after 40 hours at 70 °C. This complex is hence part of the select group of systems that enable the hydrophosphination of strictly nonactivated alkenes [62–65], which are notoriously difficult substrates for hydrophosphination reactions; on the other hand, the transformation of cyclohexene and *trans*-stilbene could not be accomplished. Kinetic investigations conducted using **19** as the precatalyst and styrene and

$HPPh_2$ as the substrates showed that the experimental rate law obeyed Eq. (2):

$$R_{HP} = k \cdot [HPPh_2]^0 \cdot [styrene]^1 \cdot [\mathbf{19}]^1 \tag{4.2}$$

It therefore matched that described for the imino-anilido calcium precatalyst [{N^N}CaN$(SiMe_3)_2$·(THF)] (**7**) [46]. On the whole, the series of complexes **19–22** represent to date the most versatile and efficient calcium hydrophosphination precatalysts. The combination of these unusual amidine-amidopyridinato ligands with barium could potentially return hydrophosphination precatalysts of unmatched efficacy. Before this, rationalization of the ligand structure–reactivity relationship would be necessary; indeed, although the different ligands in **19–22** undeniably played a key role in the study, no discussion on the exact role of the ligand substituents was provided in this otherwise thorough investigation.

4.3.2 Precatalysts with Oxygen-Based Ligands

Compared to nitrogen-based ligands, the number of hydrophosphination precatalysts incorporating alkoxides or phenolates is rather limited. Besides, these complexes have on the whole displayed lower efficiency in catalysis. They all afford selectively anti-Markovnikov addition products.

The Ca, Sr, and Ba complexes [{LO^N^O_2}AeN$(SiMe_3)_2$·$(THF)_p$] (Ae = Ca, p = 0, **23**; Ae = Sr, p = 1, **24**; Ae = Ba, p = 0, **25**; Figure 4.7) were shown to catalyze the addition of $HPPh_2$ onto styrene, affording partial conversion of the 1 : 1 mixture of substrates (50 equiv. vs. the precatalyst; 60 °C, 18.5 hours) to the linear

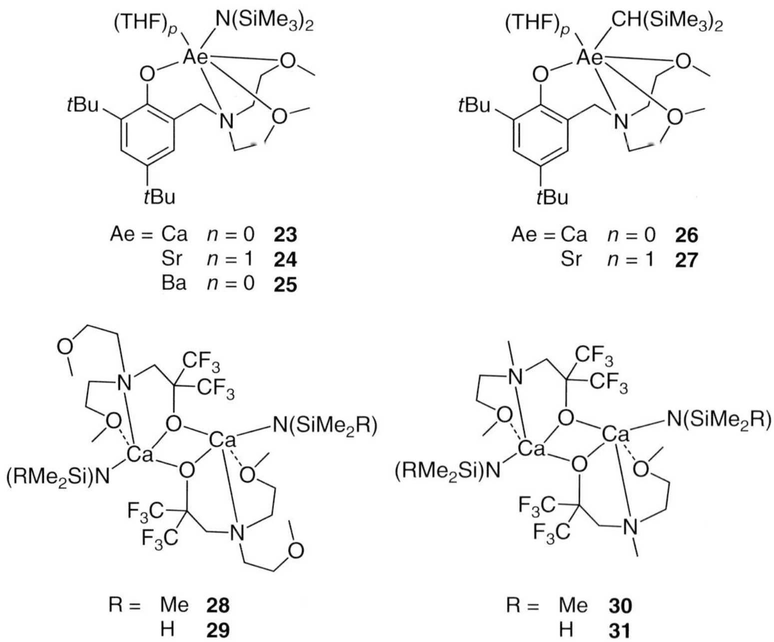

Figure 4.7 Alkaline earth complexes stabilized by phenolato and alkoxo ancillary ligands for the addition of $HPPh_2$ onto styrene [53, 66, 67].

product $PhCH_2CH_2PPh_2$ with perfect selectivity [53]. As observed for other precatalysts, the catalytic activity increased in the order Ca(12%) < Sr (26%) < Ba (46%). The behavior of **23–25** essentially matched that of the β-diketiminato complexes **4–6** under identical experimental conditions and proved inferior to **7–12**. The alkyl complexes [{LO^N^O$_2$}AeCH$(SiMe_3)_2$·(THF)$_p$] (Ae = Ca, p = 1, **26**; Ae = Sr, p = 0, **27**; Figure 4.7) demonstrated comparable activity and full anti-Markovnikov selectivity to those of the amido derivatives **23** and **24** and matched also that of the Yb(II) congener [66].

The dinuclear calcium amido-fluoroalkoxides [{RO^N^O$_2$}CaN$(SiMe_2R')]_2$ (R′ = Me, **28**; H **29**; Figure 4.7) possessing a noncoordinated methoxy dangling side arm and the related complexes [{RO^N^O}CaN$(SiMe_2R')]_2$ (R′ = Me, **30**; H **31**) also catalyzed the fully regioselective formation of $PhCH_2CH_2PPh_2$ from styrene and $HPPh_2$ [67]. Structural investigations indicated that precatalysts **29** and **31** exhibited intramolecular Ca· · ·H–Si stabilizing anagostic interactions. Compound **28** turned out to be insufficiently stable to be included in the reactivity studies. Complexes **29–31** on the contrary featured high activity during reactions carried out at 60 °C with a precatalyst loading of 1 mol%. The turnover numbers and frequencies measured for reactions performed in benzene-d_6 or with neat substrates (TON up to 400, TOF up to 50 $mol_{subst}/(mol_{Ca}$ h)) matched those reported with other calcium-based precatalysts. Besides, these precatalysts exhibited the same catalytic performances when higher substrate loadings were used, i.e. 400 equiv of each substrate vs. Ca. The two precatalysts **29** and **31** possessing the stabilizing $N(SiMe_2H)_2$ moiety were somewhat more efficacious than **30**. They respectively demonstrated TOF values of 1.8, 1.4, and 1.0 $mol_{subst}/(mol_{Ca}$ h) for nonoptimized, discriminating reactions performed over 24 hours with $[styrene]_0/[HPPh_2]_0/[Ca]_0 = 50:50:1$.

Harder and coworkers prepared tetranuclear strontium and barium siloxide/amide clusters (Figure 4.8), obtained by one-pot reaction of substituted anilines with $(Me_2SiO)_3$ and the alkaline earth amides [Sr{N$(SiMe_3)_2\}_2$·$(THF)_2$] and [Ba{N$(SiMe_3)_2\}_2$·$(THF)_2$] that displayed some interesting reactivity in hydrophosphination, yielding anti-Markovnikov addition products [68].

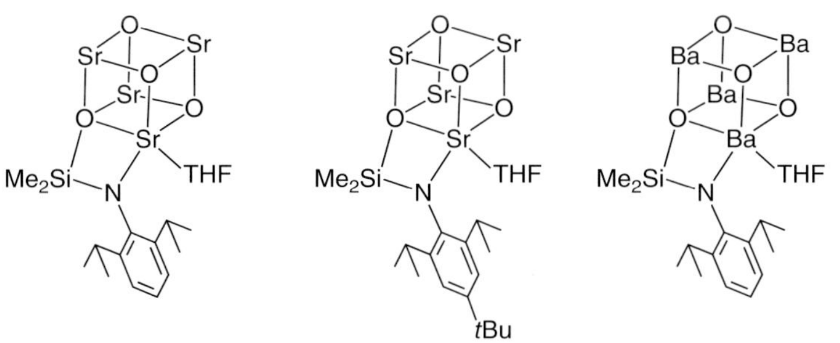

(32) **(33)** **(34)**

Figure 4.8 Strontium and barium siloxide/amide clusters used for intermolecular hydrophosphination catalysis. Only one –*N*(aryl)–$SiMe_2$– and one THF molecule depicted per cluster for clarity [68].

Catalysis took place in benzene-d_6 at 70 °C. Catalyst **34** converted α-Me-styrene, albeit the reaction required nine days to reach completion. 1,1-Diphenylethylene, a substrate neglected until then in other studies with group 2 metal catalysts, was quantitatively transformed by **32** and **34** in, respectively, 60 and 36 hours. Overall, the reaction rates were found to increase in the order α-Me–styrene <1,1-diphenylethylene < styrene. The Sr complexes proved to be substantially less active than other heteroleptic strontium catalysts $[\{L_n\}SrN(SiMe_3)_2{\cdot}(D)_p]$ and $[\{L_n\}SrCH(SiMe_3)_2{\cdot}(D)_p]$ bearing a bulky ancillary ligand disclosed in earlier studies. Besides, the strontium complex **32** was also less active than its barium analog 34. The spectator ligand in **32–34** assumed an important solubilizing function. The authors legitimately postulated a mechanism in which the siloxide/amide dianionic moiety behaved as a cooperative ligand. In such a scenario, deprotonation of the relatively acidic $HPPh_2$ (pK_a = 21.7) by the *N*-aryl side-arm (pK_a aniline = 30.7) is followed by phosphide/alkene addition, after which the intermediate is protonated again to release the final addition product and regenerate the initial complex.

4.4 Hydrophosphination of Carbodiimides

Beyond alkene hydrophosphination, Hill and coworkers also reported that $[\{BDI^{Dipp}\}CaN(SiMe_3)_2{\cdot}(THF)]$ (**1**) and, by extension, $[\{BDI^{Dipp}\}CaPPh_2{\cdot}(THF)]$ (**2**), enabled the hydrophosphination of carbodiimides to generate phosphaguanidines [69]. Moreover, they found that $[Ca\{N(SiMe_3)_2\}_2{\cdot}(THF)_2]$,

Table 4.5 Intermolecular hydrophosphination of carbodiimides catalyzed by Ae precatalysts [69].

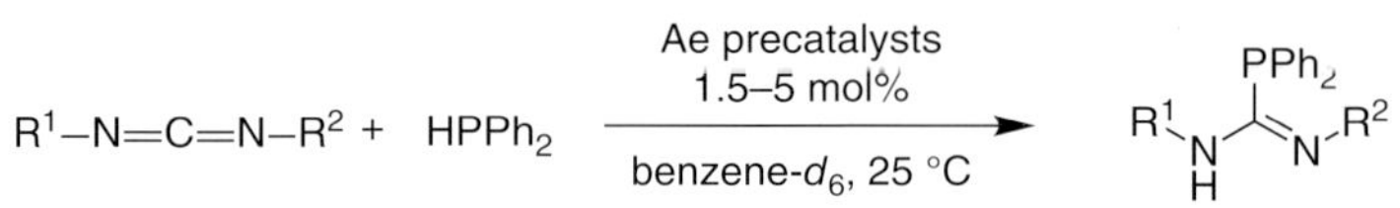

*R*¹	*R*²	Precatalyst	Loading	*t* (h)	Yield (%)
Cy	Cy	$[\{BDI^{Dipp}\}CaN(SiMe_3)_2{\cdot}(THF)]$ (**1**)	1.5	28	85
		$[Ca\{N(SiMe_3)_2\}_2{\cdot}(THF)_2]$	2	2	>95
		$[Ca\{N(SiMe_3)_2\}_2]_2$	3	4	>95
		$[Sr\{N(SiMe_3)_2\}_2{\cdot}(THF)_2]$	2	0.5	>99
		$[Ba\{N(SiMe_3)_2\}_2{\cdot}(THF)_2]$	2	0.5	>95
*i*Pr	*i*Pr	$[\{BDI^{Dipp}\}CaN(SiMe_3)_2{\cdot}(THF)]$ (**1**)	1.5	6	>99
		$[Ca\{N(SiMe_3)_2\}_2]_2$	2	1	>95
		$[Sr\{N(SiMe_3)_2\}_2{\cdot}(THF)_2]$	2	0.5	>99
*t*Bu	*t*Bu	$[Sr\{N(SiMe_3)_2\}_2{\cdot}(THF)_2]$ (60 °C)	5	16	0
*t*Bu	Et	$[Sr\{N(SiMe_3)_2\}_2{\cdot}(THF)_2]$	2	0.25	>95
p-tol	*p*-tol	$[Sr\{N(SiMe_3)_2\}_2{\cdot}(THF)_2]$	2	12	68

$[Ca\{N(SiMe_3)_2\}_2]_2$, $[Sr\{N(SiMe_3)_2\}_2{\cdot}(THF)_2]$, and $[Ba\{N(SiMe_3)_2\}_2{\cdot}(THF)_2]$ not only catalyzed this reaction but also proved more effective than the heteroleptic systems **1** and **2** (Table 4.5). The reactions took place under mild conditions in benzene-d_6, generally at room temperature with a catalyst loading of 1.5–5.0 mol%. Catalysis was rapid with $HPPh_2$; however, attempts to use the less acidic substrate $HPCy_2$ led to very slow reactions. The catalytic activity was found to be greater for Sr and Ba than for Ca. The reaction was sensitive to steric effects, as no turnover was observed for bulky substrates such as *t*Bu-N=C=N-*t*Bu. Notably, the rates were also comparatively low for carbodiimides with aryl substituents such as *p*-tolyl.

The identification of a number of reactive intermediates by nuclear magnetic resonance (NMR) and XRD methods, in particular the product of insertion of carbodiimide in the Ca–P of complex **2**, led the authors to propose the catalytic cycle described in Scheme 4.6. The rate-limiting step was thought to consist of carbodiimide insertion in the Ca—P bond.

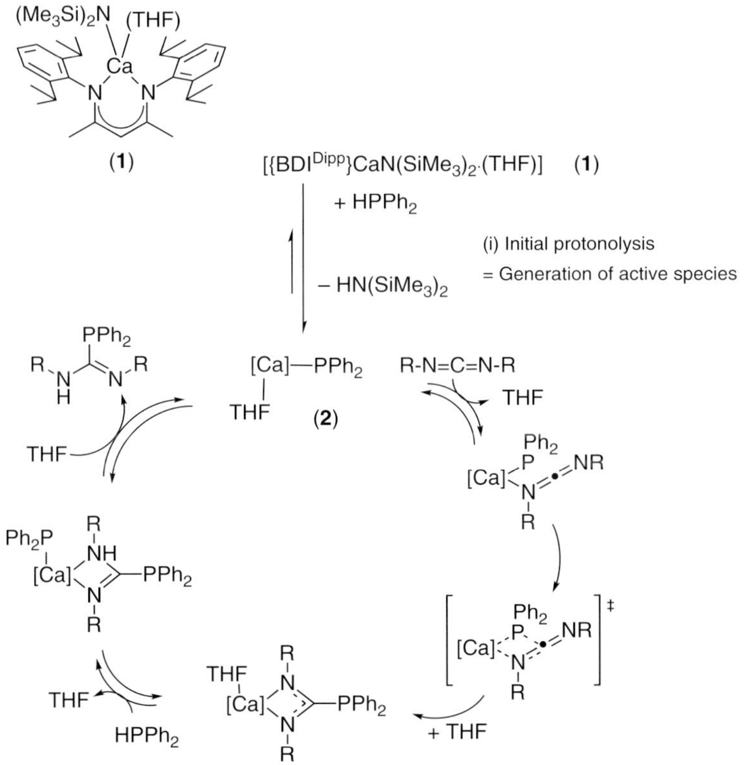

Scheme 4.6 Proposed catalytic cycle for carbodiimide hydrophosphination promoted by $[\{BDI^{Dipp}\}CaN(SiMe_3)_2{\cdot}(THF)]$ (**1**) [69].

4.5 Miscellaneous Reactions

Compared to hydrophosphination catalysis, a very limited number of studies dealing with catalyzed hydrophosphinylation and hydrophosphonylation reactions have been published. They are presented in the following.

4.5.1 Hydrophosphinylation of Alkenes and Enones

In a short publication released in 2005, Han and Zhan described the utilization of alkoxy-magnesium halides of the type $[ROMgCl]_p$ to catalyze the highly 1,2-regioregular and stereoselective addition of enantiopure phenylphosphinate onto the activated vinyldiethylphosphite, methyl acrylate, and acrylonitrile [70]. Representative examples are displayed in Scheme 4.7. The role of the metal and the mechanism of the reaction were not discussed. Side products such as $(Et_2O)_2P(O)(CH_2)_2P(O)Ph(OR)$ (R = Et, *i*Pr) resulting from deleterious transesterifications were found in small amounts (3–5%) for small Mg-alkoxides (R = Et, *i*Pr).

		t (h)	Yield (%)	R_P/S_P
EtOMgCl	(5 mol%)	24	87	94/6
*i*PrOMgCl	(5 mol%)	24	97	96/4
*t*BuOMgCl	(5 mol%)	24	79	99/1
( )MonOMgCl	(10 mol%)	12	00	00/1

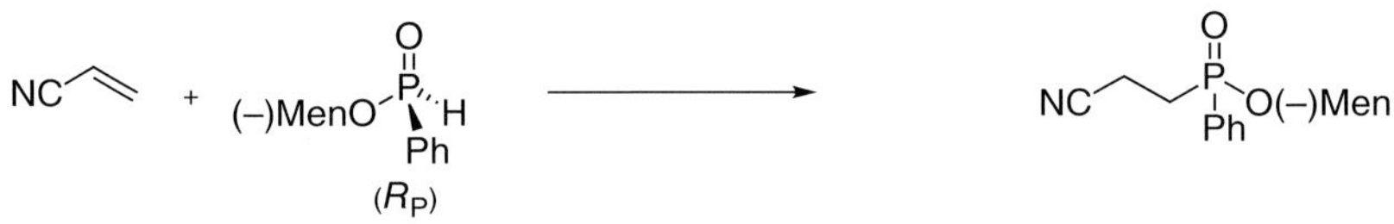

		t (h)	Yield (%)	R_P/S_P
(–)MenOMgCl	(10 mol%)	12	74	99/1

		t (h)	Yield (%)	R_P/S_P
(–)MenOMgCl	(10 mol%)	12	64	96/4

(Men = menthyl)

Scheme 4.7 Magnesium-mediated stereospecific addition of phenylphosphinates onto activated alkenes [70].

Ishihara and coworkers reported in 2013 on the highly enantioselective addition of phosphorus-based nucleophiles onto enones catalyzed by the combination of magnesium dialkyl and chiral 1,1′-binaphtol [71]. The 1,4-*hydrophosphinylation* of α,β-unsaturated esters with diarylphosphine oxides (aka phospha-Michael addition, which evolves through the formation of a transient enolate) and the 1,2-*hydrophosphonylation* of α,β-unsaturated ketones with dialkylphosphites both proceeded through cooperative Brønsted/Lewis acid–base catalysis mediated by *in situ* generated chiral magnesium(II) binaphtholate aqua species. The reactions carried out with $MgBu_2$ (10 mol%) and (*R*)-1,1′-binaphtol (10–20 mol%) in THF or toluene generally required 2–24 hours, but proceeded at low temperatures, in the range −40 to 0 °C. The addition of water (10 mol%) was necessary to observe catalytic turnover, as poor yields and stereoselectivity were obtained otherwise. A very large range of substrates was explored and could be scaled up to 10 g of the product. The different enantioselective reactions enabled by the three-component combination $MgBu_2$/(*R*)-1,1′-binaphtol/H_2O are depicted in Scheme 4.8.

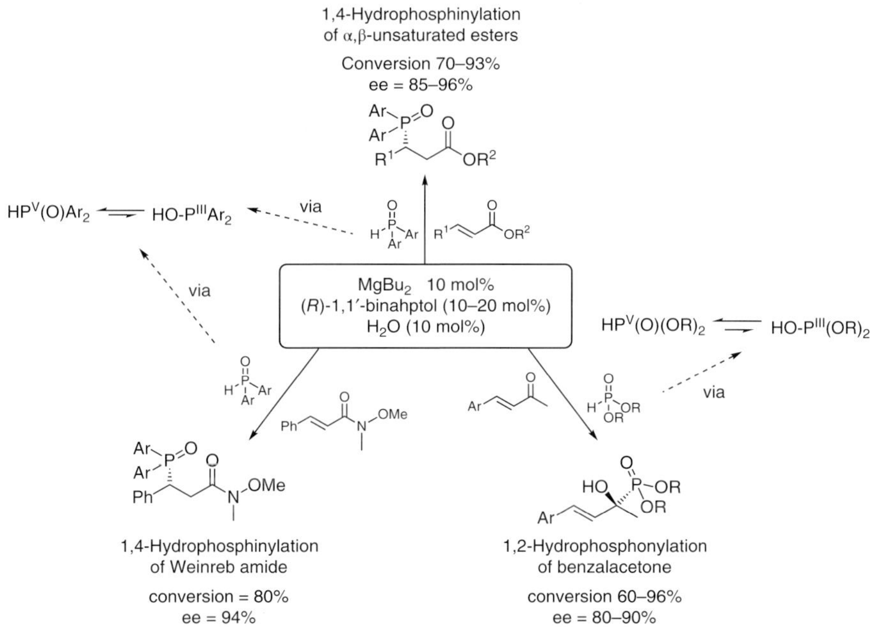

Scheme 4.8 Reaction scope for the ternary catalytic system $MgBu_2$/(*R*)-1,1′-binaphtol/H_2O [71].

Finally, a number of s-block metal catalysts, e.g. potassium and calcium diarylphosphinites, have been shown by the Westerhausen group to efficiently mediate the hydrophosphinylation of heterocumulenes, *iso*(thio)cyanates, and alkynes with $Ar_2P(O)$ or $Mes_2P(O)$ [72–75]. These results are detailed in Chapter 5.

4.5.2 Hydrophosphonylation of Aldehydes and Ketones

Sarazin and Carpentier demonstrated in 2012 that the simple alkaline earth bis(amides) [Ca{N(SiMe$_3$)$_2$}$_2$·(THF)$_2$], [Sr{N(SiMe$_3$)$_2$}$_2$·(THF)$_2$], and [Ba{N(SiMe$_3$)$_2$}$_2$·(THF)$_2$] rapidly catalyzed the addition of dialkylphosphites onto aldehydes and ketones [32]. Of interest, the catalytic activity of these homoleptic precatalysts devoid of bulky ancillary ligand equaled that of the heteroleptic complexes such as [{N^N}AeN(SiMe$_3$)$_2$·(THF)$_p$] (Ae = Ca, p = 1, 7; Ae = Sr, p = 2, **8**; Ae = Ba, p = 2, **9**; see Figure 4.4) or the ubiquitous [{BDI^{Dipp}}CaN(SiMe$_3$)$_2$·(THF)] (**1**). These Ca–Ba catalysts did not afford stereoselective reactions. On the other hand, they proved much more active than the magnesium ones. The selective addition of dialkylphosphites to aldehydes or ketones generated tertiary or quaternary phosphonates, respectively (Scheme 4.9); the reactions could be performed in neat substrates. With aldehydes, full conversion was observed within a few minutes with 0.02 mol% of catalyst at room temperature. With the less electrophilic, nonactivated ketones, unprecedented TOF values as high as 1200–1500 mol$_{subst}$/(mol$_{Ae}$ min) were achieved. The reaction was very sensitive to steric effects. Hence, for arylketones,

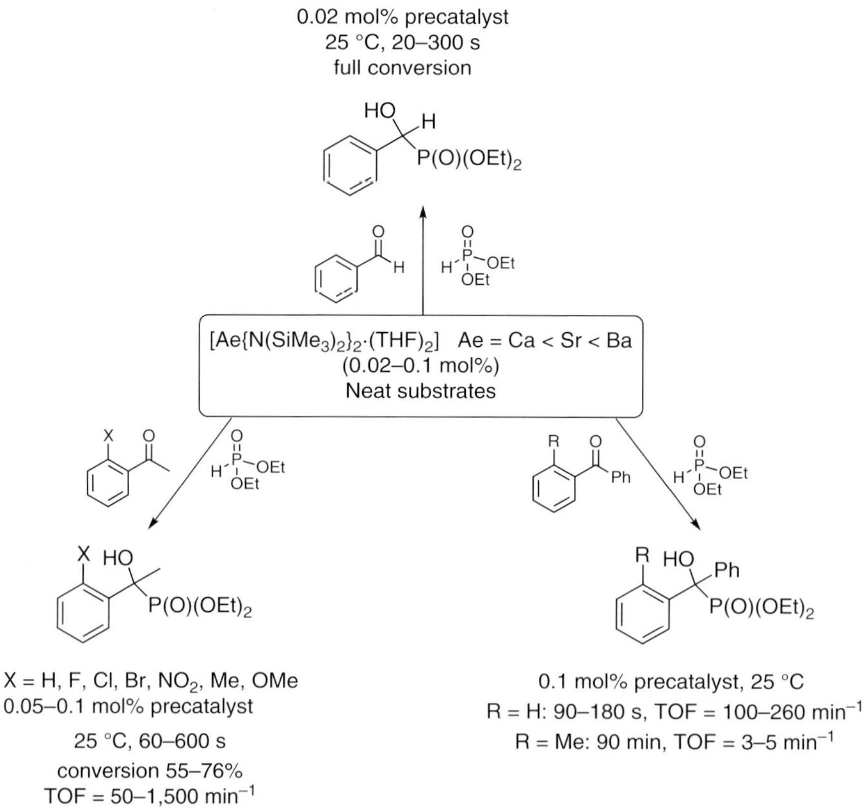

Scheme 4.9 Ca, Sr, and Ba precatalysts in the hydrophosphonylation of aldehydes and ketones with diethylphosphite [32].

the activity dropped for large substituents in ortho position of the aromatic ring. On the other hand, the efficiency of the catalyst was mostly unaffected by electronic considerations on the substrates. Under other identical experimental conditions, for a series of homologous precatalysts, reaction rates increased according to Ca < Sr < Ba.

4.6 Summary and Conclusions

Although alkali metals have seldom been considered to devise catalysts for hydrophosphination, hydrophosphonylation, and hydrophosphinylation reactions, the past 10–15 years have witnessed the development of many alkaline earth catalytic systems, chiefly based on the large calcium, strontium, and barium. Catalysis for *intermolecular* alkene hydrophosphination has received the most attention by a considerable margin; only a handful of studies have focused on hydrophosphonylation and hydrophosphinylation. Besides, the *intramolecular* hydrophosphination of phosphinoalkenes has not been investigated with Ca–Ba; this is the opposite scenario to that experienced with lanthanide(III) catalytic systems.

Since 2007 and Hill's pioneering discovery that [{BDI^{Dipp}}CaN(SiMe$_3$)$_2$·(THF)] (**1**) catalyzed the intermolecular hydrophosphination of alkenes and carbodiimides, many heteroleptic complexes, most commonly carrying a bulky nitrogen-based ligand, have emerged as competent precatalysts. The scope has been extended from calcium to the larger alkaline earths, strontium and barium, and it has been reported on several occasions that for a series of homologous complexes, the catalytic activity in alkene hydrophosphination (and hydrophosphonylation of aldehydes/ketones) was enhanced substantially upon descending group 2: Ca < Sr < Ba. This observation was also made on several other occasions for other types of catalyzed reactions (e.g. for alkene hydroamination, ring-opening polymerization of cyclic esters, and dehydrocoupling catalysis [40–43, 76]), although cases where the activity follows the reverse trend, Ba < Sr < Ca, are also known, e.g. for the intramolecular hydroamination of aminoalkenes. These trends, far from being erratic, are now well rationalized, notably through the invaluable assistance of computational studies.

Alkaline earth metals have delivered some of the most active and productive catalysts known to date for intermolecular alkene hydrophosphination, at the same time as being very regioselective and, most often, chemoselective. In particular, they consistently deliver anti-Markovnikov addition products with excellent selectivity. By contrast, late transition metal catalysts can afford both Markovnikov and anti-Markovnikov products. One of the main advantages of alkaline earths over platinum, palladium, or nickel resides in their abundance, low cost, and low toxicity; however, Ae complexes are much more air- and moisture-sensitive than the other metals and hence require strictly controlled reaction conditions. In addition, alkaline earth precatalysts are suited to relatively mildly activated alkenes such as vinyl arenes and conjugated dienes, while late transition metals often, albeit not always, require the presence of strongly electron-withdrawing groups as in acrylonitrile and Michael acceptors. In this context, both types of catalytic systems can be seen as complementary, thus enabling the assembly of primary and secondary phosphines onto a broad range of alkenes to yield a variety of new phosphines. One range of substrates that remain reluctant to this day is nonactivated alkenes, e.g. 1-hexene. Even if some advances were recently unveiled, notably through the works of Waterman [62, 65] and Trifonov [59], catalyzed reactions involving these substrates are still very slow and require harsh conditions. There is no doubt that progress in the years to come will have to address this particular issue. In addition, enriching the range of phosphines that can be used as substrates in hydrophosphination, currently mostly limited to $HPPh_2$, $PhPH_2$, and $HPCy_2$, will also constitute a main target. Other persisting challenges that should be tackled include stereoselectivity in intermolecular hydrophosphination, which cannot be accomplished for now with alkaline earths or, for this matter, with other oxophilic metals. These challenging areas will require efforts from the organometallic and organic communities, but considering the importance of P-containing materials, there can be little doubt that tremendous rewards are awaiting the imaginative chemists.

List of Abbreviations

Ae	alkaline earth
CN	coordination number
Cy	cyclohexyl
D	donor solvent (e.g. THF)
Dipp	2,6-diisopropylphenyl
Et	ethyl
$\{L_n\}^-$	monoanionic ancillary ligand
Men	menthyl
Mes	2,4,6-trimethylphenyl = mesityl
Ph	phenyl
py	pyridine
TOF	turnover frequency
p-tol	*p*-tolyl

References

1 Kolodiazhnyi, O.I. (2015). *Top. Curr. Chem.* 360: 161–236.

2 Gilheany, D.G. and Mitchell, C.M. (1990). *The Chemistry of Organophosphorus Compounds*, vol. 1 (ed. F.R. Hartley), 151–190. Chichester, England: Wiley.

3 Georgiadis, D. and Dive, V. (2015). *Top. Curr. Chem.* 360: 1–38.

4 Virieux, D., Volle, J.-N., Bakalara, N., and Pirat, J.-L. (2015). *Top. Curr. Chem.* 360: 39–114.

5 Wiemer, A.J. and Wiemer, D.F. (2015). *Top. Curr. Chem.* 360: 115–160.

6 Beletskaya, I.P. and Kazankova, M.A. (2002). *Russ. J. Org. Chem.* 38: 1391–1430.
7 Glueck, D.S. (2007). *Synlett* 17: 2627–2634.
8 Glueck, D.S. (2010). *Top. Organomet. Chem.* 31: 65–100.
9 Hudson, H.R. (1990). *The Chemistry of Organophosphorus Compounds*, vol. 1 (ed. F.R. Hartley), 399–401. Chichester, England: Wiley.
10 Dankowski, M. (1990). *The Chemistry of Organophosphorus Compounds*, vol. 1 (ed. F.R. Hartley), 496–499. Chichester, England: Wiley.
11 Rauhut, M.M., Currier, H.A., Semsel, A.M., and Wystrach, V.P. (1961). *J. Org. Chem.* 26: 5138–5145.
12 Dubois, D.L., Myers, W.H., and Meek, D.W. (1975). *J. Chem. Soc., Dalton Trans.*: 1011–1015.
13 Rauhut, M.M., Hechenbleikner, I., Currier, H.A. et al. (1959). *J. Am. Chem. Soc.* 81: 1103–1107.
14 Habib, M., Trujillo, H., Alexander, C.A., and Storhoff, B.N. (1985). *Inorg. Chem.* 24: 2344–2349.
15 Pringle, P.G. and Smith, M.B. (1990). *J. Chem. Soc. Chem. Commun.*: 1701–1702.
16 Wicht, D.K. and Glueck, D.S. (2001). *Catalytic Heterofunctionalization* (eds. A. Togni and H. Grützmacher), 143–170. Weinheim, Germany: Wiley-VCH.
17 Tanaka, M. (2004). *Top. Curr. Chem.* 232: 25–54.
18 Alonso, F., Beletskaya, I.P., and Yus, M. (2004). *Chem. Rev.* 104: 3079–3160.
19 Greenberg, S. and Stephan, D.W. (2008). *Chem. Soc. Rev.* 37: 1482–1489.
20 Beletskaya, I.P., Ananikov, V.P., and Khemchyan, L.L. (2011). *Catal. Met. Complexes* 37: 213–264.
21 Pullarkat, S.A. and Leung, P.-H. (2013). *Top. Organomet. Chem.* 43: 145–166.
22 Rosenberg, L. (2013). *ACS Catal.* 3: 2845–2855.
23 Koshti, V., Gaikwad, S., and Chikkali, S.H. (2014). *Coord. Chem. Rev.* 265: 52–73.
24 Pullarkat, S.A. (2016). *Synthesis* 48: 493–503.
25 Crimmin, M.R., Barrett, A.G.M., Hill, M.S. et al. (2007). *Organometallics* 26: 2953–2956.
26 Al-Shboul, T.M.A., Görls, H., and Westerhausen, M. (2008). *Inorg. Chem. Commun.* 11: 1419–1421.
27 Al-Shboul, T.M.A., Pálfi, V.K., Yu, L. et al. (2011). *J. Organomet. Chem.* 696: 216–227.
28 Al-Shboul, T.M.A., Görls, H., Krieck, S., and Westerhausen, M. (2012). *Eur. J. Inorg. Chem.*: 5451–5455.
29 Younis, F.M., Krieck, S., Al-Shboul, T.M.A. et al. (2016). *Inorg. Chem.* 55: 4676–4682.
30 Crimmin, M.R., Arrowsmith, M., Barrett, A.G.M. et al. (2009). *J. Am. Chem. Soc.* 131: 9670–9685.
31 Arrowsmith, M., Crimmin, M.R., Barrett, A.G.M. et al. (2011). *Organometallics* 30: 1493–1506.
32 Liu, B., Carpentier, J.-F., and Sarazin, Y. (2012). *Chem. Eur. J.* 18: 13259–13264.

33 Bellini, C., Carpentier, J.-F., Tobisch, S., and Sarazin, Y. (2015). *Angew. Chem. Int. Ed.* 54: 7679–7683.
34 Bellini, C., Orione, C., Carpentier, J.-F., and Sarazin, Y. (2016). *Angew. Chem. Int. Ed.* 55: 3744–3748.
35 Wixey, J.S. and Ward, B.D. (2011). *Chem. Commun.* 47: 5449–5451.
36 Nixon, T.D. and Ward, B.D. (2012). *Chem. Commun.* 48: 11790–11792.
37 Zhang, X., Emge, T.J., and Hultzsch, K.C. (2012). *Angew. Chem. Int. Ed.* 51: 394–398.
38 Chisholm, M.H., Gallucci, J., and Phomphrai, K. (2003). *Chem. Commun.*: 48–49.
39 Crimmin, M.R., Casely, I.J., and Hill, M.S. (2005). *J. Am. Chem. Soc.* 127: 2042–2043.
40 Harder, S. (2010). *Chem. Rev.* 110: 3852–3876.
41 Barrett, A.G.M., Crimmin, M.R., Hill, M.S., and Procopiou, P.A. (2010). *Proc. R. Soc. A* 466: 927–963.
42 Crimmin, M.R. and Hill, M.S. (2013). *Top. Organomet. Chem.* 45: 191–242.
43 Hill, M.S., Liptrot, D.J., and Weetman, C. (2016). *Chem. Soc. Rev.* 45: 972–988.
44 Douglass, M.R., Stern, C.L., and Marks, T.J. (2001). *J. Am. Chem. Soc.* 123: 10221–10238.
45 Ward, B.J. and Hunt, P.A. (2017). *ACS Catal.* 7: 459–468.
46 Liu, B., Roisnel, T., Carpentier, J.-F., and Sarazin, Y. (2013). *Chem. Eur. J.* 19: 13445–13462.
47 Tobisch, S. (2014). *Chem. Eur. J.* 20: 8988–9001.
48 Hu, H. and Cui, C. (2012). *Organometallics* 31: 1208–1211.
49 Izod, K., Clegg, W., and Liddle, S.T. (2000). *Organometallics* 19: 3640–3643.
50 Fedushkin, I.L., Petrovskaya, T.V., Bochkarev, M.N. et al. (2001). *Angew. Chem. Int. Ed.* 40: 2474–2477.
51 Harder, S. (2004). *Angew. Chem. Int. Ed.* 43: 2714–2718.
52 Liu, B., Roisnel, T., Maron, L. et al. (2013). *Chem. Eur. J.* 19: 3986–3994.
53 Liu, B., Roisnel, T., Carpentier, J.-F., and Sarazin, Y. (2012). *Angew. Chem. Int. Ed.* 51: 4943–4946.
54 Avent, A.G., Crimmin, M.R., Hill, M.S., and Hitchcock, P.B. (2005). *Dalton Trans.*: 278–284.
55 Brinkmann, C., Barrett, A.G.M., Hill, M.S., and Procopiou, P.A. (2012). *J. Am. Chem. Soc.* 134: 2193–2207.
56 Anga, S., Carpentier, J.-F., Panda, T.K. et al. (2016). *RSC Adv.* 6: 57835–57843.
57 He, M., Gamer, M.T., and Roesky, P.W. (2016). *Organometallics* 35: 2638–2644.
58 Basalov, I.V., Yurova, O.S., Cherkasov, A.V. et al. (2016). *Inorg. Chem.* 55: 1236–1244.
59 Lapshin, I.V., Yurova, O.S., Basalov, I.V. et al. (2018). *Inorg. Chem.* 57: 2942–2952.
60 Rad'kov, V., Dorcet, V., Carpentier, J.-F. et al. (2013). *Organometallics* 32: 1517–1527.
61 Radkov, V., Roisnel, T., Trifonov, A. et al. (2016). *J. Am. Chem. Soc.* 138: 4350–4353.

62 Ghebreab, M.B., Bange, C.A., and Waterman, R. (2014). *J. Am. Chem. Soc.* 136: 9240–9243.
63 Geer, A.M., Serrano, A.L., de Bruin, B. et al. (2015). *Angew. Chem. Int. Ed.* 54: 472–475.
64 Bange, C.A., Ghebreab, M.B., Ficks, A. et al. (2016). *Dalton Trans.* 45: 1863–1867.
65 Bange, C.A., Conger, M.A., Novas, B.T. et al. (2018). *ACS Catal.* 8: 6230–6238.
66 Basalov, I.V., Liu, B., Roisnel, T. et al. (2016). *Organometallics* 35: 3261–3271.
67 Roşca, S.-C., Roisnel, T., Dorcet, V. et al. (2014). *Organometallics* 33: 5630–5642.
68 Freitag, B., Stegner, P., Thum, K. et al. (2018). *Eur. J. Inorg. Chem.*: 1938–1944.
69 Crimmin, M.R., Barrett, A.G.M., Hill, M.S. et al. (2008). *Organometallics* 27: 497–499.
70 Han, L.-B. and Zhan, C.-Q. (2005). *J. Org. Chem.* 70: 10121–10123.
71 Hatano, M., Horibe, T., and Ishihara, K. (2013). *Angew. Chem. Int. Ed.* 52: 4549–4553.
72 Härling, S., Greiser, J., Al-Shboul, T.M.A. et al. (2013). *Aust. J. Chem.* 66: 1264–1273.
73 Härling, S.M., Görls, H., Krieck, S., and Westerhausen, M. (2016). *Inorg. Chem.* 55: 10741–10750.
74 Härling, S.M., Krieck, S., Görls, H., and Westerhausen, M. (2017). *Inorg. Chem.* 56: 9255–9263.
75 Härling, S.M., Fener, B.E., Krieck, S. et al. (2018). *Organometallics* **37**, 4380–4386.
76 Sarazin, Y. and Carpentier, J.-F. (2016). *Chem. Rec.* 16: 2482–2505.

5

H—N and H—P Bond Addition to Alkynes and Heterocumulenes

Sven Krieck and Matthias Westerhausen

Institut für Anorganische und Analytische Chemie, Friedrich-Schiller-Universität, August-Bebel-Strasse 2, D-07743, Jena, Germany

5.1 Introduction

The hydrofunctionalization (hydroelementation) reaction, i.e. the addition of H—E bonds (E = N, P) across multiple bonds such as alkenes, alkynes, cumulenes, and their heteroatom-substituted congeners, represents a very environmentally benign atom-economic process. Exemplarily, the hydrofunctionalization of an alkyne is depicted in Eq. (5.1) yielding *E*- and *Z*-isomers. However, strong and electron-rich Lewis bases have to interact with electron-rich multiple bonds causing electrostatic repulsion impeding this reaction. Furthermore, intermolecular hydrofunctionalization reactions are entropically highly disfavored. Significantly different energies of H—E bonds and π-bonds additionally hinder a straightforward addition. Therefore, catalyst systems are required to tackle these challenges.

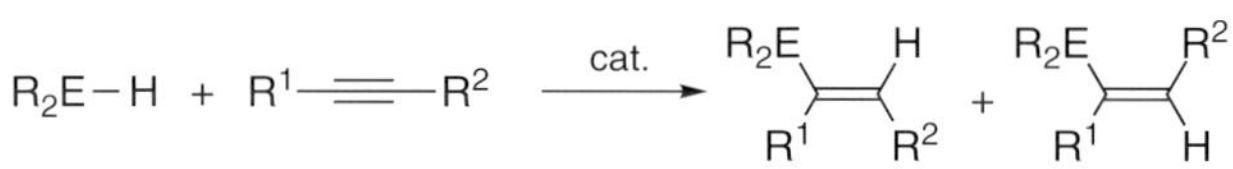

(5.1)

There exist diverse strategies to overcome the high activation barrier [1]. Activation of the multiple bonds can be achieved via side-on π-bonding at transition metals. Oxidative addition of amines or phosphanes to transition metals activates the counterpart of this reaction. Very early transition metals are able to form imides of the type [M]=ER. The s-block metals easily form amides and phosphanides of the type A-ER_2 and Ae$(ER_2)_2$ (A = alkali metal; Ae = alkaline earth metal; E = N, P) as well as phosphinites of the types A-OPR_2 and Ae$(OPR_2)_2$. Some of these s-block metals (especially sodium, potassium, magnesium, and calcium) are ideal candidates because of their nontoxicity, global abundance, and cost-efficient availability [2]. In recent years, the interest in s-block metal catalysts has gained tremendously in attention mainly because of increased environmental and ecological awareness [3–5]. The electropositive nature of these elements leads to highly polar M—E bonds and a strong nucleophilic character of the heteroatom E. Furthermore, the ions are Lewis acids with the strength depending

Early Main Group Metal Catalysis: Concepts and Reactions,
First Edition. Edited by Sjoerd Harder.

on their charge and radius. Thus, π-bonds interact with these metal ions leading to close contact with the nucleophilic ER_2 anions, a precondition for the hydrofunctionalization reaction.

This overview focuses on the intermolecular addition of amines (hydroamination; E = N; R = H, alkyl, aryl), phosphanes (hydrophosphanylation, hydrophosphination; E = P; R = alkyl, aryl), phosphane oxides and sulfides (hydrophosphorylation; E = PO, PS; R = alkyl, aryl), and phosphonates (hydrophosphonylation; E = PO; R = OR′) across alkynes and heterocumulenes, mediated by s-block metal catalysts (Scheme 5.1). Even though diverse overviews highlight certain aspects of these hydrofunctionalization reactions [6–8], a comprehensive review on s-block metal-mediated hydrofunctionalization of alkynes and heterocumulenes is justified by the increasing importance of such procedures.

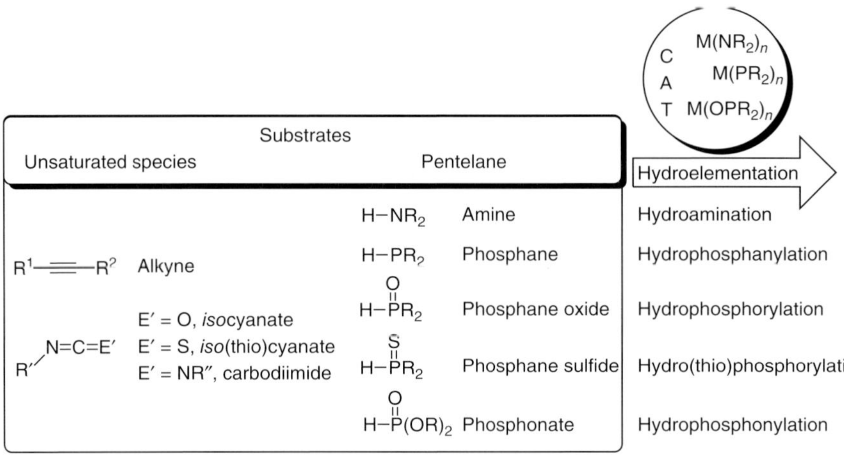

Scheme 5.1 Hydrofunctionalization (hydroelementation) reactions classified by the variety of substrates.

5.2 Hydroamination

Hydroamination of alkynes commonly yields mixtures of *E*- and *Z*-isomeric alkenylamines. The reaction can be catalyzed by strong bases converting the amine to a more reactive amide, which is able to attack the multiple bond. First investigations on s-block metal-catalyzed hydroamination of alkynes have been performed by the Reppe group [9, 10] who have used potassium or potassium hydroxide as catalysts that metalate the amines. The use of KOH/DMSO (dimethyl sulfoxide) mixtures enhances the reactivity and *N*-heterocyclic compounds (*N*-vinylpyrroles, enamines, and enaminones) are accessible [11]. $CsOH{\cdot}H_2O$ also represents a suitable catalyst for this hydroamination commonly yielding *E*/*Z*-isomeric mixtures [12].

Quite recently, the hydroamination of alkynes has been studied with s-block metal catalysts that are soluble and stable in common organic solvents in an anaerobic and anhydrous atmosphere allowing homogeneous reaction

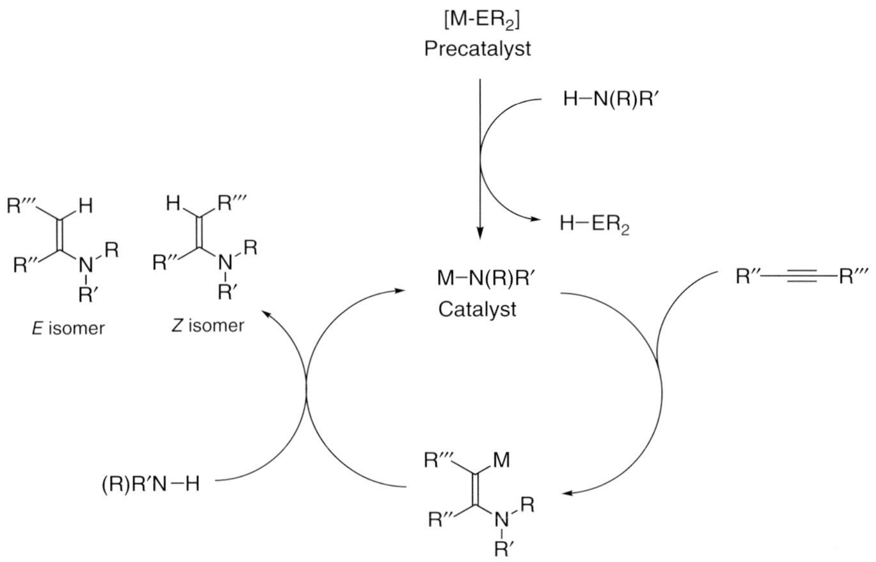

Scheme 5.2 Typical catalytic cycle for the hydroamination of alkynes yielding mixtures of *E*- and *Z*-isomers of alkenylamines after addition of the M—N bond onto a C≡C triple bond and subsequent protonation of the organometallic intermediate by an amine (E = N, CH; M = s-block metal).

conditions. A general catalytic cycle is depicted in Scheme 5.2. The precatalyst has to be transferred into the catalytically active species that adds onto an alkyne unit. Thereafter, protonation yields the *E*- and *Z*-isomeric alkenylamines and regains the catalyst.

5.2.1 Hydroamination with Secondary Amines

The influence of several factors on the suitability of alkaline earth metal-based catalysts has been studied for the hydroamination reaction of phenylacetylene and diphenylacetylene with piperidine as shown in Eq. (5.2) [13]. The hydroamination of phenylacetylene fails presumably because of deprotonation of the acetylenic C—H bond forming, especially in the case of strontium catalysts, insoluble acetylides. For the addition of piperidine across diphenylacetylene, the catalysts $[(thf)_2Ae\{CH(SiMe_3)_2\}_2]$ (Ae = Ca, Sr) and $[Sr\{N(SiMe_3)_2\}_2]_2$ have been tested in various solvents. The far most effective catalyst is $[(thf)_2Sr\{CH(SiMe_3)_2\}_2]$, leading to a quantitative conversion within two hours in tetrahydrofuran (THF) at 60 °C. The calcium congener is significantly less reactive, whereas $[Sr\{N(SiMe_3)_2\}_2]_2$ shows nearly no reactivity under these conditions.

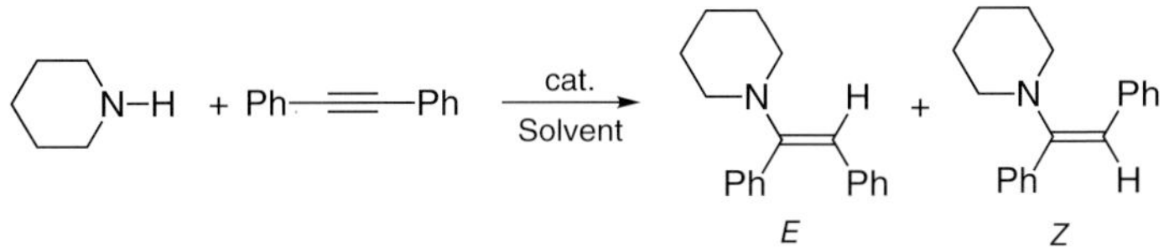

(5.2)

Besides the importance of the precatalyst, the solvent influences the activity of the catalyst and the ratio of the *E*- and *Z*-isomers. No conversion can be found in dioxane, 1,2-dimethoxyethane, and 2-methyltetrahydrofuran. THF represents the favored solvent, whereas in benzene, hexane, diethyl ether, and tetrahydropyran, significantly slower conversion rates are observed. Depending on these solvents, *E*/*Z*-isomeric ratios vary between 91 : 9 (THF) and 60 : 40 (benzene and hexane) with the *E*-isomer always being the dominant species [13]. These studies underline the importance of the choice of solvent and precatalyst.

The hydroamination of phenylbut-1-yne-3-ene with piperidine (Eq. (5.3)) requires a less reactive catalyst such as $[Sr\{N(SiMe_3)_2\}_2]_2$ yielding the cumulene derivative 1-phenyl-4-piperidylbuta-1,2-diene [13], whereas the more reactive dialkylstrontium complex $[(thf)_2Sr\{CH(SiMe_3)_2\}_2]$ leads to intensely colored reaction solutions and a complex mixture of products. The formation of these colored intermediates already shows that the organometallic species is delocalized before protonation by another piperidine molecule.

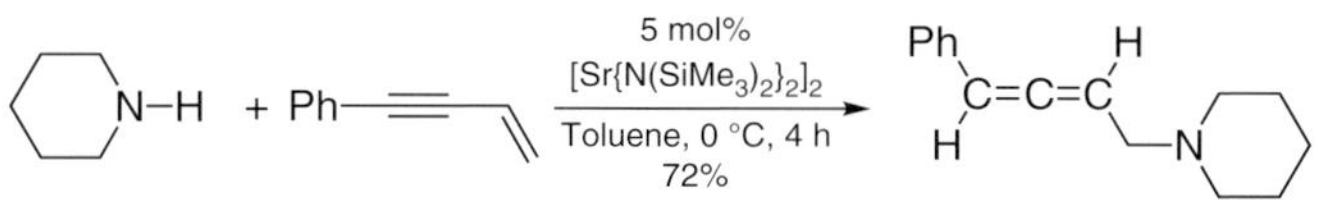

(5.3)

Hydroamination of butadiynes with secondary *N*-alkyl- and *N*-arylanilines requires highly reactive catalysts. Thus, the addition of diphenylamine across diphenylbutadiyne is neither mediated by $KNPh_2$ nor by $Ca(NPh_2)_2$, but heterobimetallic $K_2Ca(NPh_2)_4$ promotes this hydroamination reaction as depicted in Eq. (5.4) [14]. Enhanced nucleophilicity of the secondary amines and their deprotonated congeners is achieved by use of *N*-alkylanilines; hydroamination of diphenylbutadiyne with *N*-*iso*propylaniline (R = *i*Pr, R′ = C≡C—Ph) is mediated by homometallic $Ca[N(iPr)Ph]_2$ and KO*t*Bu; however, heterobimetallic $K_2Ca[N(iPr)Ph]_4$ shows a higher activity and quantitative conversion occurs in THF at 65 °C within one hour with a catalyst load of 5 mol%. The sodium-based catalyst NaN(*i*Pr)Ph is significantly less active in this application.

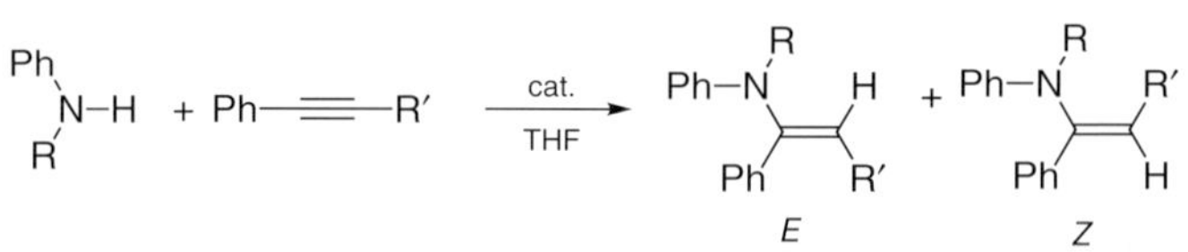

(5.4)

The heterobimetallic catalysts employed the above precipitate as thf solvates. Contrary to this finding, $[K_2Ca\{N(H)Dipp\}_4]$ crystallizes without ligated ether molecules, and consequently, this complex can be weighed, handled, and stored under an inert atmosphere without aging due to loss of ethereal coligands [15]. The hydroamination of the second far less reactive C≡C triple bond needs long reaction times of several days or even weeks and a heterobimetallic potassium calciate catalyst [16]. Regardless of the substituents at the *N*-bound aryl group (Aryl = Ph, C_6H_4-4-Me, C_6H_4-4-F), isomeric mixtures are obtained as shown in Eq. (5.5).

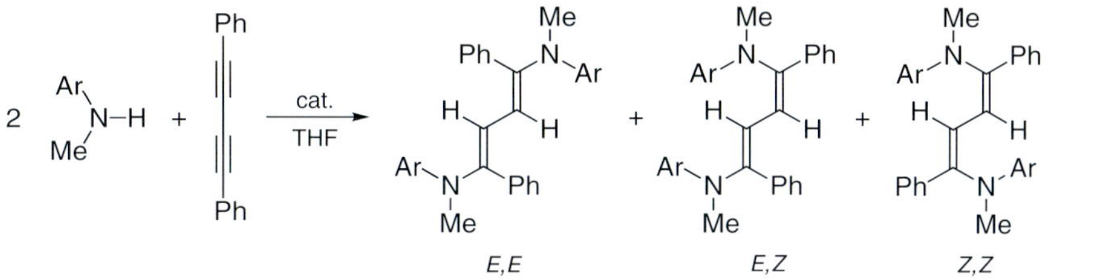

(5.5)

During the hydroamination of diphenylbutadiyne with HN(*i*Pr)Ph in boiling THF, using the catalyst KN(*i*Pr)Ph, a side product forms in small quantities via *ortho*-C—H activation of a phenyl group at the butadiyne moiety [14] (Eq. (5.6)). The first reaction step is the addition of KN(*i*Pr)Ph at an alkyne unit followed by a second addition step to another diphenylbutadiyne molecule. Thereafter, C—H activation, cyclization, and protonation sequences yield the naphthalin derivative depicted in Eq. (5.6) and the molecular structure in Figure 5.1.

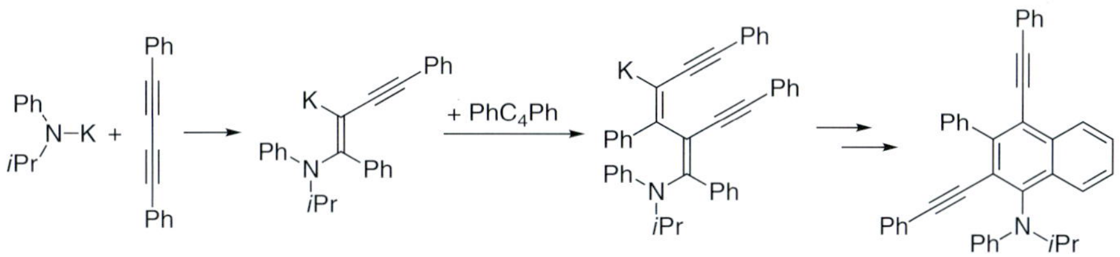

(5.6)

The different reactivities of the C≡C triple bonds in Ph—C≡C—C≡C—Ph give rise to the question whether 1,2-diaminoethanes are able to form cyclic butadiene derivatives. The reaction of 1,2-dianilinoethane with diphenylbutadiyne yields the 1 : 1 addition product and in a subsequent step the 1 : 2 hydroamination product. In agreement with earlier investigations, this hydroamination is neither promoted by the pure potassium complex $[(thf)_3K_2\{1,2\text{-}(PhN)_2C_2H_4\}]$ nor by the calcium congener $[(thf)_5Ca_2\{1,2\text{-}(PhN)_2C_2H_4\}_2]$ alone but a mixture of these complexes with a potassium:calcium ratio of 2 : 1 gives a quantitative conversion [17]. However, cyclic compounds have not been observed, but acyclic

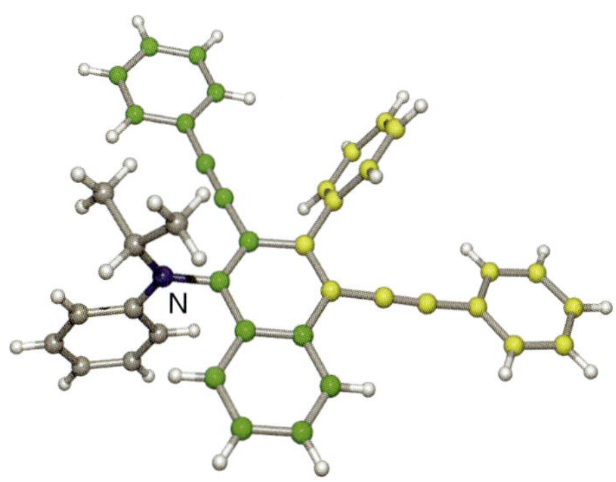

Figure 5.1 Stick-and-ball presentation of the hydroamination product of diphenylbutadiyne with *N*-diisopropylaniline containing 2 equiv of butadiyne (yellow and green) and 1 equiv of amine (C, gray; N, blue).

enyne derivatives are formed in a successive manner as *E*- and *Z*-isomeric mixtures as depicted in Eq. (5.7). The *Z,Z*-isomer is only sparingly soluble in toluene and precipitates during the hydroamination process; however, at daylight, *E*/*Z*-isomerization takes place and a mixture of all possible isomers is formed within a few hours. This outcome is remarkable because the entropically favored intramolecular cyclohydroamination is highly disfavored compared to a second intermolecular hydroamination step. This finding clearly verifies that the butydiyne and the bis(amino)ethane exhibit graded reactivities for the two alkynes as well as the two N—H functionalities.

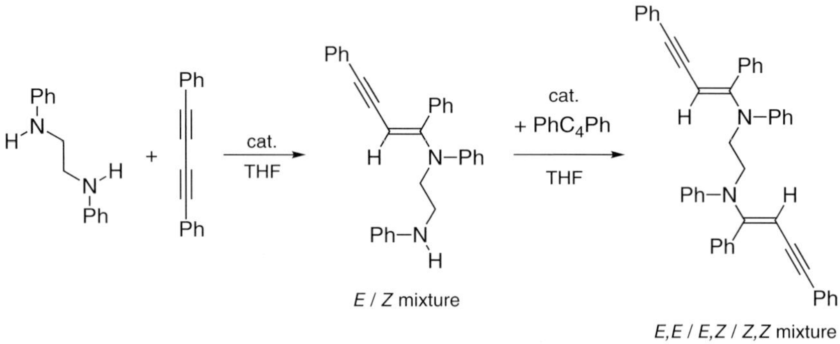

(5.7)

Enhanced reactivity of the multiple bonds eases hydroamination and less reactive catalysts are able to mediate N—H addition. Thus, hydroamination of organic *iso*cyanates succeeds with $[Ae\{N(SiMe_3)_2\}_2]_2$ (Ae = Ca, Sr, Ba) [18]. Carbodiimides [19] and *iso*thiocyanates [20] (Eq. (5.8)) can be functionalized via hydroamination with propargylamines employing sodium hydroxide as a catalyst.

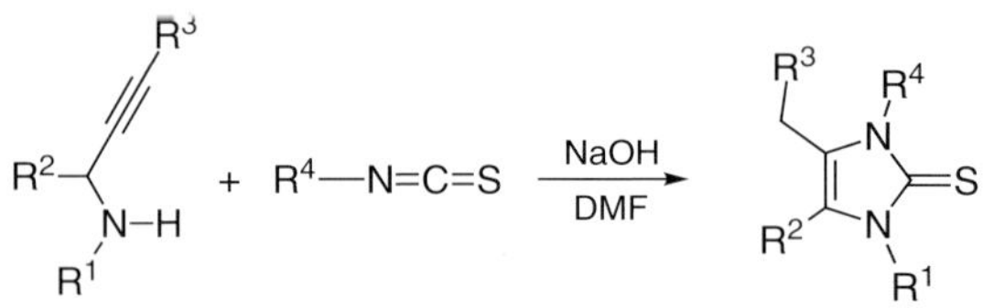

(5.8)

Responsible for the straightforward hydroamination of heterocumulenes such as (thio)cyanates and carbodiimides with “simple” catalyst systems is the positively polarized central carbon atom by neighboring electronegative atoms such as N, O, and S, easing the nucleophilic attack of strong Lewis bases at this site.

5.2.2 Hydroamination with Primary Amines

Hydroamination of butadiynes with primary amines offers the possibility of a double hydroamination with the second one being an intramolecular hydroamination. Thus, the reaction of diphenylbutadiyne with substituted anilines in boiling THF for six days yields *N*-aryl-2,5-diphenylpyrroles according to Eq. (5.9) with yields around 80% [21], regardless of the substitution pattern of the *N*-bound aryl substituent. Based on the investigations with secondary

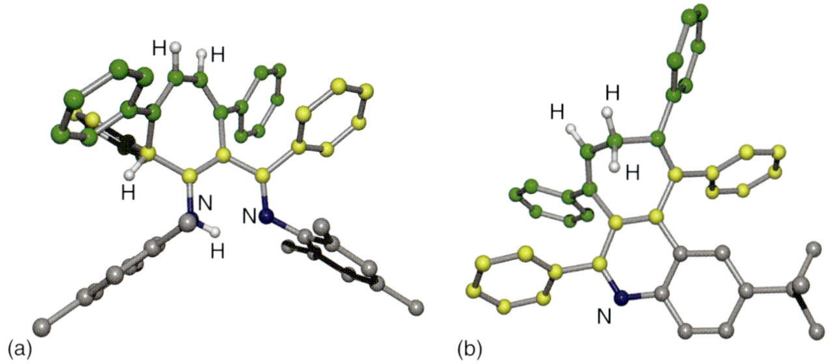

Figure 5.2 Comparison of the products of hydroamination of diphenylbutadiyne with substituted (a) and unsubstituted (b) arylamines in ortho-position at the N-bound substituent and distinction of the initial diphenylbutadiyne building blocks via different colors (the diynes are shown in yellow and green and the arylamines in gray [C atoms] and blue [N atom]. Only those H atoms are drawn [light gray] that are involved in chemical transformations or are bound at a nitrogen atom.).

amines, the hydroamination of the second alkyne moiety requires quite drastic reaction conditions (long reaction time in refluxing THF). Under these reaction conditions, the second intramolecular hydroamination of diphenylbutadiyne with primary arylamines proceeds smoothly.

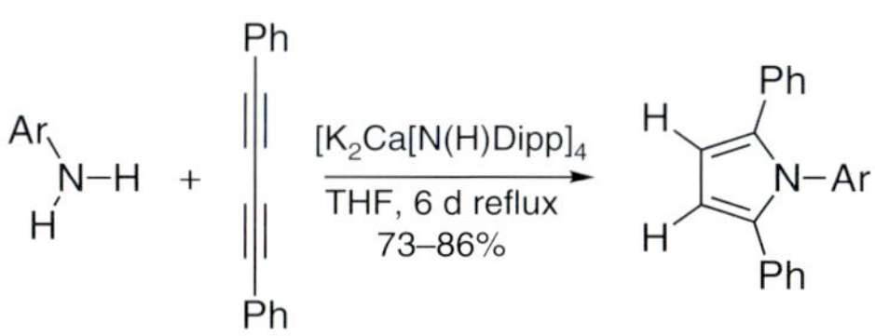

(5.9)

If this hydroamination of diphenylbutadiyne with arylamines is performed at room temperature, surprisingly different compounds are obtained. The outcome of this reaction strongly depends on the substituents in ortho-position of the *N*-bound aryl substituent. As observed earlier, the second alkyne unit of diphenylbutadiyne reacts much slower with amines. Therefore, other pathways become competitive and the intermolecular reaction with another equivalent of diphenylbutadiyne turns to the major reaction finally yielding quinoline derivatives as shown in Eq. (5.10) [21]. During the reaction sequence also, *ortho*-C—H activation at the *N*-bound aryl group takes place, finally yielding a quinoline derivative with an annelated seven-membered ring (Figure 5.2a).

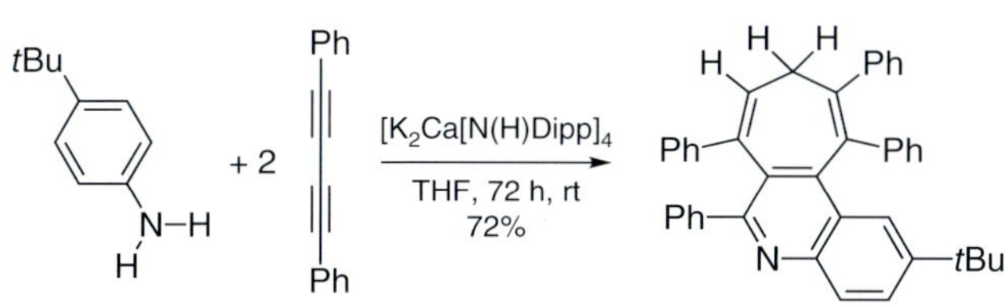

(5.10)

Blocking the ortho-position with methyl groups and use of mesitylamine (2,4,6-trimethylaniline, Mes-NH_2) enforce another reaction pathway, which

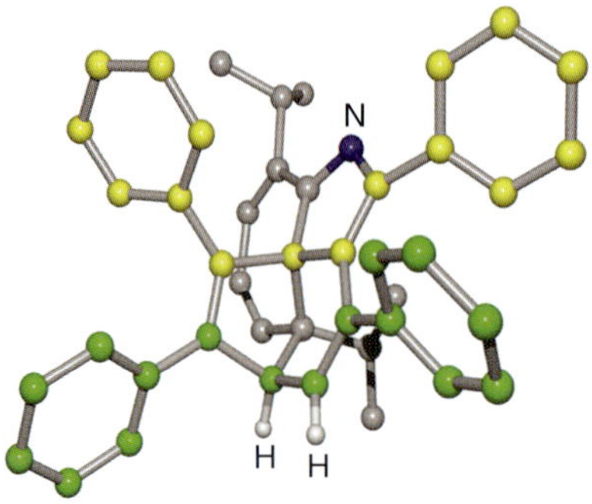

Figure 5.3 Stick-and-ball presentation of the hydroamination product of diphenylbutadiyne with 2,6-diisopropylaniline (only the formerly N-bound H atoms are depicted), containing 2 equiv of butadiyne (yellow and green) and 1 equiv of amine (C, gray; N, blue).

is depicted in Eq. (5.11) [21]. Regardless of the applied stoichiometry, a cyclohepta-1,3,5-triene derivative (Figure 5.2b) forms within three days at room temperature in THF, mediated by 5 mol% of $[K_2Ca\{N(H)Dipp\}_4]$.

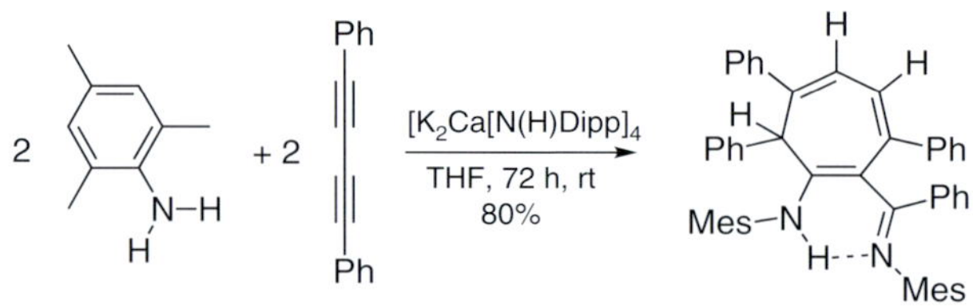

(5.11)

Enhancement of steric strain and the use of 2,6-diisopropylaniline (Dipp-NH_2) leads to another tetracyclic product also based on a seven-membered cycle as shown in Eq. (5.12) [22]. The seven membered ring is the extended *N*-bound aryl group and forms via the insertion of a carbon atom of an alkyne moiety between the *ipso*- and an *ortho*-carbon atom of the *N*-bound Dipp group, recognizable at the *iso*propyl substituted carbon atoms of this Dipp substituent (Figure 5.3).

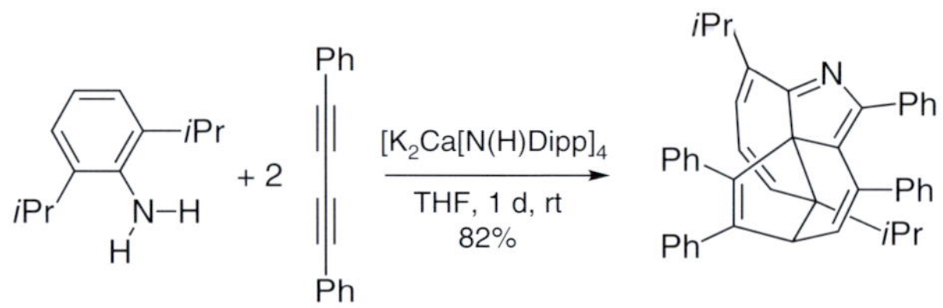

(5.12)

5.2.3 Proposed Mechanisms for the Hydroamination of Butadiynes

The activity of the s-block metal catalyst has to be adapted to the reactivity and accessibility of the alkyne functionality. The amines show increasing reactivity in hydroamination reactions of alkynes as shown in the following row:

diphenylamine < alkylaniline < dialkylamine < aniline

with the primary anilines exhibiting a deviating reactivity than the secondary congeners. Furthermore, reactivity of the catalysts can be enhanced by using heterobimetallic potassium calciates rather that homometallic potassium or calcium amide complexes. This finding suggests that these reactions are heterobimetallic processes in which actions of both metals, Ca and K, are important.

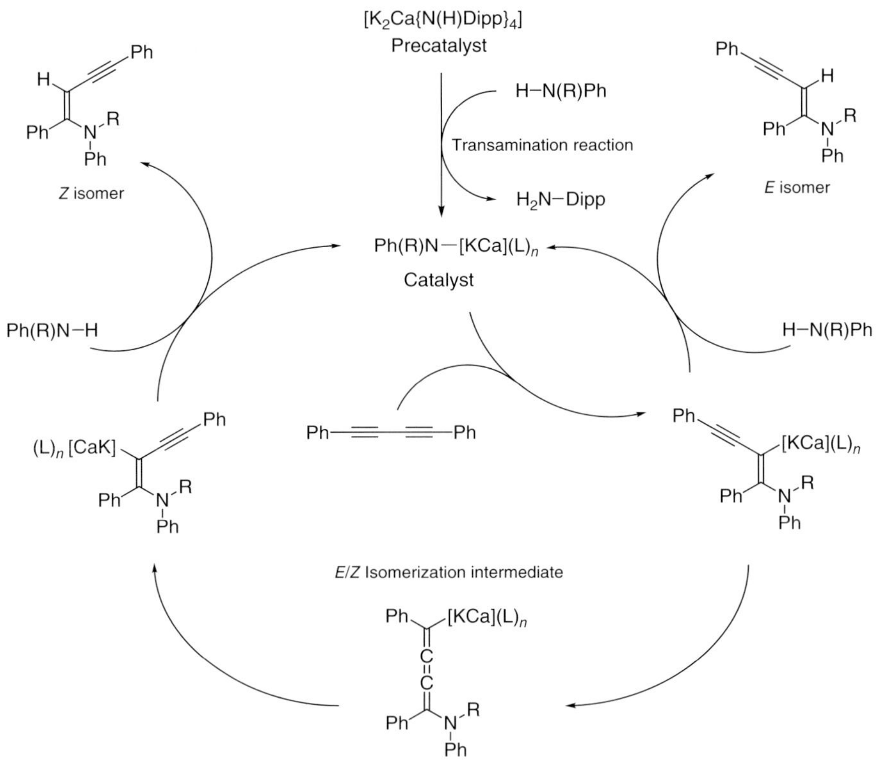

Scheme 5.3 In a typical catalytic hydroamination of butadiynes, the precatalyst is transformed into the catalytically active species via a transamination step. Thereafter, the catalytic hydrofunctionalization cycle is shown as well as the isomerization process finally yielding an *E/Z*-mixture of the anti-Markovnikov products.

A typical catalytic cycle is depicted in Scheme 5.3. In the precatalyst (e.g. $[K_2Ca\{N(H)Dipp\}_4]$), at least one of the amido groups is exchanged by the amine (transamination reaction), leading to the catalytically active calciate shown as $L_nCa[N(R)Ph]$. The cis-addition of the Ca—N bond across an alkyne moiety yields a 1-aminoalkene-2-ide, which can either be protonated by the amine substrate or isomerize via a cumulene intermediate. Because of these alternative reaction pathways, *E*- and *Z*-isomeric mixtures are obtained.

The use of primary amines extends the variety of reaction pathways significantly because intra- and intermolecular addition reactions occur depending on the reaction conditions. Starting from the catalytic active species $L_nCa[N(H)Ar]$, the cis-addition at a C≡C bond (giving **A**) is the initial step. Again, *E*/*Z*-isomerization to intermediate **B** is feasible with the *Z*-isomer allowing an intramolecular addition and the formation of pyrrole derivative **C**, which is easily protonated by amines; this reaction sequence is shown in Scheme 5.4.

The high reactivity of Ca—C bonds [23] of the intermediate **B′** also enables addition to another equivalent of diphenylbutadiyne as shown in the reaction Scheme 5.5. Two different pathways are conceivable, namely via a diradical intermediate (leading to the Bergman cyclization, **F–H**) or via a strained intermediate cycloheptatetraene derivative (**I** and **J**). In either case, the reactive radical site in **G** and the strained cumulene unit in **J** are in close contact with the *ortho*-H

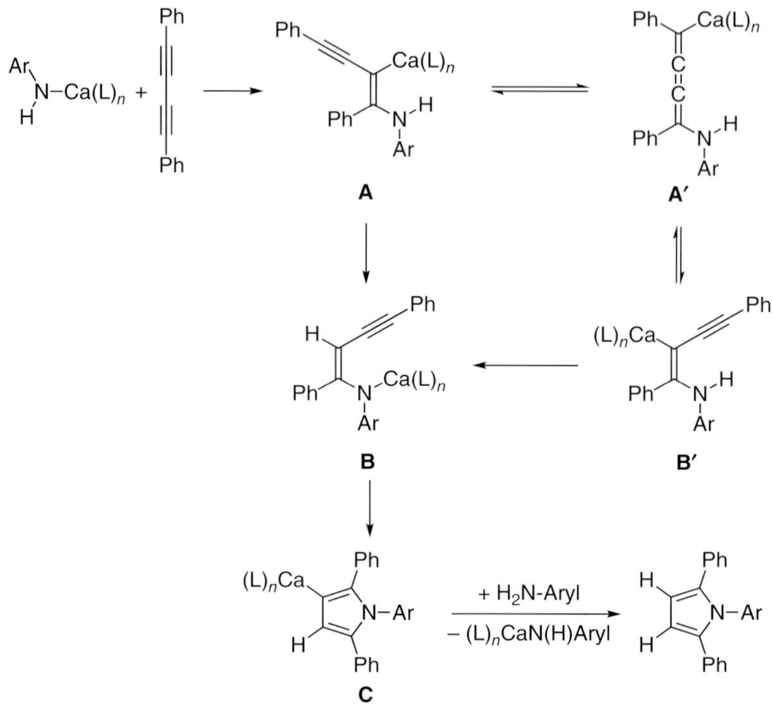

Scheme 5.4 Proposed reaction mechanism for the calcium-mediated formation of *N*-aryl-2,5-diphenylpyrroles.

atom of the *N*-bound aryl group. Cyclization via C—C bond formation forms a diamagnetic compound without ring strain.

If the *ortho*-C—H moiety is blocked by methyl groups, another pathway is feasible (Eq. (5.13)). The radical or the cycloheptatetraene attacks another equivalent of primary amine, yielding the product that is stabilized by a strong N—H· · ·N bridge and shielded by rather bulky mesityl substituents at the nitrogen bases that show a hindered rotation around the N—C bonds giving rise to six chemically different methyl groups in the ^{1}H NMR spectrum.

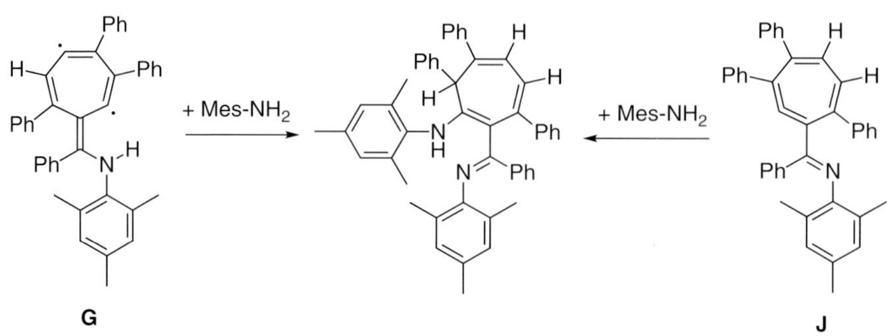

(5.13)

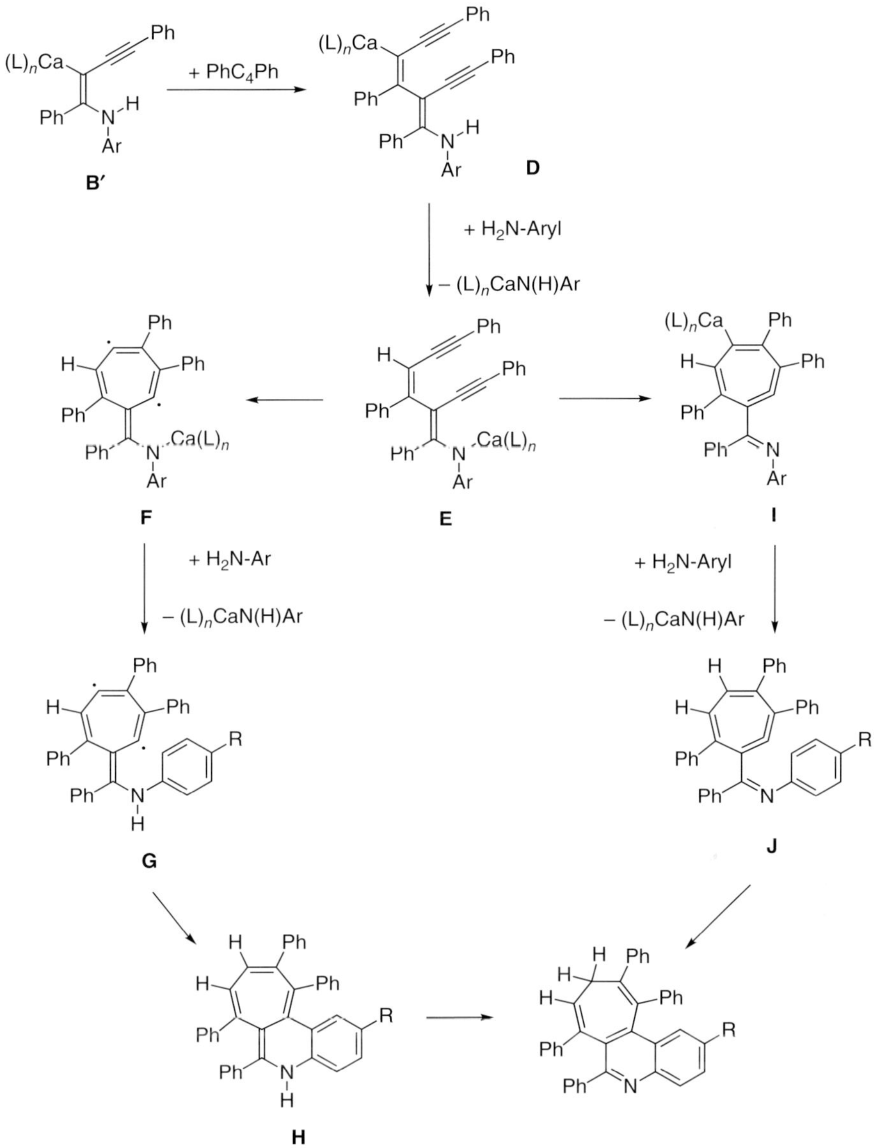

Scheme 5.5 Proposed mechanisms for the cyclization cascades via a radical mechanism (Bergman-type cyclization, left) or via acycloheptatetraene intermediate (right).

Enhancement of bulkiness by *iso*propyl groups in ortho-positions of the *N*-bound aryl groups hinders the reaction of **G** and **J** with another equivalent of 2,6-diisopropylaniline. This situation leads to an intramolecular relaxation and the product depicted in Eq. (5.14) is formed leading to ring expansion of the *N*-bound Dipp substituent.

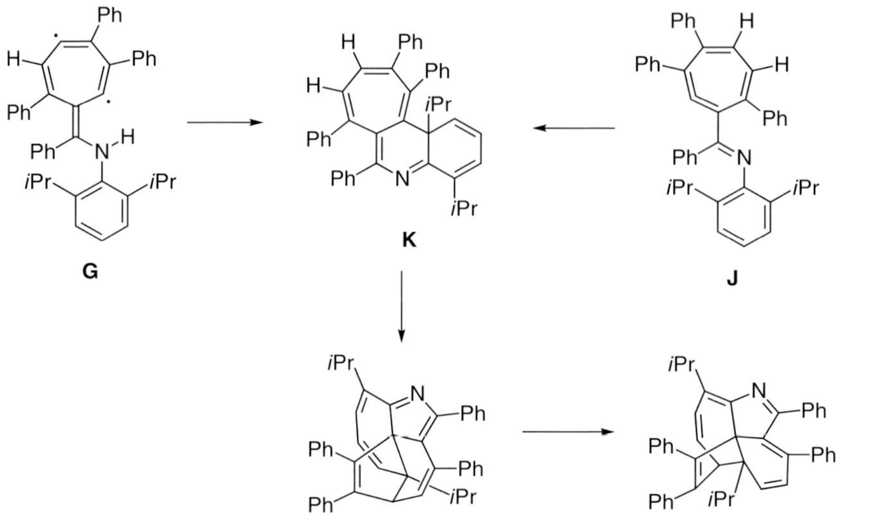

(5.14)

A subsequent slow rearrangement process finally releases steric strain and yields the end product within several days as shown in the reaction sequence (5.14). This final stabilization process can be followed by nuclear magnetic resonance (NMR) spectroscopy.

5.3 Hydrophosphanylation (Hydrophosphination)

In general, hydrophosphanylation proceeds less troublesome than the addition of amines across alkynes. Already in 1966, it has been reported that the stoichiometric reaction of $LiPPh_2$ with diphenylacetylene gives highly colored solutions in the absence of protic reagents [24]. However, in the presence of primary or secondary amines, vinylphosphanes are obtained with good yields. It is remarkable that the presence of secondary amines (diethylamine or *N*-methylaniline) leads to preferred formation of the *E*-isomer with a yield of 90%, whereas primary amines (e.g. butylamine) allow the isolation of the *Z*-isomer with a yield of 80% according to Eq. (5.15).

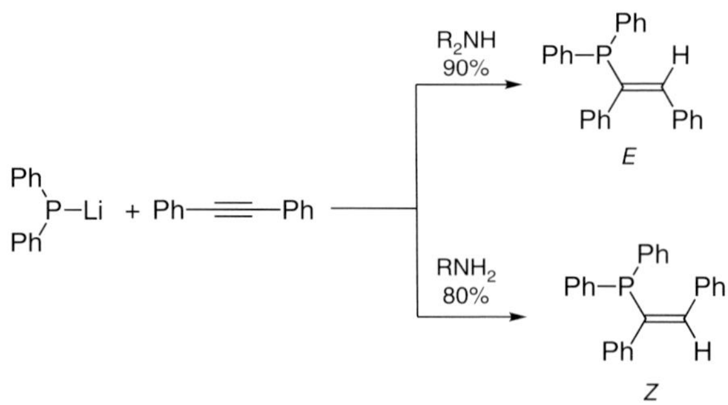

(5.15)

A catalyst- and solvent-free hydrophosphanylation of alkenes and alkynes has been reported very recently [25]. In a typical protocol, equimolar amounts of diphenylphosphane and of arylalkynes as well as alkylacetylenes have been reacted in an argon atmosphere overnight at 70 °C with isolated yields typically being above 60%. In this procedure, the anti-*Markovnikov* product $R(H)C{=}C(H)PPh_2$ of the alkenyl-diphenylphosphanes forms. More moderate reaction conditions can be achieved using metal-based catalysts for the hydrophosphanylation of alkynes. The air-stable borane adducts of secondary diaryl- and diarylphosphanes can be deprotonated with sodium hydride and added at room temperature to internal alkynes such as diarylacetylene and alkylarylacetylene in dimethylacetamide. After quenching with aqueous 5M NH_4Cl, the alkenylphosphanes are commonly obtained with excellent yields as shown in Eq. (5.16). Again, anti-*Markovnikov* products are isolated with the major component being the *E*-isomer.

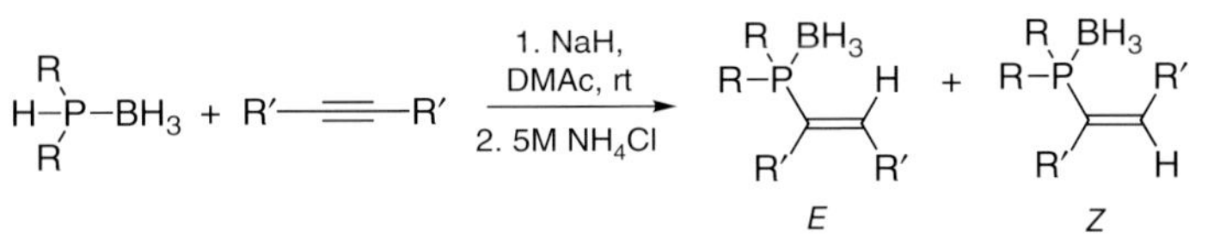

(5.16)

The reaction of 2 equiv of diphenylphosphane with 1 equiv of diarylacetylene in the presence of 0.1 equiv KO*t*Bu at room temperature in THF yields the *E*-isomer of alkenyl-diphenylphosphane and various amounts of 1,2-bis(diphenylphosphanyl)-1,2-diarylethane (Eq. (5.17)). There has been no evidence for the formation of the *Z*-isomer of alkenyl-diphenylphosphane [26]. For R = R′ = Ph, a ratio of 27 : 73 has been determined for the singly and doubly hydrophosphanylated products, whereas for the *meta*-tolyl substituents at the alkyne, the bis(diphenylphosphanyl)ethane derivative is formed in more than 95% yield.

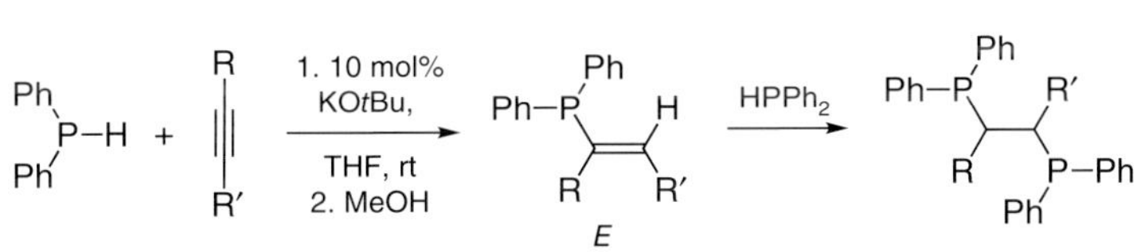

(5.17)

The accessibility of calcium complexes has evoked investigations with respect to their catalytic activity in hydrophosphanylation reactions. The straightforward strategy is to employ $[(thf)_4Ca(PPh_2)_2]$ in the addition of diphenylphosphane across alkynes because this calcium complex is soluble in ethereal solvents, easily accessible and monomeric in solution [27]. This complex (6 mol%) catalyzes the hydrophosphanylation of diphenylacetylene with diphenylphosphane at room temperature in THF and a complete conversion is achieved in less than two hours [28]. In a similar procedure, 1-phenylpropyne is hydrophosphanylated according to Eq. (5.18) giving only one regioisomer [29].

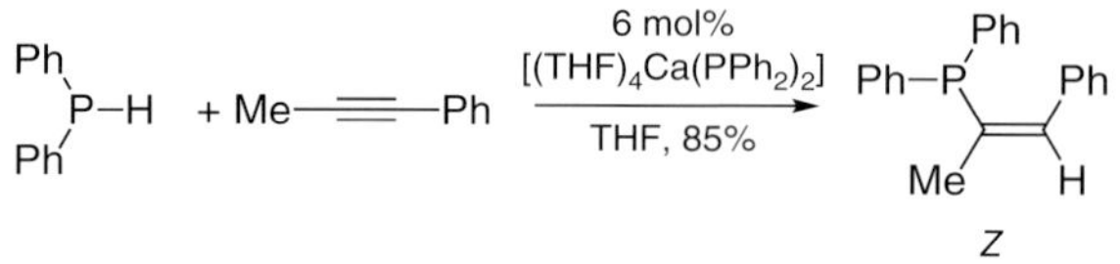

(5.18)

Calcium complexes with oligodentate anionic aza-ligands also proved to represent valuable catalyst systems for the hydrophosphanylation of alkynes [30, 31]. Surprisingly, the nature of the catalyst and different reaction conditions strongly influence the ratio of *E*- and *Z*-isomers as depicted in Scheme 5.6. Hu and Cui [31] have shown in their study that in an initial reaction, diphenylphosphane is deprotonated by the bis(trimethylsilyl)amido anion, yielding the phosphanido–calcium complex and hexamethyldisilazane. Furthermore, the calcium complex **B** with the tridentate aza-ligand generally favors formation of *Z*-isomeric alkenyl-diphenylphosphanes, whereas the similar ytterbium complex shifts the ratio in most cases in favor of the *E*-isomer.

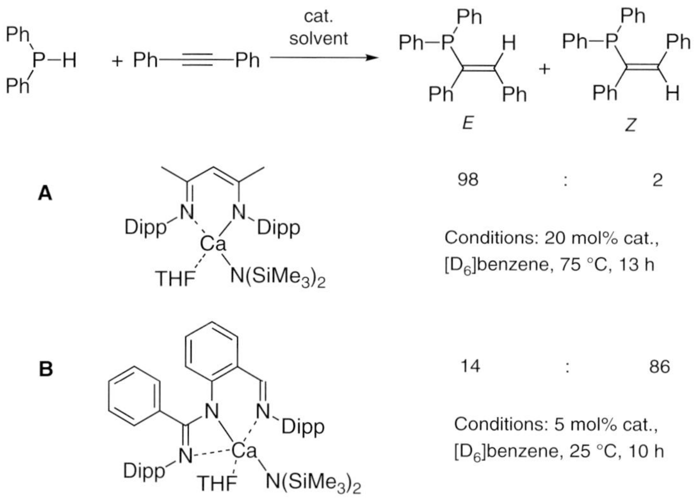

Scheme 5.6 Hydrophosphanylation of diphenylacetylene with diphenylphosphane in [D$_6$]benzene employing calcium-based catalysts with oligodentate aza-ligands, leading to significantly different *E*/*Z*-isomer distributions.

Primary phenylphosphane can also be added to diphenylacetylene using a calcium-based catalyst carrying amidinato and bis(trimethylsilyl)amido ligands [32]; again, *E*-isomeric alkenyl-diphenylphosphane is the major product.

The outcome of the hydrophosphanylation of disubstituted butadiynes with diphenylphosphane in THF, mediated by catalytic amounts of [(thf)$_4$Ca(PPh$_2$)$_2$], strongly depends on the substituents of the butadiyne. In all cases, a twofold addition and the formation of bis(diphenylphosphanyl)butadienes with different constitutions are observed. A straightforward conversion to the 1,4-bis(diphenylphosphanyl)buta-1,3-dienes is found for dimethyl- and dimesitylbutadiyne as shown in Eq. (5.19); however, a mixture of all possible stereoisomers is observed. For the catalytic hydrophosphanylation of other butadiynes (R = *t*Bu, Ph, and $SiMe_3$) with $HPPh_2$ also, 1,2-bis(diphenylphosphanyl)buta-1,3-diene and 1,4-bis(diphenylphosphanyl)-buta-1,2-diene are obtained. Many of the molecular structures of the products have been authenticated by X-ray diffraction methods, allowing the assignment of the NMR resonances and the quantification of the product distribution.

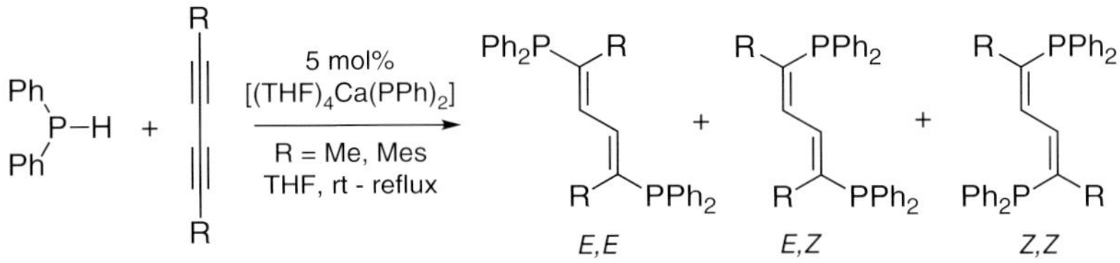

(5.19)

In order to explain the diversity of the hydrophosphanylation products, a mechanism has been proposed according to Scheme 5.7. Quantum chemical studies verify that liberation of one thf ligand to create a vacant coordination site only costs 25 kJ/mol [29]. Addition of the Ca—P bond across an alkyne moiety yields anti-Markovnikov intermediate **A** with a very reactive and heteropolar Ca—C bond allowing isomerization via cumulene intermediates **B** and **D**. Protonation with diphenylphosphane (formation of **A′**, **B′**, **C′**, and **D′**) regains the catalyst $Ca(PPh_2)_2$. A subsequent addition to another unsaturated unit leads to the formation of various products shown in the right column of Scheme 5.3. The cumulene intermediates **B′** and **D′** show two pathways depending on the attack of the catalyst at the middle or the outer C=C double bond. The cumulene intermediates **B** and **D** allow isomerization processes as also discussed for the hydroamination reactions.

According to this mechanism and in analogy to the finding that organoytterbium complexes promote hydrophosphanylation of cumulenes [33], the catalyzed addition of secondary phosphanes to carbodiimides has also been studied (Eq. (5.20)). Small amounts (1–3 mol%) of alkali metal bis(trimethylsilyl)amides are sufficient to promote this addition reaction at diisopropylcarbodiimide at room temperature [34]. Also, the alkaline earth metal bis[bis(trimethylsilyl)amides] of calcium, strontium, and barium have been tested as catalysts (1.5–3 mol%) for the conversion of carbodiimides at 25 °C in [D_6]benzene according to Eq. (5.20) [35]. With increasing size and electropositive character of the alkaline earth metal, the reaction proceeds faster. In general, the conversion is quantitative at room temperature within 0.5 up to four hours (R = *i*Pr, *c*Hex); however, the shielded di(*tert*-butyl)carbodiimide (R = *t*Bu) shows no reaction even at 60 °C for 16 hours. Slight reduction of the bulkiness of the *N*-bound alkyl groups and use of Et-N=C=N-*t*Bu with 5 mol% of $[(thf)_2Sr\{N(SiMe_3\}_2]$ again lead to complete conversion at room temperature. Diarylcarbodiimides seem to be less accessible to hydrophosphanylation and a yield of 68% is obtained for di(*para*-tolyl)carbodiimide at room temperature after 12 hours, using 2 mol% of $[(thf)_2Sr\{N(SiMe_3\}_2]$. Contrary to this finding,

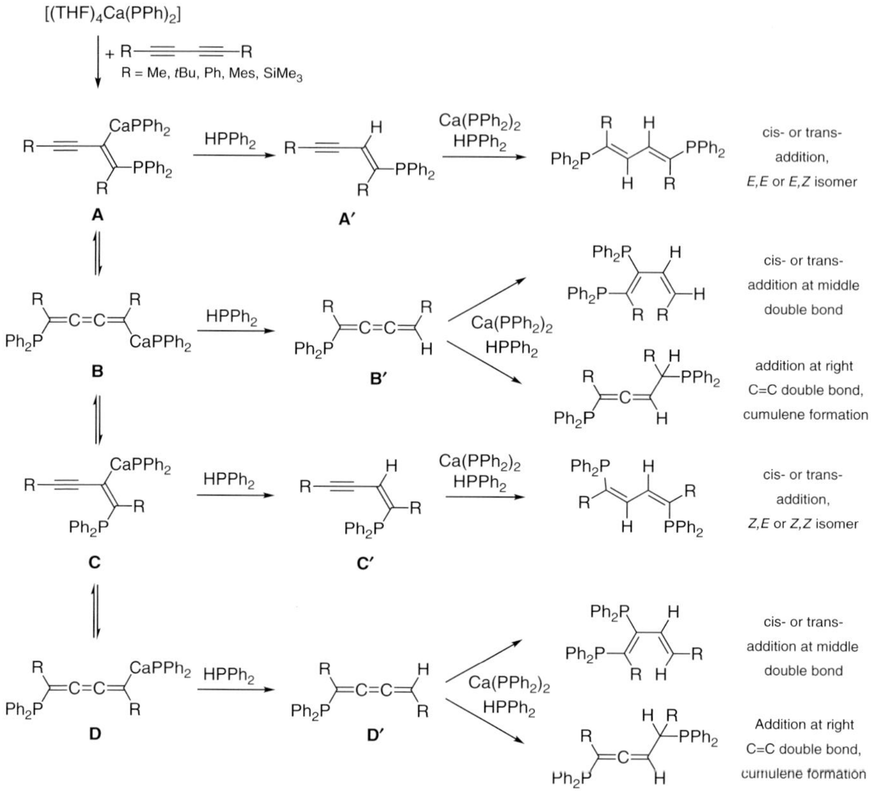

Scheme 5.7 Proposed mechanism of the hydrophosphanylation of butadiyne derivatives with diphenylphosphane catalyzed by $[(thf)_4Ca(PPh_2)_2]$. After cis-addition of a Ca—P bond to an alkyne fragment, isomerization of the intermediate is shown in the left column. The middle column contains the protonated intermediates that yield the final products after another hydrophosphanylation step.

the phospha(III)guanidinates of calcium are very poor catalysts and incomplete hydrophosphanylation of diisopropyl- and di(cyclohexyl)carbodiimide is observed at room temperature in THF [36].

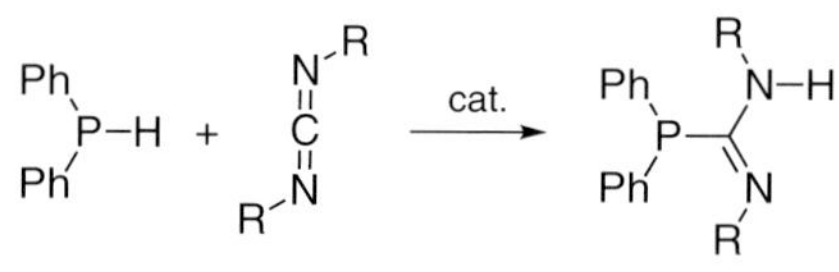

(5.20)

5.4 Hydrophosphorylation and Hydrophosphonylation

Oxidation of phosphanes to phosphane oxides and sulfides enhances the acidity of the P-bound hydrogen atom. A twofold microwave-assisted addition of diphenylphosphane oxide across R—C≡C—R can be achieved without solvent and in the absence of catalysts at 120 °C for R = COOMe [37], whereas conversion of diphenylacetylene (R = Ph) and 4-octyne (R = *n*Pr) with $Ph_2P(O)H$ requires metal catalysis. Hydrothiophosphanylation of R—C≡C—H also yields the addition products at 70 °C in air overnight under catalyst-free and solvent-free conditions [25] as depicted in Eq. (5.21). Typical yields are above 90% (R = Ph, C_6H_4–2-Cl, C_6H_4–4-OMe, C_6H_4–4-CO_2Me, and *n*Hex) with the exception of *para*-tolyl (78%), *para*-trifluoromethylphenyl (30%), and cyclohexyl substituted acetylenes (60%). The initial step is the oxidation of diphenylphosphane with sulfur and the subsequent addition of thus formed diphenylphosphane sulfide to the alkyne derivatives.

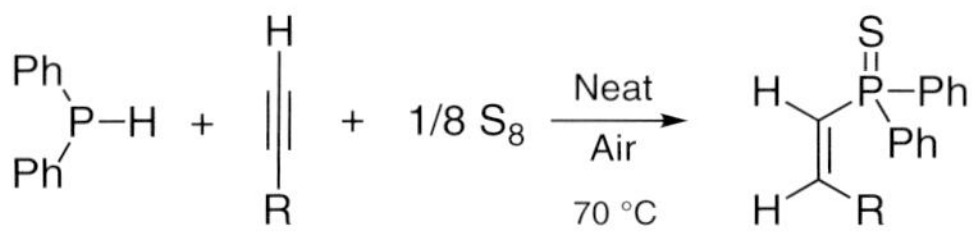

(5.21)

Twofold hydrophosphorylation of *n*-alkyl- and arylacetylenes with diphenylphosphane oxide can be achieved using catalytic amounts of LiO*t*Bu as shown in Eq. (5.24) [38]. This catalysis also proceeds in the presence of a radical scavenger supporting an ionic reaction mechanism. Sodium and KO*t*Bu also represent highly active catalysts, whereas K_2CO_3 and $NaHCO_3$ proved to be unable to promote this twofold hydrophosphorylation reaction. Heterogeneous reaction conditions using solid KOH/Al_2O_3 catalysts also promote a double addition of diethyl phosphonate at ethylpropiolate at room temperature within five minutes [39] as depicted in Eq. (5.23); however, both $(EtO)_2P(O)$ groups bind to the same carbon atom.

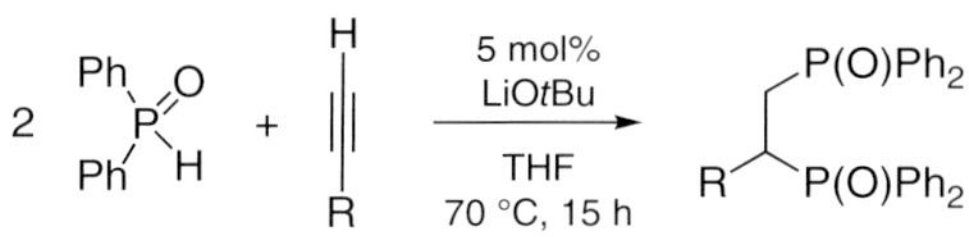

(5.22)

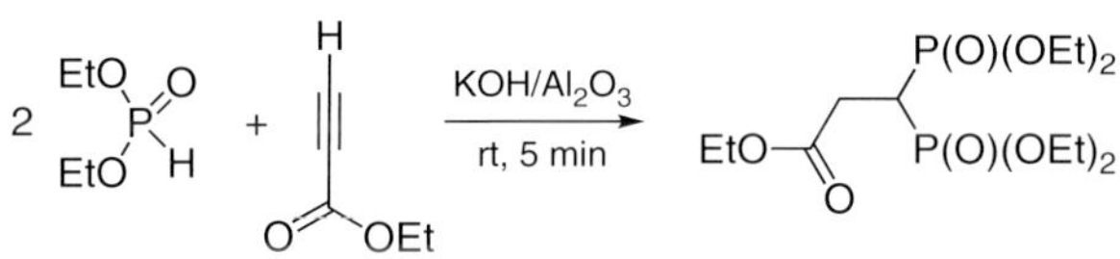

(5.23)

A comparison of the reactivities of diphenylphosphane, diphenylphosphane oxide, and sulfide as well as diethyl phosphonate in the addition of the P—H bond across the alkyne functionality catalyzed with KOH verifies an increasing reactivity in the following order [40]:

$(EtO)_2P(O)H < Ph_2P(O)H < Ph_2P(S)H < Ph_2PH$

Furthermore, Koenig and coworkers have also demonstrated that anionic and radical activation is suitable to promote the addition of P—H bonds onto alkynes. In the presence of oxygen, a radical mechanism seems to be very plausible. Oxophosphorylation of arylacetylenes with 2 equiv of diphenylphosphane oxide is performed in the presence of 10–20 mol% of a base (AOH, A_2CO_3, Et_3N; A = alkali metal) in an oxygen atmosphere at 60 °C in THF, DMSO, or a solvent mixture of water and dimethylformamide (DMF). The highest yields are achieved in solvent mixtures of H_2O and DMF with a ratio of 1 : 8 at approximately 60 °C with catalytic amounts of LiOH according to Eq. (5.24) [41]. The performance of this reaction in THF at 50 °C shows that LiOH gives higher yields than the heavier alkali metal hydroxides, whereas the use of catalytic amounts of alkali metal carbonates in a THF solution at 50 °C gives yields of only 12–16%.

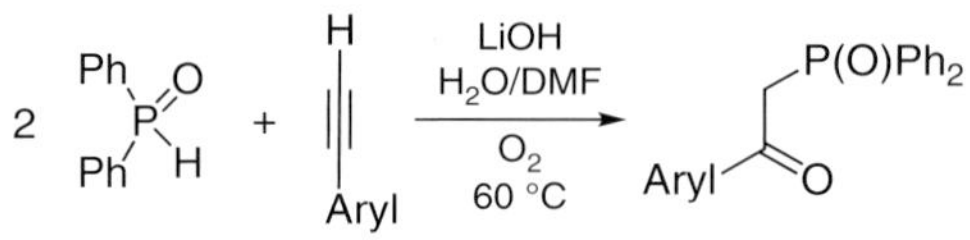

(5.24)

The mechanistic proposal is summarized in Scheme 5.8. Oxygen oxidizes diphenylphosphane oxide forming the radical cation $[Ph_2P\text{-}OH]^+$ and a superoxide ion O_2^-. After deprotonation, the radical $[Ph_2PO]$ adds onto the alkyne that

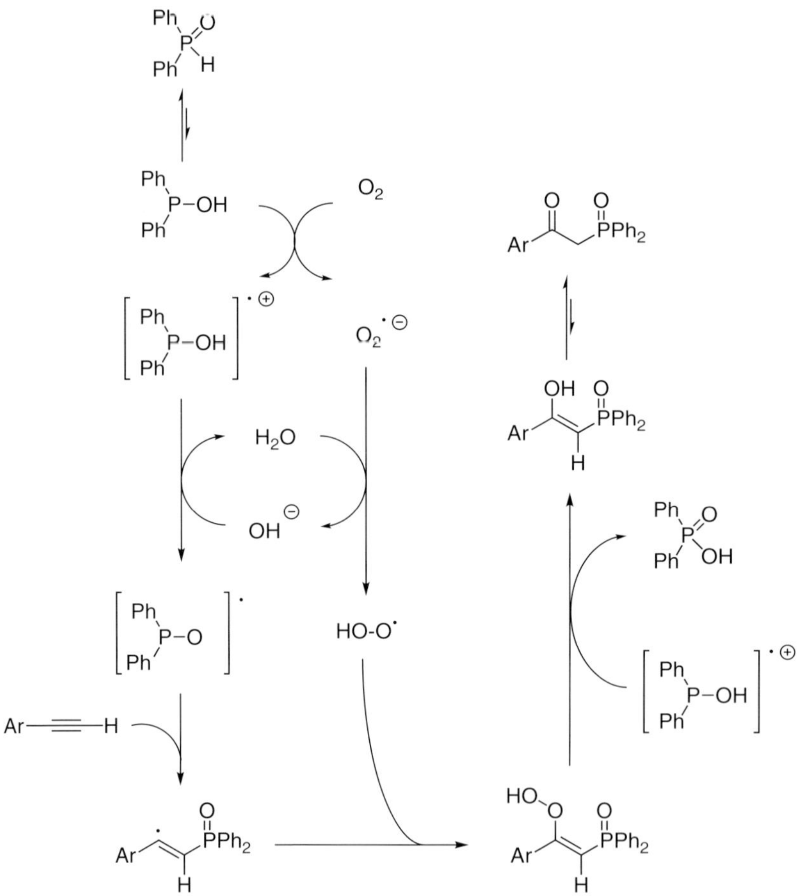

Scheme 5.8 Proposed mechanism for the base-catalyzed oxophosphosphorylation of arylacetylenes in aqueous media, assuming radical intermediates. After oxidation and deprotonation, the diphenylphosphinite radical adds onto the alkyne functionality, which is trapped by the HO_2 radical, finally yielding the product 2-diphenylphosphoryl-1-arylethanone and the by-product diphenylphosphinic acid.

is trapped by the intermediate HO—O radical. A final tautomerization gives the isolated product as shown in Eq. (5.24). This radical mechanism is supported by several experiments. In the presence of radical scavengers or in the absence of oxygen, no conversions are observed. In the absence of bases or without water, the yields drop dramatically. The use of labeled $^{18}O_2$ mainly yields compounds with labeled carbonyl groups, whereas labeled water $H_2{}^{18}O$ only leads to the formation of $Ph_2P(O)^{18}OH$.

Deprotonation of diarylphosphane oxides yields the diarylphosphinites of the type $Ar_2P—O—M$ with the metal ions bound at the oxygen atom [42–44], making the electron pair at the phosphorus atom available for a nucleophilic attack at unsaturated systems. Potassium dimesitylphosphinite acts as a catalyst in the addition of dimesitylphosphane oxide onto alkynes (Eq. (5.25)) [45].

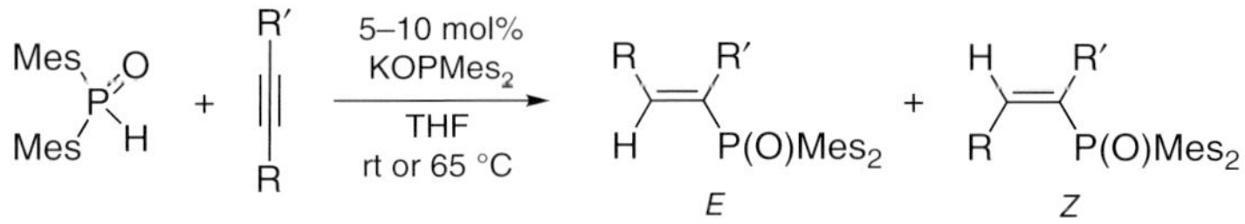

(5.25)

The success of this catalytic conversion strongly depends on the substituents R and R′. Phenyl, ester, and triisopropylsilyl groups facilitate the hydrophosphorylation of alkynes, whereas alkyl and benzyl groups suppress this reaction. If R is an aryl group, *E/Z*-isomerization occurs quite easily in solution, hampering the isolation of one specific isomer. This finding can be explained by charge delocalization, weakening the C=C double bond of the alkenyl moiety (Scheme 5.9).

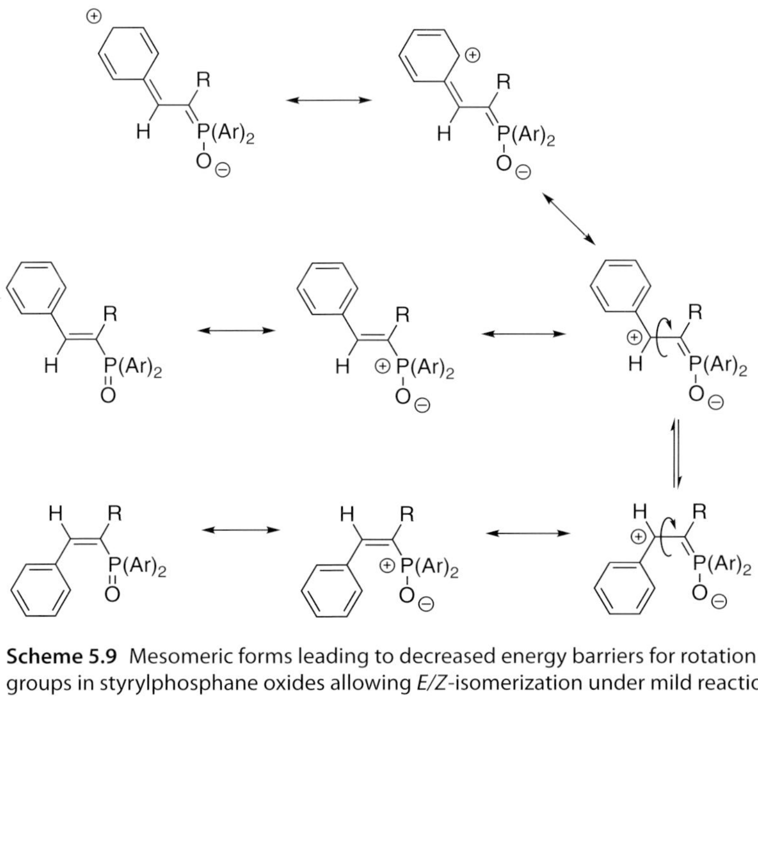

Scheme 5.9 Mesomeric forms leading to decreased energy barriers for rotation of the alkenyl groups in styrylphosphane oxides allowing *E/Z*-isomerization under mild reaction conditions.

With R = $SiMe_3$, consumption of the substrates can be monitored by ^{31}P NMR spectroscopy; however, the expected hydrophosphorylation product is not isolated, but its degradation products R′—C≡C—H and Mes_2P—O—$SiMe_3$ according to Eq. (5.26) [45].

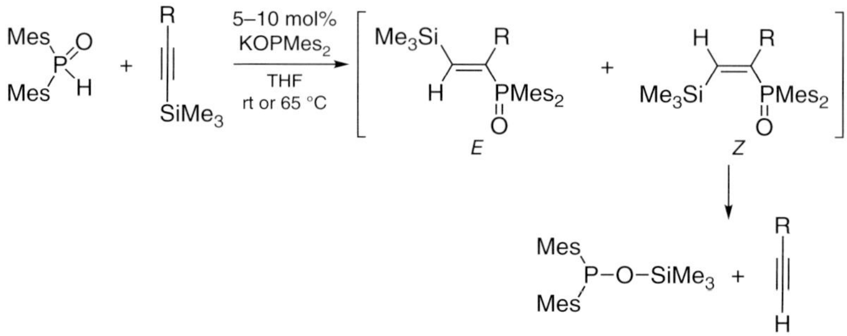

(5.26)

The hydrophosphorylation of organic *iso*(thio)cyanates R—N=C=E (R = alkyl, aryl; E = O, S) with diphenyl- and dimesitylphosphane oxide is mediated by calcium- [44, 46] and potassium-based catalysts [43, 44], as shown in Eq. (5.27). Because of the fact that $Ph_2P(O)H$ and $Ca(PPh_2)_2$ form a protonation/deprotonation equilibrium with Ph_2PH and $Ca(OPPh_2)_2$ [46], a catalyst system containing the phosphinite anions seems to be advantageous to avoid competition between hydrophosphanylation and hydrophosphorylation of the cumulenes.

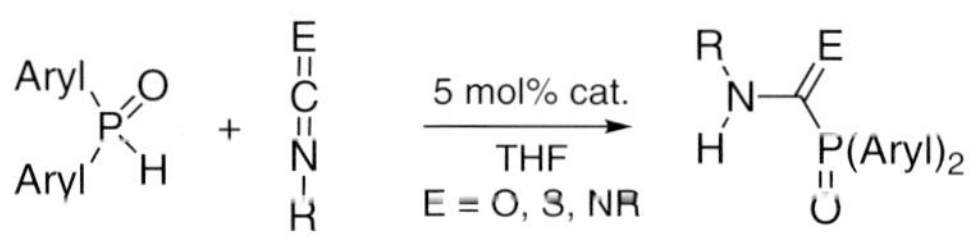

(5.27)

Several factors influence the efficiency of the catalyst in these hydrophosphorylation processes. Diphenylphosphane oxides add readily onto heterocumulenes, whereas the addition of dimesitylphosphane oxide is significantly slower, allowing kinetic studies [43]. Thus, diphenylphosphane oxide reacts with diisopropylcarbodiimide within half an hour in THF using 5 mol% of potassium diarylphosphinite, whereas dimesitylphosphane oxide is inert toward this substrate under similar reaction conditions. Furthermore, the influence of the hardness of the catalyst has been investigated. Potassium ions are significantly softer than the isoelectronic calcium ions. The solid-state structures differ significantly. The calcium complex crystallizes monomeric as $[(thf)_4Ca\{O\text{-}P(Ar)_2\}_2]$ [42], whereas the potassium congener precipitates as a tetranuclear cage compound of the type $[(thf)K\{O\text{-}P(Ar)_2\}]_4$ with central K_4O_4 heterocubane units [43]. Therefore, both complexes have been reacted with 18-crown-6 ether to stabilize mononuclear complexes of the type $[(18C6)Ca\{O\text{-}P(Ar)_2\}_2]$ and $[(18C6)K\{O\text{-}P(Ar)_2\}]$ [44]. In addition and based on the experience that heterobimetallic potassium calciates have shown an enhanced activity in hydroamination reactions, also the heterodinuclear calciate $[(L)K_2Ca\{O\text{-}P(Ar)_2\}_4]$ (L = thf, $Ar_2P(O)H$) has been included in these hydrophosphorylation experiments. In this study, the potassium complexes show a higher reactivity than the calcium congeners. Furthermore, the coordination of 18-crown-6 reduces the reactivity significantly because the hemilabile thf ligands are substituted by a tightly bound hexadentate ether shielding the metal centers much more effectively. Very similar catalytic activities have been observed for the calcium and calciate complexes $[(thf)_4Ca\{O\text{-}P(Ar)_2\}_2]$ and $[(L)K_2Ca\{O\text{-}P(Ar)_2\}_4]$. Regardless of the catalyst, an initial induction period has been observed. This finding suggests that the catalytic species forms during this time from the precatalyst systems and a breakdown of the tetranuclear potassium complex has been proposed for $[(thf)K\{O\text{-}P(Ar)_2\}]_4$ [44].

Challenges arise from the tendency of the phosphinite complexes to degrade into the diarylphosphanide and diarylphosphinate ions as depicted in Eq. (5.28) [43, 44]. In contrast to the phosphanides $M—P(Ar)_2$, which are soluble in ethereal solvents (and can be monitored by ^{31}P NMR spectroscopy), the phosphinates $M—O—P(O)(Ar)_2$ are only very sparingly soluble because these latter anions occupy bridging positions between the metal ions leading to large aggregates. Quenching of the reaction solution of diphenylphosphane sulfide with dicyclohexylcarbodiimide, catalyzed with 5 mol% of potassium diphenylthiophosphinite, shows resonances for $HPPh_2$, $HP(S)Ph_2$, and $HS\text{-}P(S)Ph_2$ besides the product in the ^{31}P NMR spectrum.

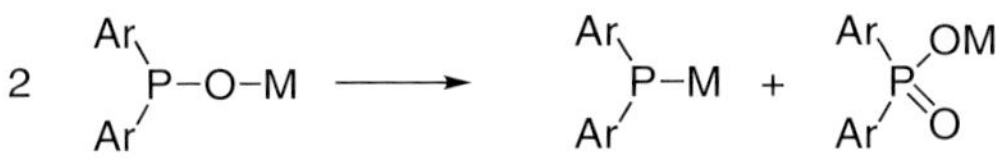

(5.28)

These experiments show that hydrophosphorylation reaction require active catalysts that must be accessible by the substrates. Shielding by bulky *P*-bound aryl groups or tightly bound multidentate ether bases hinders the catalytic conversion. Furthermore, electron-withdrawing groups at the multiple bond systems raise the conversion rates.

5.5 Summary and Conclusions

Hydrofunctionalization (hydroelementation) of alkynes and other π-systems is generally acid, base, or metal catalyzed; however, in selected cases, radical mechanisms are also discussed. Diverse metal catalysts represent suitable activators for the π-systems and/or amines, phosphanes, phosphane oxides, and sulfides. In a typical catalytic cycle (Scheme 5.10), several reaction steps finally yield the addition products. The initial reaction is the attack of the amide (hydroamination), phosphanide (hydrophosphanylation and hydrophosphination), or phosphinite (hydrophosphorylation) at the alkyne or cumulene, leading to an intermediate with a highly reactive metal–carbon bond. This organometallic complex deprotonates the protic substrate (amine, phosphane, phosphane oxide, and sulfide) regaining the catalyst and forming the product.

Amines are strong bases and very weak Brønsted acids with pK_a values strongly depending on solvent (solvation effects) and metal ions [47];

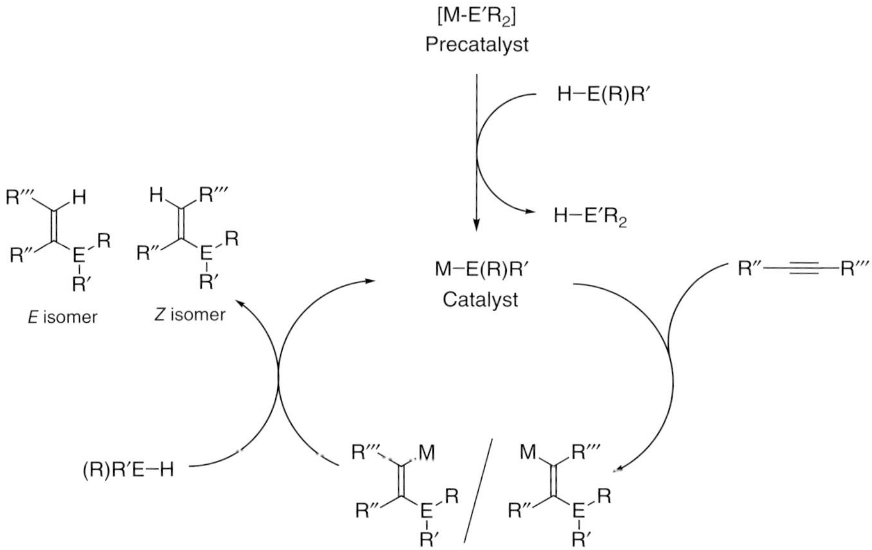

Scheme 5.10 Generic catalytic cycle for the hydrofunctionalization (hydroelementation) of alkynes with substrates containing H—E bonds [E, E′ = N (hydroamination), P (hydrophosphanylation), PO (hydrophosphorylation), and PS (hydrothiophosphorylation)].

nevertheless, characteristic pK_a values lie around 30. Substitution of N by the homologous element P enhances the acidity significantly. Thus, aniline and *N*-methylaniline exhibit pK_a values of 30.6 and 29.5, respectively, whereas phenyl- and *P*-methyl-phenylphosphane have pK_a values of 22.4 and 26.7 [48]. Further acidity enhancement is observed for the phosphane oxides (pK_a values for $Ph_2P(O)H$ 20.6, Ph(Me)P(O)H 23.9, and $Me_2P(O)H$ 26.9) [48] and sulfides (pK_a value of $Ph_2P(S)H$ 12.8) [40]. For catalytic hydrophosphanylation reactions, increased Brønsted acidity accelerates the protonation step, but reduced Lewis basicity requires more drastic conditions for the nucleophilic attack at the multiple bond systems.

Diverse factors influence the accomplishment of the s-block metal-mediated hydrofunctionalization reactions of alkynes and heterocumulenes such as carbodiimides and *iso*(thio)cyanates.

(i) The *stability of the catalysts* is guaranteed for the s-block metal amides and phosphanides allowing the preparation of stock solutions of these reagents. The alkali and alkaline earth metal phosphinites are significantly less stable and disproportionate into soluble phosphanide and sparingly soluble phosphinate that precipitates and draws the equilibrium toward the degradation products. Therefore, freshly prepared phosphinite solutions give the best results in hydrofunctionalization reactions.

(ii) The *metal ions* play a crucial role because charge and radius determine their hardness and hence electrostatic attraction between these ions and the Lewis basic π-systems. Furthermore, the metal–carbon bonds of the intermediately formed organometallics are extremely reactive and deprotonate the substrate.

(iii) *Solvent* plays an important (and often underestimated) role, which has to act as a donor (Lewis base) to the metal ions ensuring solubility in organic solvents; however, these bases must be hemilabile to open coordination sites for binding and activation of substrates. Multidentate Lewis bases like 18-crown-6 ethers block the accessibility of the metal ions.

(iv) The hydroelementation reaction at alkynes is *regioselective* and generally yields anti-Markovnikov products. However, this reaction lacks stereoselectivity and *E*- and *Z*-isomers are obtained. Often, subsequent *E/Z*-isomerization processes impede separation and isolation of specific isomers. After addition of an E—M bond of the catalytically active species, M—ER_2 across a C≡C triple bond of butadiynes, cumulene intermediates of the type R′(R_2E)C=C=C=C(M)R′ interconvert the *E*- and *Z*-isomers of R′(R_2E)C=C(M)—C=C—R′, which are protonated thereafter by H—ER_2.

(v) The *substituents at the substrates* (protic component and π-bases) exert influence not only via their electronic properties (electron pushing and withdrawing effects, inductive properties) but also the bulkiness of the groups strongly affects the rate of conversion. Heavily shielded substrates can even completely suppress hydrofunctionalization reactions.

(vi) The *rate-determining step* is the approach of an anionic species (amide, phosphanide, phosphinite, or thiophosphinite) to an electron-rich π-system of alkenes, alkynes, cumulenes, and the heteroatom-substituted congeners. The protonation of the thus intermediately formed organometallics is fast and therefore the pK_a values of the substrates play a minor role. The proton source may play a role for the ratio of the *E*- and *Z*-isomeric product distribution [24]; however, this influence has not been studied in detail for catalytic reactions.

(vii) These hydrofunctionalization reactions have to be performed in an *inert and anhydrous atmosphere* to exclude radical reactions and interference of radical and ionic reaction steps. It has been proposed that diphenylphosphane oxide reacts with oxygen in the presence of catalytic amounts of bases such as LiOH or NEt_3 yielding intermediately the proposed radical cation $[Ph_2P(OH)]^+$ and the superoxide anion [41]. In an oxygen atmosphere, the radical pathway represents the dominant reaction.

In summary, the hydrofunctionalization of π-systems (alkenes, alkynes, cumulenes, and their heteroatom-substituted congeners such as imines, nitriles, isocyanates, carbodiimides, and others) is a beneficial atom-economic reaction that can be performed with complexes containing the nontoxic and globally abundant metals such as potassium and calcium. These two essentially isoelectronic metal ions represent a soft and large alkali metal cation and a significantly harder Lewis acidic alkaline earth metal cation for catalytic requirements in interaction with ligands of different hardness and size. In general, the approach of the electron-rich substrates (amides, phosphanides, and phospinites on the one side and C≡C triple bonds on the other) represents the rate-determining step, strongly depending on steric and electrostatic repulsion, the basicity and nucleophilicity, as well as to a minor extent on pK_a values.

5.6 Acknowledgments

We thank the coworkers who developed and advanced the fascinating field of hydrofunctionalization of π-systems with environmentally benign metal catalysts and who are cited in the reference list. Furthermore, we acknowledge the valuable support of the NMR service platform (www.nmr-jena.de) of the Faculty of Chemistry and Earth Sciences for often time-consuming NMR experiments for kinetic studies. Mass spectrometric experiments were performed by the mass spectrometry (MS) platform (www.ms.uni-jena.de).

5.7 Abbreviations

A	alkali metal, group I element
Ae	alkaline earth metal, group II element
Ar	aryl
*c*Hex	cyclohexyl
Dipp	2,6-diisopropylphenyl
DMAc	dimethylacetamide
DMF	dimethylformamide
Et	ethyl
*i*Pr	*iso*propyl
M	metal without specification to a certain group
Me	methyl
Mes	2,4,6-trimethylphenyl, mesityl
*n*Hex	*n*-hexyl
*n*Pr	*n*-propyl
Ph	phenyl
*t*Bu	*tert*-butyl
THF/thf	tetrahydrofuran (solvent/ligated molecule)

References

1 Togni, A. and Grützmacher, H. (eds.) (2001). *Catalytic Heterofunctionalization: From Hydroamination to Hydrozirconation*. Weinheim, Germany: Wiley-VCH.

2 Krieck, S. and Westerhausen, M. (2017). *Inorganics* 5: 17.

3 Harder, S. (2010). *Chem. Rev.* 110: 3852–3876.

4 (a) Barrett, A.G.M., Crimmin, M., Hill, M.S., and Procopiou, P.A. (2010). *Proc. R. Soc. A* 466: 927–963. (b) Crimmin, M.R. and Hill, M.S. (2013). *Top. Organomet. Chem.* 45: 191–242. (c) Hill, M.S., Liptrot, D.J., and Weetman, C. (2016). *Chem. Soc. Rev.* 45: 972–988.

5 (a) Ananikov, V.P. and Tanaka, M.T. (eds.) (2013). *Hydrofunctionalization* (Top. Organomet. Chem., vol. 43). Heidelberg: Springer. (b) Westerhausen, M., Krieck, S., Langer, J. et al. (2013). *Coord. Chem. Rev.* 257: 1049–1066. (c) Rodriguez-Ruiz, V., Carlino, R., Bezzenine-Lafollee, S. et al. (2015). *Dalton Trans.* 44: 12029–12059. (d) Trifonov, A.A., Basalov, I.V., and Kissel, A.A. (2016). *Dalton Trans.* 45: 19172–19193. (e) Bezzenine-Lafollee, S., Gil, R., Prim, D., and Hannedouche, J. (2017). *Molecules* 22: 1901/1–1901/29.

6 Recent reviews on hydroamination: (a) Reznichenko, A.L. and Hultzsch, K.C. (2013). *Top. Organomet. Chem.* 43: 51–114. (b) Reznichenko, A.L., Nawara-Hultzsch, A.J., and Hultzsch, K.C. (2014). *Top. Curr. Chem.* 343: 191–260. (c) Yim, J.C.-H. and Schafer, L.L. (2014). *Eur. J. Org. Chem.*: 6825–6840. (d) Vo, C.-V.T. and Bode, J.W. (2014). *J. Org. Chem.* 79: 2809–2815. (e) Huang, L., Arndt, M., Goossen, K. et al. (2015). *Chem. Rev.* 115: 2596–2697. (f) Coman, S.M. and Parvulescu, V.I. (2015). *Org. Process Res. Dev.* 19: 1327–1355. (g) Villa, M. and von Wangelin, A.J. (2015). *Angew. Chem. Int. Ed.* 54: 11906–11908. (h) Weiße, M., Zille, M., Jacob, K. et al. (2015). *Chem. Eur. J.* 21: 6511–6522. (i) Mahdi, T. and Stephan, D.W. (2015). *Chem. Eur. J.* 21: 11134–11142. (j) Arrowsmith, M. (2015). *The Lightest Metals: Science and Technology from Lithium to Calcium* (ed. T.P. Hanusa), 255–280. Chichester: Wiley. (k) Isaeva, V.I. and Kustov, L.M. (2016). *Top. Catal.* 59: 1196–1206. (l) Michon, C., Abadie, M.-A., Medina, F., and Agbossou-Niedercorn, F. (2017). *J. Organomet. Chem.* 847: 13–27. (m) Patel, M., Saunthwal, R.K., and Verma, A.K. (2017). *Acc. Chem. Res.* 50: 240–254. (n) Lepori, C. and Hannedouche, J. (2017). *Synthesis* 49: 1158–1167. (o) Kalck, P. and Urrutigoity, M. (2018). *Chem. Rev.* 118: 3833–3861.

7 Selected reviews on hydrophosphanylation (hydrophosphination): (a) Delacroix, O. and Gaumont, A.C. (2005). *Curr. Org. Chem.* 9: 1851–1882. (b) Waterman, R. (2009). *Dalton Trans.*: 18–26. (c) Julienne, D., Delacroix, O., and Gaumont, A.-C. (2010). *Curr. Org. Chem.* 14: 457–482. (d) Kawaguchi, S.-i. and Ogawa, A. (2013). *Synlett* 24: 2199–2215. (e) Rosenberg, L. (2013). *ACS Catal.* 3: 2845–2855. (f) Pullarkat, S.A. and Leung, P.-H. (2013). *Top. Organomet. Chem.* 43: 145–166. (g) Koshti, V., Gaikwad, S., and Chikkali, S.H. (2014). *Coord. Chem. Rev.* 265: 52–73. (h) Bange, C.A. and Waterman, R. (2016). *Chem. Eur. J.* 22: 12598–12605. (i) Gusarova, N.K., Cherysheva, N.A., and Trofimov, B.A. (2017). *Synthesis* 49: 4783–4807.

8 Selected reviews on hydrophosphorylation and hydrophosphonylation: (a) Montchamp, J.-L. (2005). *J. Organomet. Chem.* 690: 2388–2406. (b) Coudray, L. and Montchamp, J.-L. (2008). *Eur. J. Org. Chem.*: 3601–3613. (c) Albrecht, L., Albrecht, A., Krawczyk, H., and Jørgensen, K.A. (2010). *Chem. Eur. J.* 16: 28–48. (d) Demmer, C.S., Krogsgaard-Larsen, N., and Bunch, L. (2011). *Chem. Rev.* 111: 7981–8006. (e) Ali, T.E. and Abdel-Kariem, S.M. *ARKIVOC* 2015: 246–287. (f) Dondoni, A. and Marra, A. (2015). *Org. Biomol. Chem.* 13: 2212–2215. (g) Shaikh, R.S., Ghosh, I., and König, B. (2017). *Chem. Eur. J.* 23: 12120–12124. (h) Chen, L. (2018). *Synthesis* 50: 440–469.

9 Reppe, W. (1956). *Liebigs Ann. Chem.* 601: 81–138.

10 Chekulaeva, I.A. and Kondrat'eva, L.V. (1965). *Russ. Chem. Rev.* 34: 669–680.

11 (a) Tarasova, O.A., Mal'kina, A.G., Mikhaleva, A.I. et al. (1994). *Synth. Commun.* 24: 2035–2037. (b) Settambolo, R., Mariani, M., and Caiazzo, A. (1998). *J. Org. Chem.* 63: 10022–10026. (c) Verma, A.K., Joshi, M., and Singh, V.P. (2011). *Org. Lett.* 13: 1630–1633. (d) Joshi, M., Patel, M., Tiwari, R., and Verma, A.K. (2012). *J. Org. Chem.* 77: 5633–5645.

12 Tzalis, D., Koradin, C., and Knochel, P. (1999). *Tetrahedron Lett.* 40: 6193–6195.

13 Brinkmann, C., Barrett, A.G.M., Hill, M.S., and Procopiou, P.A. (2012). *J. Am. Chem. Soc.* 134: 2193–2207.

14 Glock, C., Görls, H., and Westerhausen, M. (2012). *Chem. Commun.* 48: 7094–7096.

15 Glock, C., Görls, H., and Westerhausen, M. (2011). *Dalton Trans.* 40: 8108–8113.

16 Younis, F.M., Krieck, S., Görls, H., and Westerhausen, M. (2016). *Dalton Trans.* 45: 6241–6250.

17 Ziemann, S., Krieck, S., Görls, H., and Westerhausen, M. (2018). *Organometallics* 37: 924–933.

18 Barrett, A.G.M., Boorman, T.C., Crimmin, M.R. et al. (2008). *Chem. Commun.*: 5206–5208.

19 Jia, J.-H., Yu, C., Xu, M. et al. (2015). *Synthesis* 47: 3473–3478.

20 Ranjan, A., Yerande, R., Wakchaure, P.B. et al. (2014). *Org. Lett.* 16: 5788–5791.

21 Younis, F.M., Krieck, S., Görls, H., and Westerhausen, M. (2015). *Organometallics* 34: 3577–3585.

22 Glock, C., Younis, F.M., Ziemann, S. et al. (2013). *Organometallics* 32: 2649–2660.

23 (a) Hanusa, T.P. (2000). *Coord. Chem. Rev.* 210: 329–367. (b) Westerhausen, M. (2001). *Angew. Chem. Int. Ed.* 40: 2975–2977. (c) Alexander, J.S. and Ruhlandt-Senge, K. (2002). *Eur. J. Inorg. Chem.*: 2761–2774. (d) Westerhausen, M., Gärtner, M., Fischer, R. et al. (2007). *Chem. Eur. J.* 13: 6292–6306. (e) Westerhausen, M., Gärtner, M., Fischer, R., and Langer, J. (2007). *Angew. Chem. Int. Ed.* 46: 1950–1956. (f) Westerhausen, M. (2009). *Z. Anorg. Allg. Chem.* 635: 13–32. (g) Buchanan, W.D., Allis, D.G., and Ruhlandt-Senge, K. (2010). *Chem. Commun.* 46: 4449–4465. (h) Westerhausen, M., Langer, J., Krieck, S., and Glock, C. (2011). *Rev. Inorg. Chem.* 31: 143–184. (i) Torvisco, A. and Ruhlandt-Senge, K. (2011). *Inorg. Chem.* 50: 12223–12240. (j) Westerhausen, M., Langer, J., Krieck, S. et al. (2013). *Top. Organomet. Chem.* 45: 29–72. (k) Westerhausen, M., Koch, A., Görls, H., and Krieck, S. (2017). *Chem. Eur. J.* 23: 1456–1483.

24 Aguiar, A.M. and Archibald, T.G. (1966). *Tetrahedron Lett.* 7: 5541–5546.

25 Moglie, Y., González-Soria, M.J., Martin-Garcia, I. et al. (2016). *Green Chem.* 18: 4896–4907.

26 Bookham, J.L., Smithies, D.M., Wright, A. et al. (1998). *J. Chem. Soc., Dalton Trans.*: 811–818.

27 (a) Gärtner, M., Görls, H., and Westerhausen, M. (2007). *Z. Anorg. Allg. Chem.* 633: 2025–2031. (b) Crimmin, M.R., Barrett, A.G.M., Hill, M.S. et al.

(2007). *Inorg. Chem.* 46: 10410–10415. (c) Langer, J., Al-Shboul, T.M.A., Younis, F. et al. (2011). *Eur. J. Inorg. Chem.*: 3002–3007.
28 Al-Shboul, T.M.A., Görls, H., and Westerhausen, M. (2008). *Inorg. Chem. Commun.* 11: 1419–1421.
29 Al-Shboul, T.M.A., Pálfi, V.K., Yu, L. et al. (2011). *J. Organomet. Chem.* 696: 216–227.
30 Crimmin, M.R., Barrett, A.G.M., Hill, M.S. et al. (2007). *Organometallics* 26: 2953–2956.
31 Hu, H. and Cui, C. (2012). *Organometallics* 31: 1208–1211.
32 Basalov, I.V., Yurova, O.S., Cherkasov, A.V. et al. (2016). *Inorg. Chem.* 55: 1236–1244.
33 Takaki, K., Koshoji, G., Komeyama, K. et al. (2003). *J. Org. Chem.* 68: 6554–6565.
34 (a) Zhang, W.-X., Nishiura, M., and Hou, Z. (2006). *Chem. Commun.*: 3812–3814. (b) Zhang, W.-X., Nishiura, M., Mashiko, T., and Hou, Z. (2008). *Chem. Eur. J.* 14: 2167–2179.
35 Crimmin, M.R., Barrett, A.G.M., Hill, M.S. et al. (2008). *Organometallics* 27: 497–499.
36 Al-Shboul, T.M.A., Volland, G., Görls, H., and Westerhausen, M. (2009). *Z. Anorg. Allg. Chem.* 635: 1568–1572.
37 Stone, J.J., Stockland, R.A., Reyes, J.M. et al. (2005). *J. Mol. Catal. A Chem.* 226: 11–21.
38 Yoshimura, A., Saga, Y., Sato, Y. et al. (2016). *Tetrahedron Lett.* 57: 3382–3384.
39 Albouy, D., Laspéras, M., Etemad-Moghadam, G., and Koenig, M. (1999). *Tetrahedron Lett.* 40: 2311–2314.
40 Semenzin, D., Etemad-Moghadam, G., Albouy, D. et al. (1997). *J. Org. Chem.* 62: 2414–2422.
41 Zhong, W., Tan, T., Shi, L., and Zeng, X. (2018). *Synlett* 29: 1379–1384.
42 Al-Shboul, T.M.A., Volland, G., Görls, H. et al. (2012). *Inorg. Chem.* 51: 7903–7912.
43 Härling, S.M., Görls, H., Krieck, S., and Westerhausen, M. (2016). *Inorg. Chem.* 55: 10741–10750.
44 Härling, S.M., Krieck, S., Görls, H., and Westerhausen, M. (2017). *Inorg. Chem.* 56: 9255–9263.
45 Härling, S.M., Fener, B.E., Krieck, S. et al. (2018). *Organometallics* 37 https://doi.org/10.1021/acs.organomet.8b00368.
46 Härling, S., Greiser, J., Al-Shboul, T.M.A. et al. (2013). *Aust. J. Chem.* 66: 1264–1273.
47 See for example: (a)Velinov, G. and Georgieva, M. (1982). *Anal. Chim. Acta* 140: 339–341. (b) Ahlbrecht, H. and Schneider, G. (1986). *Tetrahedron* 42: 4729–4741. (c) Streitwieser, A., Facchetti, A., Xie, L. et al. (2012). *J. Org. Chem.* 77: 985–990.
48 Li, J.-N., Liu, L., Fu, Y., and Guo, Q.-X. (2006). *Tetrahedron* 62: 4453–4462.

6

Early Main Group Metal-Catalyzed Hydrosilylation of Unsaturated Bonds

Sjoerd Harder

University Erlangen-Nürnberg, Inorganic and Organometallic Chemistry, Egerlandstrasse 1, 91058 Erlangen, Germany

6.1 Introduction

Hydrosilylation, the formal addition of a Si—H bond across an unsaturated bond, is without doubt the most important reaction in silicon chemistry. It converts a large variety of π-bonds, including C=C, C≡C, C=O, C=N, or C≡N, in valuable industrial products ranging from bulk materials such as silicone polymers, rubbers, lubricants, or coatings to fine chemicals such as pharmaceuticals [1]. Given this industrial importance, from the early 1960s onward, academia and industry embarked on a long journey to explore hydrosilylation catalysis. The results of these studies have been captured in many review articles and book chapter contributions. For further background information, articles that present the most recent progress are recommended [2–8]. The enormous wealth of information and the many different mechanisms that have been proposed demonstrate that hydrosilylation catalysis has many different faces. This is not only because of a large variety of different catalysts but also because of the fact that the hydrosilylation mechanism is strongly dependent on the nature of the substrate, catalyst, and silane. First catalysts were based on Pt, but over the years, a variety of transition metal complexes were shown to be active. Most recently, advancement has been made in Lewis acid and s-block metal-catalyzed hydrosilylation. This chapter starts with the history of hydrosilylation catalysis, gives a short overview of general hydrosilylation catalysis, and continues with latest developments in the search for cheaper and more abundant catalysts, followed by a systematic survey of alkene, ketone, and imine hydrosilylation using s-block metal catalysts.

6.2 Historical Development

The first catalytic alkene hydrosilylation was simply catalyzed by a radical source such as AcOOAc [9]. Its proposed mechanism is relatively straightforward, including the formation of a radical Si species followed by addition to an alkene (Scheme 6.1). This classical radical process also led, especially for activated

Early Main Group Metal Catalysis: Concepts and Reactions,
First Edition. Edited by Sjoerd Harder.

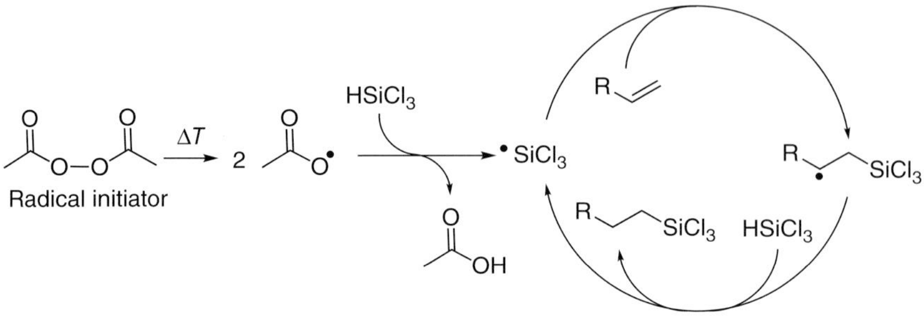

Scheme 6.1 First alkene hydrosilylation by a radical mechanism.

alkenes such as styrene, to alkene polymerization. A major breakthrough was achieved in 1957 by the introduction of Speier's hexachloroplatinic acid catalyst $[PtCl_6^{2-}][H_3O^+]_2$ [10]. The latter catalyst showed very high activities, which were topped by the even more active Karstedt catalyst introduced by the General Electric Company in 1973 [11]. The mechanistic details for Pt-catalyzed alkene hydrosilylation are described by the Chalk-Harrod cycle (Scheme 6.2) [12]. It involves oxidative addition of the Si—H bond to form a Pt(II) complex with a hydride and silanide ligand. The original mechanism proposes insertion of the alkene in the Pt—H bond followed by reductive elimination. Alternatively, Perutz discussed a mechanism in which the alkene inserts in the Pt—SiR_3 bond [13]. The nature of the insertion step depends very much on the metal: the Perutz mechanism was proposed for a Rh catalyst.

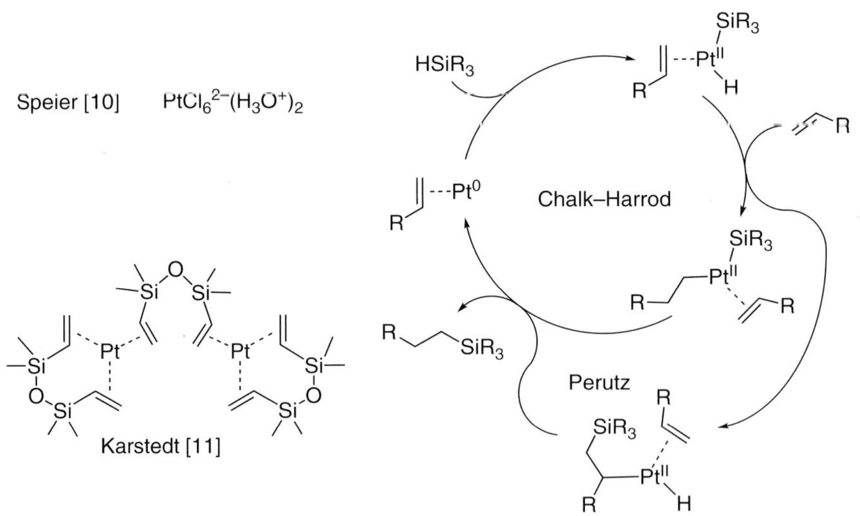

Scheme 6.2 The Chalk–Harrod mechanism for platinum-catalyzed alkene hydrosilylation.

The Karstedt catalyst has been used in the industry for nearly half a century and is still today the benchmark in hydrosilylation catalysis [8]. Research on Pt-catalyzed alkene hydrosilylation reached a very high level of sophistication.

There are, however, some drawbacks. The Pt catalysts are known to cause undesired side reactions to an extent that depends strongly on the substrates (Scheme 6.3). For substituted alkenes, regioselectivity, i.e. Markovnikov (branched) or anti-Markovnikov (linear) product formation, is an issue. Other problems involve dehydrogenative silylation, hydrogenation, olefin isomerization or polymerization, and silane redistribution reactions, giving rise to various hydrosilylation products. It is thought that the formation of platinum black (small metal particles) is responsible for these side reactions. Apart from chemoselectivity problems, the instability of the platinum price is also highly undesirable. For the most important applications (curing of silicone coatings and rubbers), the Pt catalyst is trapped in the product and cannot be easily recovered. The silicone industry consumes c. 6 tons of Pt annually, which is c. 3% of the world production [2]. This is the reason why, after nearly half a century, research on catalyst improvement (activity, selectivity) is still booming. Part of this research is also directed toward nonprecious metal catalysts.

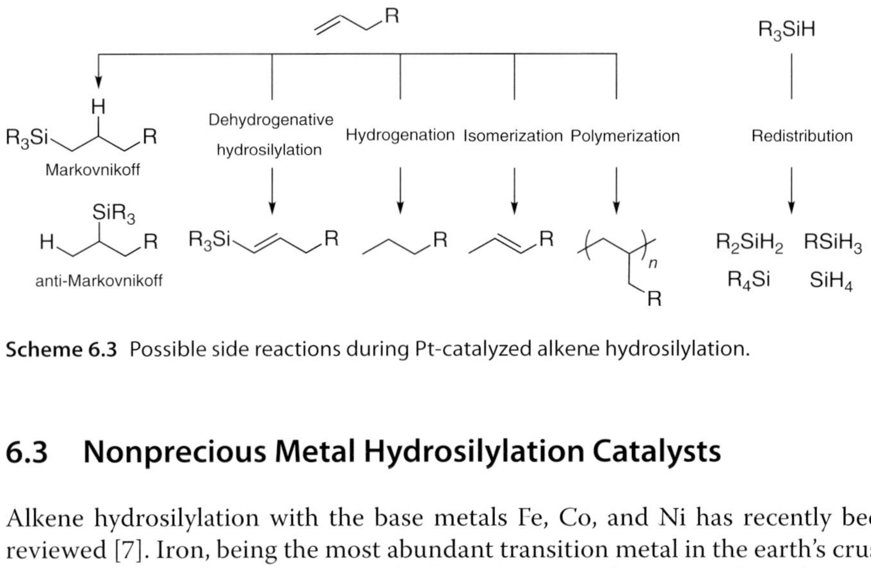

Scheme 6.3 Possible side reactions during Pt-catalyzed alkene hydrosilylation.

6.3 Nonprecious Metal Hydrosilylation Catalysts

Alkene hydrosilylation with the base metals Fe, Co, and Ni has recently been reviewed [7]. Iron, being the most abundant transition metal in the earth's crust, found already application in alkene hydrosilylation catalysis since the early 1960s. The simple Fe^0 complex $Fe(CO)_5$ was used as a catalyst in the hydrosilylation of α-olefins to give the anti-Markovnikov product [14]. The first step in the mechanism is catalyst activation by dissociation of CO forming $Fe(CO)_4$, a process that needs either elevated temperatures (>100 °C) or photoirradiation. Side products were consistently observed. Because of complicated redox processes at the Fe center, which can have oxidation states between −II and +VI, the mechanism is not fully understood. The iron center is also prone to undergo one-electron redox processes, instead of the two-electron transformations for the Karstedt catalyst. However, use of noninnocent ligands, which can deliver or take up electrons, also enables for Fe valuable two-electron processes. A prime example is Chirik's imino(pyridine) iron complex (Scheme 6.4) [15]. This compound was found to catalyze the hydrosilylation of terminal alkenes with a remarkably high

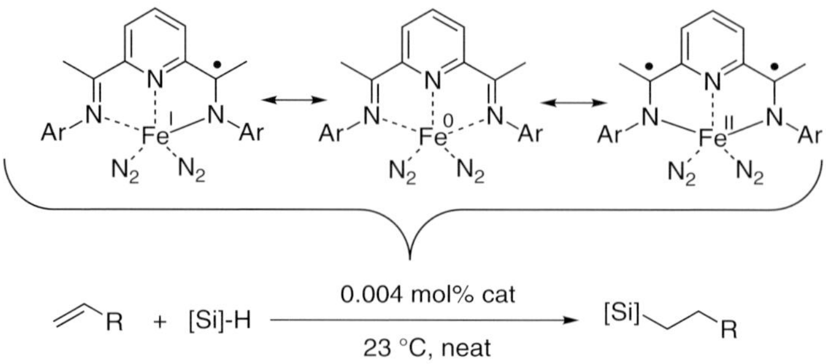

Scheme 6.4 Alkene hydrosilylation catalyzed by a Fe^0 complex with noninnocent ligands.

TON (25 000) and TOF (100 000 h^{-1}), i.e. an activity that is comparable to that of the Karstedt catalyst.

Apart from numerous investigations on other base metal catalysts [7], Ru seems to be a most promising alternative for low-cost hydrosilylation catalysis. The metal is known for excellent functional group tolerance and within the platinum group metals (Ru, Os, Rh, Ir, Pd, and Pt) by far the most economical. Tilley and coworkers introduced a cationic Ru catalyst for alkene hydrosilylation, which operates through a unique mechanism that involves a Ru-silylene species (Scheme 6.5) [16].

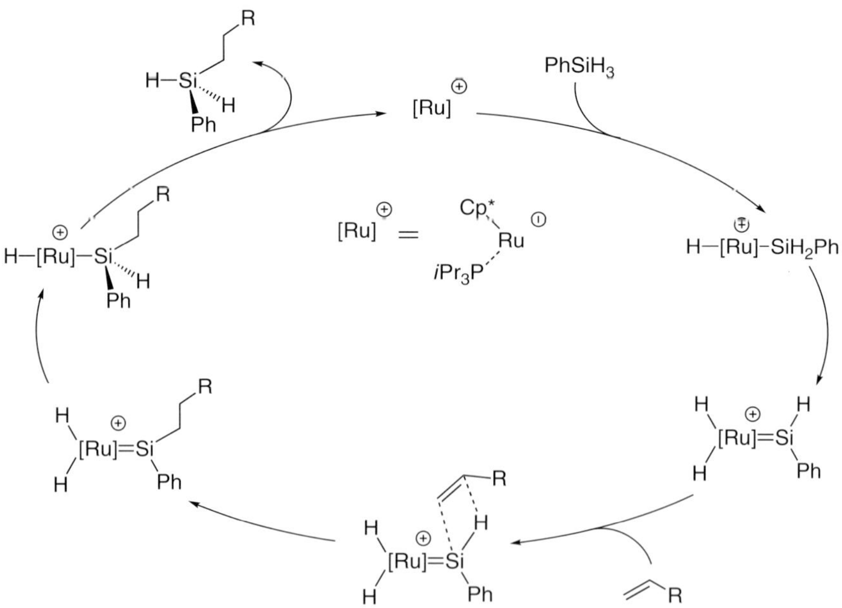

Scheme 6.5 Catalytic cycle for olefin hydrosilylation mediated by a Ru–silylene complex.

The turn of this century also saw further development in fully metal-free Lewis acid-catalyzed alkene hydrosilylation: Parks and Piers reported the hydrosilylation of aromatic aldehydes, ketones, and esters with $B(C_6F_5)_3$ [17].

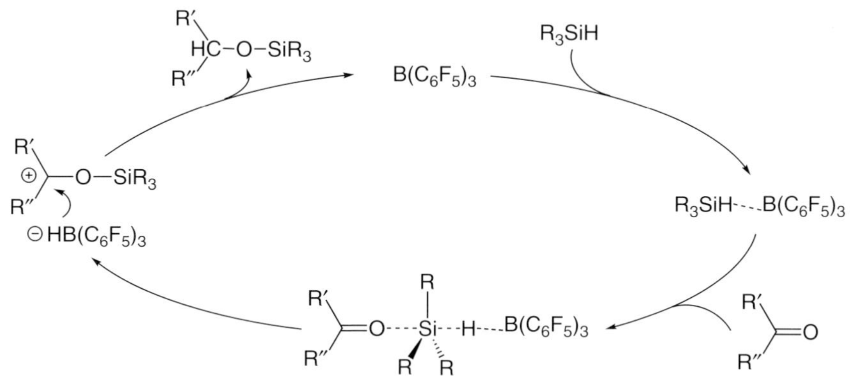

Scheme 6.6 Catalytic cycle for the $B(C_6F_5)_3$-catalyzed hydrosilylation of carbonyl compounds.

This conversion follows a unique nucleophilic–electrophilic mechanism in which the borane abstracts H− from the silane, assisted by nucleophilic attack of the ketone at Si (Scheme 6.6). The intermediate cation–anion pair $[R_3Si\text{-}OC(R)R'^+][HB(C_6F_5)_3^-]$ subsequently reacts by transferring a hydride from the borate to the electron poor C. This cooperative mechanism, in which borane and ketone team up to enable the Si to B hydride transfer, has been unambiguously confirmed by the Oestreich group. They demonstrated an inversion of configuration for chiral silanes [18]. A similar ionic outer-sphere mechanism [19] was also established for Brookhart's cationic pincer Ir hydrido catalyst for ketone hydrosilylation [20]. The $B(C_6F_5)_3$-catalyzed ketone hydrosilylation was extended to alkene hydrosilylation by Rubin et al. [21] and quite recently also highly Lewis acidic cationic Al complexes were found to be efficient catalysts [22]. Early transition metals and the lanthanides catalyze alkene hydrosilylation [23]. Their catalytic activity is not based on redox chemistry but relies on activation by a combination of a highly electrophilic metal and a strongly polarized metal–carbon bond. Therefore, it was anticipated that group 2 metal reagents may also be active in alkene hydrosilylation. The beginning of this century saw the start of hydrosilylation catalysis using highly polarized s-block organometallics, which will be discussed in the following sections of this chapter. Most of the work is related to alkene hydrosilylation and minor reports describe the reduction of ketones. For imine hydrosilylation, preliminary results are discussed.

6.4 C=C Bond Hydrosilylation with s-Block Metal Catalysts

Harder and coworkers published in 2006 the first early main group metal catalysts for alkene hydrosilylation [24]. This initial set of catalysts is based on the highly abundant, nontoxic, and environmentally benign, alkali and alkaline earth metals K, Ca, and Sr. The catalysts **1–4** (Scheme 6.7) that have been tested are based on the 2-dimethylamino-α-trimethylsilyl-benzyl (DMAT), which is sterically protected and electronically stabilized by the benzylic trimethylsilyl

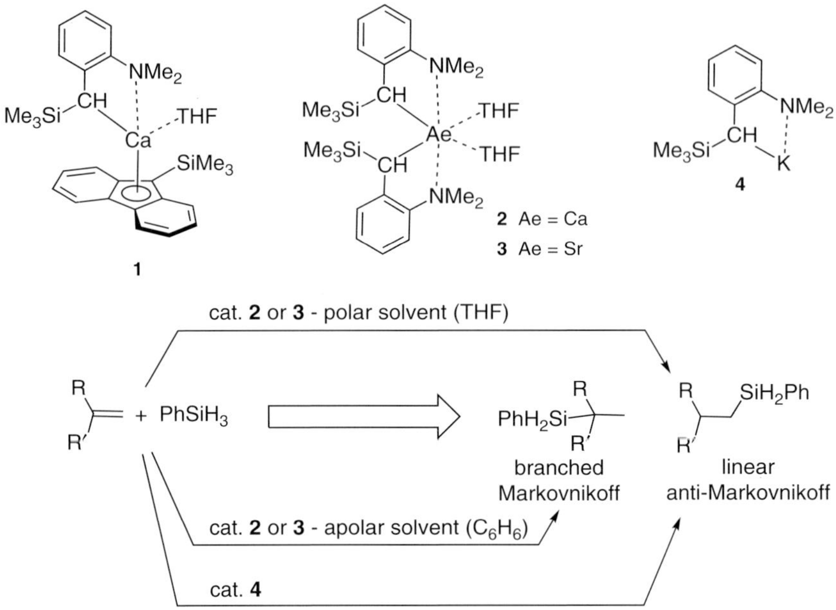

Scheme 6.7 Alkene hydrosilylation with Ca, Sr, and K catalysts (**1**–**4**). The regioselectivity is controlled either by metal or solvent choice.

substituent, while the dimethylamino substituent attributes to stability by intramolecular coordination.

Initially, the hydrosilylation of 1,1,-diphenylethylene (DPE) with $PhSiH_3$ was investigated. The reason for using DPE is its documented insensitivity toward polymerization. The heteroleptic complex **1** only gave stoichiometric conversion but $(DMAT)_2Ca \cdot (THF)_2$ (**2**) gave with a catalyst loading of 2.5 mol% overnight quantitative conversion to the branched Markovnikov product. Not only α-methylstyrene but also styrene was quantitatively converted to the branched product without traces of the linear anti-Markovnikov isomer, a regioselectivity that is rarely achieved with transition metal catalysts. The Sr catalyst $(DMAT)_2Sr \cdot (THF)_2$ (**3**) showed the same selectivity but was found to be an order of magnitude more active. Surprisingly, the conversion of styrene, an alkene highly sensitive to polymerization, was smooth and fast and without polystyrene side products. In contrast to DPE and α-methylstyrene, which were hydrosilylated at 50 °C, styrene was fully converted already within five minutes at 20 °C. This indicates that hydrosilylation is much faster than the potentially competing polymerization side reaction. The catalyst loadings could generally be lowered to 2.5 mol% for DPE down to 0.5 mol% for styrene. Also, the less reactive secondary silane $Ph(Me)SiH_2$ gave good conversion.

Apart from styrenic substrates, one of the double bonds in cyclohexadiene could be hydrosilylated, but the remaining C=C bond was left fully intact. This suggests that the substrate scope is limited to conjugated (activated) alkenes, an assumption that was confirmed by the inactivity of cyclohexene, allylbenzene, and norbornene. This restriction may be a disadvantage, but could be fortuitous in cases where selectively only one out of two conjugated double bonds should be converted, a challenge that is often difficult to meet in transition metal catalysis.

Interestingly, hydrosilylation of DPE with the analog K catalyst (DMAT)K (**4**) gave the pure linear anti-Markovnikov product. Considering that this may be related to the higher polarity of the C—K bond vs. that of the C—Ca or C—Sr bonds, reactions with the Ca (**2**) and Sr (**3**) catalysts were run in solvents of increasing polarity: $C_6H_6 < Et_2O <$ THF. Although the reaction in the apolar solvent benzene gave only the branched product, in THF, exclusively the linear isomer was formed. In the somewhat less polar Et_2O, a branched/linear ratio of 75/25 was observed. This clearly shows that at least two mechanisms operate next to each other and that solvent polarity controls their mutual preference.

The proposed mechanism for solvent and metal-dependent alkene hydrosilylation with s-block metal catalysts is shown in Scheme 6.8. The cycle starts with catalyst initiation: the reaction between the [M]-DMAT species and $PhSiH_3$ generates the metal hydride species [M]-H and $PhSiH_2$-DMAT, which unequivocally could be characterized by ^{1}H NMR. Attempted isolation of (DMAT)CaH from reaction of $(DMAT)_2Ca \cdot (THF)_2$ with $PhSiH_3$ in C_6D_6 failed. Although nuclear magnetic resonance (NMR) monitoring showed clean conversion, attempted crystallization gave $(DMAT)_2Ca \cdot (THF)_2$. This suggests that larger hydride-rich clusters, $[(DMAT)_{<1}CaH_{>1}]_n$, remain in the solution. Further cooling of the mother liquor repeatedly gave $(DMAT)_2Ca \cdot (THF)_2$, which shows that clusters in solution keep growing and become more hydride-rich. The "true" catalyst is therefore likely a series of larger hydride-rich clusters. Using a bulky ß-diketiminate ligand, a discrete dimeric Ca hydride complex (**5**) has been isolated [25]. From the reaction of $Ca[N(SiMe_3)_2]_2$ with $PhSiH_3$, larger Ca hydride clusters have also been identified [26].

The *in situ* formed metal hydride species reacts further with the alkene. In this addition step, the regioselectivity is determined. The addition of polar metal hydrides to conjugated double bonds will always give a resonance-stabilized intermediate. Either benzylic or allylic reaction intermediates are formed, whereas primary unstabilized alkyl carbanions are avoided. This explains the very high regioselectivity observed. Subsequent reaction with $PhSiH_3$ gives the Markovnikov product. The recognition that the "true" catalyst is a metal hydride species suggested that alkene hydrosilylation may also be mediated simply by CaH_2. Freshly ground finely divided CaH_2, however, did not catalyze hydrosilylation of DPE, but the more reactive KH did, giving the anti-Markovnikov product.

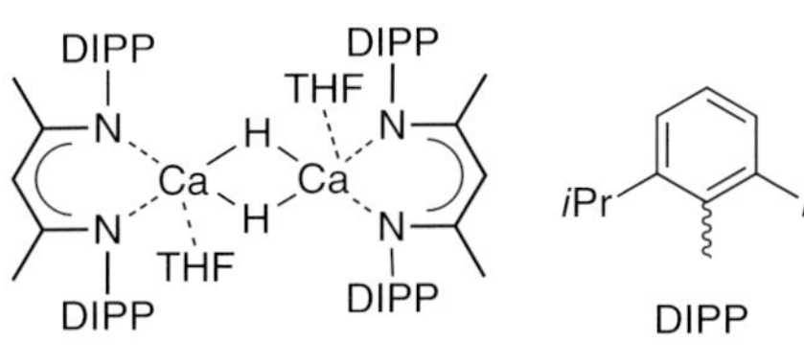

5

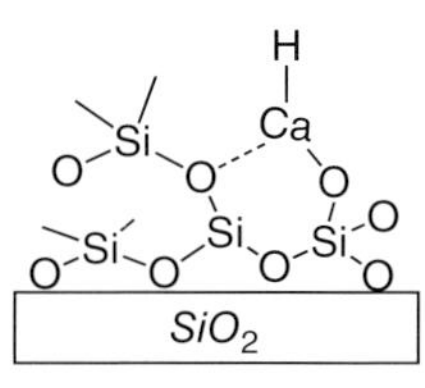

6

The formation of the anti-Markovnikov product under polar conditions may be explained by switching from a hydride cycle to an ion-pair or silanide

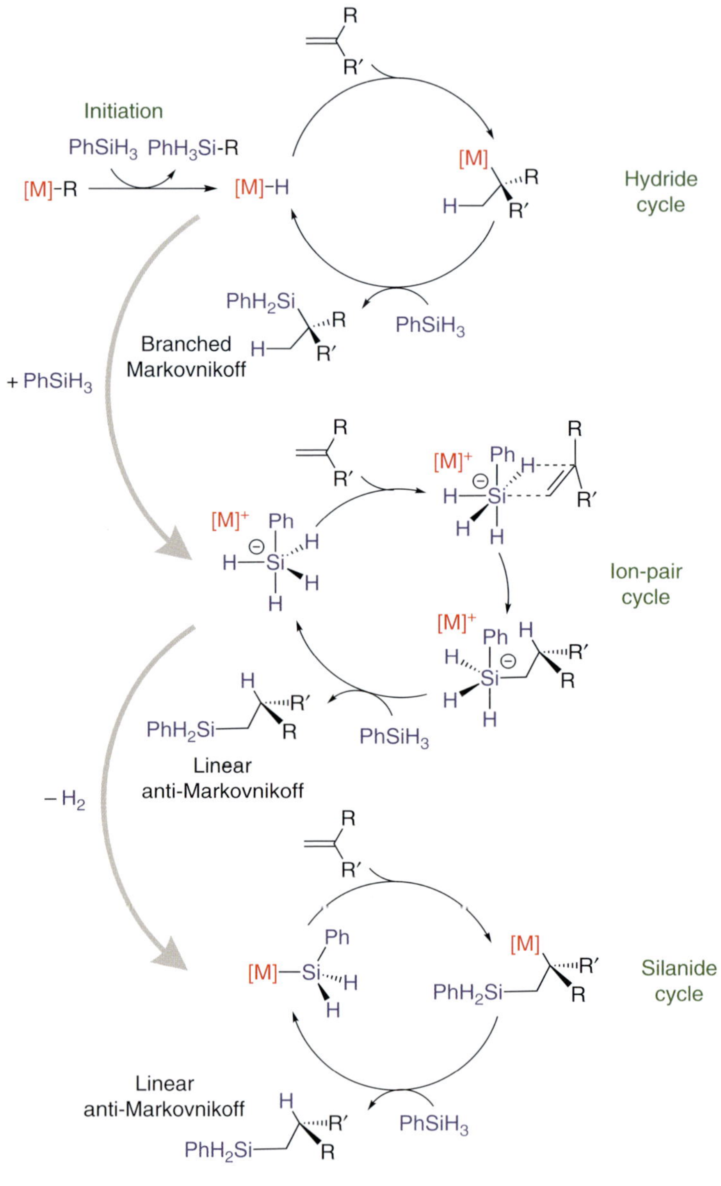

Scheme 6.8 Initiation and possible catalytic cycles for alkene hydrosilylation with early main group metal catalysts.

cycle. Reaction of [M]-H with $PhSiH_3$, instead of alkene, generates a hypervalent silicate, which is the entrance point for the ion-pair cycle. Silicates with electron-withdrawing alkoxy substituents, e.g. $(i\text{PrO})_4SiH^- K^+$, are known to be powerful reducing agents [27]. It is not clear whether a $PhSiH_4^-$ M^+ species is stable, but the crystal structure of the less hydride-rich complex, $[Ph_2SiH_3^-][K^+(18\text{-crown-}6)]$, has been reported [28]. The transition state for a concerted alkene addition to a silicate involves a six-coordinate Si atom. Similar to that in alkene hydroboration, the alkene substituents would be arranged in such a way to avoid steric repulsion, giving selectively the anti-Markovnikov product. This mechanism is also reminiscent of the well-established ketone hydrosilylation catalyzed by K^+F^- [29]. In the last step of the cycle, a hydride is transferred to $PhSiH_3$ regenerating the silicate catalyst $PhSiH_4^-$ M^+.

Alternatively, the formation of the linear product could also be explained by a silanide cycle, which is closely related to the hydride cycle but features the metal silanide [M]-SiH_2Ph as the active catalyst. The latter could be formed by release of H_2 from $PhSiH_4^-$, reducing the oxidation state of Si from +IV to +II, a reductive elimination at Si that has been previously postulated by Corriu [30]. Another route for silanide formation could be the direct deprotonation of $PhSiH_3$ by the metal reagent. Similar to that in the hydride cycle, addition of [M]-SiH_2Ph to the alkene is highly stereoselective. It gives only the resonance-stabilized intermediate that explains the exclusive formation of the linear anti-Markovnikov product.

Because the ion-pair cycle involves concerted Si–H/alkene addition and the silanide cycle a stepwise conversion, one can discriminate between the two possibilities by choosing a cyclic alkene substrate. The hydrosilylation of 1-Ph-cyclohexene with $PhSiH_3$ gave 1-Ph-2-$PhSiH_2$-cyclohexane exclusively as its cis-diastereomer from which a trans-addition of the silane can be concluded. A converted ion-pair cycle would give cis-addition, resulting in the trans-diastereomer. It is therefore most likely that the formation of the linear product proceeds through a silanide cycle. Metal-dependent switching between hydride and silanide cycles has also been discussed for transition metal-catalyzed alkene hydrosilylation [13].

These first pioneering studies clearly showed that s-block metal-catalyzed alkene hydrosilylation is unique in the sense that the regioselectivity can be steered by either metal or solvent choice. The very high selectivities obtained are related to the stabilization of the intermediate by charge delocalization. This is also the reason why isolated alkenes failed to react. Although this restriction may be a disadvantage, it is also a fluke that allows for selective mono-hydrosilylation of dienes.

Following these first studies, also the activity of a discrete calcium hydride complex (**5**) in the hydrosilylation of DPE with $PhSiH_3$ was investigated [31]. Independent of the solvent used, polar or apolar, the product distribution was mainly biased to the branched silane (Markovnikov) with only traces of the linear isomer. Catalysis in THF gave a high fraction of alkene hydrogenation. This clearly demonstrates that the spectator ligand can strongly influence activities and selectivities in alkene hydrosilation.

Alkene hydrosilylation with Ca hydride grafted to a SiO_2 surface (**6**) gave the same selectivity as the homogeneous Ca catalyst **2** [32]. Because the activities were in some cases much lower, catalysis proceeds presumably in a heterogeneous manner at the surface and not homogeneously in solution.

In 2014, the Okuda group introduced the Ca silanide catalyst $Ca(SiPh_3)_2 \cdot (THF)_4$ (**7**) as a catalyst for the hydrosilylation of styrene substrates with the tertiary silane Ph_3SiH (Scheme 6.9) [33]. Reactions were performed without solvent at 60 °C (2.5 mol% cat loading) and full conversion was relatively slow: $Ph_2C{=}CH_2$ (62 hours) and $Ph(Me)C{=}CH_2$ (2 hours). Styrene was polymerized and 1-octene did not react under these conditions. Variation

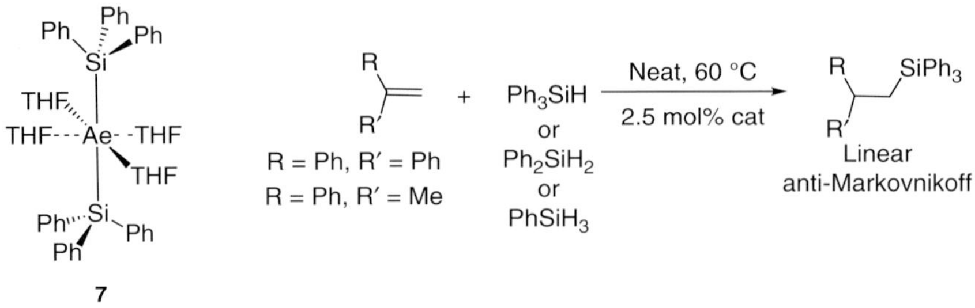

Scheme 6.9 Alkene hydrosilylation with Ca silanide catalyst **7** gave the anti-Markovnikov product.

of the silane from Ph_3SiH to Ph_2SiH_2 and $PhSiH_3$ shortened the reaction times. Most importantly, the products obtained were in all cases the linear anti-Markovnikov silane. In contrast, the Ca catalyst $(DMAT)_2Ca \cdot (THF)_2$ (**2**) exclusively gave the Markovnikov product (at least under apolar conditions). This implies that preformed Ca silanide catalysts follow the silanide mechanism (Scheme 6.8), whereas dibenzylcalcium catalysts follow the hydride route.

Okuda and coworkers continued their investigations with the syntheses of a set of alkali metal silanides: $Ph_3SiLi \cdot$ (12-crown-4), $Ph_3SiNa \cdot$ (15-crown-5), and $Ph_3SiK \cdot$ (18-crown-6)(THF) [34]. Both the monomeric Li and K complexes converted DPE and Ph_3SiH in the anti-Markovnikov product (2.5 mol% cat, 80 °C, 16–20 hours). Exchange of the crown ether by THF elongated the reaction times. Silane reactivity increased along the row: $Ph_3SiH < Ph_2SiH_2 < PhSiH_3$, but primary and secondary silanes led to catalyst decomposition and incomplete conversion. The catalysts were not active in the hydrosilylation of styrene, 1-octene, and cyclohexene. In a follow-up report [35], they provide mechanistic evidence for the silanide mechanism (Scheme 6.8, bottom). The stoichiometric reaction between $KSiPh_3$ and DPE at 25 °C led to immediate formation of $Ph_3SiCH_2CPh_2K$, which was structurally characterized as its crown ether solvate. Reaction of this product with Ph_3SiH gave the product $Ph_3SiCH_2C(H)Ph_2$ and $KSiPh_3$ but clearly needed a higher reaction temperature of 60 °C, indicating that this σ-bond metathesis reaction is the rate-determining step. Indeed, the reaction with Ph_3SiD needed considerably longer reaction times and a kinetic isotope effect of $k_H/k_D = 3.1$ has been estimated. This demonstrates that the Si—H bond is cleaved in the slow step. The latter σ-bond metathesis step reached only 40% conversion and is likely an equilibrium. This was confirmed by reaction of $Ph_3SiCH_2C(H)Ph_2$ with $KSiPh_3$, giving a similar mixture of compounds.

Most recently, the Okuda group introduced a set of alkali metal amides with an anionic aza-crown ether-type ligand (Me_3TACD^-, Scheme 6.10) [36]. The metal amides $M(Me_3TACD)$ (M = Li, Na, and K) form aggregated species that are soluble in pentane and react very fast with Ph_2SiH_2 to give the hypervalent silicates **8**–**10** that have been structurally characterized. Although the Li complex **8** does not show Si—H· · ·Li interactions, stabilizing secondary interactions are evident in the Na and K complexes **9** and **10** (the latter K complex also shows Ph· · ·K interactions). The Li complex (**8**) is already at 20 °C highly active in the hydrosilylation of DPE, styrene, and even 4-MeO-styrene, giving in all cases the linear anti-Markovnikov product. On the other hand, the Na (**9**) and K (**10**) complexes do not catalyze the alkene hydrosilylation and instead react with DPE

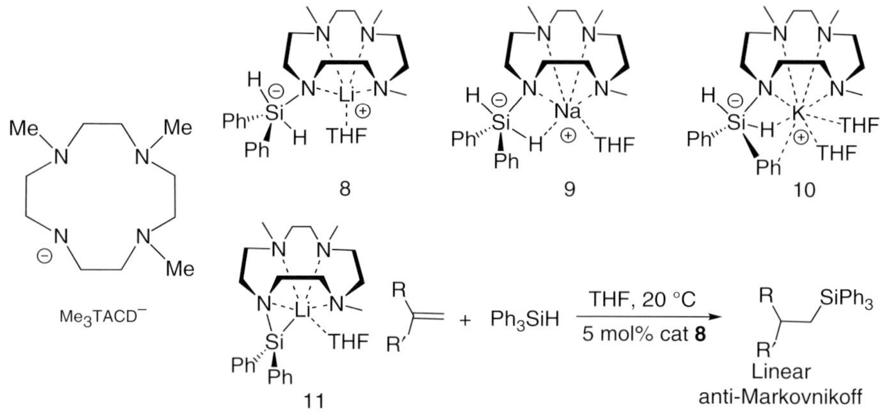

Scheme 6.10 The Me_3TACD^- ligand and alkene hydrosilylation with the hypervalent hydridosilicate catalysts **8**.

transferring a hydride to give $MeCPh_2M$. The mechanism for DPE hydrosilylation with Li complex **8** is unclear but may involve the Li silanide species **11**.

6.5 C=O Bond Hydrosilylation with s-Block Metal Catalysts

Quite some work has been reported on the valuable conversion of C=O double bonds to alcohols. Catalytic hydrosilylation is an attractive alternative to the direct hydrogenation with H_2. In 1997, Kobayashi et al. reported the reduction of ketones with polymethylhydrosiloxane (PMHS): [—OSi(H)MeO—]$_n$ [37]. This is a type of silicon oil with reducing properties and is the most economical silane [38]. The method is based on the use of K^+F^- as a catalyst (Scheme 6.11). The very

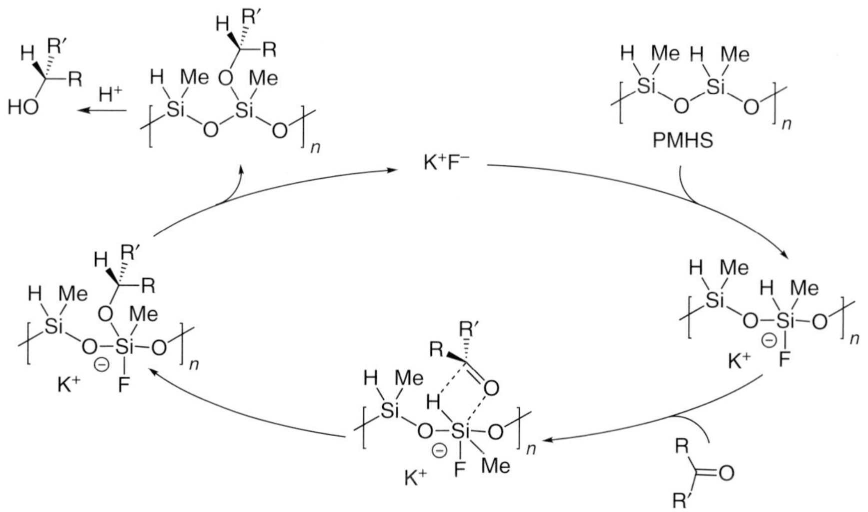

Scheme 6.11 Potassium fluoride-catalyzed ketone hydrosilylation with polymeric PMHS.

high affinity of Si for F^- leads to silicate formation, which increases the hydridic nature of the silane. Ketone addition proceeds through a transition state with a hypervalent six-coordinate Si center. The siloxane-bound alcoholates are eliminated by acidic hydrolysis.

This reactivity is based on silicate formation, and over the years, many variations have been reported. Hosomi and coworkers described the reduction of esters with $(MeO)_3SiH$ and catalytic quantities of MeOLi [39]. The electron-withdrawing MeO substituents facilitate the formation of the reducing silicate $(MeO)_4SiH^-Li^+$. The Kagan group reported asymmetric catalytic ketone reduction with $(MeO)_3SiH$ using the mono-lithium salt of 1,1′-bi-2-naphtol (Li-BINOL) as a chiral enantiopure catalyst [40]. Although there are many possible side reactions, the catalytic conversion of ketones in their respective chiral alcohols is quite clean and reached ee's up to 93%. Beller and coworkers reported the reduction of ketones and aldehydes using $PhSiH_3$ or the less expensive silane $(EtO)_2Si(Me)H$ and KO*t*Bu, NaO*t*Bu, or powdered NaOH as a catalyst [41], whereas the Nolan group reduced ester and tertiary amides in alcohols and amines, respectively, using KOH as a catalyst [42]. The Nikonov group reduced ketones and esters using KO*t*Bu and KOH as catalysts and PMHS as a reducing agent [43]. They made valuable contributions to mechanistic understanding showing that the base catalyst breaks down the PMHS polymer in smaller silanes such as H_2SiMe_2 or H_3SiMe that react further with the base to highly reducing silicates. Most recently, the Guo group described a practical method for aldehyde and ketone reduction using low-cost PMHS and K_2CO_3 [44].

The above-described catalysis with simple bulk bases such as NaOH and K_2CO_3 or soluble bases such as KO*t*Bu proceeds in all cases via a mechanism in which silicates $R_4SiH^-M^+$ are proposed to be the reactive species. Depending on the nature of the base, the metal cation, the solvent, and the silane, these silicates may form tightly bound contact ion pairs or solvent-separated ion pairs (Scheme 6.12). The influence of the metals on the reactivity of these silicates is especially for solvent-separated ion pairs small or negligible. It is therefore questionable whether it is possible to control C=O hydrosilylation with an

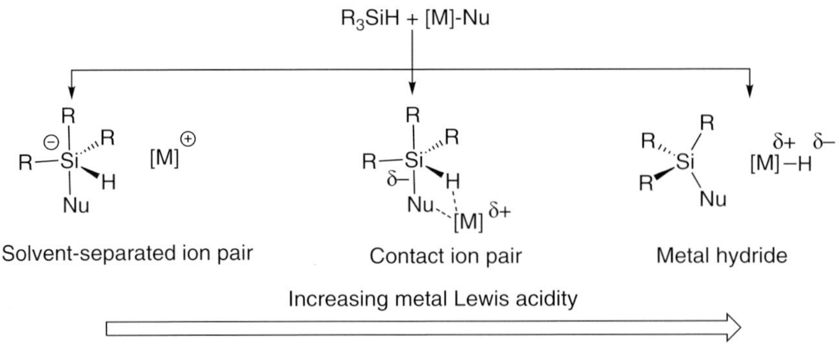

Scheme 6.12 Reaction of a nucleophilic metal reagent with silane may follow different pathways.

organometallic catalyst that can be tuned by ligand variation. For metals with a high Lewis acidity, however, formation of a metal hydride species may be possible in which case the catalyst could be controlled by ligand design.

In 2008, the Harder group investigated ketone hydrosilylation with the hydride-rich silane $PhSiH_3$ using the well-defined ß-diketiminate calcium hydride catalyst **5** (Scheme 6.13) [45]. In a first set of reactions, the selectivity of the calcium hydride to ketone addition was investigated. For reaction of PhC(O)Ph with **5**, only addition is possible, but α-hydrogen-containing ketones are prone to enolization and further aldol condensation (Scheme 6.13). Especially for acetone, cyclohexanone, or dibenzylketone, substantial enolization was found. Surprisingly, in catalytic ketone hydrosilylation, a very high selectivity for addition was observed. This means that there is a mismatch in selectivity between stoichiometric and catalytic reactions.

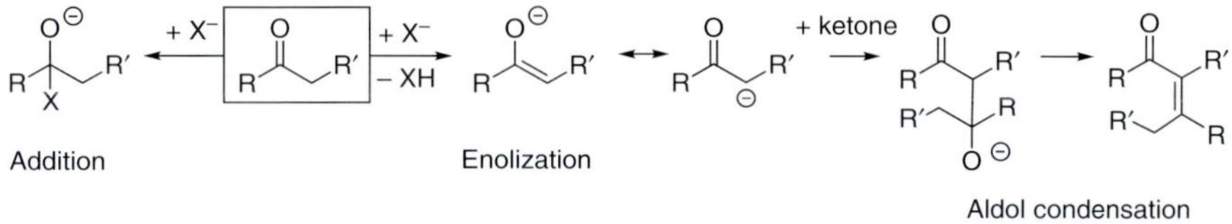

Stoichiometric reaction: $[L\text{-}Ca\text{-}H]_2$ (**5**) + ketone → $[L\text{-}Ca\text{-}OR]_2$

Catalytic hydrosilylation: $PhSiH_3$ + ketone →(cat. **5**) $PhSi(OR)_2H$

	Addition	Enol	Aldol	Addition/enol
PhC(O)Ph	100	0	0	100/0
PhC(O)Me	85	15	0	100/0
MeC(O)Me	0	76	24	—
Cyclohexanone	58	35	7	96/4
$PhCH_2C(O)CH_2Ph$	68	32	0	100/0
Adamantone	100	0	0	100/0

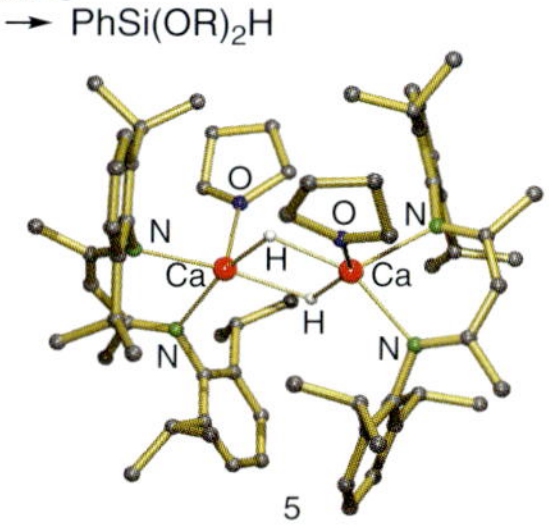

5

Scheme 6.13 Possible reactions of ketones with X⁻ nucleophile (or base). Differences in selectivity between the stoichiometric reaction of **5** with ketones and catalytic ketone hydrosilylation.

Ketone hydrosilylation with $PhSiH_3$ was found to proceed smoothly (1.25 mol% **5**, benzene, 50 °C), giving mainly the double hydrosilylation product $PhHSi(OR)_2$, even when excess silane was used. This clearly shows that the intermediate $PhH_2Si(OR)$ is a more reactive silylation reagent than $PhSiH_3$. This may be explained by its enhanced electrophilicity or by coordination of the Si—O—R unit to the metal facilitating intramolecular conversion. The triply hydrosilylation product $PhSi(OR)_3$ was formed only in traces (c. 2%), which is likely because of steric crowding around the Si center. The following reaction times for full conversion were found: PhC(O)Ph 15 hours, PhC(O)Me 34 hours, cyclohexanone 3 hours, $PhCH_2C(O)CH_2Ph$ 34 hours, and adamantone 0.2 hours; i.e. reaction times for Ph-containing substrates are considerably longer.

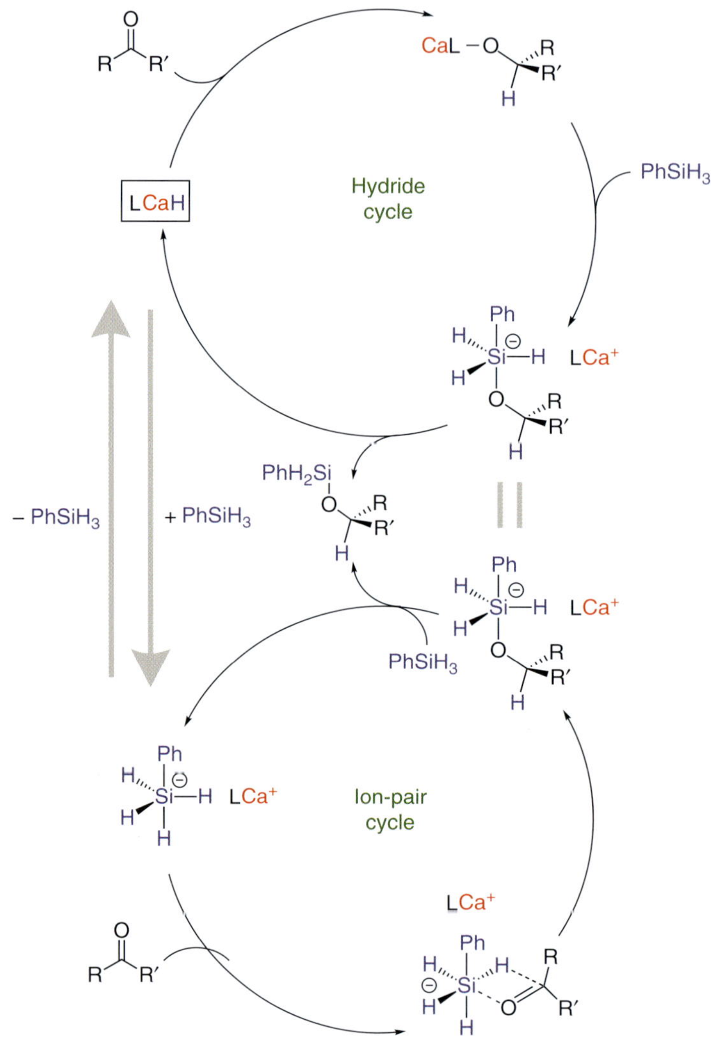

Scheme 6.14 Tentative mechanisms for ketone hydrosilylation with a calcium hydride catalyst. The hydride cycle is connected to the ion-pair cycle by an equilibrium and by sharing a common intermediate.

A tentative mechanism for ketone hydrosilylation is shown in Scheme 6.14. Starting with a Ca hydride species, two cycles are possible: a metal hydride cycle or an ion-pair cycle. The metal hydride cycle follows the common protocol of hydride to ketone addition succeeded by an alkoxide/hydride exchange with $PhSiH_3$.

Alternatively, the metal hydride catalyst could first react with $PhSiH_3$ to give the hypervalent silicate $[LCa^+][PhSiH_4^-]$, which, through a transition state with a six-coordinate Si center and hydride transfer, releases the product. Both cycles are connected through an equilibrium and a common intermediate and could also operate in interplay. It is, however, most likely that ketone hydrosilylation proceeds mainly through the ion-pair cycle. This conclusion is based on the discrepancy between the selectivity of the stoichiometric reaction between **5** and a series of ketones (addition vs. enolization) and that in the catalytic ketone hydrosilylation (Scheme 6.13). Ketones that show substantial enolization, such as dibenzylketone, gave a hydrosilylation product with exclusive hydride addition. This clearly rules out a calcium hydride intermediate. Depending on ketone, silane, metal, or solvent, it is not excluded that both mechanisms are operative.

In 2015, Hill and coworkers reported a special example of C=O hydrosilylation: the Mg hydride catalyzed reduction of CO with $PhSiH_3$ [46]. The stoichiometric reaction of the ß-diketiminate Mg hydride complex **12** with CO gave the double addition product **16** (Scheme 6.15). It is proposed that this reaction starts with a 1,1-addition of hydride to CO giving the mixed formyl-hydride dimer **13**. A second 1,1-addition would form **14** with the bridging dianion: H_2CO^{2-}. Although unusual, a Ca complex of a related isoelectronic dianion is known: $[Ca^{2+}][Ph_2CNPh^{2-}]$ [47]. Highly reactive **14** could react with a second CO to form **15**, which after a 1,2-H shift gives the final product **16**, which has been fully characterized. The intermediacy of the formyl species **13** is supported by NMR monitoring. Authors claim that the formation of the formyl intermediate **13** is also demonstrated by the hydrosilylation of CO with $PhSiH_3$ using **12** as a catalyst (10 mol%, toluene, 60 °C): $CO + 3\ PhSiH_3 \rightarrow PhH_2SiCH_3 + PhH_2SiOSiH_2Ph$ (^{13}C labeling showed that the CH_3 group originates from CO). A mechanism in which Mg hydride complex **12** is the catalyst has been proposed (Scheme 6.16). The cycle involves a series of steps in which the Mg hydride catalyst is repeatedly eliminated and inserted. In the light of the aforementioned ketone hydrosilylation by Ca hydride catalyst **5**, which most likely proceeds through an ion-pair mechanism (Scheme 6.14), it is questionable whether the cycle indeed involves the intermediacy of the Mg hydride species. The mismatch between the stoichiometric reaction of **12** with CO, which proceeds already at −60 °C to give **16** [46], and the catalytic CO hydrosilylation, which is even at +60 °C a very slow transformation (after three weeks only 15% $PhSiH_3$ is converted), makes

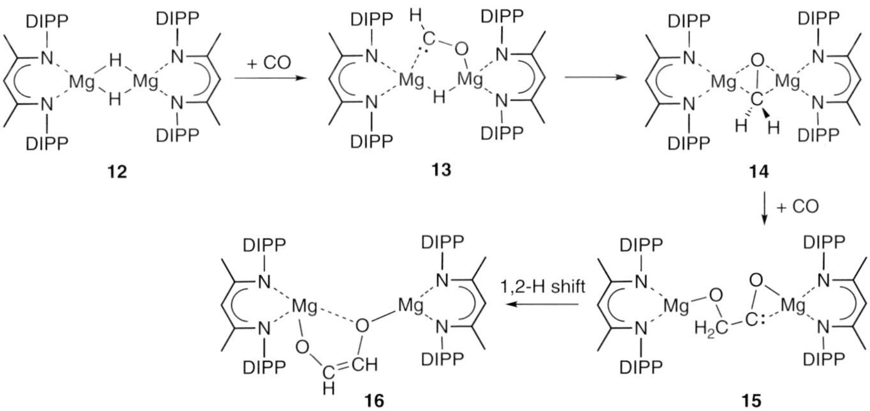

Scheme 6.15 Stoichiometric reaction of Mg hydride complex **12** with CO giving homologation and formation of **16**.

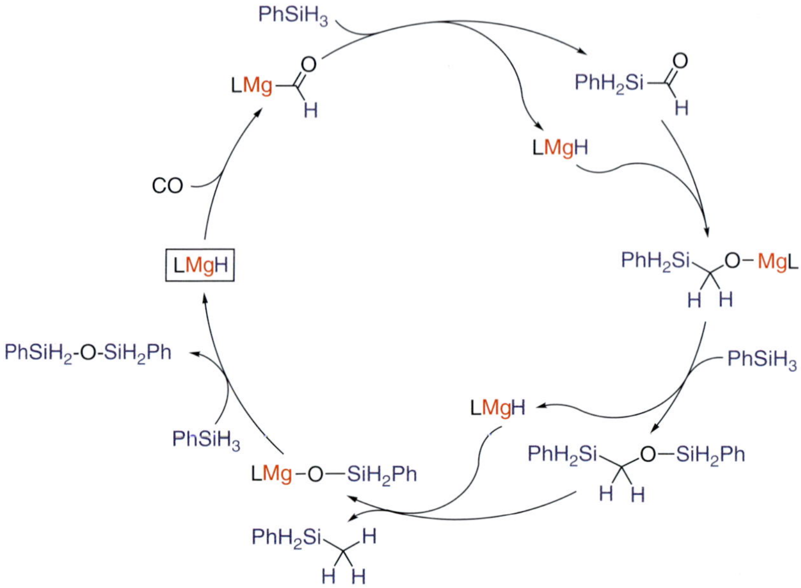

Scheme 6.16 Reported mechanism for Mg hydride-catalyzed CO hydrosilylation [46].

it likely that hydrosilylation catalysis follows a mechanism that does not involve a Mg hydride intermediate. Given the fast reaction of **12–16**, a metal hydride mechanism, as shown in Scheme 6.16, would also likely produce the product $PhH_2Si—OC(H)=C(H)O—SiH_2Ph$, which has not been detected. The formation of the silicate $[LMg^+][PhSiH_4^-]$, which could act as a catalyst, is a valid alternative. Insertion of CO in a Si—H bond followed by insertion of the resulting formyl carbene in a Si—H bond would also explain the lack of CO homologation products. The complexity of these reactions leaves at this stage only room for speculation.

Sadow and coworkers reported in 2015 another special case of C=O hydrosilylation: the Mg catalyzed hydrosilylation of enones such as methyl methacrylate to give silyl ketene acetals (Scheme 6.17) [48]. The catalyst is based on the anionic

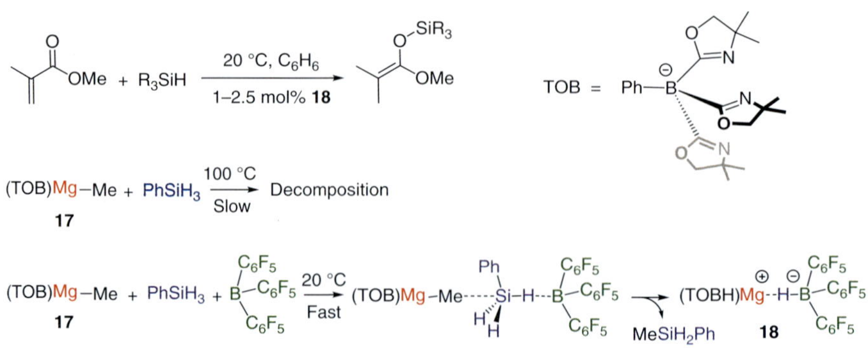

Scheme 6.17 Hydrosilylation of methyl methacrylate with an Mg hydridoborate catalyst (**18**).

tridentate *tris*-oxazolinyl-borato ligand (TOB). Under harsh conditions (100 °C), (TOB)MgMe (**17**) reacts with $PhSiH_3$, but only decomposition is observed. In contrast, reaction of (TOB)MgMe (**17**) with $PhSiH_3$ in the presence of the Lewis acid $B(C_6F_5)_3$ is fast and proceeds at 20 °C. This is likely because of the activation of the Si—H bond by B coordination. The Mg hydridoborate complex **18** was found to be an efficient catalyst for enone hydrosilylation, converting a range of silanes (including tertiary silanes); temperatures: 30–80 °C, reaction times: 0.5–8 hours. Complex (TOB)MgMe (**17**) is not catalytically active and instantaneously polymerizes methyl methacrylate. Although the precise mechanism is hitherto unknown, the anion $HB(C_6F_5)_3^-$ is essential to catalysis.

6.6 C=N Bond Hydrosilylation with s-Block Metal Catalysts

Although various transition metal and lanthanide catalysts have been developed for imine hydrosilylation [49], reports on early main group metal-mediated reactions are rare. This section discusses the hydrosilylation of pyridine substrates, which could be seen as the reduction of a special C=N bond and discloses first results in imine hydrosilylation using Ca, Sr, and K catalysts.

The well-known NADPH/NADP$^+$ redox cycle demonstrates that the pyridine/1,4-dihydropyridine interconversion is essential to life. The reagent 1,4-dihydropyridine is also an important reducing agent in organocatalysis, delivering 1 equiv of H_2 while forming semiaromatic pyridine [50]. The reverse reaction, reduction and dearomatization, of pyridine is still a very challenging reaction that requires potent reducing reagents. The addition of s-block metal hydrides to pyridine substrates is known since the early work by Ashby et al. who showed that reaction of MgH_2 with pyridine gave a mixture of 1,2- and 1,4-pyridides, abbreviated as 1,2-DHP and 1,4-DHP (Scheme 6.18a) [51]. Heating this mixture led to exclusive isomerization to 1,4-DHP, demonstrating that 1,2-DHP is the kinetic product while 1,4-DHP is thermodynamically more stable. Also, pyridine reduction with $LiAlH_4$ gave mixtures of 1,2- and 1,4-products [52]. The poor regioselectivity was confirmed by the stoichiometric reaction of the ß-diketiminate Mg hydride complex **12** with pyridine that initially gave a mixture of 1,2- and 1,4-DHP products, which after heating evolved to pure 1,4-DHP (Scheme 6.18a) [53]. These dearomatized pyridide mixtures react as disguised metal hydrides, and Mulvey and coworkers showed that the reaction of *n*BuLi with pyridine also provides such a metal hydride surrogate [54]. This observation opens up a range of catalytic transformations. Although Mg pyridide complexes provide catalytic activity in pyridine hydroboration [53], attempted pyridine hydrosilylation failed [55]. In 2015, Harder and coworkers showed that the heavier, more reactive, Ca hydride complex **5** is an efficient catalyst for hydrosilylation of pyridine substrates with $PhSiH_3$ (Scheme 6.18b) [56]. In contrast to earlier observations, the addition of the Ca hydride complex **5** to pyridine proceeds with exclusive 1,2-selectivity (Scheme 6.18a). This facile reaction proceeded already at 20 °C, but the heteroleptic product is sensitive to

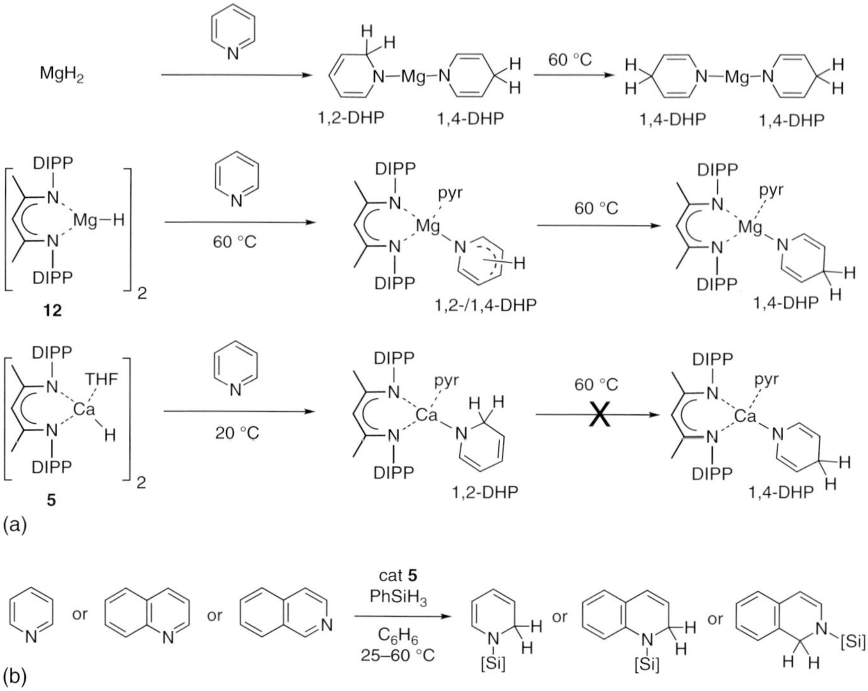

Scheme 6.18 (a) Stoichiometric addition of pyridine to metal hydrides. (b) Hydrosilylation of pyridine substrates with Ca hydride catalyst **5**.

ligand scrambling and could not be crystallized. Heating a mixture of hetero- and homoleptic products, however, did not give the expected 1,2-DHP → 1,4-DHP isomerization. DFT calculations demonstrated that, although the ß-diketiminate Ca-1,4-DHP complex is 4 kcal/mol more stable, the barrier for this isomerization is extremely high (75 kcal/mol, B3PW91/6-311G(d,p)). For the analog ß-diketiminate Mg-1,2-DHP complex, a barrier for isomerization of only 14.8 kcal/mol was calculated. The surprising kinetic stability of the Ca-1,2-DHP complex was exploited in hydrosilylation catalysis. Using Ca hydride catalyst **5**, pyridine, quinoline, and isoquinoline have been reduced by hydrosilylation with $PhSiH_3$ (5 mol% **5**, C_6D_6, 25–60 °C, 24 hours, 80–90% conversion). The catalyst loading could be lowered to 0.12 mol%. All substrates showed highly selective 1,2-reduction. This method therefore nicely complements the highly selective Ru-catalyzed conversion of pyridines to 1,4-substituted products [57].

In contrast to these observation, Okuda and coworkers found that $Ca(SiPh_3)_2 \cdot (THF)_4$ (**7**) reacted with pyridine to form $Ca(1,4\text{-DHP})_2 \cdot (pyr)_n$ and 4-Ph_3Si-pyridine [58]. It is likely that this 1,4-selective pyridine reduction proceeds via selective hydride transfer from the pyridide intermediate 4-Ph_3Si-$(C_5H_5N)^-$ to the para-position of pyridine to give 1,4-DHP. It was also found that $Mg(1,4\text{-DHP})_2 \cdot (pyr)_4$ is a slow catalyst for pyridine hydrosilylation (10 mol% cat, 80 °C, 72 hours, 50% conversion) giving a mixture of 1,2- and 1,4-products. These results compare well with slow pyridine hydrosilylation with another amido Mg hydride catalyst [59].

There is hardly precedence for the hydrosilylation of imines with early main group metal catalysts. Hosomi and coworkers reported a rare case of enantioselective imine reduction with $(MeO)_3SiH$ using a mixture of *n*BuLi and chiral alcohol as the catalyst (20 mol%) and reached ee's up to 65% [60]. The Harder group launched a more comprehensive investigation on imine hydrosilylation with the K, Ca, and Sr catalysts **2–4**, earlier used for alkene hydrosilylation (Scheme 6.7) [61]. A variety of aldimine substrates were reduced by $PhSiH_3$ in C_6D_6 while monitoring conversion with ^1H NMR (Scheme 6.19). Using catalyst loadings of 2.5–5 mol% and reaction times between five minutes and 24 hours, generally, full conversion was achieved. An aryl group at C is beneficial for fast conversion. This may be explained by C=N bond activation by conjugation with the aryl substituent. A Ph substituent at N slows down conversion, which is likely because of the formation of the resonance-stabilized metal amide [M]-N(Ph)CH_2R, an intermediate that only slowly reacts with $PhSiH_3$. Alkyl substituents at N are clearly more advantageous for efficient catalysis. The reaction is also surprisingly tolerant to functional groups as demonstrated by full conversion of 4-R-C_6H_4-C(H)=N*t*Bu with R = Me, Cl, and MeO. The Sr catalyst **3** was found to be substantially more active than Ca catalysts **2**, while K catalyst **4** generally was the slowest. The primary silane $PhSiH_3$ was found to be more reactive than the secondary silane Ph_2SiH_2. This is in agreement with the observation that $PhSiH_3$ is generally only mono-silylated to $PhH_2SiN(R)R'$. Double-silylated products such as $PhHSi[N(R)R']_2$ were rarely detected. The tertiary electron-poor silane $(EtO)_3SiH$ did not react at all.

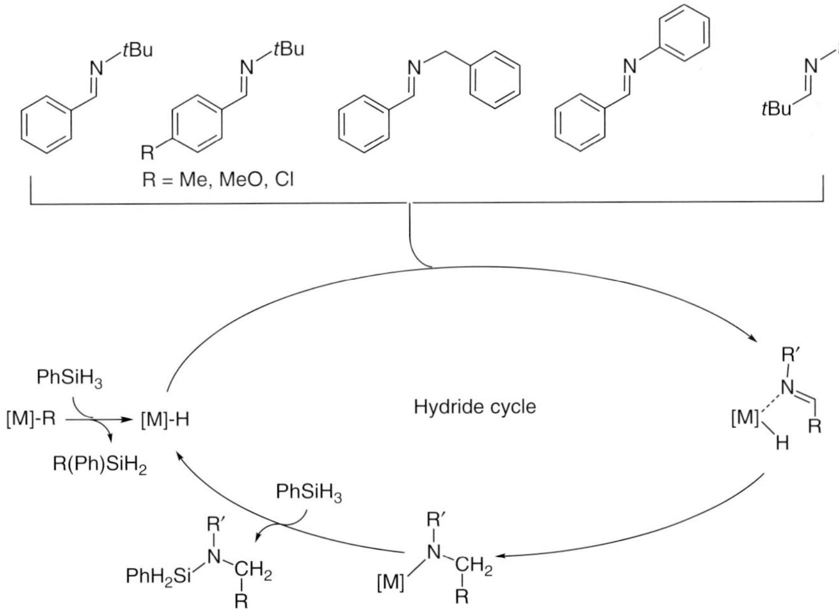

Scheme 6.19 Catalytic hydrosilylation of aldimines with K, Ca, and Sr catalysts **2–4** (See Scheme 6.7).

The mechanism likely follows the metal hydride cycle (Scheme 6.19). This is supported by stoichiometric reactions showing that the well-defined Ca hydride complex **5** cleanly reacted with imines to give Ca amide products, which were subsequently converted with $PhSiH_3$ back to **5**. DFT calculations on the hydrosilylation of *t*BuN=C(H)Ph with $PhSiH_3$ and the model catalyst CaH_2 confirm the possibility for a hydride cycle. The activation barrier for hydride/imine addition was calculated to be $\Delta H = 7.6$ kcal/mol (B3PW91/6-311++G**, benzene solvent correction), whereas the barrier for subsequent reaction of the Ca amide intermediate with $PhSiH_3$ was found to be somewhat higher ($\Delta H = 13.3$ kcal/mol), suggesting that the second step is rate determining.

6.7 Conclusions

s-Block metal-catalyzed hydrosilylation of unsaturated double bonds is a welcome addition to established transition metal-catalyzed methods. One major advantage is the very high selectivity for Markovnikov alkene hydrosilylation products, which is still a rarely reported challenge [3]. Notable is also the possibility to steer the selectivity either by solvent or metal choice. The latter feature originates from the strong influence of solvents on early main group metal chemistry. Based on the observations, a plausible mechanistic explanation of this behavior has been proposed. It is likely that at least two different catalytic cycles operate next to each other. There is, however, still plenty of room for comprehensive studies on this subject. There is especially a lack of computational studies.

One drawback of early main group metal-catalyzed alkene hydrosilylation is its limitation to conjugated (activated) double bonds, which may also be a blessing if one strives for selective mono-hydrosilylation of dienes. The stronger polarized C=O and C=N bonds do not need activation by conjugation. Hydrosilylation of ketones likely follows a pathway in which hypervalent silicates are the main reducing species. Imine hydrosilylation is still in its infancy, but preliminary results suggest that it most likely proceeds through a metal hydride cycle. Major future challenges are the enantioselective hydrosilylation of double bonds. Apart from hydrosilylation of unsaturated bonds, the recent discovery of C—H bond activation by dehydrogenative silylation using KO*t*Bu as a simple catalyst [62] is another breakthrough that opens up new possibilities for the early main group metals. It is anticipated that the rapidly increasing interest in group 2 metal chemistry will stimulate further developments in s-block metal-catalyzed hydrosilylation.

List of Abbreviations

Ae	alkaline earth
DPE	1,1-diphenylethylene
PMHS	polymethylhydrosiloxane
DMAT	2-dimethylamino-α-trimethylsilyl-benzyl
BINOL	1,1′-bi-2-naphtol
TOB	*tris*-oxazolinyl-borato

References

1 Ojima, I., Li, Z., and Zhu, J. (1998). *The Chemistry of Organic Silicon Compounds* (eds. Z. Rappoport and Y. Apeloig). New York: Wiley.
2 Shimada, S. and Nakajima, Y. (2015). *RSC Adv.* 5: 20603–20616.
3 Zaranek, M. and Pawluc, P. (2018). *ACS Catal.* 8: 9865–9876.
4 Obligacion, J.V. and Chirik, P. (2018). *Nat. Rev. Chem.* 2: 15–34.
5 Nikonov, G.I. (2017). *ACS Catal.* 7: 7257–7266.
6 Lipke, M.C., Liberman-Martin, A.L., and Tilly, T.D. (2017). *Angew. Chem. Int. Ed.* 56: 2260–2294.
7 Du, X. and Huang, Z. (2017). *ACS Catal.* 7: 1227–1243.
8 Troegel, D. and Stohrer, J. (2011). *Coord. Chem. Rev.* 255: 1440–1459.
9 Sommer, L.H., Pietrusza, E.W., and Whitmore, F.C. (1947). *J. Am. Chem. Soc.* 69: 188.
10 Speier, J.L., Webster, J.A., and Barnes, G.H. (1957). *J. Am. Chem. Soc.* 79: 974–979.
11 Karstedt, B.D. (1973). General Electric Company, Patent: US3775452A.
12 Chalk, A.J. and Harrod, J.F. (1965). *J. Am. Chem. Soc.* 87: 16–21.
13 Duckett, S.B. and Perutz, R.H. (1992). *Organometallics* 11: 90–98.
14 (a) Freidlina, R.K., Chukovskaya, E.C., Tsao, I., and Nesmeyanov, A. (1960). *Dokl. Akad. Nauk SSSR* 132: 374–377. (b) Nesmeyanov, A.N., Freidlina, R.K., Chukovskaya, E.C. et al. (1962). *Tetrahedron* 17: 61–68.
15 Tondreau, A.M., Atienza, C.C.H., Weller, K.J. et al. (2012). *Science* 335: 567–570.
16 Glaser, P.B. and Tilley, T.D. (2003). *J. Am. Chem. Soc.* 125: 13640–13641.
17 Parks, D.J. and Piers, W.E. (1996). *J. Am. Chem. Soc.* 118: 9440–9441.
18 Rendler, S. and Oestreich, M. (2008). *Angew. Chem. Int. Ed.* 47: 5997–6000.
19 Iglesias, M., Fernández-Alvarez, F., and Oro, L.A. (2014). *ChemCatChem* 6: 2486–2489.
20 Yang, J. and Brookhart, M. (2007). *J. Am. Chem. Soc.* 129: 12656–12657.
21 Rubin, M., Schwier, T., and Gevorgyan, V. (2002). *J. Org. Chem.* 67: 1936–1940.
22 Jakobsson, K., Chu, T., and Nikonov, G.I. (2016). *ACS Catal.* 6: 7350–7356.
23 (a) Harrod, J.F. and Yun, S.S. (1987). *Organometallics* 6: 1381–1387. (b) Fu, P.-F., Brard, L., Li, Y., and Marks, T.J. (1995). *J. Am. Chem. Soc.* 117: 7157–7168.
24 Buch, F., Brettar, J., and Harder, S. (2006). *Angew. Chem. Int. Ed.* 45: 2741–2745.
25 Harder, S. and Brettar, J. (2006). *Angew. Chem. Int. Ed.* 45: 3474–3478.
26 Maitland, B., Wiesinger, M., Langer, J. et al. (2017). *Angew. Chem. Int. Ed.* 56: 11880–11884.
27 Chuit, C., Corriu, R.J.P., Reye, C., and Young, J.C. (1993). *Chem. Rev.* 93: 1371–1448.
28 Bearpark, M.J., McGrady, G.S., Prince, P.D., and Steed, J.W. (2001). *J. Am. Chem. Soc.* 123: 7736–7737.

29 Boyer, J., Corriu, R., Perz, R., and Reye, C. (1981). *Tetrahedron* 37: 2165–2171.
30 Brefort, J.L., Corriu, R., Guérin, C., and Henner, B. (1989). *J. Organomet. Chem.* 370: 9–15.
31 Ruspic, C. and Harder, S. (2007). *Inorg. Chem.* 46: 5320–5326.
32 Gauvin, R.M., Buch, F., Delevoye, L., and Harder, S. (2009). *Chem. Eur. J.* 15: 4382–4393.
33 Leich, V., Spaniol, T.P., Maron, L., and Okuda, J. (2014). *Chem. Commun.* 50: 2311–2314.
34 Leich, V., Lamberts, K., Spaniol, T.P., and Okuda, J. (2014). *Dalton Trans.* 43: 14315–14321.
35 Leich, V., Spaniol, T.P., and Okuda, J. (2016). *Organometallics* 35: 1179–1182.
36 Schuhknecht, D., Leich, V., Spaniol, T.P., and Okuda, J. (2018). *Chem. Eur. J.* 24: 13424–13427.
37 Kobayashi, Y., Takahisa, E., Nakano, M., and Watatani, K. (1997). *Tetrahedron* 53: 1627–1634.
38 Lipowitz, J. and Bowman, S.A. (1973). *Aldrichim. Acta* 6: 1–6.
39 Hojo, M., Murakami, C., Fujii, A., and Hosomi, A. (1999). *Tetrahedron Lett.* 40: 911–914.
40 Schiffers, R. and Kagan, H.B. (1997). *Synlett*: 1175–1178.
41 Addis, D., Zhou, S., Das, S. et al. (2010). *Chem. Asian J.* 5: 2341–2345.
42 Fernandez-Salas, J.A., Manzini, S., and Nolan, S.P. (2013). *Chem. Commun.* 49: 9758–9760.
43 Revunova, K. and Nikonov, G.I. (2014). *Chem. Eur. J.* 20: 839–845.
44 Zhao, X.-X., Zhang, P., and Guo, Z.-X. (2017). *ChemistrySelect* 2: 7670–7677.
45 Spielmann, J. and Harder, S. (2008). *Eur. J. Inorg. Chem.*: 1480–1486.
46 Anker, M.D., Hill, M.S., Lowe, J.P., and Mahon, M.F. (2015). *Angew. Chem. Int. Ed.* 54: 10009–10011.
47 Buch, F. and Harder, S. (2007). *Organometallics* 26: 5132–5135.
48 Lampland, N.L., Pindwal, A., Neal, S.R. et al. (2015). *Chem. Sci.* 6: 6901–6907.
49 Li, B., Sortais, J.-B., and Darcel, C. (2016). *RSC Adv.* 6: 57603–57625.
50 Ouellet, S.G., Walji, A.M., and MacMillan, D.W.C. (2007). *Acc. Chem. Res.* 40: 1327–1339.
51 (a) Ashby, E.C. and Goel, A.B. (1977). *J. Am. Chem. Soc.* 99: 310–311. (b) Ashby, E.C. and Goel, A.B. (1977). *J. Chem. Soc. Chem. Commun.*: 169–169.
52 Lansbury, P.T. (1961). *J. Am. Chem. Soc.* 83: 429–434.
53 Arrowsmith, M., Hill, M.S., Hadlington, T. et al. (2011). *Organometallics* 30: 5556–5559.
54 Mulvey, R.E., Dunbar, L., Clegg, W., and Horsburgh, L. (1996). *Angew. Chem., Int. Ed. Engl.* 35: 753–755.
55 Hill, M.S., Kociok-Köhn, G., MacDougall, D.J. et al. (2011). *Dalton Trans.* 40: 12500–12509.
56 Intemann, J., Bauer, H., Pahl, J. et al. (2015). *Chem. Eur. J.* 21: 11452–11461.
57 (a) Gutsulyak, D.V., van der Est, A., and Nikonov, G.I. (2011). *Angew. Chem. Int. Ed.* 50: 1384–1387. (b) Königs, C.D.F., Klare, H.F.T., and Oestreich, M. (2013). *Angew. Chem. Int. Ed.* 52: 10076–10079.

58 Lemmerz, L.E., Spaniol, T.P., and Okuda, J. (2018). *Dalton Trans.* 47: 12553–12561.
59 Fohlmeister, L. and Stasch, A. (2016). *Chem. Eur. J.* 22: 10235–10246.
60 Nishikori, H., Yoshikara, R., and Hosomi, A. (2003). *Synlett*: 561–563.
61 Elsen, H., Harder, S., manuscript in preparation.
62 Banerjee, S., Yang, Y.-F., Jenkins, I.D. et al. (2017). *J. Am. Chem. Soc.* 139: 6880–6887.

7

Early Main Group Metal Catalyzed Hydrogenation

Heiko Bauer and Sjoerd Harder

Universität Erlangen-Nürnberg, Department Chemie und Pharmazie, Chair für Inorganic and Organometallic Chemistry, Egerlandstraße 1, 91058 Erlangen, Germany

7.1 Introduction

Unsaturated double bonds such as C=C, C=N, or C=O consist of a strong stable σ-bond in conjunction with a much weaker π-interaction. The latter forms the basis for a rich chemistry and allows for the addition of a wide range of substrates. Hydrogenation, the addition of H_2 to a double bond, is probably one of the best investigated transformations. As the product is lower in energy than the educts, the hydrogenation of double bonds with H_2 is a thermodynamically favorable reaction. Nevertheless, a catalyst is needed for this exergonic chemical transformation. Simple frontier orbital theory demonstrates that all HOMO/LUMO of the double bond containing substrate and H_2 result in an overlap integral of zero (Figure 7.1) meaning that the four-membered ring transition state for the synchronous addition of H_2 to the double bond is symmetry forbidden [1]. The high activation barrier is overcome by a catalyst promoting the heterolytic cleavage of H_2 through the interaction of orbitals of the metal with the hydrogen orbitals. As d-orbitals of the transition metals show perfect symmetry for the interaction with the HOMO and LUMO of H_2, most catalytic hydrogenation reactions are based on transition metal catalysts.

The first catalytic hydrogenation was already discovered in 1874 by Wilde, who reported the catalytic conversion of acetylene and H_2 into ethane using "platinum black" (finely divided Pt powder) [2]. This early example of heterogeneous hydrogenation was the starting point for the discovery of further heterogeneous hydrogenation catalysts. Sabatier developed around 1900 a procedure for the catalytic hydrogenation of ethylene with H_2 and finely divided nickel at surprisingly low temperatures of 30–45 °C. Further improvements on catalytic hydrogenation of acetylene with Co, Fe, and Cu catalysts were ultimately acknowledged with the 1912 Nobel Prize in chemistry [3]. At the beginning of the twentieth century, Murray Raney developed his famous Raney nickel and showed that his finely powdered nickel is an active heterogeneous catalyst for the hydrogenation of C=N double and C≡N triple bonds. Other well-known heterogeneous

Early Main Group Metal Catalysis: Concepts and Reactions,
First Edition. Edited by Sjoerd Harder.

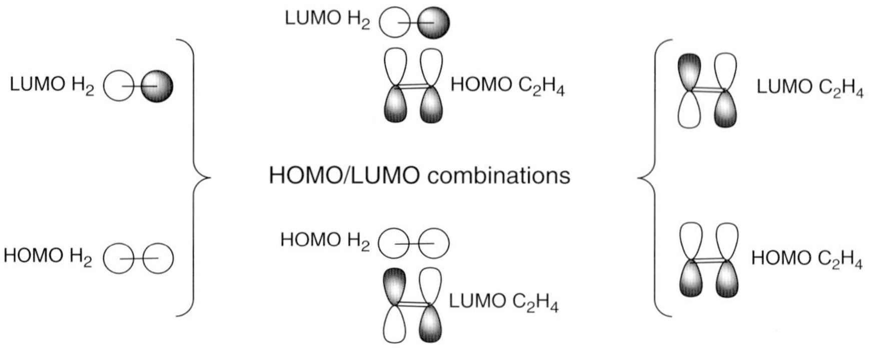

Figure 7.1 Representation of the HOMOs and LUMOs of H_2, C_2H_4, and unproductive orbital interactions for the concerted addition of H_2 to C_2H_4.

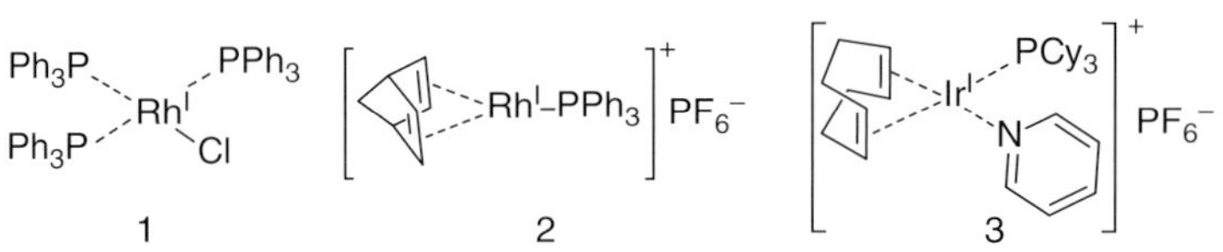

Figure 7.2 Early homogeneous transition metal hydrogenation catalysts.

catalysts, which are still used today, are, for example, platinum dioxide or Pt and Pd on carbon [4].

In conjunction with the rapidly developing field of transition metal complex chemistry in the early 1960s, also homogeneous catalysis started to develop rapidly. The very first catalysts for homogeneous hydrogenation are based on late transition metals. An early example is $RhCl(PPh_3)_3$ (**1**), introduced by Wilkinson as a precatalyst for the hydrogenation of olefins (Figure 7.2). The catalytically active species was reported to be $RhCl(PPh_3)_2$, which is able to hydrogenate olefins at room temperature and normal pressure [5]. Only a few years later in 1976, Osborn and Schrock developed a further catalyst, which is still in use. The rhodium complex $[Rh(nbd)PPh_3{}^+][PF_6{}^-]$ (nbd = norbornadiene) (**2**) (Figure 7.2) was synthesized by exchanging the chlorine atom for neutral nbd, thus creating a cationic rhodium species that led to significantly increased activities in the catalytic hydrogenation of olefins [6]. A third example of a highly active olefin hydrogenation catalyst is the cationic iridium complex $[Ir(cod)PCy_3(pyr)^+][PF_6{}^-]$ (cod = cyclooctadiene, pyr = pyridine) (**3**) (Figure 7.2) introduced by Crabtree in 1979 [7].

These pioneering hydrogenation catalysts incorporate expensive and rare transition metals. In contrast, enzymes in nature are generally based on more abundant, less noble metals. The catalytic center in hydrogenases, enzymes that regulate both the uptake and production of H_2 in microorganisms, was generally believed to be based on Fe and/or Ni. The early 1990s saw the surprising discovery of a metal-free enzyme (Hmd: H_2-forming methylene-tetrahydromethanopterin) that catalyzed hydrogenation without the need for Ni or Fe–S clusters [8]. This spectacular breakthrough in hydrogenation catalysis was later unmasked. The cofactor of Hmd was found to contain Fe as could be shown unequivocally by determination of its molecular structure [8]. Although not purely organocatalytic [9–11], it still presents an important milestone in the trend toward less toxic and cheaper metals in hydrogenation catalysis.

Apart from using the more economical first row transition metals there is also increasing interest in using redox-inactive main group or lanthanide metals in hydrogenation catalysis. First attempts in the homogeneously catalyzed hydrogenation of C=C double bonds without transition metals started already in the 1960s. Slaugh et al. could show that $LiAlH_4$ or alkaline metal hydrides and alkaline earth metal hydrides [12, 13] are able to hydrogenate alkenes. However, the harsh conditions required (35 mol% cat, 190 °C, 80 bar H_2) and poor performance did not encourage any further research and clearly favored the transition metal catalysts. Thus, it took nearly 30 years until the development of non-transition metal catalysts started to take off. Marks and coworkers in 1985 proved, in a mechanistic study, that lanthanides are very active hydrogenation catalysts [14]. The oxidation state of the lanthanide metal does not change during this transformation and therefore a mechanism different from the oxidative addition/reductive elimination protocol, typical for transition metal catalyzed hydrogenations, was proposed. Lanthanide catalyzed alkene hydrogenation proceeds via an olefin/metal hydride insertion, followed by a metal alkyl/H_2 σ-bond metathesis. The catalytic procedure could be extended to asymmetric hydrogenation [15] or hydrogenation of C=N double bonds [16]. On a similar note, Buchwald and coworker reported on the asymmetric hydrogenation of imines with chiral redox-inactive titanocene catalysts [17]. These methods started a change from precious late transition metal catalysts toward cheaper, more abundant, early transition metal catalysts.

The last decade has seen the rapid development of a new concept in hydrogenation catalysis: Stephan introduced organocatalytic metal-free hydrogenation using frustrated Lewis pairs (FLPs) [18–21]. The key to this chemistry is the rupture of the H—H bond by a Lewis base/acid combination that for steric reasons does not form a Lewis pair. Starting with catalytic imine hydrogenation [18] (Figure 7.3), this concept was soon extended to the diastereoselective

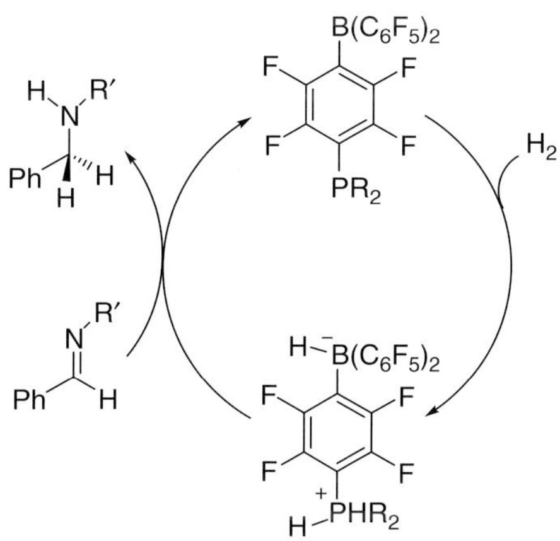

Figure 7.3 Catalytic cycle for metal-free imine hydrogenation.

hydrogenation of imines [19] and to FLP-mediated alkene hydrogenation [20]. In imine hydrogenation the imine substrate itself may play the role of the Lewis base, thereby simplifying the catalyst to the Lewis acid $B(C_6F_5)_3$ [21]. Wang et al. introduced highly Lewis acidic hydridoboranes as catalysts for alkene hydrogenation [22]. This unusual process starts with alkene hydroboration by $HB(C_6F_5)_2$, which is followed by cleavage of the alkyl-$B(C_6F_5)_2$ bond by H_2. Although this process does not need a Lewis base, the reaction conditions (140 °C, 72–120 hours) and catalyst loadings (20 mol%) are not very attractive when compared to FLP hydrogenation catalysis.

Apart from metal-free hydrogenation catalysis, there is also a growing interest in the use of d^{10}-metals such as Zn [23] and a revival of late main group metals such as Al [24]. Major breakthroughs, however, were made by using highly earth-abundant and nontoxic alkaline earth metals. The dogma that efficient alkene hydrogenation is only possible with transition metal catalysts was for the first time broken in 2008 by the organocalcium-catalyzed hydrogenation of conjugated alkenes by the Harder group [25]. As alkaline earth metals do not dispose of partially filled d-orbitals, binding of the substrate can occur only via an electrostatic interaction in which the alkene is activated by polarization induced by the Lewis acid. This chapter describes the first steps and the rapid development of main group metal catalyzed hydrogenation reactions, which is mainly dominated by group 2 metal catalysts. It exclusively focuses on the challenging reduction with the bulk commodity H_2. Addition of the much more polar, highly reactive, hydridoboranes ($R_2B^{\delta+}$-$H^{\delta-}$) and hydridosilanes ($R_3Si^{\delta+}$-$H^{\delta-}$) to unsaturated double bonds is not part of this chapter. This contribution zooms in on the many pitfalls but also focuses on the major successes and future challenges. This chapter is divided in three categories that describe C=C, C=N, and C=O bond hydrogenation, respectively.

7.2 Hydrogenation of C=C Double Bonds

The challenging isolation of the first Ca hydride complex (Figure 7.4 (**4**)) could be seen as the starting point in the development of alkaline earth metal catalyzed hydrogenation chemistry [26]. This complex with the constitution $[(DIPPnacnac)CaH{\cdot}(THF)]_2$ (DIPPnacnac = $HC[(CMe)N(C_6H_3\text{-}2,6\text{-}i\text{Pr})]_2$) is stabilized by a large sterically demanding β-diketiminate ligand that prevents ligand exchange reactions via the Schlenk equilibrium. The latter equilibrium, which is well known for Grignard reagents, describes the ligand exchange between heteroleptic complexes L^1–Ae–L^2 (Ae = alkaline earth metal) to give homoleptic complexes L^1–Ae–L^1 and L^2–Ae–L^2 carrying equal ligands at the metal. The isolation of a Ca hydride complex LCaH, in which L is a spectator ligand, is fully dependent on successful suppression of ligand scrambling. The Schlenk equilibrium would lead to formation of highly insoluble $(CaH_2)_\infty$ salts, thus shifting the equilibrium completely to the side of the homoleptic species. Hitherto, the factors that influence these Schlenk equilibria are not very well understood; however, it is clear that strongly coordinating multidentate ligands

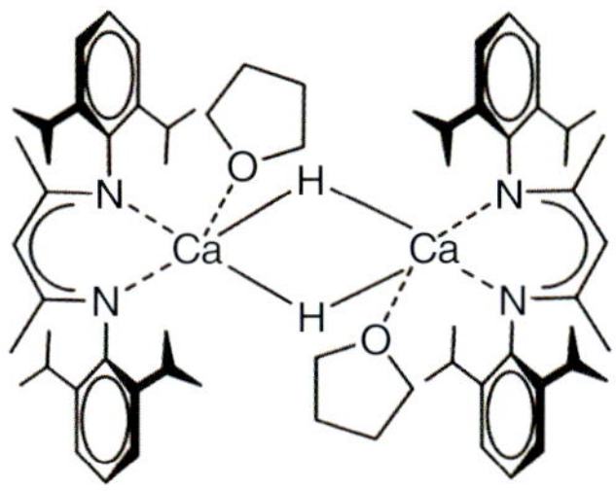

4

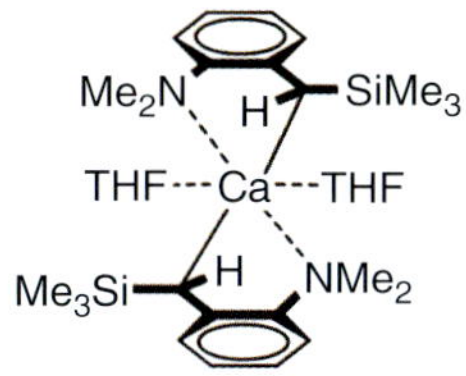

5

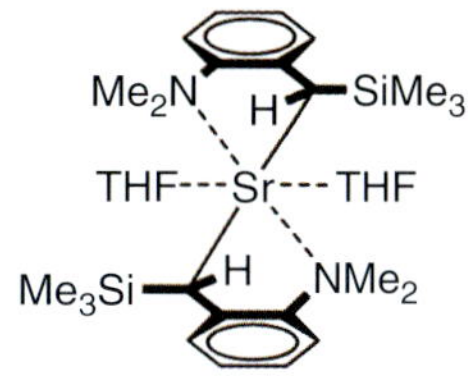

6

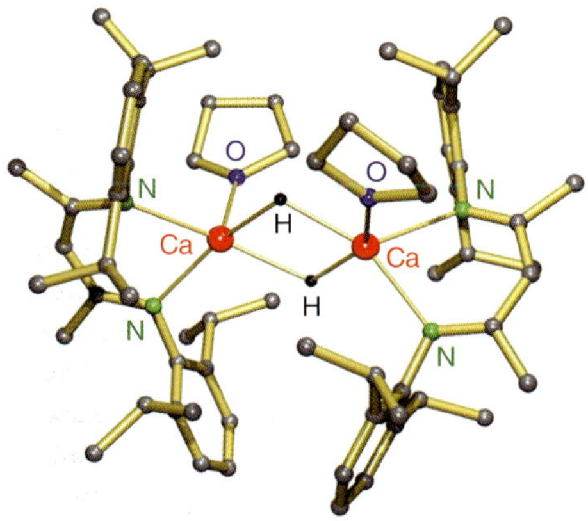

Figure 7.4 Catalysts for the hydrogenation of alkenes: [(DippNacNac)CaH·(THF)]$_2$ (**4**), (DMAT)$_2$Ca·(THF)$_2$ (**5**), and (DMAT)$_2$Sr·(THF)$_2$ (**6**), and the crystal structure of **4**.

are crucial for the isolation of Ca hydride complexes. Since Schlenk equilibria are faster for the bigger metals Sr and Ba, the recent isolation of the first Sr [27] and Ba hydride [28] complexes turned out to be even more challenging.

Just after the isolation of the first Ae hydride complex, catalytic application in the hydrogenation of alkenes was reported by Harder et al. [25]. Besides the calcium hydride complex [(DIPPnacnac)CaH·(THF)]$_2$ (**4**) also the benzyl complexes (DMAT)$_2$Ca·(THF)$_2$ (**5**) and (DMAT)$_2$Sr·(THF)$_2$ (**6**) (DMAT = 2-dimethylamino-α-trimethylsilyl-benzyl) (Figure 7.4) were introduced as (pre)catalysts for alkene hydrogenation. The calcium hydride complex **4** is able to convert styrene to ethylbenzene at room temperature with 5 mol% loading within 15 hours. The reaction proceeds at 20 bar H_2 pressure but besides the main product ethylbenzene also 19% of styrene oligomers are formed. The proposed mechanism (Figure 7.5), which was verified by stoichiometric conversion and characterization of the intermediates, shows that alkene polymerization is competitive with alkene hydrogenation. After exclusive transfer of the hydride to the terminal C of styrene a resonance-stabilized benzylcalcium complex is formed. This reaction is highly regioselective: the alternative product, a primary alkylcalcium complex, is not resonance stabilized. The benzylcalcium intermediate subsequently reacts with H_2 through a σ-bond metathesis mechanism, which is well established in lanthanide chemistry [14] but had previously not been demonstrated for group 2 metals. The latter

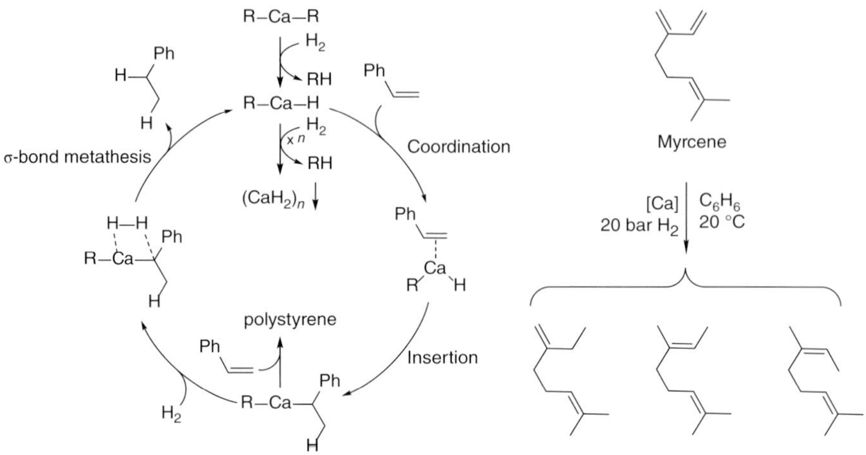

Figure 7.5 Mechanism for the calcium-catalyzed hydrogenation of styrene and the conversion of myrcene leaving the isolated C=C bond unaffected.

conversion, which formally could be viewed as a deprotonation of H_2 by a benzylcalcium reagent, results in product formation and recovery of the Ca hydride catalyst. Formation of oligomers can be explained by a competing side reaction: styrene polymerization. Instead of reaction of the benzylcalcium intermediate with H_2, the intermediate could react with the other substrate present, i.e. styrene. Given the well-established use of benzylcalcium reagents as initiators in styrene polymerization [29], it is a major challenge to prevent this side reaction. Oligomeric products can partially be prevented by using a higher H_2 pressure, which accelerates the σ-bond metathesis route. Formation of oligomers is also substrate and catalyst dependent. Catalytic hydrogenation with the Ca hydride complex **4** could also be achieved for 1,1-diphenylethylene (49% 1,1-diphenylethane) and α-methylstyrene (60% isopropylbenzene) without observation of oligomers but only uncomplete conversion was observed at an elevated temperature of 60 °C. The catalytic performance could be considerably improved using the homoleptic dibenzyl calcium complex $(DMAT)_2Ca{\cdot}(THF)_2$ (Figure 7.4) (**5**). Despite the potential formation of insoluble CaH_2 salts, hydrogenation of styrene was complete after 15 hours at room temperature with only half of the catalyst loading (2.5 mol%) and decreased oligomer formation (15%). The analog strontium complex $(DMAT)_2Sr{\cdot}(THF)_2$ gave a comparable catalytic performance. Exchanging the benzene solvent with a more polar medium such as THF accelerated the catalytic reaction and decreased the reaction time to 3.5 hours with only 8% oligomer formation. This effect could be further increased by addition of hexamethylphosphoramide (HMPA), a highly polar cosolvent. In this case, complete conversion was obtained after 1.5 hours with only 4% oligomer formation. The rate-enhancing effect of a polar reaction medium can be explained by its general ability to stabilize polar transition states. It may, however, also be related to its ability to keep in situ generated species such as nascent "CaH_2" in solution. Reactions with finely ground CaH_2 in the presence of a HMPA cosolvent were, however, not successful. Harder et al. also tested simple alkali metal catalysts [25]. While catalytic quantities of BuLi/TMEDA gave essentially no hydrogenation of 1,1,-diphenylethylene (DPE), (DMAT)K converted DPE nearly quantitatively to 1,1′-diphenylethane (only 3% oligomerization), but high H_2 pressures up to 100 bar were necessary. In this case, however, solid KH could also simply be used as a catalyst.

The Ca-catalyzed alkene hydrogenation was applied to the reduction of myrcene, an ingredient of many natural ethereal oils, containing two conjugated and one isolated C=C double bond. The products (Figure 7.5) showed that only one of the conjugated bonds was hydrogenated. It seemed therefore that the method is limited to the use of activated alkenes in which the C=C bond is conjugated either with a Ph ring or another C=C bond, an assumption that was found not to be true (*vide infra*). The fact is that conjugated double bonds react faster than isolated double bonds. The ease of hydrogenation of conjugated double bonds can be explained by the stability of the intermediates formed, i.e. either benzyl or allylcalcium complexes in which the negative charge is resonance stabilized. The limitation of the substrate scope to conjugated double bonds may be seen as a curse or a blessing. The Ca-catalyzed hydrogenation of cyclohexadiene gave exclusively cyclohexene as the product. Such very high selectivity for a single double bond reduction is not easily reached in transition metal catalyzed diene hydrogenation.

Ten years after the first alkene hydrogenation with group 2 metal catalysts, Hill and coworkers reported a THF-free catalyst **4** [30]. This calcium hydride dimer is much more Lewis acidic and was able to react with isolated alkenes such as ethylene and 1-hexene. The latter is the basis for 1-hexene reduction with H_2 but the catalytic transformation has to be carried out at room temperature in order to prevent ligand exchange to homoleptic species and nucleophilic attack of the solvent, which makes this transformation very slow (C_6D_6, 10 mol% cat., 25 °C, 99%, 21 days) [30].

There are not many ligands that are able to stabilize Ca hydride complexes against ligand distribution reactions. In 2012, Okuda and coworkers isolated Ca hydride complexes using the strongly coordinating tetradentate monoanionic ligand Me_3TACD^- (1,4,7-trimethyl-1,4,7,10-tetraazacyclododecane). A cationic complex of composition $(Me_3TACD)_3Ca_3H_2^+$ was isolated and structurally characterized (**7**, Figure 7.6) [31]. This complex was shown to be an active catalyst in the hydrogenation of 1,1-diphenylethylene, DPE (Figure 7.7). With

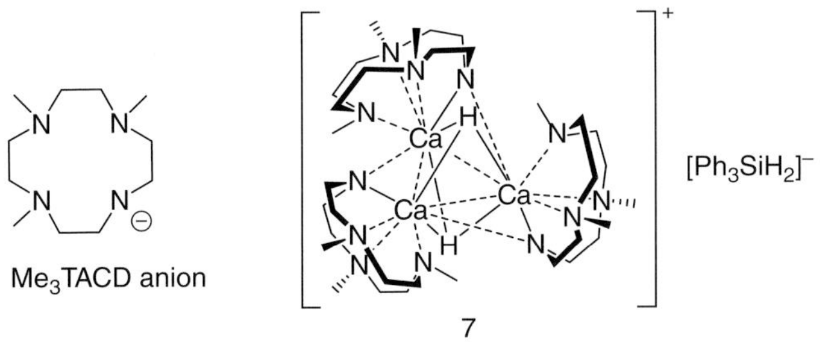

Figure 7.6 A cationic Me_3TACD-stabilized calcium hydride complex (**7**).

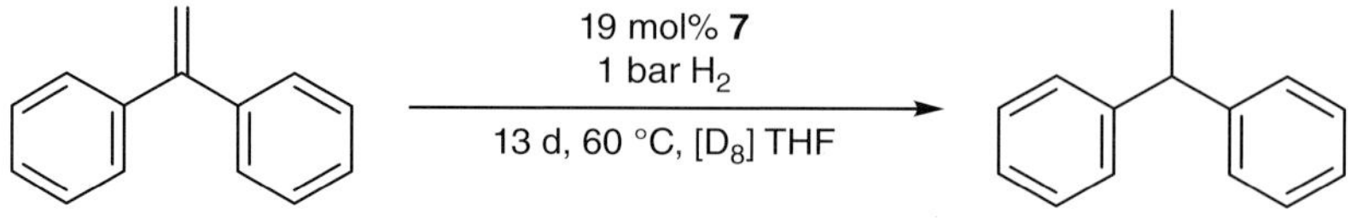

Figure 7.7 Catalytic hydrogenation of 1,1-diphenylethylene with a Me_3TACD-stabilized cationic calcium hydride complex (**7**).

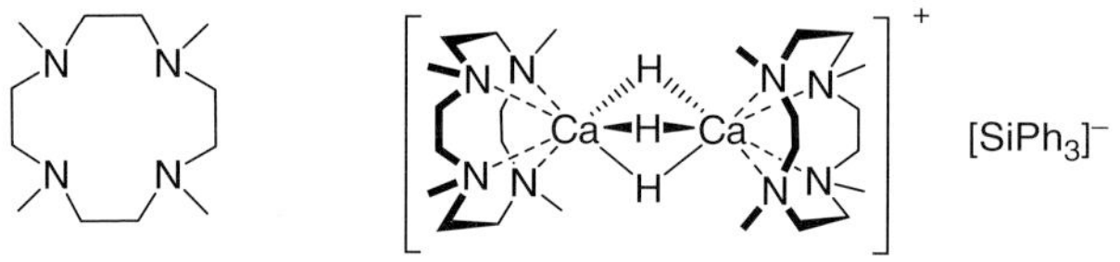

Me_4TACD

8

Figure 7.8 Me_4TACD-stabilized cationic calcium hydride complex (8).

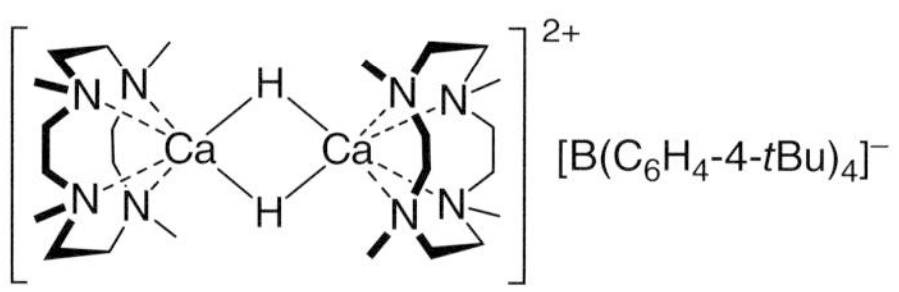

9

Figure 7.9 The Me_4TACD-stabilized dicationic calcium hydride complex (9).

a catalyst loading of 19 mol% full conversion to 1,1-diphenylethane could be observed at 60 °C and 1 bar H_2 pressure after 13 days. Catalysis proceeds at a surprisingly low pressure of 1 bar, and no oligomeric by-products were observed (oligomerization of DPE is known to be difficult [29]). In contrast, it suffers from a high catalyst loading and very long reaction times. It was suggested that the low activity may originate from the poor solubility of the cationic catalyst: complex 7 is quite insoluble even in THF in which the catalysis was performed.

Complexes with increased solubility could be obtained using the neutral tetradentate ligand Me_4TACD [32]. The resulting dinuclear complex of composition $(Me_4TACD)_2Ca_2H_3^+$ (**8**) formed in the reaction of $Ca(SiPh_3)_2$ with H_2 in the presence of Me_4TACD (Figure 7.8) and was found to be soluble in THF. Further development and different synthetic approaches led to the dicationic calcium hydride complex $(Me_4TACD)_2Ca_2H_2^{2+}$ that was isolated as its borate salt (**9**) (Figure 7.9). The latter dicationic complex showed a remarkable activity in the hydrogenation of alkenes, especially with less or nonactivated alkenes [33]. The activated alkene styrene could be hydrogenated at room temperature with 2.5 mol% catalyst loading at 1 bar H_2 (Table 7.1). The low reaction temperature significantly reduced formation of oligomeric by-products to a mere 5%, which is especially noteworthy considering the ease of styrene oligomerization. More importantly, it was found that the catalyst could also fully hydrogenate isolated double bonds such as that in $Me_3SiCH{=}CH_2$. While the latter alkene is still substantially polarized by the electropositive Si substituent, it was shown that also 1-hexene could be reduced. This more challenging unactivated alkene needs an increased catalyst loading of 5 mol% and a somewhat higher reaction temperature of 60 °C. Reaction times for nonactivated alkenes

Table 7.1 Cationic calcium hydride catalyzed hydrogenations of alkenes with 1 bar H_2.

Loading (mol %)	Substrate	Temperature (°C)	Time (h)	Conversion (%)
2.5	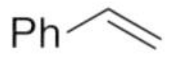	25	10	92
2.5	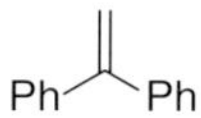	25	6	98
5	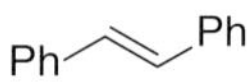	60	24	96
5	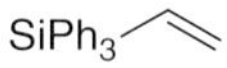	60	16	97
5	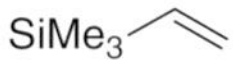	60	36	98
5	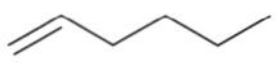	60	24	95
5	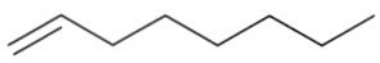	60	24	95
5	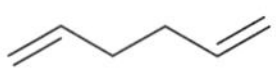	60	36	91
5		60	36	95
5	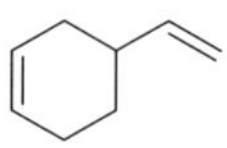	60	36	97[a)]
5		60	24	0
10	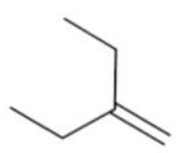	80	24	0

a) Only hydrogenation of the vinyl double bond.

such as 1-hexene, 1-octene, vinylcyclohexene, or other dienes were found to be generally longer (24–36 hours). Although some derivatives such as cyclohexene or 2-ethyl-1-butene could not be hydrogenated with the cationic calcium catalyst, the dicationic complex **9** showed an unprecedented performance. It was reasoned that the 2+ charge on the complex is critical in imparting sufficient electrophilicity to the electropositive calcium center and is crucial for the activation of isolated olefinic bonds by metal–alkene coordination. As the nucleophilicity of the hydride ligands is decreased by the 2+ charge, it seems that electrophilic substrate activation is more important than nucleophilicity. The proposed mechanism for the hydrogenation of alkenes with cationic calcium complexes follows the earlier reported cycle shown in Figure 7.5. A major difference is the reversibility of the insertion step of 1-alkenes into the Ca—H bond. Stoichiometric reactions of the Ca hydride complex with unactivated 1-alkenes showed that the intermediate Ca alkyl species could not be observed or isolated [33]. It was proposed that this Ca alkyl intermediate is unstable toward ß-hydride elimination. However, the small part of Ca alkyl reagent formed in this equilibrium is highly reactive, thus enabling catalysis.

Most recently, the Harder group demonstrated that very simple homoleptic Ae amide reagents, AeN''_2 ($N'' = N(SiMe_3)_2$), can effectively be used in catalytic imine hydrogenation (*vide infra*) [34]. This rather unexpected observation meant that these easily accessible catalysts may be useful in alkene hydrogenation

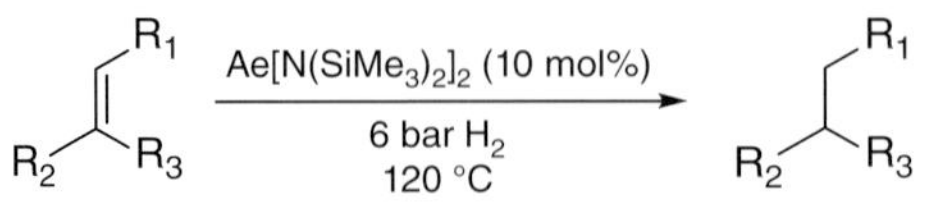

Figure 7.10 Catalytic hydrogenation of alkenes with alkaline earth metal amides AeN''_2 (Ae = Ca, Sr, Ba; N″ = $N(SiMe_3)_2$).

Table 7.2 Alkaline earth metal amide catalyzed hydrogenation of alkenes.

Catalyst	mol %	Substrate	*P* (bar)	*T* (°C)	Time (h)	Product	%
CaN''_2	10	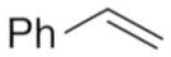	6	80	1.5	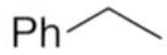	99
CaN''_2	10		6	120	0.5		99
CaN''_2	10		12	80	1		99
CaN''_2	10		1	80	24		97
CaN''_2	1		6	120	24–48		99[a)]
MgN''_2	10		6	120	24		99[a)]
SrN''_2	10		6	120	0.25		99
BaN''_2	10		6	120	0.25		99
BaN''_2	10	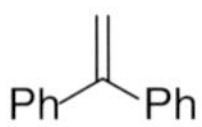	6	120	0.75	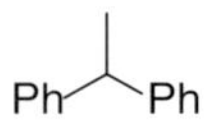	99
BaN''_2	10	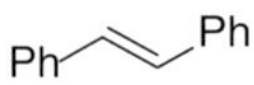	6	120	0.5	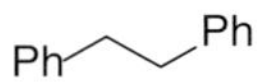	99
BaN''_2	10	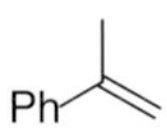	6	120	0.25	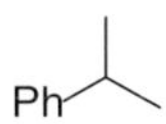	99
BaN''_2	10	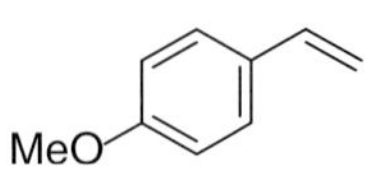	6	120	0.25	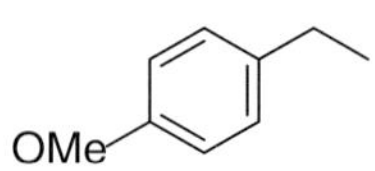	99
BaN''_2	10		6	120	0.5	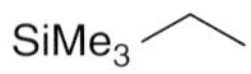	99
BaN''_2	10		6	120	24	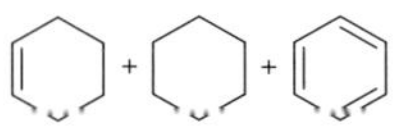 76 1 23	99
BaN''_2	10		6	120	24	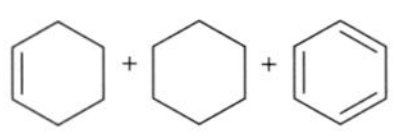 50 30 20	50
BaN''_2	10		6	120	24	 29 39 32	99
BaN''_2	10	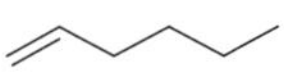	6	120	24	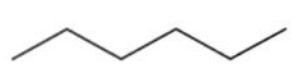	54[b)]
$(DMAT)_2Ca(THF)_2$	10		6	120	20		99

a) C. 5% oligomers formed by thermal styrene polymerization.
b) Isomerization to 2-hexene is observed as a side reaction.

as well. Indeed, a range of catalysts with Ae = Ca, Sr, and Ba smoothly hydrogenated activated alkenes such as styrene (Figure 7.10, Table 7.2) [35]. The activity increases along the row Mg < Ca < Sr < Ba. Using 10 mol% of BaN''_2, hydrogenation of styrene to ethylbenzene was completed within 15 minutes (120 °C, 6 bar H_2). The strontium amide catalyst SrN''_2 showed a similar fast conversion but CaN''_2 is slightly slower (full conversion after 30 minutes). With

MgN''_2 full conversion needed 24 hours. Increase in pressure from 6 to 12 bar showed only minor acceleration. An increase in temperature from 80 to 120 °C decreased the reaction times by a factor of c. 2–3. Using only 1 mol% of catalyst is possible but this extended the reaction times to one to two days. In contrast to alkene hydrogenation using $(DMAT)_2Ae{\cdot}(THF)_2$ (Ae = Ca, Sr) [25], addition of THF inhibited the catalysis completely.

The unique feature of these Ae amide catalysts is their ability to suppress the formation of oligomeric by-products. The key to directing the catalytic cycle toward hydrogenation is the presence of small quantities of $HN(SiMe_3)_2$ formed in the catalyst initiation step (Figure 7.11). The competitive alkene polymerization, which may partially be avoided by applying high H_2 pressure, is now completely shut off by a very fast reaction of the benzyl (or alkyl) metal intermediate with the relatively Brønsted acidic $HN(SiMe_3)_2$ ($pK_a = 25.8$ [36]). This reaction is much faster than deprotonation of the much less acidic H_2 ($pK_a \approx 49$ [37]). Only for sensitive monomers (e.g. styrene) and in those cases where longer reaction times (>24 h) or higher temperature (120 °C) were needed (e.g. for the slower Mg catalyst or at low catalyst loadings) c. 5% oligomers formed by thermally induced styrene polymerization. The free amine $HN(SiMe_3)_2$ generated in the catalyst initiation step is therefore not just a by-product but essentially traps the highly reactive metal benzyl intermediates, thus preventing polymerization. However, increasing the concentration of free

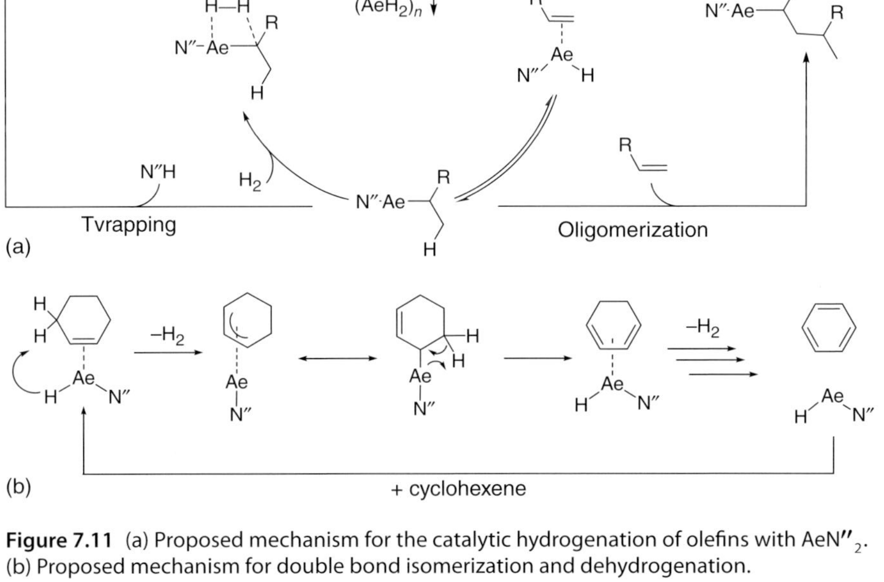

Figure 7.11 (a) Proposed mechanism for the catalytic hydrogenation of olefins with AeN''_2. (b) Proposed mechanism for double bond isomerization and dehydrogenation.

amine by deliberately adding excess $HN(SiMe_3)_2$ slowed down the conversion significantly. A higher $HN(SiMe_3)_2$ concentration shifts the proposed equilibrium to the catalytically inactive species: $AeN''_2 + H_2 \rightleftarrows HAeN''_2 + N''H$. The optimal concentration of free amine N″H is therefore a compromise between prevention of oligomerization and catalyst activity.

Especially the most reactive BaN''_2 catalyst shows a broad scope and was found to hydrogenate a range of styrenic substrates (1,1-diphenylethylene, 1,2-diphenylethylene, α-methyl styrene, and *p*-methoxy styrene) in less than one hour (120 °C, 6 bar). The C=C double bonds of these styrenic substrates are activated by the adjacent phenyl group. As mentioned previously, Okuda's cationic Ca hydride complexes were found to hydrogenate isolated C=C double bonds, such as that in 1-hexene, but this conversion remains challenging. Generally, higher catalyst loadings and longer reaction times are needed. Apart from that a side reaction such as the isomerization of 1-hexene to 2-hexene was found. Since the latter 1,2-substituted alkene could not be further reduced, the yields for hydrogenation are never quantitative.

It was found that especially the most active Ae amide catalyst, BaN''_2, is also able to hydrogenate isolated alkenes showing a surprisingly outstanding performance. For example, trimethylsilyl ethylene was completely reduced within 30 minutes. While the C=C bond in $Me_3SiCH{=}CH_2$ is not conjugated it is to some extent activated since the intermediate carbanion after hydride attack, $Me_3Si(Et)CH^-$, is stabilized by negative hyperconjugation. The isolated double bonds in norbornadiene could also be reduced, giving a mixture of norbornene, nortricyclene, and norbornane. Although the C=C bond in norbornadiene is partially activated by through-space homoconjugation, the observation of norbornane suggests that also truly unactivated C=C bonds may be reduced. Indeed, the BaN''_2 catalyst also hydrogenates the more challenging substrate 1-hexene. The poor conversion of 54% is due to partial isomerization, which, we propose, proceeds through a classical mechanism: deprotonation of 1-hexene gives the 1-*n*Pr-allyl anion, which is protonated by either H_2 or $HN(SiMe_3)_2$ to give 2-hexene. The latter 1,2-disubstituted alkene is inert toward further hydrogenation. This is in line with the observation of Okuda and coworkers that cyclohexene could not be hydrogenated using cationic Ca catalyst **9** [33]. The superb activity of BaN''_2 in hydrogenation catalysis is underscored by the fact that both cyclohexadiene and cyclohexene could be fully converted. However, besides cyclohexane also benzene has been observed. The formation of benzene is formally a dehydrogenation reaction that can be explained by successive deprotonation and ß-hydride elimination reactions (Figure 7.11(b)).

The previously mentioned myrcene hydrogenation by $(DMAT)_2Ca{\cdot}(THF)_2$ (**5**) (see Figure 7.5) suggested that only conjugated double bonds can be reduced. It was found, however, that the same catalyst could quantitatively convert 1-hexene into hexane without alkene isomerization (Table 7.2). This experiment clearly shows that there is a metal-dependent balance between substrate deprotonation, leading to isomerization, and hydride addition, giving reduction. The more important conclusion is that hydrogenation of isolated alkenes is not restricted to cationic Ca catalysts. The ease of C=C double bond reduction seems to be determined more by the alkene substitution pattern. As in transition metal catalyzed alkene hydrogenation, an increasing number of alkene substituents lowers the reactivity of the double bond. This explains why the tri-substituted isolated C=C bond in myrcene is inert toward Ca-catalyzed hydrogenation but 1-hexene can be successfully converted. The simple Ba amide complex BaN″$_2$ is the first group 2 metal catalyst that can also convert 1,2-substituted alkenes such as norbornene or cyclohexene.

7.3 Hydrogenation of C=N Double Bonds

The catalytic hydrogenation of C=N double bonds with molecular hydrogen is relatively unattended and remains a challenge. Harder and coworkers recently demonstrated that aldimines can be hydrogenated under mild conditions (80 °C, 1–6 bar H_2) using very simple, easily available Ae amide catalysts (AeN″$_2$; N″ = $N(SiMe_3)_2$), a discovery that was rather unexpected [34].

Looking at a hypothetical reaction mechanism for the Ae-catalyzed hydrogenation of imines (Figure 7.12), which is related to the catalytic cycle for olefin hydrogenation (Figure 7.5), the difficulties become immediately apparent. The last step of the catalytic cycle (σ-bond metathesis) is formally a deprotonation of H_2. While this deprotonation step was shown to be possible in olefin hydrogenation, it seemed questionable whether it would work in imine hydrogenation. In the latter catalytic cycle H_2 has to be deprotonated by an amide intermediate, which is a much weaker base than the Ca benzyl or alkyl intermediates formed during alkene hydrogenation. Considering the large difference in the pK_a values of H_2 (pK_a ≈ 49 [37]) and amines (pK_a ≈ 35) or N″H (pK_a = 25.8 [36]), efficient and rapid deprotonation of H_2 by the much weaker amide bases should be unlikely. In contrast to expectation, a range of catalysts AeN″$_2$ (Ae = Mg, Ca, Sr, Ba) showed catalytic activity in the reduction of aldimines to amines

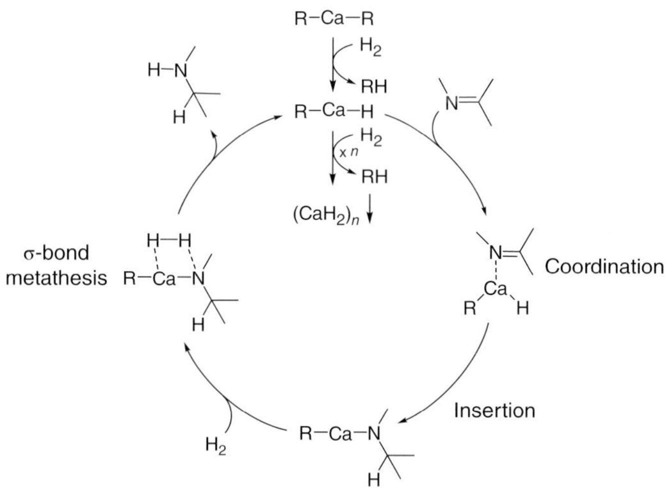

Figure 7.12 Hypothetical catalytic cycle for imine hydrogenation with a Ca catalyst.

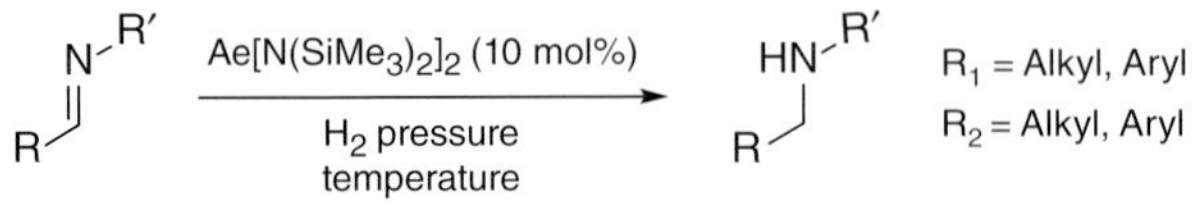

Figure 7.13 Catalytic hydrogenation of aldimines with alkaline earth metal amide catalysts AeN″$_2$.

Table 7.3 Alkaline earth metal mediated hydrogenations of Ph(H)C = N*t*Bu.

Catalyst (mol %)	H_2 (bar)	Temperature (°C)	Time (h)[a]
MgN″$_2$ (10)[b]	6	80	19.0
CaN″$_2$ (10)[b]	6	80	3.0
CaN″$_2$ (10)[b]	6	120	0.5
CaN″$_2$ (10)[b]	12	80	2.25
SrN″$_2$ (10)[b]	6	80	1.25
BaN″$_2$·(THF)$_2$ (10)[b]	6	80	0.75
(DMAT)$_2$Ca·(THF)$_2$ (10)[c]	6	80	<0.25

a) Reaction time for >99% conversion.
b) N″ = N(SiMe$_3$)$_2$.
c) DMAT = 2-dimethylamino-α-trimethylsilyl-benzyl.

(Figure 7.13). The catalyst CaN″$_2$ resulted in, after three hours at 80 °C and 6 bar H_2 pressure, complete hydrogenation of the imine Ph(H)C = N*t*Bu, used in this study as the benchmark substrate (Table 7.3). Doubling the pressure slightly increased the conversion speed but influences of pressure were generally found to be small. Catalytic conversion, however, is strongly dependent on the temperature. The strongly polar solvent THF was found to be detrimental to conversion. This is likely due to metal THF ligation, which blocks coordination and the subsequent reduction of the imine. Also the influence of the amine N″H, liberated during catalyst initiation, was studied. Increasing its concentration led to slower conversion. Catalyst initiation can be considered an acid–base equilibrium and therefore any increase in the N″H concentration shifts the equilibrium to the side of the homoleptic amide complex AeN″$_2$, thus decelerating catalysis. Liberation of the inhibitor N″H during catalyst initiation can be avoided by using the catalyst (DMAT)$_2$Ca·(THF)$_2$ (**5**), which would give DMAT-H after catalyst initiation. The low acidity of DMAT-H makes the initiation reaction a unidirectional process, and consequently very fast and quantitative conversion is observed. Functional groups such as Cl- or MeO-substituents in para-position of the phenyl group in Ph(H)C = N*t*Bu are tolerated. Although various combinations of alkyl or aryl substituents at C or N are allowed, the substrate scope is currently limited to aldimines. Ketimines, a much more challenging group of substrates, could not be hydrogenated to date. The AeN″$_2$ catalysts become more active descending down the group of Ae metals: Mg < Ca < Sr < Ba. The latter Ba amide catalyst, BaN″$_2$·(THF)$_2$, showed a superior performance despite the presence of inhibiting THF ligands.

The catalyst loadings are generally high and could only be lowered to 5 mol%. This major drawback is explained by the formation of larger Ae hydride clusters. In the first step, the precatalyst AeN″$_2$ reacts with H_2 to HAeN″ and N″H. Although this step is quite endothermic, aggregation and ligand exchange of HAeN″ may give larger clusters of the form $Ae_x(H)_y(N'')_{2-y}$, a process that is exothermic. The formation of larger aggregates is therefore the driving force for the generation of catalytically active metal hydride species. Aggregation to larger species, however, also lowers the concentration of the active catalyst present in solution, which may explain the need for higher catalyst loadings. The existence of larger metal hydride clusters could be observed by Diffusion-Ordered-Spectroscopy (DOSY) NMR spectroscopy in which measurement of diffusion speed provides information on particle size. Clusters with molecular weight up to 7500 could be detected. Further evidence for the presence of bigger aggregates could be obtained from the recent discovery of well-defined Ca, Sr, and Ba hydride clusters by Harder and coworkers (Figure 7.14) [27, 28].

Theoretical studies on the mechanism of imine hydrogenation with the CaN″$_2$ as the catalyst have been reported [34]. Three conceptually different reaction mechanisms were systematically evaluated (Figure 7.15). Mechanism A represents the classical metal hydride route, which presupposes the formation of a metal hydride species. Mechanism B involves a bifunctional catalyst, which, similar to the Shvo or Noyori catalysts [38], contains a hydridic $H^{\delta-}$ and a protic $H^{\delta+}$ for dual imine attack. This bifunctional catalyst is the first intermediate in the formation of HCaN″ and is probably short-lived. Mechanism C avoids the formation of a high-energy metal hydride intermediate. This mechanism, which is similar to that proposed by Berkessel et al. for ketone hydrogenation with a KO*t*Bu catalyst [39], involves a six-membered ring transition state. Attempts to optimize the six-membered ring transition states proposed for B and C gave in all cases four-membered ring structures. Either a transition state for hydride-imine attack (B') or for Ca hydride formation (C') was found. This shows that the metal hydride mechanism A is the only true mechanism. The energies in the catalytic

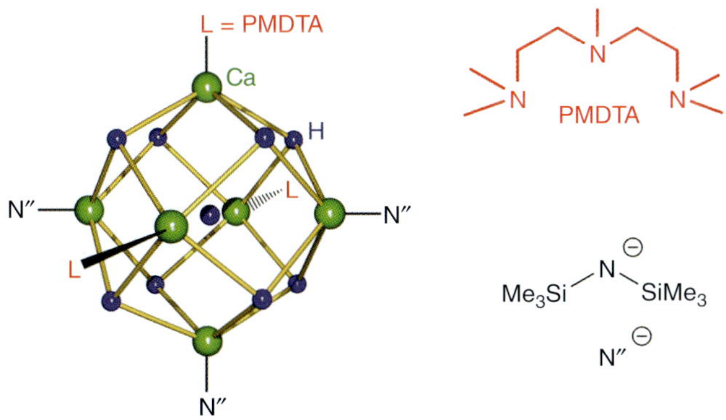

Figure 7.14 A complex Ca hydride cluster isolated from the reaction of CaN″$_2$ and $PhSiH_3$ in the presence of PMDTA.

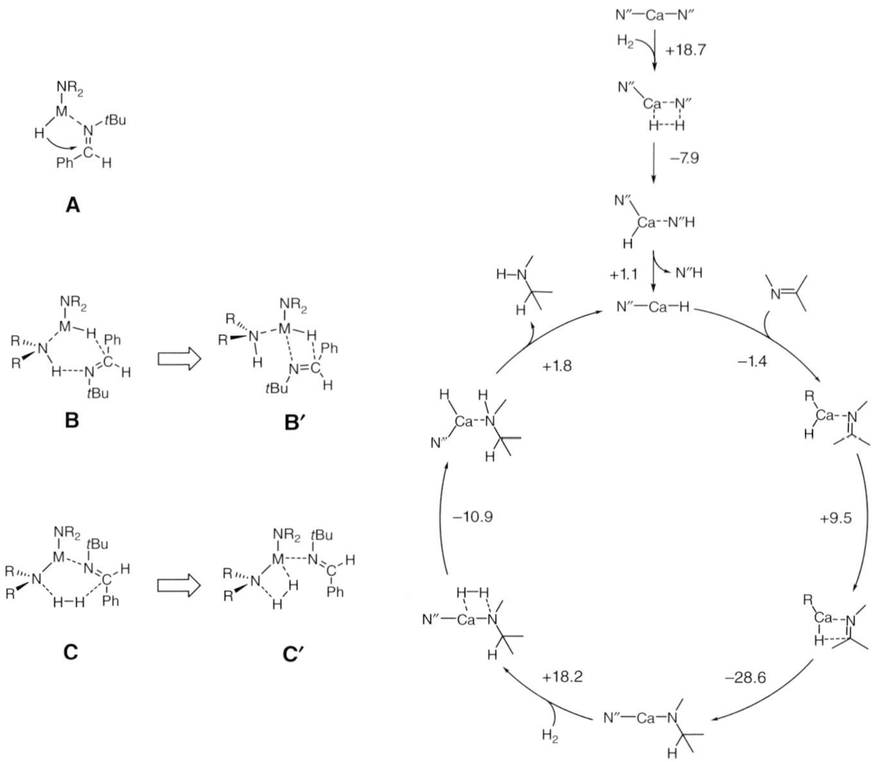

Figure 7.15 Calculated pathways for imine hydrogenation (**A**–**C**) and the proposed mechanism for the catalytic hydrogenation of imines with CaN″$_2$ and H_2; energies given as ΔG (80 °C, 6 bar) in kcal/mol at the level M06/6-311++G(d,p)//M06/6-31++G(d,p).

cycle (Figure 7.15) clearly demonstrate that the deprotonation of H_2 is the critical step (the activation energy for this step is c. 18 kcal/mol).

This method could not be extended to group 1 metal catalysts. The amide catalysts LiN″, NaN″, and KN″ (10 mol%) were tested in the catalytic hydrogenation of Ph(H)C = N*t*Bu (80 °C, 6 bar) but showed only low conversion (up to 29% after 24 hours for KN″). This shows that the higher Lewis acidity of the 2+ alkaline earth cations is essential for an efficient catalysis.

Although the metal amide catalysts CaN″$_2$ are very simple, easily accessible metal complexes, later work by the Harder group showed that imine hydrogenation can also be accomplished using catalytic quantities of $LiAlH_4$, an even simpler commercially available bulk commodity [40]. Under very harsh working conditions, $LiAlH_4$ was previously shown to be an active catalyst for alkyne reduction [12, 13]. Under much milder conditions it was applied as a catalyst for the reduction of alkenes and aldehydes/ketones by the much more reactive hydridoboranes [41]. Imine reduction by the bulk commodity H_2 using the simple catalyst $LiAlH_4$ (down to 2.5 mol%) under mild conditions (1 bar H_2 and 85 °C) was, however, unprecedented. Selected results are summarized in Table 7.4. An increase in pressure accelerates the conversion implying that

Table 7.4 $LiAlH_4$ catalyzed hydrogenations of Ph(H)C = N*t*Bu.

Catalyst (mol %)	H_2 (bar)	Temperature (°C)	Time (h)[a]
$LiAlH_4$ (10)	1	85	2
$LiAlH_4$ (5)	1	85	15
$LiAlH_4$ (2.5)	1	85	66
$LiAlH_4$ (5)	5	85	6
$LiAlH_4$ (5)	1	65	6 [39%]
$LiAlH_4$ (5)	1	100	6 [51%]

a) Reaction time for >99% conversion or, in case no full conversion was reached, time for conversion given in square brackets.

the reaction with H_2 is rate limiting. The reaction needs a higher temperature of 80 °C, but at 100 °C, however, hardly any conversion is found. This can be attributed to thermal decomposition of $LiAlH_4$ to Li_3AlH_6, $Al^{0,}$ and H_2, which lowers the catalyst loading. The reaction cannot be catalyzed by a mixture of LiH and AlH_3. Alternative reducing agents such as RedAl ($LiAlH_2(OCH_2CH_2OMe)_2$) and $NaAlH_4$ decelerated the hydrogenation noticeably whereas $NaBH_4$ is inactive. This indicates a mechanism in which heterobimetallic cooperativity is important. The preliminary mechanism is based on NMR observations, a crystal structure of a potential intermediate, preliminary density functional theory (DFT) calculations, and analogies to the cycle shown in Figure 7.15 (Figure 7.16). Although the activities and scope of $LiAlH_4$-mediated imine hydrogenation are inferior to the previously discussed group 2 metal amide catalysis, its simplicity has far-reaching consequences. Classical imine-to-amine reduction uses stoichiometric quantities of $LiAlH_4$ and ether solvents. The final hydrolysis step is sometimes uncontrolled and risky (especially at a larger scale) and workup can be tedious due to formation of considerable quantities of Li and Al salts as a waste product. In contrast, catalytic imine reduction using H_2 and small amounts of $LiAlH_4$ is a solvent-free 100% atom-economical process that does not require tedious workup and produces much less waste. Since the proposed catalyst consists of different metal cations and different anions, there is plenty of room for catalyst optimization. It is likely that the activity, substrate scope, and perhaps even chemo- and stereoselectivities will be optimized further.

7.4 Hydrogenation of C=O Double Bonds

Interest in transition metal free hydrogenation catalysis was documented already in the 1960s when Walling and Bollyky chose KO*t*Bu as a strong base to hydrogenate benzophenone in *tert*-butanol (Figure 7.17) [42]. The development of the catalytic hydrogenation of C=O bonds with simple main group metal bases was based on two observations. First, it was found that bases catalyze H/D exchange between D_2 and H_2O. This implies the unusual chemical equilibrium shown in

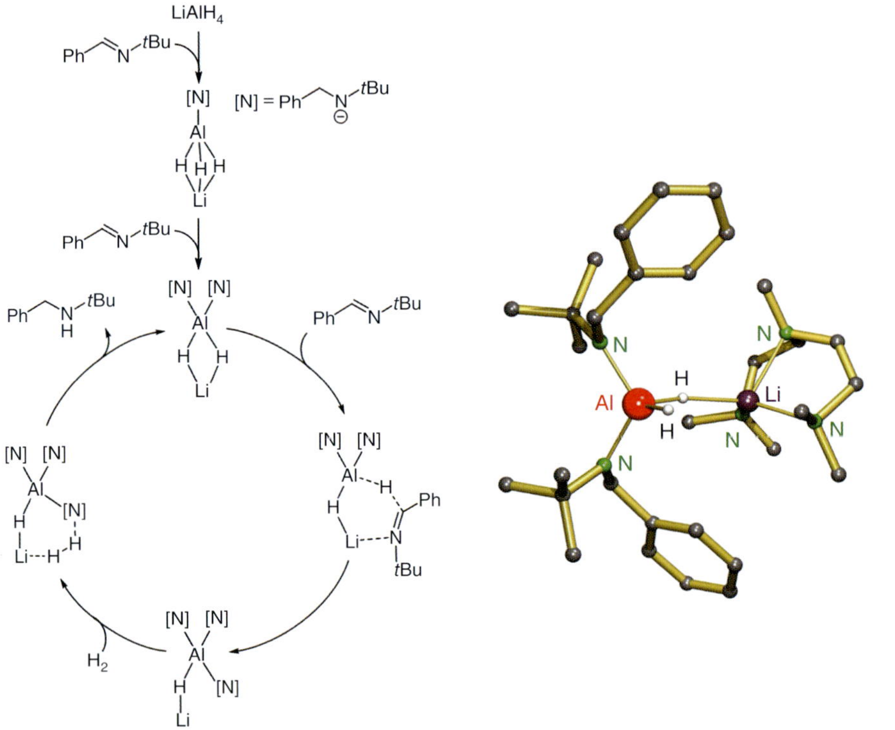

Figure 7.16 Proposed mechanism for the catalytic hydrogenation of imines with $LiAlH_4$ and H_2. The crystal structure of a possible catalyst is shown: $[N]_2AlH_2Li{\cdot}PMDTA$.

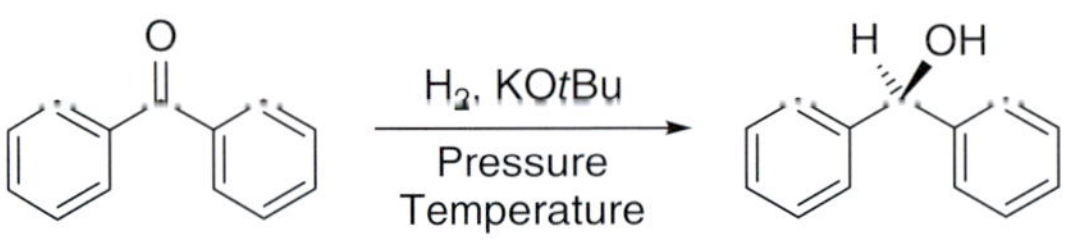

Figure 7.17 Catalytic hydrogenation of benzophenone with KO*t*Bu.

Eq. (7.1), which most unfavorably lies strongly to the side of D_2 [43]. Second, organic molecules can be dehydrogenated by strong bases such as in the Varrentrapp reaction [44]. This shows that under suitable conditions (e.g. high H_2 pressure) base-catalyzed reverse hydrogenation catalysis might be feasible and this led Walling et al. to propose an alkoxide-mediated ketone hydrogenation along the lines of Eqs. (7.3)–(7.4). Deprotonation of H_2 by an alkoxide would be highly unusual but a small percentage of active hydride may be sufficient to catalyze hydrogenation.

$$D_2 + OH^- \rightleftharpoons D^- + HOD \tag{7.1}$$

$$RO^- + H_2 \rightleftharpoons ROH + H^- \tag{7.2}$$

$$H^- + R_2C{=}O \rightleftharpoons R_2CH{-}O^- \tag{7.3}$$

$$R_2CHO^- + ROH \rightleftharpoons R_2CHOH + RO^- \tag{7.4}$$

It was indeed found that with the very simple alkali metal catalyst KO*t*Bu good conversion of benzophenone to benzhydrol can be reached, provided that harsh reaction conditions are applied (150–200 °C, 100 bar H_2). In general, reactions times are long and the catalyst loadings are high (Table 7.5). With a catalyst loading of 20 mol%, the reduction of benzophenone is completed within 25 hours (135 bar, 210 °C). Faster conversion at lower temperature can be achieved with a threefold excess of KO*t*Bu. Among the different solvents tested in this hydrogenation, diethylene glycol dimethyl ether (diglyme) was found to be most effective. The reason for the acceleration of the hydrogenation in diglyme could be the better solvation of the cation, making $t\mathrm{BuO}^-$ a stronger base. The substrate scope of this method is limited to stable, non-enolizable ketones. In order to improve the method and reach broader applicability, Berkessel et al. investigated the mechanism of this catalytic hydrogenation of ketones in detail [39] and could expand the scope to ketones such as PhC(O)*t*Bu, $(t\mathrm{Bu})_2\mathrm{CO}$, and 2,2,5,5-tetramethylcyclopentanone. The latter aliphatic substrate showed a lower conversion than aromatic substrates. The fast conversion of Ph ring containing ketones may be attributed to a K^+–Ph π interaction but may also be due to activation of the C=O bond by conjugation with the Ph ring. In contrast to the original procedure, Berkessel was not able to obtain any conversion in diglyme as solvent. A decrease in the reaction temperature also gave decreased reaction rates, as observed earlier. Kinetic investigations show that the reduction is first order in substrate, base, and H_2. Different alkaline metals have been tested, showing an increasing activity along the line $\mathrm{Li} \ll \mathrm{Na} \ll \mathrm{K} \approx \mathrm{Rb} < \mathrm{Cs}$, which follows the covalence of the metal–oxygen bond. Experiments with deuterium revealed no effect on the kinetic course of the hydrogenation, thus excluding a kinetic isotope effect. Using D_2 for the hydrogenation of Ph_2CO gave only 60% deuteration incorporation in the final product benzhydrol. This indicates a base-catalyzed H/D exchange between D_2 and the solvent *t*BuOH, which could be confirmed in the absence of substrate. The experimental observations led to the conclusion that a mechanism related to Noyori's Ru-catalyzed asymmetric hydrogenation of ketones [45] could be assumed. This mechanism is based on a six-membered intermediate in which H_2 is activated by bridging between the alkoxide and the ketone C moieties (Figure 7.18). This transition state circumvents the highly endothermic formation of KH from KO*t*Bu and H_2. Most recently, however, calculations have shown that this process is not feasible.

Table 7.5 KO*t*Bu catalyzed hydrogenations of benzophenone with H_2 in *t*BuOH.

Catalyst (mol %)	H_2 (bar)	Temperature (°C)	Time (h)[a]	Conversion (%)
KO*t*Bu (20)	135	210	25	98
KO*t*Bu (20)	102	153	50.5	47
KO*t*Bu (300)	96	150	14.5	98
KO*t*Bu (20)	100	170	18	52[a]
KO*t*Bu (300)	78	130	5	98[a]

a) Diglyme as solvent.

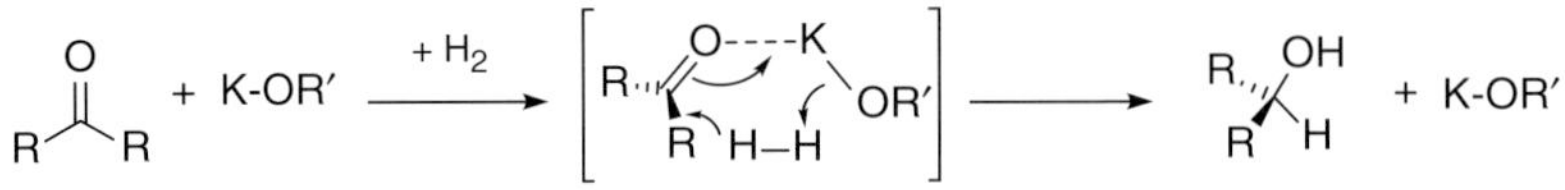

Figure 7.18 Base-catalyzed hydrogenation of ketones.

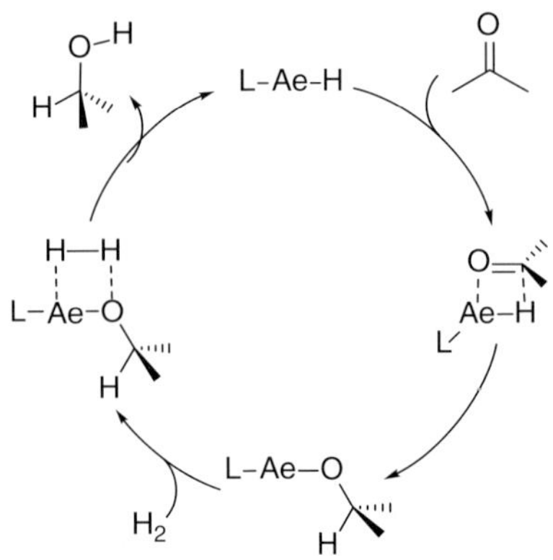

Figure 7.19 Possible mechanism for the catalytic hydrogenation of ketones with alkaline earth metals.

Instead, a two-step mechanism is proposed. In the first stage of the reaction KH is formed, which in the second step reacts with the ketone [46]. This indicates that, in contrast to chemical intuition, indeed a highly reactive short-living K hydride species can be formed by deprotonation of H_2 with the very weak base KO*t*Bu. A mechanism for the potential, hitherto unobserved, hydrogenation of ketones with Ae metal catalysts could therefore be similar to that of imine hydrogenation (Figure 7.19).

7.5 Summary and Perspectives

Although catalytic hydrogenation of unsaturated bonds is likely one of the most investigated processes, s-block metal catalyzed conversions are still a big challenge and hitherto underdeveloped. Compared to transition metal catalyzed hydrogenation, the scope and the activity of early main group metal complexes toward H—H bond activation are still limited.

The use of Ca, Sr, or even K catalysts in alkene hydrogenation broke the dogma that transition metals are needed for hydrogenation catalysis. The initial results were far from being competitive with commercially available transition metal hydrogenation catalysts. The scope seemed to be limited to activated alkenes. Conjugated double bonds, such as those in styrene or cyclohexadiene, are much easier targets for hydrogenation. This is due to the higher stability of the benzyl or allyl intermediates along the catalytic pathway. Other drawbacks are the low functional group tolerance and the observation of alkene oligomerization as an undesired side reaction (especially for alkenes sensitive to polymerization, e.g. styrene). Synthesis and isolation of cationic Ca hydride complexes gave impetus to the field. These cationic catalysts generally showed a somewhat higher reactivity but catalysis can be complicated by the low solubility of these species. The dicationic dimeric complex $[(Me_4TACD)_2Ca_2H_2]^{2+}$ (**9**) showed very high activity and allowed for hydrogenation of unactivated (nonconjugated) alkenes such as 1-hexene; however, internal (doubly substituted) C=C bonds as in cyclohexene were unaffected. It was postulated that the cationic nature of **9** makes the metals more Lewis acidic and therefore more efficient in activating the C=C bond for hydrogenation. Later observations showed that also neutral Ca catalysts such as $(DMAT)_2Ca{\cdot}(THF)_2$ (**5**) are able to hydrogenate isolated double bonds, e.g. 1-hexene. As part of a general trend to use simpler catalysts, the readily available group 2 metal amides (AeN''_2, $N'' = N(SiMe_3)_2$) have been investigated in alkene hydrogenation. Using metal amide complexes as precatalysts in alkene hydrogenation has a major advantage. During the initiation step the free amine is formed as a side product: $AeN''_2 + H_2 \rightarrow HAeN'' + N''H$. The latter amine is relatively acidic and rapidly traps any Ae alkyl intermediate, thus suppressing alkene oligomerization as an undesired side reaction. Apart from their simplicity and facile synthetic access, these catalysts have another advantage. The heaviest barium amide complex BaN''_2 showed an unsurpassed reactivity. Preliminary catalytic tests demonstrated that it not only converts isolated alkenes such as 1-hexene but also doubly substituted alkenes such as cyclohexene or norbornene. Therefore, unactivated double bonds may be more difficult targets in hydrogenation catalysis but also neutral Ae metal complexes are able to transform these functionalities. It seems that the number of substituents at the double bond influences the reactivity. As in transition metal catalyzed alkene hydrogenation, the reactivity of alkenes decreases with the number of substituents. While activated double bonds with two substituents can be hydrogenated easily (*cf*. DPE or cyclohexadiene), doubly substituted unactivated double bonds are a challenge that so far has only been met by the Ba catalyst BaN''_2. The hydrogenation of cyclohexadiene led to the formation of benzene. The observation of this *de*hydrogenation product suggested that the catalytic cycle for hydrogenation is fully reversible, which would pave the way for transfer hydrogenation catalysis. Indeed, during the writing of this chapter the first successful alkene transfer hydrogenation using 1,4-cyclohexadiene as a proton source and simple AeN''_2 catalysts was reported [47]. DFT calculations support a mechanism that proceeds through an intermediate with an unstabilized Meisenheimer anion ($C_6H_7^-$) that rapidly transfers a hydride to the metal (Figure 7.20). This reactivity provides an alternative pathway to formation of the metal hydride catalyst. Catalytic activity for BaN''_2 is generally substantially higher than for CaN''_2 and apart from activated alkenes (styrene, diphenylethylene) also the isolated double bond in 1-hexene could be cleanly reduced without isomerization.

Simple catalysts such as AeN''_2 were also shown to be active in imine hydrogenation with an increasing reactivity down the group $Mg < Ca < Sr < Ba$. The method is limited to aldimines but in some cases even the commercially available bulk metal hydride reagent $LiAlH_4$ could be used in catalytic quantities down to 2.5 mol% and at relatively mild reaction conditions (80 °C and 1 bar H_2) [40]. The latter heterobimetallic catalyst offers great potential for variation and catalyst optimization by choosing the best combination of s- and p-block metals.

Ketone hydrogenation is a much less developed field. This is due to the very difficult formation of metal hydride species by the reaction of metal alkoxides (an

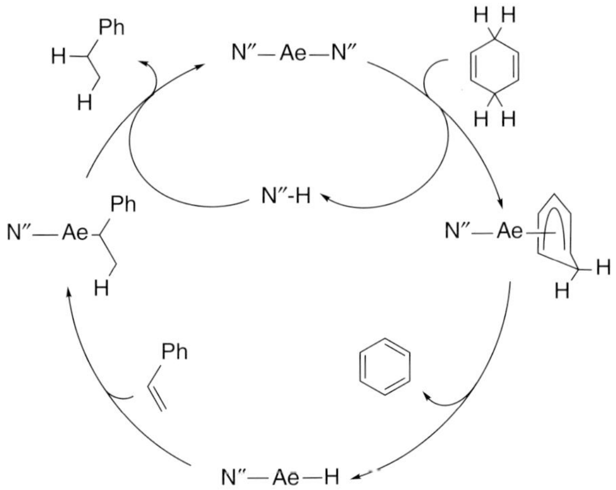

Figure 7.20 Alkene transfer hydrogenation with AeN″$_2$ catalysts exemplified by the reduction of styrene using 1,4-cyclohexadiene as a proton source.

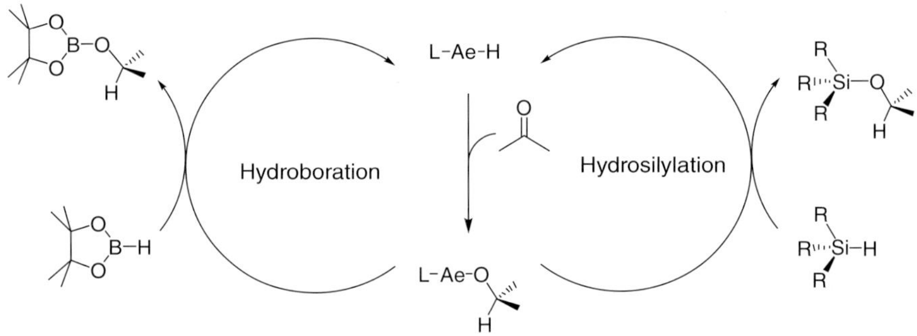

Figure 7.21 Reduction of ketones by catalytic hydroboration and hydrosilylation.

intermediate in the catalytic cycle) with molecular H_2. Consequently, harsh reaction conditions are hitherto always a prerequisite. For the reduction of ketones it is much easier to use the more polar silanes (R_3SiH) as a reducing reagent (Figure 7.21) as has been demonstrated by Harder et al. [48]. Apart from hydrosilylation, also hydroboration using polar hydridoboranes (HBpin) can be easily accomplished as has been reported by Hill and coworkers (Figure 7.20) [49]. The driving force in these catalytic cycles is the very high oxophilicity of B or Si.

All main breakthroughs in early main group metal catalyzed hydrogenation catalysis have been reported in the twenty-first century, which shows that this is still a very young field with full potential for future investigations. It would be an exciting goal to include electron transfer or redox processes in hydrogenation catalysis. Main group metals are not known for a wide variety of oxidation states but the recent isolation of Mg(I) complexes (Figure 7.22) by Jones and coworkers [50] may provide opportunities for such redox-based catalysis. Unlike low valent Ni^0 complexes, compounds of the type L–Mg–Mg–L so far did not show any redox reactivity with molecular H_2 [51]. Heterobimetallic complexes with polarized metal–metal bonds may be the key to the heterolytic cleavage of H_2 but the design of a catalytic cycle with an oxidative H—H bond activation step presupposes also a catalytic metal reduction step, which likely is the largest challenge. For that matter, introduction of redox non-innocent ligands may be a more fruitful approach to such chemistry.

Although the limitations of s-block metal catalyzed hydrogenation hitherto pose an open question, the increasing number of young researchers entering the field promises a rapid development. It is likely that the near future will see important breakthroughs and improvements in substrate scope and functional group tolerance.

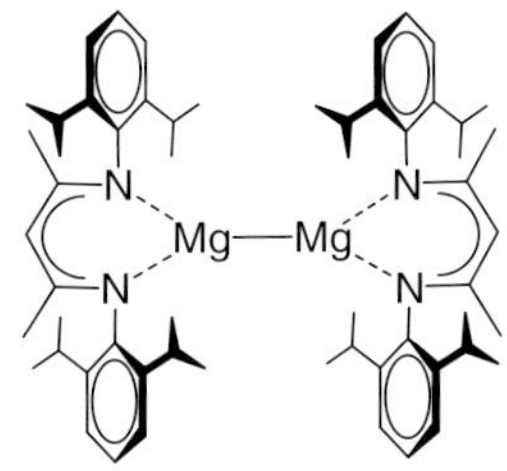

Figure 7.22 Mg(I) complex.

References

1 Steinborn, D. (2012). *Fundamentals of Organometallic Catalysis*, vol. 1, 49–50. Weinheim, Germany: Wiley-VCH.

2 von Wilde, M.P. (1874). *Chem. Ber.* 7: 352–357.

3 (a) Sabatier, P. (1913). *La Catalyse en chimie organique*. Paris, Béranger. https://gallica.bnf.fr/ark:/12148/bpt6k9615479v/f15.image.texteImage. (b) Sabatier, P. and Nye, M.J. (2004). *Chem. World* 1: 46–49.

4 Nishimura, S. (2001). *Handbook of Heterogeneous Catalytic Hydrogenation for Organic Synthesis*. Weinheim, Germany: Wiley.

5 (a) Wilkinson, G. (1966). *J. Am. Soc. (A)* 1711–1732. (b) Wilkinson, G. (1968). *Bull. Soc. Chim. Fr.* 12: 5055–5058.

6 (a) Schrock, R.R. and Osborn, J.A. (1976). *J. Am. Chem. Soc.* 98: 2134–2143. (b) Schrock, R.R. and Osborn, J.A. (1976). *J. Am. Chem. Soc.* 98: 2143–2147. (c) Schrock, R.R. and Osborn, J.A. (1976). *J. Am. Chem. Soc.* 98: 4450–4455.

7 Crabtree, R.H. (1979). *Acc. Chem. Res.* 12: 331–337.

8 (a) Zirngibl, C., Hedderich, R., and Thauer, R.K. (1990). *FEBS Lett.* 261: 112–116. (b) Thauer, R.K., Klein, A.R., and Hartmann, G.C. (1996). *Chem. Rev.* 96: 3031–3042. (c) Berkessel, A. (2001). *Curr. Opin. Chem. Biol.* 5: 486–490. (d) Shima, S. and Thauer, R.K. (2007). *Chem. Rec.* 7: 37–46.

9 (a) Lyon, E.J., Shima, S., Burrman, G. et al. (2004). *Eur. J. Biochem.* 271: 195–204; (b) Korbas, M., Vogt, S., Meyer-Klaucke, W. et al. (2006). *J. Biol. Chem.* 281: 30804–30813.

10 (a) Shima, S., Pilak, O., Vogt, S. et al. (2008). *Science* 321: 572–575; (b) Hiromoto, T., Ataka, K., Pilak, O. et al. (2009). *FEBS Lett.* 583: 585–590.

11 (a) Wright, J.A., Turrell, P.J., and Pickett, C.J. (2010). *Organometallics* 29: 6146–6156. (b) Kalz, K.F., Brinkmeier, A., Dechert, S. et al. (2014). *J. Am. Chem. Soc.* 136: 16626–16634.

12 Slaugh, L.H. (1966). *Tetrahedron* 22: 1741–1746.

13 Slaugh, L.H. (1967). *J. Org. Chem.* 32: 108–113.

14 Jeske, G., Lauke, H., Mauermann, H. et al. (1985). *J. Am. Chem. Soc.* 197: 8111–8118.

15 Haar, C.M., Stern, C.L., and Marks, T.J. (1996). *Organometallics* 15: 1765–1784.

16 Obora, Y., Ohta, T., Stern, C.L., and Marks, T.J. (1997). *J. Am. Chem. Soc.* 119: 3745–3755.

17 Willoughby, C.A. and Buchwald, S.L. (1994). *J. Am. Chem. Soc.* 116: 11703–11714.

18 Chase, P.A., Welch, G.C., Jurca, T., and Stephan, D.W. (2007). *Angew. Chem. Int. Ed.* 46: 8050–8053.

19 Heiden, Z.M. and Stephan, D.W. (2011). *Chem. Commun.* 47: 5729–5731.

20 Greb, L., Oña-Burgos, P., Schirmer, B. et al. (2012). *Angew. Chem. Int. Ed.* 51: 10164–10168.

21 (a) Stephan, D.W. (2010). *Chem. Commun.* 436: 8526–8533. (b) Hounjet, L.J. and Stephan, D.W. (2014). *Org. Process Res. Dev.* 18: 385–391. (c) Stephan, D.W. (2015). *Acc. Chem. Res.* 48: 306–316. (d) Stephan, D.W. and Erker, G. (2015). *Angew. Chem. Int. Ed.* 54: 6400–6441. (e) Stephan, D.W. (2016). *Science* 354: 1248–1256.

22 Wang, Y., Chen, W., Lu, Z. et al. (2013). *Angew. Chem. Int. Ed.* 52: 7496–7499.

23 (a) Jochmann, P. and Stephan, D.W. (2013). *Angew. Chem. Int. Ed.* 52: 9831–9835. (b) Jochmann, P. and Stephan, D.W. (2014). *Chem. Eur. J.* 20: 8370–8378.

24 Hatnean, J.A., Thomson, J.W., Chase, P.A., and Stephan, D.W. (2014). *Chem. Commun.* 50: 301–303.

25 Spielmann, J., Buch, F., and Harder, S. (2008). *Angew. Chem. Int. Ed.* 47: 9434–9438.

26 Harder, S. and Brettar, J. (2006). *Angew. Chem. Int. Ed.* 45: 3474–3478.

27 Maitland, B., Wiesinger, M., Langer, J. et al. (2017). *Angew. Chem. Int. Ed.* 56: 11880–11884.

28 Wiesinger, M., Maitland, B., Färber, C. et al. (2017). *Angew. Chem. Int. Ed.* 56: 16654–16659.

29 (a) Weeber, A., Harder, S., Brintzinger, H.-H., and Knoll, K. (2000). *Organometallics* 19: 1325–1332. (b) Feil, F., Knoll, K., and Harder, S. (2001). *Angew. Chem. Int. Ed.* 40: 4261–4264.

30 Wilson, A.S.S., Dinoi, C., Hill, M.S. et al. (2018). *Angew. Chem. Int Ed.* 57: 15500–15504.

31 Jochmann, P., Davin, J.P., Spaniol, T.P. et al. (2012). *Angew. Chem. Int. Ed.* 51: 4452–4455.

32 Leich, V., Spaniol, T.P., Maron, L., and Okuda, J. (2016). *Angew. Chem. Int. Ed.* 55: 4794–4797.

33 Schuhknecht, D., Lhotzky, C., Spaniol, T.P. et al. (2017). *Angew. Chem. Int. Ed.* 56: 12367–12371.

34 Bauer, H., Alonso, M., Färber, C. et al. (2018). *Nat. Catal.* 1: 40–47.

35 Bauer, H., Alonso, M., Fischer, C. et al. (2018). *Angew. Chem. Int. Ed.* 57: 15177–15182.

36 Fraser, R.R., Mansour, T.S., and Savard, S. (1985). *J. Org. Chem.* 50: 3232–3234.

37 Abdur-Raschid, K., Fong, T.P., Greaves, B. et al. (2000). *J. Am. Chem. Soc.* 122: 9155–9171.

38 Samec, J.S.M. and Bäckvall, J.-E. (2002). *Chem. Eur. J.* 8: 2955–2961.

39 Berkessel, A., Schubert, T.J.S., and Müller, T.N. (2002). *J. Am. Chem. Soc.* 124: 8693–8698.

40 Elsen, H., Färber, C., Ballmann, G., and Harder, S. (2018). *Angew. Chem. Int. Ed.* 57: 7156–7160.

41 (a) Pollard, V.A., Orr, S.A., McLellan, R. et al. (2018). *Chem. Commun.* 54: 1233–1236. (b) Bismuto, A., Thomas, S.P., and Cowley, M.J. (2018). *ACS Catal.* 8: 2001–2005.

42 (a) Walling, C. and Bollyky, L. (1961). *J. Am. Chem. Soc.* 83: 2968–2969. (b) Walling, C. and Bollyky, L. (1964). *J. Am. Chem. Soc.* 86: 3750–3752.

43 (a) Wirtz, K. and Bonhoeffer, K.F. (1936). *Z. Physik. Chem.* 177A: 1–6. (b) Wilmarth, W.K., Dayton, J.C., and Fluornoy, J.M. (1953). *J. Am. Chem. Soc.* 75: 4549–4553. (c) Miller, S.L. and Rittenberg, D. (1958). *J. Am. Chem. Soc.* 80: 64–65.

44 Varrentrapp, F. (1840). *Liebig's Ann.* 35: 196.

45 (a) Noyori, R., Yamakawa, M., and Hashiguchi, S. (2001). *J. Org. Chem.* 66: 7931–7944. (b) Noyori, R. and Ohkuma, T. (2001). *Angew. Chem. Int. Ed.* 40: 40–73. (c) Ohkuma, T., Koizumi, M., Ikehira, H. et al. (2000). *Org. Lett.* 2: 659–662. (d) Yamakawa, M., Ito, H., and Noyori, R. (2000). *J. Am. Chem. Soc.* 122: 1466–1478.

46 Dub, P.A., Henson, N.J., Martin, R.L., and Gordon, J.C. (2014). *J. Am. Chem. Soc.* 136: 3505–3521.

47 Bauer, H., Thum, K., Alonso, M. et al. (2019). *Angew. Chem. Int. Ed.* 58, early view.

48 Spielmann, J. and Harder, S. (2008). *Eur. J. Inorg. Chem.* 1480–1486.

49 Arrowsmith, M., Hadlington, T.J., Hill, M.S., and Kociok-Köhn, G. (2012). *Chem. Commun.* 48: 4567–4569.

50 Green, S.P., Jones, C., and Stasch, A. (2007). *Science* 318: 1754–1757.

51 Bonyhady, S.J., Jones, C., Nembenna, S. et al. (2010). *Chem. Eur. J.* 16: 938–955.

8

Alkali and Alkaline Earth Element-Catalyzed Hydroboration Reactions

Aaron D. Sadow

Iowa State University, US Department of Energy Ames Laboratory, Center for Catalysis, and Department of Chemistry, 2438 Pammel Dr, Ames IA, 50011, USA

8.1 Introduction and Overview

The addition of a boron–hydrogen bond to an unsaturated organic moiety is known as hydroboration. This class of transformations is useful for the synthesis of compounds containing boron–carbon bonds via addition to alkenes and alkynes; boron–nitrogen bonds by additions to imines, pyridines, nitriles, and carbodiimides; and boron–oxygen bonds through additions to aldehydes, ketones, and esters. The BX_2 group in these products may be substituted by hydrogen through hydrolysis (sometimes called protodeborylation), resulting in a two-step process that is a functional equivalent to hydrogenations of alkenes to alkanes, carbonyls to alcohols, or imines to amines. Hydroboration of esters results in C—O bond cleavage, whereas hydroboration of amides provides amines via C=O cleavage. Alternatively, oxidation of boronate esters ($R–BX_2$) provides alcohols, and RBX_2 are also very useful for C—C bond forming reactions (i.e. Suzuki coupling).

Diborane (B_2H_6), diethylborane, 9-borabicyclo[3.3.1]nonane (9-BBN), and related boranes react spontaneously with alkenes and other unsaturated functional groups in useful, catalyst-free transformations. Applications of hydroboration in organic synthesis have been reviewed in detail [1, 2]. Additions of boranes to alkenes and alkynes occur with syn or cis selectivity, respectively, and provide anti-Markovnikov products. These reactions contrast and complement methods that afford Markovnikov products, such as acid-mediated hydrations. The proposed mechanism for hydroboration of alkenes by three-coordinate boranes invokes the unoccupied p orbital on boron as a Lewis acid acceptor that interacts with the filled π-orbital on the alkene, followed by transfer of hydrogen from boron to carbon via a four-center transition state (Figure 8.1a). Alternatively, ketones and aldehydes are reduced by borohydride reagents, such as sodium borohydride. $NaBH_4$ is considerably less reactive toward esters, amides, carbamates, and imines. The boron center in anionic borohydrides is four-coordinate, and thus, borohydrides are proposed to react with carbonyls as nucleophilic hydride transfer reagents (Figure 8.1b).

Early Main Group Metal Catalysis: Concepts and Reactions,
First Edition. Edited by Sjoerd Harder.

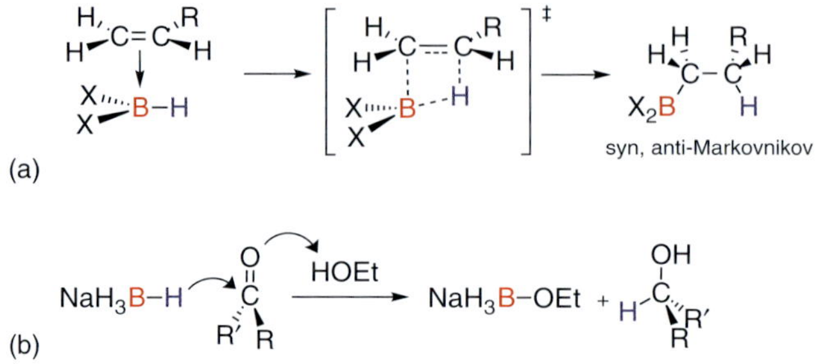

Figure 8.1 Generally proposed mechanisms for catalyst-free (a) hydroboration and (b) borohydride reduction.

Nonetheless, direct hydroboration reactions are often limited by competing pathways. For example, reactions of unsaturated fatty esters give mixtures resulting from ester reduction and hydroboration. Instead, catalyzed additions of less reactive 4,4,5,5-tetramethyl-1,3,2-dioxaborolane (pinacolborane, HBpin) and 1,3,2-benzodioxaborole (catecholborane, HBcat) may kinetically dictate reaction features, including chemoselectivity, regioselectivity, stereochemistry, and/or enantioselectivity [3]. Rhodium-based molecular catalysts have received considerable attention. Some catalysts, such as $RhCl(PPh_3)_3$ (Wilkinson's catalyst), mediate the hydroboration reaction of catecholborane and alkenes in the presence of carbonyl groups [4]. Other catalysts, such as trialkylphosphine-coordinated rhodium, afford trans-selective hydroborations of alkynes with catecholborane or pinacolborane in the presence of triethylamine [5]. Early transition metal (e.g. group 4) [6–9], first-row metal [10–17], lanthanide complexes [18–23], and even group 13 and 14 compounds [24–26] catalyze hydroborations of alkenes and carbonyls.

Many of these reactions are readily studied by ^{11}B NMR (nuclear magnetic resonance) spectroscopy, which distinguishes boron centers in three-coordinate boranes, four-coordinate neutral boranes, and four-coordinate borates by chemical shift [27]. Naturally abundant ^{11}B (80%) is a quadrupolar nucleus ($I = 3/2$), thus lower symmetry species (BX_3) give broader signals than higher symmetry compounds (BX_4). In addition, $^1J_{BH}$ coupling, even in broad three-coordinate boranes, reveals the number of hydrogen atoms bonded to the boron center. For example, HBpin appears as a doublet at 28.5 ppm ($^1J_{BH} = 174$ Hz) [8]. The ^{11}B NMR spectrum of HBcat contains a doublet at 28.8 ppm (benzene-d_6, $^1J_{BH} = 193$ Hz) [28].

An early review on transition-metal-catalyzed hydroboration has been supplemented with an analysis of recent studies on additions of boranes and polar unsaturated moieties, such as carbonyls [29, 30]. In addition, recent reviews summarized the catalytic chemistry of main group elements in reductions [31] and group 2 reagents in catalysis [32]. Main group metal catalysts provide alternative reaction pathways to direct hydroborations with BH_3, borohydride reductions, and transition-metal-catalyzed additions. Trends of hydroboration reactivity with these catalysts tend to follow (perhaps) expected patterns, with aldehydes and ketones typically showing higher numbers of catalytic turnover

(TON = moles product per moles catalyst) and faster rates than esters, imines and pyridines, and amides. The present chapter highlights recent developments in group 1 and group 2 catalyzed hydroborations.

Because alkylboranes and BH_3·tetrahydrofuran (THF) react with carbonyls, alkenes, and other unsaturated functionality in the absence of catalysts [1, 2], side reactions of HBpin and HBcat with metal centers to form such species could lead to B–H additions occurring by uncatalyzed pathways. Transition metal complexes, such as Cp_2TiCl_2, not only act as catalysts for $[BH_4]^-$ additions to alkenes but also react to generate B_2H_6 [33]. Moreover, Cp_2ZrH_2 and HBcat react to form $Cp_2Zr(BH_4)_2$ [34]. A calcium hydride reacts with HBcat to give $[H_2Bcat]^-$, $B_2(cat)_3$, and B_2H_6 [28]. Likewise, $(Cp^*{}_2LaH)_2$ and HBpin react to form $Cp_2La\{\kappa^2\text{-}O(BH_3)\text{–}(CMe_2)_2\text{–}OBpin\}$, which could also produce B_2H_6 [19]. Related rearrangements have been noticed in the presence of magnesium compounds [35, 36]; interestingly, these two examples come from studies on pyridine or imine hydroboration. Thus, contamination of HBpin and HBcat can generate BH_3; ^{11}B NMR of starting materials and products is critical to avoid mischaracterizing hydroboration processes.

In addition, some ambiguity exists in the literature with regard to the need for catalysts in hydroboration reactions. In the initial report of HBpin, terminal alkynes, internal alkynes, and alkenes react with excess HBpin (2 equiv) in methylene chloride at room temperature [3]. In that study, esters, nitriles, and alkyl halide functional groups are inert to HBpin. Aldehydes and HBpin are also reported to react under solvent-free conditions [37], and many early studies suggested slow additions at elevated temperatures.

8.2 Thermodynamic Considerations

8.2.1 Hydroboration, Hydrosilylation, and Hydrogenation

Because the overall transformation of hydrosilylation and hydroboration, followed by hydrolysis, is functionally equivalent to hydrogenation, it is worth comparing the thermodynamic constraints of each reaction. The enthalpy for hydrogenation ($\Delta H_{\text{rxn-hydrogenation}}$) of acetophenone as a representative ketone at 298 K is benchmarked as a favorable process by c. −35 kJ/mol, using ΔH_f values for acetophenone and 1-phenylethanol (Figure 8.2) [38]. The requisite ΔH_f values for borane and corresponding hydroboration products, as well as silane and hydrosilylation products, are not available. Instead, $\Delta H_{\text{rxn-hydroboration}}$ can be approximated by the reported homolytic bond dissociation enthalpies (BDEs) for B—H and B—O bonds. The $\Delta H_{\text{rxn-hydrogenation}}$ using BDEs for C=O and H—H in the reactants and C—O, C—H, and O—H bonds in products is estimated to be −28 kJ/mol in an approximation that, nonetheless, provides some confidence in this approach for estimation of $\Delta H_{\text{rxn-hydroboration}}$. Note that the enthalpy values for reduction of C=O and for the formation of the C—H bond will be equivalent, and the energy difference between E—H and E—OR will determine the relative reaction thermodynamics.

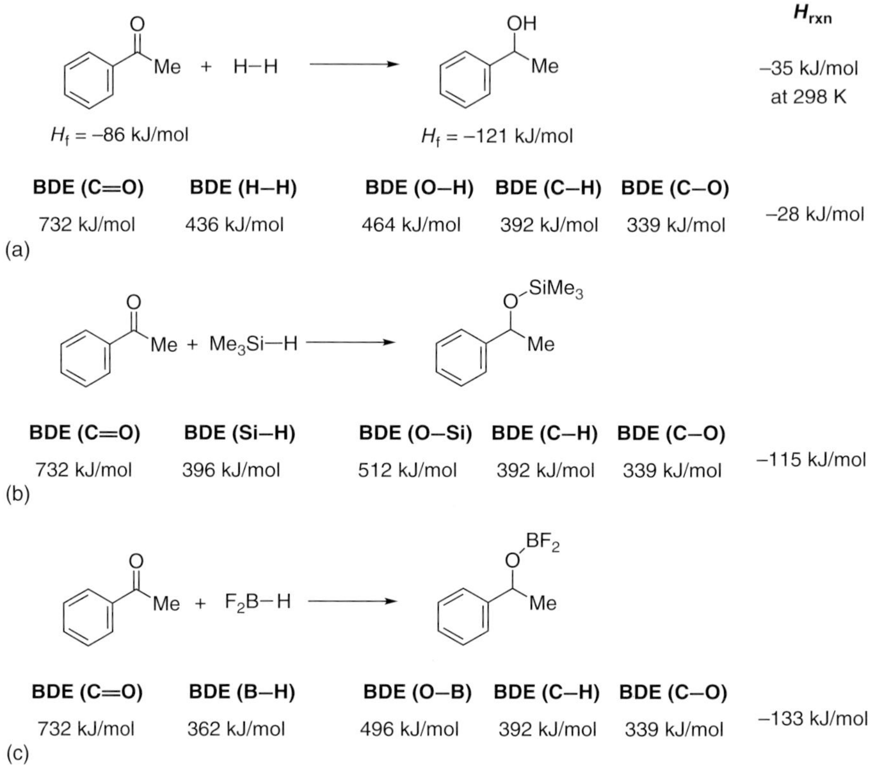

Figure 8.2 Estimated ΔH_{rxn} for (a) hydrogenation, (b) hydrosilylation, and (c) hydroboration of acetophenone.

The B—H BDE for BH and BH_2 in polyhedral boranes was reported by Glockler to be 67 and 71 kcal/mol (280 and 297 kJ/mol), respectively [39]. Higher values of 74.6 kcal/mol (312 kJ/mol) for H_2B—H and 86.5 kcal/mol (362 kJ/mol) for F_2B—H are derived from gas-phase ΔH_f [40, 41]. Alternatively, computational studies report the B—H BDE in HBF_2, HBcat, and HBpin as remarkable 108.6 kcal/mol (454 kJ/mol), 111 kcal/mol (466 kJ/mol), and 109.5 kcal/mol (458 kJ/mol), respectively [42–44]. The corresponding experimental boron–oxygen bond in F_2B—O is 118 kcal/mol (496 kJ/mol) [40]. Thus, hydroboration of a ketone by HBF_2 (as an analogue of HBpin or HBcat) is estimated to be exothermic by −31.7 kcal/mol (−133 kJ/mol), which is significantly more than the value estimated for hydrogenation. The O—B BDE in RO—Bpin is not reported, although the B—O bond in MeOBpin is calculated to be 141 kcal/mol (590 kJ/mol). Despite the significantly inequivalent absolute values for B—H and B—O BDEs evaluated by experimental and computational methods, the differences in B—O and B—H BDEs measured experimentally for XBF_2 and determined by computational methods for XBcat are in fact similar, being 134 kJ/mol (32 kcal/mol) and 124 kJ/mol (30 kcal/mol), respectively.

The present analysis employs the experimental BDE values, and a weaker thermochemical potential of 10 kJ/mol leads to similar conclusions for relative energies of hydroboration compared to hydrogenation and hydrosilylation.

The values of silicon–hydrogen and silicon–oxygen BDEs in $Me_3Si—H$ and $Me_3Si—OH$ of 90.3 kcal/mol (378 kJ/mol) and 128 kcal/mol (536 kJ/mol), respectively [45], lead to the estimate of $\Delta H_{\text{rxn-hydrosilylation}}$ to be −37.6 kcal/mol (−158 kJ/mol). Alternatively, the Si—H BDE of 94.7 kcal/mol (396 kJ/mol) and silicon–oxygen bond enthalpy in $Me_3Si—OEt$ is 122 kcal/mol (512 kJ/mol) [40], which corresponds to the $\Delta H_{\text{rxn-hydrosilylation}}$ value of −27.5 kcal/mol (−115 kJ/mol). Thus, both hydroboration and hydrosilylation are considerably more exothermic than hydrogenation.

This analysis of enthalpy does not consider the effect of entropy or corresponding free energy. ΔS_{rxn} will be negative because the addition reaction decreases the number of molecules. Reactions performed at higher temperatures feature increasing ΔG_{rxn} and smaller equilibrium constants. Thus, low-temperature reaction pathways are essential to allow thermodynamically favorable operating conditions, and these can be enabled by catalysis. Still, more accurate values, obtained from calorimetry experiments, would be useful for better estimates of thermodynamics of hydroboration and hydrosilylation reactions, particularly as the unsaturated substrate and E—H-containing reactant is varied.

8.2.2 Thermochemistry of Metal–Oxygen Bonds and Element–Hydrogen Bonds

In reactions in which oxygenates are reduced, species containing metal–oxygen bonds are plausible intermediates. A generic catalytic cycle that features metal–alkoxide, amide, and hydrocarbyl intermediates, often proposed for metal-mediated hydrogenations (E = H), hydrosilylations (E = SiX_3), and hydroboration (E = BX_2) of carbonyls (Y = O), imines (Y = NR′), heterocycles, alkenes (Y = CR'_2), and alkynes, is shown in Figure 8.3. Clearly, the diverse

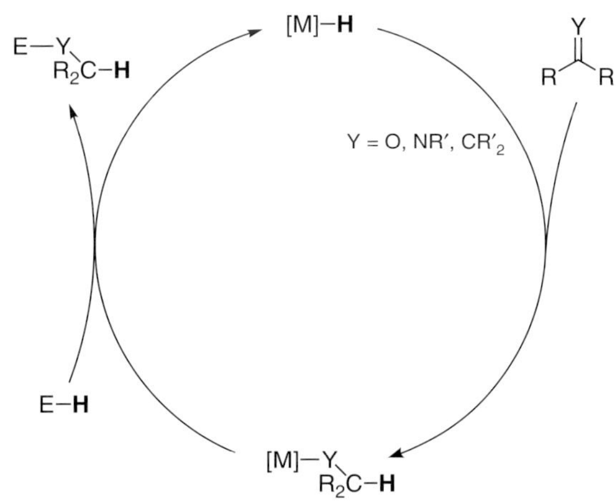

Figure 8.3 Generic two-step mechanism for metal-mediated addition reactions to unsaturated organic compounds such as carbonyls, imines, and alkenes and various derivatives, involving (i) insertion into a metal hydride and (ii) metathetical M–Y/E–H exchange.

catalysts for any one of these hydroelementations are not obligated to follow such a path. Moreover, hydrogen (H—H), silanes (H—SiX_3), and boranes (H—BX_2) are distinct reactants in terms of H—E bond polarity and dissociation enthalpy, acceptor orbitals (H—H σ*, H—SiX_3 σ*, and H—BX_2 *p*), and Lewis acidity of reagent. Thus, accessible reaction pathways would likely be inequivalent for hydrogenation, hydrosilylation, and hydroboration. Despite these caveats, the simple cycle facilitates a comparison of possible steps that are often invoked and can help assess if a particular pathway is reasonable.

In general, early transition metal compounds, lanthanides, and early main group metals all are strongly oxophilic. The zirconium–oxygen bond in (η^5-C_5Me_5)$_2$Zr(OCH_2CF_3)$_2$ is 103 kcal/mol (431 kJ/mol) and the samarium–oxygen bond is 81 kcal/mol (339 kJ/mol) in (η^5-C_5Me_5)$_2$Sm–O*t*Bu [46, 47]. These values, along with metal–hydrogen BDE values, may be used to predict viability of elementary steps in plausible catalytic cycles. Over the three reactions in Figure 8.4, the difference in BDE between E—H and E—O moieties increases for H < SiX_3 < BX_2, and only the borane reagent is sufficiently reducing for the equilibrium to favor exchange with the strong Sm—O bond. The analysis based on (η^5-C_5Me_5)$_2$Zr(OCH_2CF_3)$_2$ reveals a similar trend, although organosilanes are also competent for zirconium–hydrogen bond formation. Similar effects are expected with oxophilic group 1 and group 2 catalysts.

These estimated reaction enthalpies indicate that early metal hydrides are most easily accessed in reactions of alkoxides with boranes, in comparison with silanes and hydrogen. Any catalytic reaction involving a HBX_2 reagent (e.g. HBpin or HBcat) could involve this metathetical exchange. In fact, a number of group 1 and group 2 complexes that catalyze hydroboration of carbonyls form hydrides upon treatment with HBpin or HBcat, and catalytic cycles related to the one in Figure 8.3 are sometimes proposed, as discussed below. In other cases, selectivity, isolated complexes, and kinetic studies have ruled out this insertion/metathesis cycle and other pathways have been proposed.

(A)	Cp^*_2Sm—O*t*Bu +	H–H	⟶	Cp^*_2Sm—H +	H–O*t*Bu	
	BDE (Sm–O)	**BDE (H–H)**		**BDE (Sm–H)**	**BDE (O–H)**	
	339 kJ/mol	436 kJ/mol		217 kJ/mol	464 kJ/mol	94 kJ/mol
(B)	Cp^*_2Sm—O*t*Bu +	Me_3Si—H	⟶	Cp^*_2Sm—H +	H–O*t*Bu	
	BDE (Sm–O)	**BDE (Si–H)**		**BDE (Sm–H)**	**BDE (O–Si)**	
	339 kJ/mol	396 kJ/mol		217 kJ/mol	512 kJ/mol	6 kJ/mol
(C)	Cp^*_2Sm—O*t*Bu +	F_2B—H	⟶	Cp^*_2Sm—H +	F_2B–O*t*Bu	
	BDE (Sm–O)	**BDE (B–H)**		**BDE (Sm–H)**	**BDE (O–B)**	
	339 kJ/mol	362 kJ/mol		217 kJ/mol	496 kJ/mol	−12 kJ/mol

Figure 8.4 Estimated enthalpy for reaction of metal–alkoxide and hydrogen, trimethylsilane, and difluoroborane. Note that the BDE values for O–H, O–Si, and O–B are given for methanol, trimethylsilyl ethylether, and difluoroborate rather than the species in the balanced chemical equation.

Note that this analysis does not predict rates of M–OR/E–H exchange or rates of catalytic reduction. Reaction pathways that avoid metal–oxygen-bonded intermediates are potentially viable for carbonyl reduction, even with dihydrogen. Mechanistic analysis of catalytic hydroboration, from kinetic studies and investigations of elementary steps, might then suggest possible strategies for designing catalysts for hydrosilylations or hydrogenations.

8.3 Group 1-Catalyzed Hydroboration Reactions

8.3.1 Overview

Ketones, and especially aldehydes, react readily with HBpin and HBcat in the presence of a number of lithium, sodium, and potassium-containing organometallics, alkoxides or hydroxides, or borohydride salts. Inter- and intramolecular competition studies between ketone and aldehyde functionality typically favor conversion of the aldehyde with up to 99 : 1 selectivity.

Often, the ratios of carbonyl substrate to precatalyst and TONs, as well as the rates, as measured by turnover frequency (TOF = moles product per moles catalyst per unit time), are high. (Rates and TOF are dependent on substrate and precatalyst concentration with the proportionality related to the rate law, and comparisons must be treated carefully.) We have found that even LiCl, solubilized by ether solvents, catalyzes additions of aldehydes and HBpin. This reaction also occurs at elevated temperatures after long reaction times under catalyst-free conditions, and highly concentrated reactions proceed even more readily [37]. In that case, a second-order rate law ($\sim[\text{HBpin}]^1[\text{aldehyde}]^1$) suggests a bimolecular mechanism.

Conversions of less reactive unsaturated species, either with C=E bonds of lower polarity and carbon centers with reduced electrophilicity or of molecules lacking an overall dipole moment, are also catalyzed by alkali metal salts. For example, hydroboration of carbon dioxide is catalyzed by alkali metal complexes of the $HBPh_3$ anion [48]; the anionic species is invoked as an essential component for catalysis. Imines and aryl-substituted alkenes and alkynes react with HBpin, catalyzed by NaOH under more forcing conditions than are required for aldehydes [49]. In contrast, hydroboration of esters, amides, nitriles, and heterocycles are not readily catalyzed by group 1 metal-based species, at least with the compounds reported thus far. Overall, most of the studies suggest that the participation of both the cation and the counter anion is needed in hydroboration by alkali metal complexes. These points will be highlighted in the examples given below.

8.3.2 Base-Catalyzed Hydroborations

Alkali metal-based precatalysts for carbonyl hydroboration include sodium-*tert*-butoxide [50], sodium hydroxide, potassium *tert*-butoxide [49], *n*-butyllithium, and lithium anilide [51]. In an early study, seemingly by the common requirement of bases such as *tert*-butoxide in transition-metal-catalyzed hydroboration reactions, the addition of HBpin to aryl and dialkyl ketones is catalyzed by NaO*t*Bu

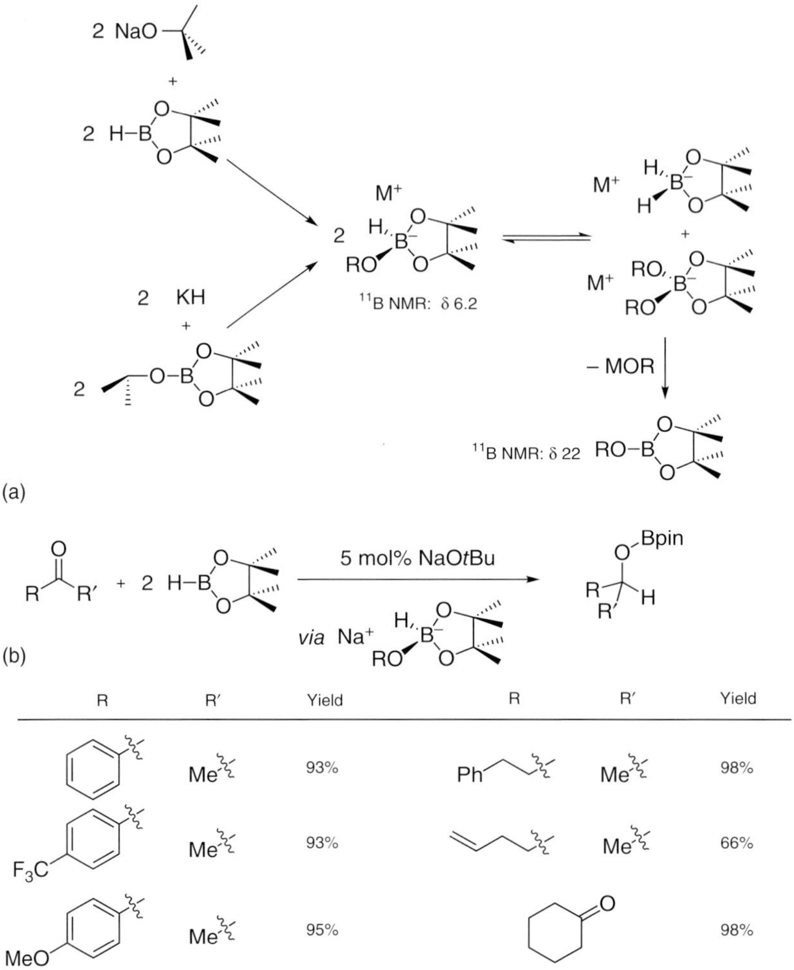

Figure 8.5 (a) Reactions that generate and consume M[RO(H)Bpin], which is a proposed catalytic species. (b) Catalytic performance of NaO*t*Bu.

(Figure 8.5; room temperature, three hours, toluene) [50]. NaO*t*Bu and HBpin react to form Na[*t*BuO(H)Bpin], which is identified in THF by a 1 : 1 : 1 : 1 quartet at −0.09 ppm ($^1J_{BH}$ = 82 Hz) in the ^{1}H NMR spectrum and a singlet in the ^{11}B NMR spectrum at 6.2 ppm. Other species are present, including $NaBH_4$ (^{11}B NMR: −41 ppm, q, $^1J_{BH}$ = 81 Hz). Na[*t*BuO(H)Bpin] is redistributed into Na[H_2Bpin] and Na[(tBuO)$_2$Bpin], and the latter species extrudes NaO*t*Bu to form *t*BuOBpin (^{11}B NMR: 22 ppm). In addition, KH and *i*-C_3H_7OBpin react to give K[*i*-C_3H_7O(H)Bpin]. Thus, a number of hydride species may be involved in these, and related, catalytic hydroborations; despite the possibilities of multiple pathways, useful synthetic protocols have been developed.

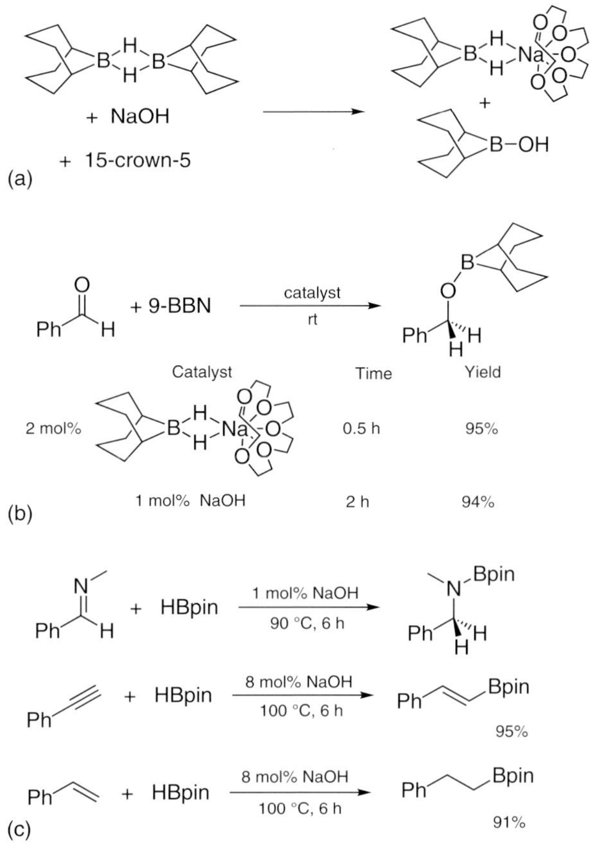

Figure 8.6 (a) Reaction of NaOH and 9-BBN gives NaH·9-BBN adduct. (b) Catalytic performance of NaOH and NaH·9-BBN adduct. (c) Representative NaOH-catalyzed hydroboration reactions.

For example, powdered NaOH catalyzes hydroboration of aldehydes, ketones, imines, vinyl arenes, and alkynyl arenes (Figure 8.6) [49]. Reactions of aldehydes require only 15 minutes at room temperature with 1 mol% NaOH, and examples range from nitro-, halide, and cyano-substituted benzaldehydes to heteroaromatic aldehydes. Ketones were allowed to react for 24 hours, while *N*-benzylidenemethylamine requires 90 °C for six hours (Figure 8.6c). The ^{11}B NMR resonances of pinacolborate products appear at 22 ppm. Styrene and X-$C_6H_4CH{=}CH_2$ (X = F, Cl, Br, and OMe) react with excess HBpin (1.4 equiv) to give anti-Markovnikov product at even higher temperatures (100 °C, six hours) than simple carbonyls and imines. The ^{11}B NMR signals of alkyl pinacolboronate esters appear at 33.8 ppm as a broad singlet. Phenylacetylene, X—$C_6H_4CH{\equiv}CH$ (X = F, Cl, Me, OMe, Me), and 2-ethynylthiophene react with HBpin to give the anti-Markovnikov and cis-addition products under the same conditions. The ^{11}B NMR signals for vinyl pinacolboronate esters appear at slightly lower frequencies (c. 30 ppm) compared with the alkyl boronates.

Most of the detailed experiments probing the reaction pathway involve benzaldehyde, 9-BBN, and NaOH. NaOH and 2 equiv of 9-BBN react, in the presence of 15-crown-5 ether, to give (15-c-5)NaH_2-9-BBN and HO-9-BBN. The former species was characterized by single-crystal X-ray diffraction and IR in the solid state (BH = 2106 cm^{-1}), and ^{11}B NMR spectroscopy in THF (17.9 ppm, *t*, $^1J_{BH}$ = 74.1 Hz). The reaction of benzaldehyde and 9-BBN is catalyzed by (15-c-5)NaH_2-9-BBN. One possible pathway for benzaldehyde hydroboration was modeled using density functional theory (DFT) methods, which explicitly considers two THF molecules coordinated with a sodium center and a universal continuum solvation model.

8.3.3 Alkali Metal Hydridoborate and Aluminate-Catalyzed Hydroboration

In the systems described above, alkali metal hydridoborate species are invoked as catalytic intermediates. Alternatively, alkali metal hydridoborate or aluminate salts may be used directly as precatalysts. In such systems, the cationic alkali metal complex and the counterion may synergistically provide enhanced reactivity or selectivity. In a notable example, [{Me_6-TREN}M][$HBPh_3$] (Me_6-TREN = $N(CH_2CH_2NMe_2)_3$) is a precatalyst for additions of HBpin to aldehyde, ketone, and carbon dioxide substrates [48]. The species are prepared sequentially from tetramethyldisilazido compounds $MN(SiHMe_2)_2$ and Me_6-TREN, followed by addition of BPh_3 in THF, or else by reaction of Me_6-TREN and M[$HBPh_3$]. The $[HBPh_3]^-$ gives a doublet resonance in the ^{11}B NMR spectra at c. −8 ppm ($^1J_{BH}$ = 78 Hz), measured in THF-d_8. Interestingly, the former reaction involving BPh_3 generates 0.5 equiv of 1,3-diazadisilacyclobutane as the by-product, but only in THF. Reaction of [{Me_6-TREN}M][$N(SiHMe_2)_2$] and BPh_3 in benzene yields [{Me_6-TREN}M][$(Me_2HSi)_2N{-}BPh_3$] as the result of boron–nitrogen bond formation [52]. Apparently, the nucleophilicity of amide is greater in benzene, whereas the hydride on silicon in the tetramethyldisilazide is more nucleophilic in THF.

These $[HBPh_3]^-$ species are highly active for the hydroboration of benzophenone. For example, a TON of 100 000 is obtained after only 1.5 hours with [{Me_6-TREN}Li][$HBPh_3$] as the precatalyst, corresponding to a TOF of 18.5 s^{-1}. Because TOF is calculated based on the time to quantitative conversion, and rates likely depend on reactant concentrations, initial rates may be even faster. The alkali metal cation appears to be engaged in the catalysis, indicated by slower conversions with [{Me_6-TREN}M][$HBPh_3$] (M = Na, K) giving TOF of 0.06 s^{-1}, while [$n$$Bu_4$N][$HBPh_3$] gives a TOF of 0.04 s^{-1}. Moreover, [{Me_6-TREN}Li][$HBPh_3$] outperforms Li[$HBPh_3$], and Me_6-TREN-coordinated lithium cation is faster compared with other ligands, following the trend Me_6-TREN (TOF > 60 000 h^{-1}) > THF (TOF = 13 300 h^{-1}) > PMDTA ~ TMEDA (TOF = 10 000 h^{-1}) > Me_4-TACD (600 h^{-1}), compared under equivalent precatalyst concentrations (PMDTA = $NMe(CH_2CH_2NMe_2)_2$; Me_4-TACD = $(N(Me)CH_2CH_2)_4$) [53].

[{Me_6-TREN}Li][$HBPh_3$] reacts with benzophenone in THF at room temperature to give [{Me_6-TREN}Li][$Ph_2HCO{-}BPh_3$]. Likewise, carbon dioxide and the complex react at room temperature in THF to give formoxytriphenylborate [{Me_6-TREN}Li][HCO_2BPh_3], whereas ethyl acetate, dimethyl acetamide, diisopropylcarbodiimide, diphenylacetylene, pyridine, thiophene, styrene, and acetonitrile are inert under these conditions. Although [{Me_6-TREN}Li][Ph_2HCO–BPh_3] crystallizes as a THF·Li adduct, THF is apparently labile. In fact, the formoxytriphenylborate species crystallizes with a Li—O—CH—O—B bridging structure (Figure 8.7a).

A THF-d_8 solution composed of a mixture of [$HBPh_3$]$^-$ and HBpin provides NMR spectra with identical chemical shifts as spectra of the individual species, and the authors interpret this result to indicate that hydridoborate and borane exchange does not occur. Such a degenerate hydrogen exchange might be expected to be faster than alkoxide transfer from [RO–BPh_3]$^-$ to HBpin. However, [{Me_6-TREN}Li][Ph_2HCO–BPh_3] and HBpin immediately react in THF-d_8 to form Ph_2HCO–Bpin and regenerate [{Me_6-TREN}Li][$HBPh_3$] [52].

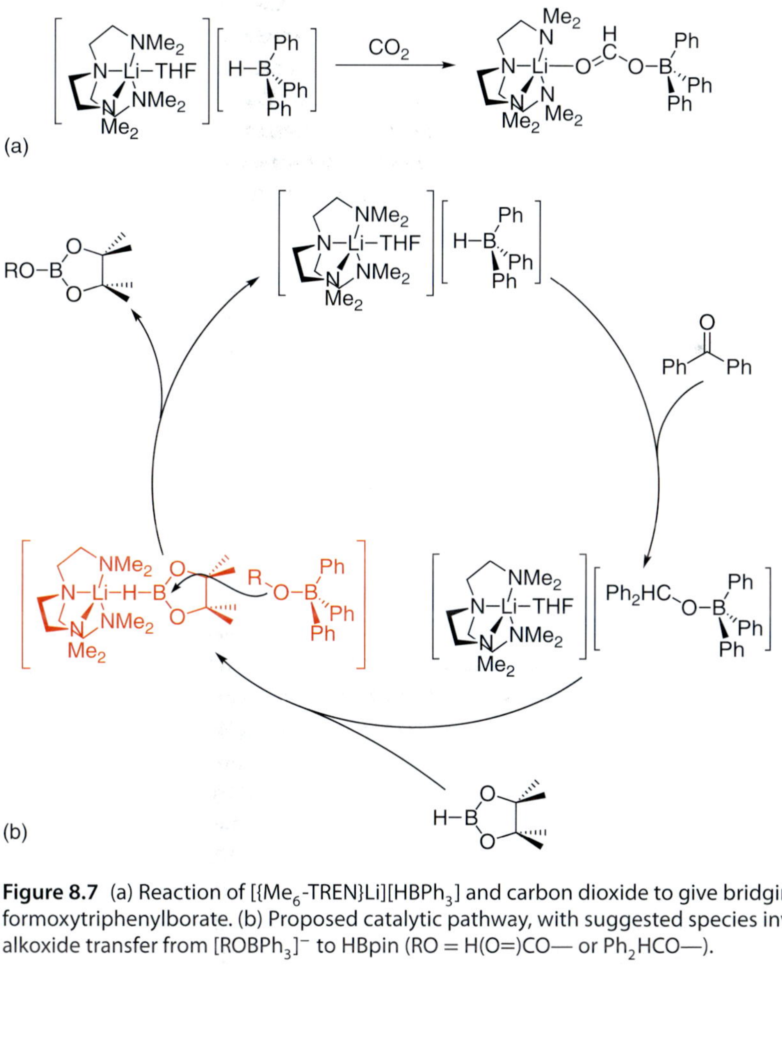

Figure 8.7 (a) Reaction of [{Me_6-TREN}Li][$HBPh_3$] and carbon dioxide to give bridging formoxytriphenylborate. (b) Proposed catalytic pathway, with suggested species involved in alkoxide transfer from [$ROBPh_3$]$^-$ to HBpin (RO = H(O=)CO— or Ph_2HCO—).

The reaction of [{Me_6-TREN}Na][Ph_2HCO–BPh_3] and HBpin is slower (c. 15 minutes), which is consistent with the relative rates of catalytic action of lithium-based precatalyst being greater than the sodium analogue. More detailed studies of rates are yet needed to assess the catalytic relevance of the reaction between [{Me_6-TREN}M][Ph_2HCO–BPh_3] and HBpin.

A few other observations, however, provide some clues. First, the reactions of benzophenone and [{Me_6-TREN}M][$HBPh_3$] are equivalently fast for M = Li, Na, and K, implying that this step does not distinguish the Li > Na ~ K catalytic performance of the series of alkali metal compounds. In addition, catalytic conversion using the labeled DBpin is c. 2.5× slower (four hours for complete conversion) than equivalent conditions with HBpin. [{Me_6-TREN}Li][BPh_4] and [{Me_6-TREN}M][$HB(C_6F_5)_3$] are not precatalysts for benzophenone hydroboration; however, a pathway initiated by {Me_6-TREN}MH or {Me_6-TREN}MOR species has not yet been ruled out.

A two-step catalytic cycle (Figure 8.7b) was proposed on the basis of the influence of alkali metal cation and its coordination environment on the catalytic rate of hydroboration of benzophenone, a similar trend in the reaction of [{Me_6-TREN}M][Ph_2HCO–BPh_3] and HBpin and slower hydroboration of benzophenone with DBpin compared to HBpin.

Related chemistry involving alkali metal aluminum hydride chemistry has recently been developed [54]. Hydroboration of styrene occurs at 110 °C under solvent-free conditions in the presence of $LiAlH_4$ (LAH) or sodium (bis-2-methoxyethoxy) aluminum hydride ($Na(MeOCH_2CH_2O)_2AlH_2$, also known as RED-Al). Aliphatic 1-alkenes, such as 1-hexene, t-butylethylene, and allylbenzene, react with HBpin to form boronate esters in 70–85% yield and selectivity for the linear product ranging from 90% to 99%. Conversions and selectivity for additions of HBpin and functionalized styrenes were in a similar range. Note, however, that 95% yield with 10% $LiAlH_4$ or RED-Al corresponds to a TON of 9.5, meaning that the TON per equivalent of hydride is 2.4 and 4.75, respectively.

The styrene addition product is also formed in lower yield (<50%) with *neutral* aluminum species diisobutylaluminum hydride (DIBAL-H) and Et_3Al·DABCO (1,4-Diazabicyclo[2.2.2]octan), as well as LiH and NaH. $LiAlH_4$ also catalyzes the addition of HBpin to acetophenone (0.5 mol% catalyst, room temperature, 30 minutes), ethyl acetate (0.5 mol% catalyst, room temperature, eight hours), and 4-trifluoromethylbenzonitrile (1 mol% catalyst, room temperature, six hours). Moreover, *t*-butylethylene and DIBAL-H react to give isolable $t\text{BuCH}_2\text{CH}_2\text{Al}i\text{Bu}_2$, while both RED-Al and $LiAlH_4$ react with alkenes to apparently give carbon–carbon-bonded products. Note that alanes and boranes readily exchange hydrogen and alkyl groups. Together, these results suggest that the alkali metal aluminum hydride species may be functioning as precursors to alanes during the hydroboration reactions. In fact, a few reports show that alane-catalyzed hydroborations operate without alkali metal cations [25, 55, 56].

In contrast, the cationic alkali metal center and its ligand affect the catalytic properties (in ketone and aldehyde hydroboration) of well-defined aluminum hydride salts, including lithium bis(hexamethyldisilazido)aluminum hydride $(THF)_3Li(\mu\text{-}H)AlH\{N(SiMe_3)_2\}_2$, its ligated derivatives $(L)LiH_2Al\{N(SiMe_3)_2\}_2$ (L = PMDTA, TMEDA, 12-crown-4), a mono-TMP aluminate (TMP = tetramethylpiperidyl) $[Li{\cdot}HAl(TMP)\textit{i}Bu_2]_2$, its ligated derivatives $[(L)Li{\cdot}HAl(TMP)\textit{i}Bu_2]$ (L = TMEDA, PMDTA, 12-crown-4, DABCO), and a sodium analogue $[(THF)Na{\cdot}HAl(TMP)\textit{i}Bu_2]_2$ [57, 58]. These species form either separated ion pairs (SIPs) or contact ion pairs (CIPs) in the solid state, whereas diffusion-ordered NMR spectroscopy (DOSY) experiments typically suggest CIP and dynamic structures in solution. The CIP structures generally contain bridging M–H–Al motifs. For example, the PMDTA-coordinated $[(L)Li{\cdot}HAl(TMP)\textit{i}Bu_2]$ containing a Li–H–Al CIP motif gives the least active hydroboration catalyst (for addition to acetophenone) in benzene, whereas an order of magnitude faster rates are obtained with the same precatalyst in THF. The authors suggest that enolate formation by acetophenone, in combination with the basic tetramethylpiperidine-bonded aluminate, may interfere with catalysis; however, benzophenone was also less efficiently converted with the PMDTA-coordinated catalyst and other TMP-containing aluminates were more effective for both acetophenone and benzophenone. In fact, the donor-free $[Li{\cdot}HAl(TMP)\textit{i}Bu_2]_2$ is generally the most effective catalyst in the series, whereas these aluminates are much less active than $[\{Me_6\text{-}TREN\}Li][HBPh_3]$.

In the disilazido series, catalytic hydroboration chemistry was only reported for the THF adduct $(THF)_3Li(\mu\text{-}H)AlH\{N(SiMe_3)_2\}_2$ (Figure 8.8a). The reaction of $(THF)_3Li(\mu\text{-}H)_2Al\{N(SiMe_3)_2\}_2$ and benzaldehyde yields a doubly inserted product, isolated and subjected to single-crystal X-ray diffraction analysis. In the product $(THF)_2Li(\mu\text{-}OCH_2Ph)_2Al\{N(SiMe_3)_2\}_2$, two alkoxide ligands bridge between lithium and aluminum centers. This addition reaction is postulated as part of a pathway for hydroboration in both disilazido- and TMP-supported aluminum catalysts. In fact, $(THF)_2Li(\mu\text{-}OCH_2Ph)_2Al\{N(SiMe_3)_2\}_2$ is catalytically competent for hydroboration of benzaldehyde.

Interestingly, $[Li{\cdot}HAl(TMP)\textit{i}Bu_2]_2$ is also a precatalyst for the hydroboration of phenylacetylene (Figure 8.8b). Here, the reactivity of $[Li{\cdot}HAl(TMP)\textit{i}Bu_2]_2$ or $(PMDTA)Li{\cdot}HAl(TMP)\textit{i}Bu_2$ involves alkyne metalation, rather than addition. Alumination of terminal alkynes to give alkynylaluminum species was proposed as part of a diketimimate dihydroalane-catalyzed alkyne hydroboration [25]. Neither $(PMDTA)Li{\cdot}HAl(TMP)\textit{i}Bu_2$ nor $\{HC(CMeNAr)_2\}AlH_2$ species are effective for hydroboration of internal alkynes. Unfortunately, the reaction of alkynylaluminate and HBpin did not provide the product of catalytic hydroboration.

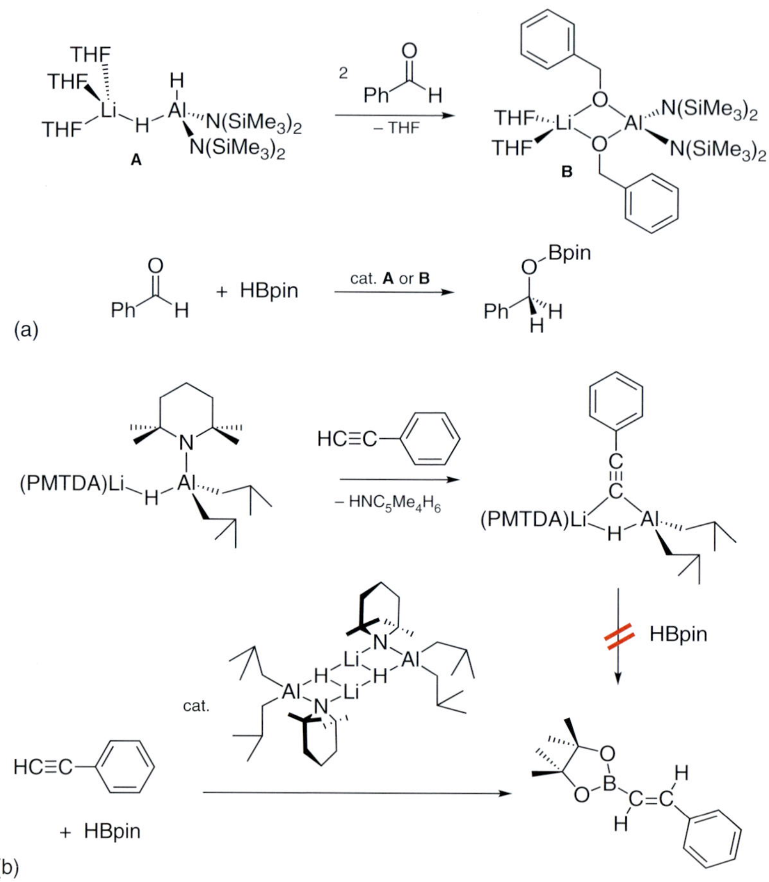

Figure 8.8 (a) Hydroboration of benzaldehyde, catalyzed by a lithium aluminum hydride, and a catalytically competent lithium aluminum alkoxide. (b) Catalytic hydroboration of phenyl acetylene and a catalytically inactive lithium aluminum acetylide.

8.4 Group 2-Catalyzed Hydroboration Reactions

8.4.1 Overview

Group 2 complexes catalyze hydroboration reactions, resulting in the addition of HBpin and carbonyl compounds. In many cases, trends and observations evoke similar principles to those outlined for group 1-catalyzed hydroborations; however, there are significant differences as well. Group 2 compounds, mainly based on magnesium, tend to also catalyze hydroboration of less electrophilic carbonyls (such as esters and amides). Additions to pyridine derivatives are also reported. The divalent nature of group 2 metals allows one L_nX-type ancillary ligand, in contrast to the requirement of L_n-type coordination to alkali metal cations. In addition, isolable metal hydride species such as those supported by β-diketiminato ligand "Nacnac" ligand (Nacnac = $HC\{C(Me){=}N(2,6\text{-}C_6H_3iPr_2)\}_2$) $[\{Nacnac\}MgH]_2$ undergo well-defined insertion reactions with unsaturated substrates, giving some experimental basis for mechanistic proposals.

8.4.2 β-Diketiminate Magnesium-Catalyzed Hydroborations

In particular, {Nacnac}Mg*n*Bu and HBpin react to form *n*Bu–Bpin (^{11}B NMR: 37 ppm) and [{Nacnac}MgH]$_2$ [35]. [{Nacnac}MgH]$_2$ and excess HBpin, however, also react to form an adduct in a crystallized product. Moreover, DOSY evidence suggests that *n*Bu–Bpin and [{Nacnac}MgH]$_2$ are interacting in solution. Nonetheless, {Nacnac}Mg*n*Bu is a precatalyst for hydroboration of pyridine and substituted derivatives into mixtures of 2- and 4-hydropyridylboranes. The substitution pattern on the pyridine affects the relative amounts of isomer formed, with the parent pyridine giving mainly 1,4-addition, while substitution in the 4-position favors 1,2-addition. For example, *N*-boryl-4-hydropyridine is the major product in hydroboration reactions performed at 80 °C (16 hours), while reactions at 35 °C give a 1 : 1 ratio of 1,2- and 1,4-adducts (after four days). Alternatively, 4-methylpyridine (4-picoline) reacts at 70 °C over 23 hours to afford an 81 : 19 mixture of 2-hydro-1-boryl-pyridine:4-hydro-1-boryl-pyridine, which was isolated as 1,2-dihydropicoline in 68% yield.

[{Nacnac}MgH]$_2$, generated from {Nacnac}Mg*n*Bu(pyridine) and PhSiH$_3$, reacts with pyridine at 60 °C to give mixtures of 2- and 4-hydropyridyl magnesium pyridine adducts (Figure 8.9a) [35]. *N*,*N*-linked and 1,4-C$_6$H$_4$-linked diketiminate ligands, which support multimetallic magnesium hydrides, also react with pyridine to afford dearomatized products [59]. The *N*,*N*-linked magnesium hydride system is dimeric [(μ-H)Mg(Nacnac–Nacnac)Mg(μ-H)]$_2$ and shows remarkable selectivity for the 2-hydropyridylmagnesium species over the 4-hydropyridyl analogue, whereas the C$_6$H$_4$-linked diketiminate reacts to give c. 1 : 1 mixtures of 2-hydro and 4-hydropyridyl products. Remarkably, the application of the *N*,*N*-linked compounds as precatalysts for pyridine hydroboration gives *N*-boryl-2- and 4-hydropyridines in a c. 1 : 3 ratio on a 0.4 mmol scale, but a 3 : 1 ratio on preparative scale (60 °C, 30 hours; Figure 8.9b). The amount of 1,2-addition product is higher in this linked system than with the parent {Nacnac}Mg*n*Bu(pyridine), but not as high as in the reaction of [(μ-H)Mg(Nacnac–Nacnac)Mg(μ-H)]$_2$ and pyridine. On the basis of this inequivalent selectivity for hydroelementation of pyridines under stoichiometric and catalytic conditions, the authors proposed a catalytic pathway invoking hydride transfer from a *N*-borohydrido-2-hydropyridyl ligand to a second pyridine.

The same precatalyst, namely {Nacnac}Mg*n*Bu, has been used for the hydroboration of aldehydes and ketones [60], aldimines and ketimines [36], nitriles [61], carbodiimides [62], and isocyanates [63]. As noted above, hydroboration of aldehydes is particularly facile. Other compounds may serve as precatalysts for aldehyde and ketone hydroboration, such as divalent magnesium dimers [{Nacnac}Mg]$_2$ [64], although it is also clear that side reactions that produce BH$_3$ may be involved via uncatalyzed additions [65]. In this context, ^{11}B NMR analysis of reaction products is essential.

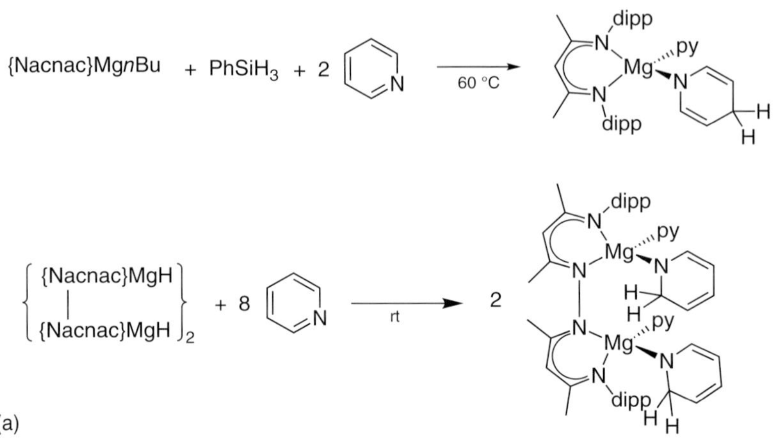

(a)

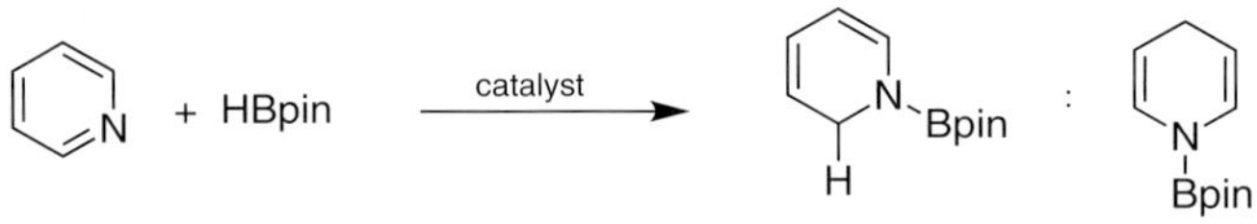

Catalyst	Temperature	Time	Yield (isolated)	Ratio
{Nacnac}Mg*n*Bu	70 °C	16 h	92(59)	37 : 63
{Nacnac}Mg*n*Bu	35 °C	96 h	81	49 : 51
{{Nacnac}MgH–{Nacnac}MgH}$_2$	60 °C	30 h	100(76)	22 : 8

(b)

Figure 8.9 Reactions pertinent to the hydroboration of pyridine. (a) Reactions of pyridine with *in situ* generated and isolated magnesium hydrides. (b) Hydroboration of pyridine catalyzed by mono-magnesium and multi-magnesium compounds.

N-benzylideneaniline, for example, undergoes quantitative hydroboration within 10 minutes. at room temperature with {Nacnac}MgBn as the pre-catalyst, showing a reactivity similar to the substrate benzaldehyde. Bulkier aldimines and bulkier ketimines, such as *N*-benzylidene-*tert*-butylamine and *N*-benzylidene-cyclohexylamine, required heating and gave lower conversions [36]. Under such conditions, B_2pin_3 (^{11}B NMR: 22 or 25 ppm) and BH_3 are formed as decomposition products of HBpin, mediated by heating in the presence of {Nacnac}Mg*n*Bu.

Interestingly, measurements of reaction kinetics for hydroboration of many of the classes of reactants listed above provide unique rate laws, suggesting that hydroboration of each class of substrate by a catalyst generated from {Nacnac}Mg*n*Bu, HBpin, and unsaturated species follows a distinct pathway. For example, hydroboration of *N*-benzylideneaniline shows second-order dependence on [imine], carbodiimides react with a slight induction period, and pyridine hydroboration shows a variable dependence of rate on [isoquinoline] [66].

Closer examination of the hydroboration of nitriles further illustrates rich mechanistic features. Kinetic studies of catalytic hydroboration of propionitrile reveal that it reacts with zero-order dependence on both [nitrile] and [HBpin]. Moreover, magnesium aldimides, formed from reaction of {Nacnac}Mg*n*Bu, HBpin, and *t*BuCN, react with an additional equivalent of HBpin to give the isolable adduct {Nacnac}Mg{N=CH*t*Bu)HBpin}. A rate law, in which the rate is proportional to only catalyst concentration (and zero order in both reactants), suggests a mechanism in which the turnover-limiting step is unimolecular. This pathway is consistent with {Nacnac}Mg{N=CH*t*Bu)HBpin} as the catalytic resting state because all the reactants are already associated in this species. Alternatively, mixtures of HBpin and aryl substituted aldiminido magnesium intermediates show only spectral signatures of the reactants, rather than a HBpin adduct. Kinetic measurements of the catalytic system further indicate the complexity of this interaction, with arylnitriles showing first-order dependence (either on [HBpin] or on [nitrile], but likely the former). Thus, the mechanisms for arylnitriles and alkylnitrile apparently follow different rate laws relating to the relative nucleophilicity of the iminidomagnesium species.

8.4.3 Tris(4,4-dimethyl-2-oxazolinyl)phenylborato Magnesium-Catalyzed Hydroboration of Ester and Amides

Magnesium centers supported by tris(4,4-dimethyl-2-oxazolinyl)phenylborate catalyze the hydroboration of esters and amides [67, 68]. The reaction stoichiometry of these two hydroboration reactions requires 2 equiv of HBpin, similar to the case described above for nitrile hydroboration. As mentioned in the Introduction section, ester hydroboration results in cleavage of the ester into two borate-substituted (RCH_2OBpin) fragments (Figure 8.10). Esters containing nitriles and nitro groups, organohalides, and cyclopropyl groups are readily converted into borate products at room temperature.

Lactones undergo ring-opening upon hydroboration. For example, lactide is split into 2 equiv of 1,2-diboratopropane. The same product may be produced by hydroboration of polylactic acid, again using $To^{M}MgMe$ as a catalyst precursor (Figure 8.11).

The cleavage of esters during hydroboration manifests itself in interesting kinetic behavior, which provides valuable mechanistic data. In particular, the ester may be cleaved before the turnover-limiting, irreversible step of the catalysis, leading to half-order dependence on [ester]. Alternatively, C—O bond cleavage could occur subsequent to the turnover-limiting step, which would give first-order dependence on ester concentration. Of course, reversible elementary steps connecting the resting state and the turnover-limiting step may create more complex (i.e. noninteger) dependence of the rate on either ester or HBpin concentrations, or both. In $To^{M}MgMe$-catalyzed hydroboration of ethyl acetate, a half-order dependence on the concentration of ester reveals that ester cleavage occurs before the turnover-limiting step. The rate is independent of [HBpin] (Eq. (8.1)):

$$\frac{-d[\mathrm{MeCO_2Et}]}{\mathrm{d}t} = k_{\mathrm{obs}}[\mathrm{To^{M}MgMe}]^1[\mathrm{HBpin}]^0[\mathrm{MeCO_2Et}]^{1/2} \quad (8.1)$$

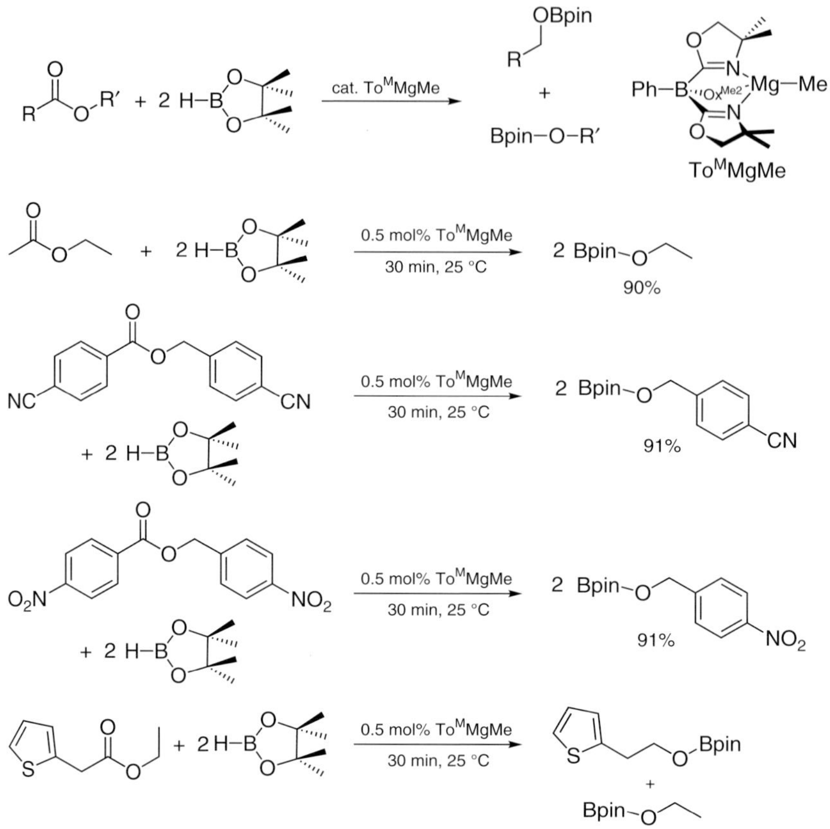

Figure 8.10 Hydroboration of representative esters, which gives C—O bond cleavage into two borates.

Ester cross-metathesis, catalyzed by To^M MgMe, provides support for the idea that ester cleavage occurs before hydroboration and that ester cleavage does not require HBpin. In addition, To^M MgMe and excess HBpin react to give To^M MgH_2Bpin, which was isolated and characterized by single-crystal X-ray diffraction. The rate dependence on ester concentration could mean that ester is catalytically split into 2 equiv of aldehyde as the rate-determining step, and then, the resulting aldehyde reacts by hydroboration in a faster (and thus kinetically invisible) process. Alternatively, the mechanism could involve reversible cleavage of ester into aldehyde and turnover-limiting hydroboration of aldehyde. The ester cross-metathesis reaction catalyzed by To^M MgMe (as the precatalyst) is fast, and exchange products were detected at room temperature. This fast exchange means that ester cleavage is even faster, and these results were interpreted to mean that the C—O bond cleavage during hydroboration of esters was not the turnover-limiting step in this system.

Insight into the mechanism for aldehyde hydroboration, instead, can contribute to interpretation of ester hydroboration kinetics. The rate law for aldehyde hydroboration, catalyzed by To^M MgMe, is zero order in both

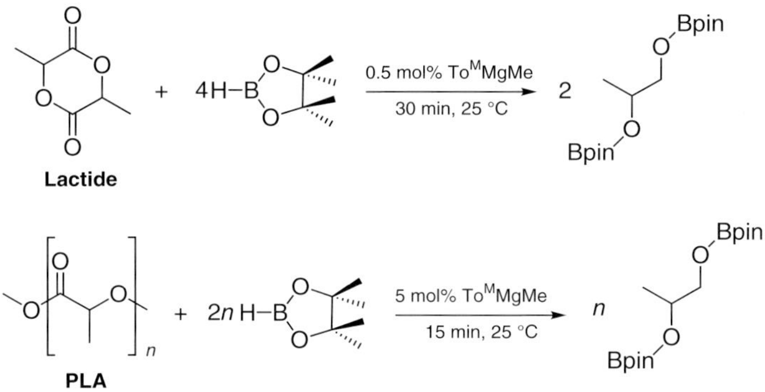

Figure 8.11 Hydroboration of cyclic esters and deconstruction of polylactic acid by hydroboration.

aldehyde and HBpin concentrations. A proposal for the aldehyde hydroboration mechanism, then, involves all three components assembled in the resting state and unimolecular extrusion of the product (Figure 8.12a). Because the same catalytic species are present and the same products are formed for both aldehyde and ester hydroboration, the latter transformation builds upon the idea that extrusion of RCH_2OBpin from $To^MMg\{(RCH_2O)_2Bpin\}$ is the turnover-limiting step, with reversible ester cleavage occurring before the same unimolecular rate-determining step (Figure 8.12b). The equilibrium of the two catalytic reactions, however, is not equivalent as a result of the effective concentration of the aldehyde. Regardless, the cumulative evidence points toward the formation of a magnesium hydridoborate, $To^MMg\{(RCH_2O)HBpin\}$, perhaps involving a bridging Mg–H–B moiety, which reacts with the carbonyl functionality. This pathway would not explicitly involve metathesis of [Mg]–OCH_2R and HBpin into [Mg]–H and RCH_2OBpin via a concerted electrocyclic transition state, even though such a step can be observed by the reaction of To^MMgOR and HBpin (in the presence of a large excess of HBpin).

The rate law for hydroboration of esters, interestingly, varies for different precatalysts, implying that not all catalysts follow the same mechanism. For example, guanidinate-supported magnesium hexamethyldisilazide, as well as bis(hexamethyldisilazido)magnesium ($Mg\{N(SiMe_3)_2\}_2$), also catalyzes this reaction [69]. The kinetic data for this $Mg\{N(SiMe_3)_2\}_2$-catalyzed hydroboration of benzyl benzoate is consistent with a second-order catalytic process, with first-order dependence on the concentration of the precatalyst and first-order dependence on *either* ester concentration *or* HBpin concentration. Such a rate law provides only a glimpse of possible pathways, although ester cleavage before the turnover-limiting step is explicitly excluded from the mechanism because the reaction rate law is not half order in benzyl benzoate. Recently, we reported that catalytic hydroboration of esters is also catalyzed by the homoleptic tris(alkyl) $La\{C(SiHMe_2)_3\}_3$ [23]. Although both magnesium and lanthanum hydroboration pathways likely occur without change in valence at the metal center (i.e. magnesium is divalent, lanthanum is trivalent) and the mechanisms might be expected to be similar, the experimental data show that the pathways have very

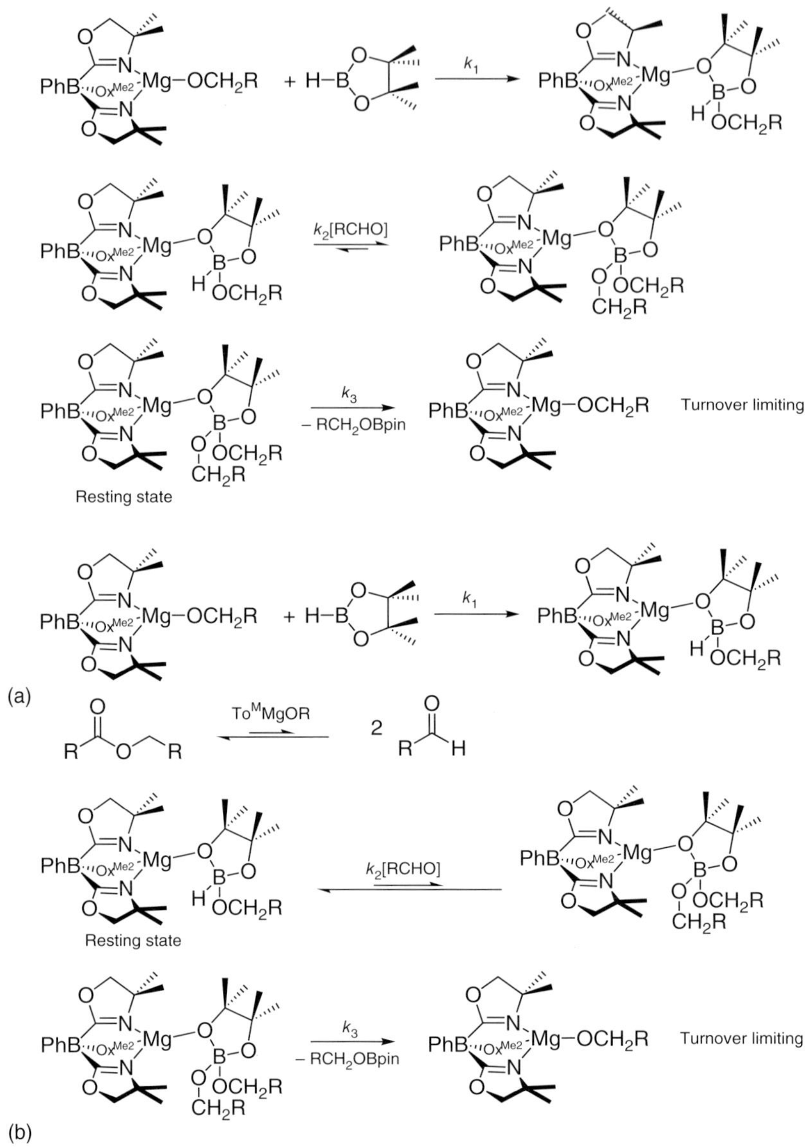

Figure 8.12 Proposed steps in To^{M}MgMe-catalyzed hydroboration of (a) aldehydes and (b) esters.

distinct features. First, lanthanum-catalyzed hydroboration of benzoate esters is third order (Eq. (8.2)):

$$\frac{-d[\mathrm{PhCO_2Me}]}{dt} = k_{\mathrm{obs}}[\mathrm{La\{C(SiHMe_2)_3\}_3}]^1[\mathrm{HBpin}]^1[\mathrm{PhCO_2Me}]^1 \qquad (8.2)$$

Thus, the C—O bond in the ester is broken after the rate-limiting step. Moreover, the reaction is first order in HBpin concentration. Independent experiments suggest that the reaction of La{C(SiHMe$_2$)$_3$}$_3$ and HBpin give lanthanum hydride species based on insertion reactions of ketones. Yet, lanthanum hydride and esters do not react in the absence of HBpin, and similar to the magnesium system, the proposed pathway involving insertion of carbonyl into borohydride is favored over the route in which ester inserts into the metal hydride bond.

Amide reduction is an important reaction for synthesis of amines, and transformations that avoid lithium aluminum hydride as a stoichiometric reductant could be useful. Hydroboration of tertiary and secondary amides is catalyzed by $To^{M}MgMe$ at room temperature, giving the amine product (equivalent to a $LiAlH_4$ reduction). For example, formyl amides are reduced to methylamine, acyl amides are reduced to ethylamine, and benzamide give benzylamine. The boron-containing by-product is pinB–O–Bpin.

8.4.4 Magnesium Triphenylborate-Catalyzed Hydroboration

The observations and kinetic measurements described above indicate that a two-step mechanism, such as the one involving insertion of the unsaturated moiety into a metal hydride and metathesis-type exchange of Mg—E and B—H bonds, is unlikely to represent the rich reactivity of main group catalysts. Thus, a magnesium triphenylborohydride, related to the lithium triphenylborohydride catalysts for aldehyde, ketone, and CO_2 hydroboration noted above, is a catalyst for hydroboration of a range of polar unsaturated species, including ethyl acetate, dimethyl acetamide, carbon dioxide, diisopropyl carbodiimide, and *t*-butylisocyanate [70]. Interestingly, dimethylsulfoxide is also reduced under hydroboration conditions.

8.4.5 Supported Catalysts for Hydroboration

A few supported catalysts, e.g. on silica, or on metal–organic frameworks (MOFs), have been reported, including a supported zirconium amide ($Zr(NMe_2)_3$@MSN; MSNs = mesoporous silica nanoparticles) [8], a titanium alkoxide catalyst supported on a UiO-67 MOF [9], and a supported magnesium species on a derivative of the MOF UiO-69 [71]. The latter magnesium catalyst is synthesized by the reaction of $MgMe_2$ and Zr_3OH sites in the secondary building units (SBUs) of the MOF to give a proposed monomeric, yet immobilized Zr_3O–Mg–Me site. This material mediates hydroboration of aldehydes, ketones, and imines, and the material may be repeatedly used for acetophenone hydroboration without loss of measured activity. As in many of the catalytic systems noted above, a magnesium hydride intermediate is proposed. In this case, reaction of the precatalyst and HBpin provides MeBpin and presumably Zr_3O–Mg–H-active catalytic sites. The rate law (Eq. (8.3)) for this reaction is more typical for ketones:

$$\frac{-d[\text{PhCOMe}]}{\text{d}t} = k_{\text{obs}}[\text{TPHN-MOF-MgMe}]^1[\text{HBpin}]^0[\text{PhCOMe}]^1 \tag{8.3}$$

The proposed mechanism involves ketone insertion into a magnesium–hydride bond, followed by σ-bond metathesis. This proposal is consistent with kinetic data, but the data do not rule out other pathways such as a magnesium borohydride adduct.

8.5 Summary and Conclusions

Magnesium complexes have been reported to catalyze a wider range of hydroboration reaction upon unsaturated substrates compared with the alkali metals, reactions of which are mainly limited to aldehydes and ketones. There are likely opportunities for catalytic applications of larger alkaline earth metals (calcium, strontium, and barium) in hydroboration. A key challenge for selective hydroborations, employing group 1 and group 2 metals (as well as other oxophilic early metal and rare earth element centers), will be preventing redistribution reactions that lead to B_2H_6. These processes may contribute to the limited use of calcium and heavier group 2 congeners in catalytic hydroborations.

References

1 Brown, H.C. (1962). *Hydroboration*. New York: W.A. Benjamin.
2 Dhillon, R.S. (2007). *Hydroboration and organic synthesis: 9-Borabicyclo [3.3.1] Nonane (9-BBN)*. Berlin; New York: Springer.
3 Tucker, C.E., Davidson, J., and Knochel, P. (1992). *J. Org. Chem.* 57: 3482–3485.
4 Männig, D. and Nöth, H. (1985). *Angew. Chem. Int. Ed. Engl.* 24: 878–879.
5 Ohmura, T., Yamamoto, Y., and Miyaura, N. (2000). *J. Am. Chem. Soc.* 122: 4990–4991.
6 He, X. and Hartwig, J.F. (1996). *J. Am. Chem. Soc.* 118: 1696–1702.
7 Oluyadi, A.A., Ma, S., and Muhoro, C.N. (2012). *Organometallics* 32: 70–78.
8 Eedugurala, N., Wang, Z., Chaudhary, U. et al. (2015). *ACS Catal.*: 7399–7414.
9 Huang, Z., Liu, D., Camacho-Bunquin, J. et al. (2017). *Organometallics* 36: 3921–3930.
10 Wu, J.Y., Moreau, B., and Ritter, T. (2009). *J. Am. Chem. Soc.* 131: 12915–12917.
11 Ely, R.J. and Morken, J.P. (2010). *J. Am. Chem. Soc.* 132: 2534–2535.
12 Semba, K., Fujihara, T., Terao, J., and Tsuji, Y. (2012). *Chem. Eur. J.* 18: 4179–4184.
13 Zhang, L., Peng, D., Leng, X., and Huang, Z. (2013). *Angew. Chem. Int. Ed.* 52: 3676–3680.
14 Obligacion, J.V. and Chirik, P.J. (2013). *Org. Lett.* 15: 2680–2683.
15 Zhang, G., Zeng, H., Wu, J. et al. (2016). *Angew. Chem. Int. Ed.* 55: 14369–14372.
16 Bagherzadeh, S. and Mankad, N.P. (2016). *Chem. Commun.* 52: 3844–3846.
17 Tamang, S.R. and Findlater, M. (2017). *J. Org. Chem.* 82: 12857–12862.
18 Harrison, K.N. and Marks, T.J. (1992). *J. Am. Chem. Soc.* 114: 9220–9221.
19 Dudnik, A.S., Weidner, V.L., Motta, A. et al. (2014). *Nat. Chem.* 6: 1100–1107.
20 Weidner, V.L., Barger, C.J., Delferro, M. et al. (2017). *ACS Catal.* 7: 1244–1247.

21 Chen, S., Yan, D., Xue, M. et al. (2017). *Org. Lett.* 19: 3382–3385.
22 Yan, D., Dai, P., Chen, S. et al. (2018). *Org. Biomol. Chem.* 16: 2787–2791.
23 Patnaik, S. and Sadow, A.D. (2019). *Angew. Chem. Int. Ed.* 58: 2505–2509.
24 Hadlington, T.J., Hermann, M., Frenking, G., and Jones, C. (2014). *J. Am. Chem. Soc.* 136: 3028–3031.
25 Yang, Z., Zhong, M., Ma, X. et al. (2016). *J. Am. Chem. Soc.* 138: 2548–2551.
26 Yang, Z., Zhong, M., Ma, X. et al. (2015). *Angew. Chem. Int. Ed.* 54: 10225–10229.
27 Kidd, R.G. (1983). *Boron-11*, vol. 2. New York: Academic Press.
28 Harder, S. and Spielmann, J. (2012). *J. Organomet. Chem.* 698: 7–14.
29 Burgess, K. and Ohlmeyer, M.J. (1991). *Chem. Rev.* 91: 1179–1191.
30 Chong, C.C. and Kinjo, R. (2015). *ACS Catal.* 5: 3238–3259.
31 Revunova, K. and Nikonov, G.I. (2015). *Dalton Trans.* 44: 840–866.
32 Hill, M.S., Liptrot, D.J., and Weetman, C. (2016). *Chem. Soc. Rev.* 45: 972–988.
33 Burgess, K. and van der Donk, W.A. (1993). *Tetrahedron Lett.* 34: 6817–6820.
34 Männig, D. and Nöth, H. (1984). *J. Organomet. Chem.* 275: 169–171.
35 Arrowsmith, M., Hill, M.S., Hadlington, T. et al. (2011). *Organometallics* 30: 5556–5559.
36 Arrowsmith, M., Hill, M.S., and Kociok-Köhn, G. (2013). *Chem. Eur. J.* 19: 2776–2783.
37 Stachowiak, H., Kaźmierczak, J., Kuciński, K., and Hreczycho, G. (2018). *Green Chem.* 20: 1738–1742.
38 Yaws, C.L. (2009). *Yaws' Handbook of Thermodynamic Properties for Hydrocarbons and Chemicals*. Houston, USA: Gulf Publishing.
39 Glockler, G. (1963). *J. Chem. Soc. Faraday Trans.* 59: 1080–1085.
40 Luo, Y.-R. (2007). *Comprehensive Handbook of Chemical Bond Energies*. Boca Raton: CRC Press.
41 Chase, M.W. (1998). *NIST-JANAF Thermochemical Tables*. Woodbury, N.Y.: National Institute of Standards and Technology.
42 Rablen, P.R., Hartwig, J.F., and Nolan, S.P. (1994). *J. Am. Chem. Soc.* 116: 4121–4122.
43 Rablen, P.R. and Hartwig, J.F. (1996). *J. Am. Chem. Soc.* 118: 4648–4653.
44 Xu, D., Shan, C., Li, Y. et al. (2017). *Inorg. Chem. Front.* 4: 1813–1820.
45 Walsh, R. (1981). *Acc. Chem. Res.* 14: 246–252.
46 Schock, L.E. and Marks, T.J. (1988). *J. Am. Chem. Soc.* 110: 7701–7715.
47 Nolan, S.P., Stern, D., and Marks, T.J. (1989). *J. Am. Chem. Soc.* 111: 7844–7853.
48 Mukherjee, D., Osseili, H., Spaniol, T.P., and Okuda, J. (2016). *J. Am. Chem. Soc.* 138: 10790–10793.
49 Wu, Y., Shan, C., Ying, J. et al. (2017). *Green Chem.* 19: 4169–4175.
50 Query, I.P., Squier, P.A., Larson, E.M. et al. (2011). *J. Org. Chem.* 76: 6452–6456.
51 Zhu, Z., Wu, X., Xu, X. et al. (2018). *J. Org. Chem.* 83: 10677–10683.
52 Osseili, H., Mukherjee, D., Beckerle, K. et al. (2017). *Organometallics* 36: 3029–3034.

53 Osseili, H., Mukherjee, D., Spaniol, T.P., and Okuda, J. (2017). *Chem. Eur. J.* 23: 14292–14298.

54 Bismuto, A., Cowley, M.J., and Thomas, S.P. (2018). *ACS Catal.* 8: 2001–2005.

55 Bismuto, A., Thomas, S.P., and Cowley, M.J. (2016). *Angew. Chem. Int. Ed.* 55: 15356–15359.

56 Jakhar, V.K., Barman, M.K., and Nembenna, S. (2016). *Org. Lett.* 18: 4710–4713.

57 Pollard, V.A., Orr, S.A., McLellan, R. et al. (2018). *Chem. Commun.* 54: 1233–1236.

58 Lemmerz, L.E., McLellan, R., Judge, N.R. et al. (2018). *Chem. Eur. J.* 24: 9940–9948.

59 Intemann, J., Lutz, M., and Harder, S. (2014). *Organometallics* 33: 5722–5729.

60 Arrowsmith, M., Hadlington, T.J., Hill, M.S., and Kociok-Kohn, G. (2012). *Chem. Commun.* 48: 4567–4569.

61 Weetman, C., Anker, M.D., Arrowsmith, M. et al. (2016). *Chem. Sci.* 7: 628–641.

62 Weetman, C., Hill, M.S., and Mahon, M.F. (2016). *Chem. Eur. J.* 22: 7158–7162.

63 Yang, Y., Anker, M.D., Fang, J. et al. (2017). *Chem. Sci.* 8: 3529–3537.

64 Li, J., Luo, M., Sheng, X. et al. (2018). *Org. Chem. Fron.* 5: 3538–3547.

65 Jones, D.D.L., Matthews, A.J.R., and Jones, C. (2019). *Dalton Trans.* Advance Article.

66 Weetman, C., Hill, M.S., and Mahon, M.F. (2016). *Polyhedron* 103: 115–120.

67 Mukherjee, D., Ellern, A., and Sadow, A.D. (2014). *Chem. Sci.* 5: 959–964.

68 Lampland, N.L., Hovey, M., Mukherjee, D., and Sadow, A.D. (2015). *ACS Catal.* 5: 4219–4226.

69 Barman, M.K., Baishya, A., and Nembenna, S. (2017). *Dalton Trans.* 46: 4152–4156.

70 Mukherjee, D., Shirase, S., Spaniol, T.P. et al. (2016). *Chem. Commun.* 52: 13155–13158.

71 Manna, K., Ji, P., Greene, F.X., and Lin, W. (2016). *J. Am. Chem. Soc.* 138: 7488–7491.

9

Dehydrocoupling and Other Cross-couplings

Merle Arrowsmith

Julius-Maximilians-Universität Würzburg, Institut für Anorganische Chemie, Am Hubland, 97074 Würzburg, Germany

9.1 Introduction

Dehydrocoupling (DHC), or the formation of main group element–element bonds through loss of dihydrogen, represents an attractively clean method for accessing new main group compounds. Two types of DHC reactions are to be distinguished: that of two identical E—H bonds to form an apolar E—E bond, or homo-DHC (Scheme 9.1a), and that of a hydridic $E^{\delta+}$—$H^{\delta-}$ and a protic $E'^{\delta-}$—$H^{\delta+}$ unit to form a polar $E^{\delta+}$—$E'^{\delta-}$ bond, or cross-DHC (Scheme 9.1b). These reactions become particularly appealing when applied to species containing multiple E—H bonds that can undergo dehydropolymerization (DHP) reactions (Scheme 9.1c,d), yielding new inorganic materials and hybrid polymers for applications in a wide range of modern technologies [1]. Although entropically favored by the generation of gaseous H_2, the ease of homo- or cross-DHC greatly depends on the relative strength of the newly formed E—E/E′ bonds vs. the original E/E′—H bonds (Table 9.1). In the p-block, the E—H bond dissociation energy (BDE) steadily decreases in each group as orbital overlap decreases, and increases from left to right across the period as the bond becomes more ionic. A similar trend applies to homodinuclear BDEs, albeit with much starker decreases down each group and increases from one group to the next. Heteronuclear BDEs increase with orbital overlap and bond polarity and are generally higher than homodinuclear ones, making cross-DHC in general thermodynamically more favorable than homo-DHC.

Although the thermal dehydrogenation of inorganic main group hydrides has been known since the late nineteenth century report of the decomposition of SiH_4 into elemental silicon and H_2 at 400 °C [3], the term "dehydrocoupling" was only coined by Sneddon and coworker in 1985 for the Pt^{II}-catalyzed homo-DHC of small boron clusters (Scheme 9.2a) [4]. In the same year, Harrod and coworkers reported the first efficient and chemoselective homo-DHP of silanes with titanocene precatalysts (Scheme 9.2b) [5], which were to shape the field for the next three decades. The success of transition metal (TM)- and lanthanide-catalyzed silane DHC [6] paved the way for a flourishing chemistry

Early Main Group Metal Catalysis: Concepts and Reactions,
First Edition. Edited by Sjoerd Harder.

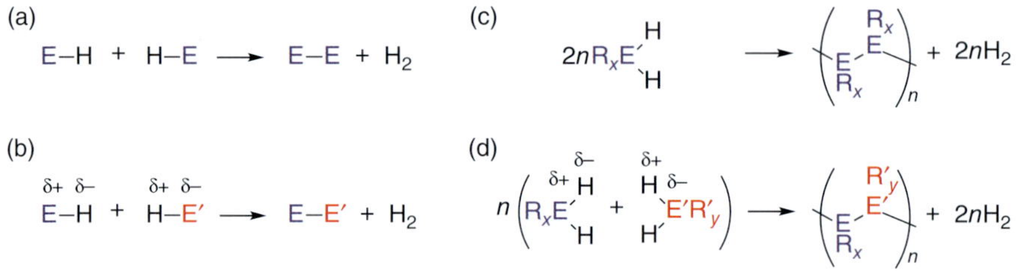

Scheme 9.1 Types of dehydrocoupling reactions: (a) Homo-dehydrocoupling; (b) cross-dehydrocoupling; (c) homo-dehydropolymerization; (d) cross-dehydropolymerization.

Table 9.1 Selected E—H, E—E, and E—E′ bond dissociation energies (BDEs) in kcal/mol.

	E—H	BDE		E—E	BDE		E—E′	BDE
Group 13	B—H	83	**Group 1**	H—H	104	**Group 13–14**	B—Si	76
Group 14	C—H	81	**Group 13**	B—B	69	**Group 13–15**	B—N	90
	Si—H	70	**Group 14**	C—C	148		B—P	83
	Ge—H	63		Si—Si	74	**Group 13–16**	B—O	193
Group 15	N—H	81		Ge—Ge	63	**Group 14–14**	C—Si	107
	P—H	71	**Group 15**	N—N	226	**Group 14–15**	Si—N	104
Group 16	O—H	103		P—P	117	**Group 14–16**	Si—O	191

Source: Luo 2007 [2]. Reproduced with permission of Taylor & Francis.

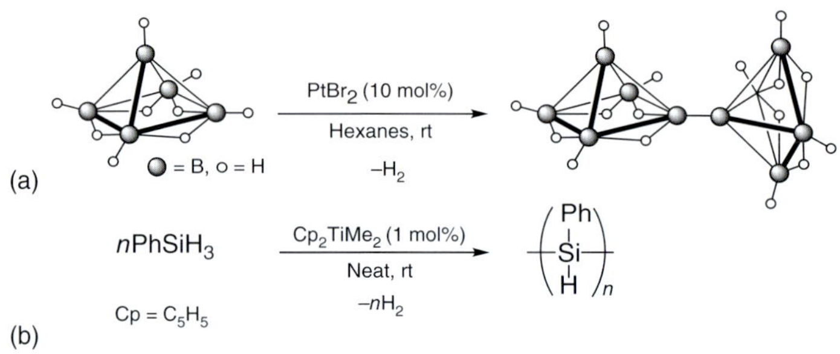

Scheme 9.2 Early examples of TM-catalyzed homo-DHC.

of group 13–15 homo-DHC [7], which will not be discussed herein, given that there are no examples of s-block-catalyzed homo-DHC to date. Although many cross-DHC reactions can be thermally induced, the use of catalysts for cross-DHC has proven especially useful for controlling reaction outcomes and accessing tailored hybrid polymers and inorganic materials based on Si–O (polysiloxane) [8], B–P (polyphosphinoborane), and B–N (polyaminoborane) linkages [9]. To date, the scope of catalytic cross-DHC further encompasses B—O, Al—N [10], C—P, C—Si, Si—N, Si—P, Si—S, Ge—P, Sn—Te, N—P, P—O, and P—S bond formation [7].

A variety of cross-DHC reaction mechanisms have been proposed over the years, depending on the nature of the substrates, metal center, and ligand

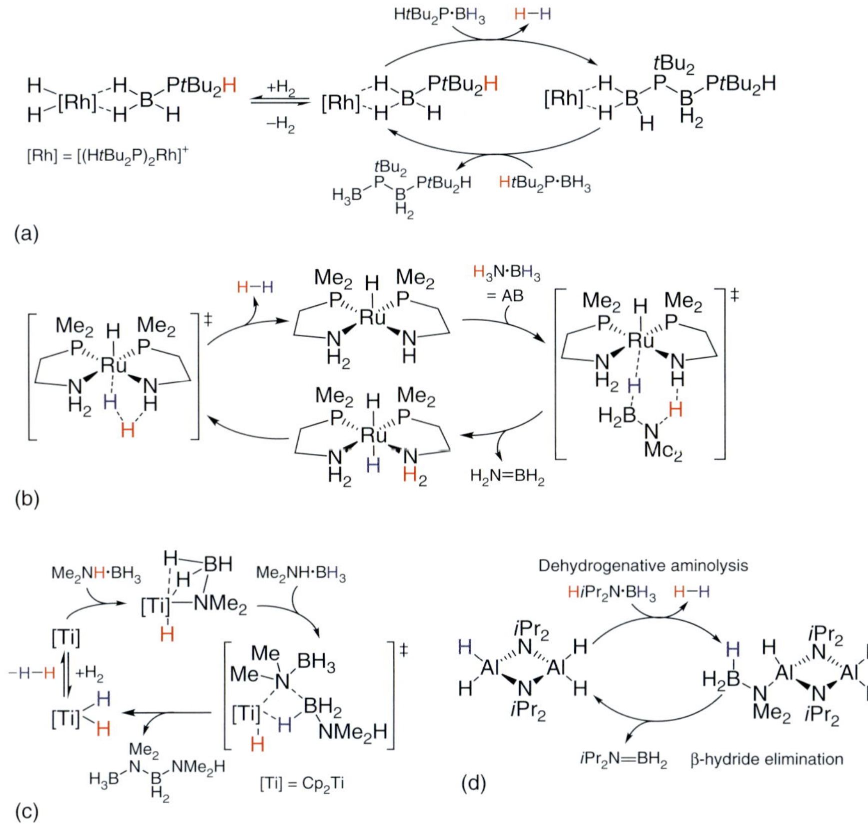

Scheme 9.3 Selection of TM-catalyzed cross-DHC mechanisms: (a) Rh^{I}-catalyzed B–H/P–H DHC; (b) Ru^{II}-catalyzed, concerted, ligand-assisted B–H/N–H DHC; (c) Ti^{II}/Ti^{IV}-catalyzed B–H/N–H DHC; (d) Al^{III}-catalyzed B–H/N–H DHC.

environment, a selection of which is shown in Scheme 9.3. Unlike TM-catalyzed homo-DHC, which often involves oxidative addition/reductive elimination or intermediates featuring TM=E double bonds, the majority of processes in TM-catalyzed cross-DHC are redox neutral [7a]. Weller's Rh^{I} cation-catalyzed DHC of $HtBu_2P{\cdot}BH_3$, for example, proceeds via the formation of a μ^2-borane adduct, which undergoes dehydrogenative B—P bond formation with an incoming $HtBu_2P{\cdot}BH_3$ molecule before release of the coupled diphosphinoborane product, the reversible formation of a Rh^{III} dihydride being an off-cycle process (Scheme 9.3a) [11]. For the DHC of ammonia borane (AB) into the aminoborane $H_2N{=}BH_2$ by the P,N-chelated Ru^{II} hydride complex, Fagnou proposed a concerted, ligand-assisted mechanism, in which the hydridic B—H bond is activated by the metal center and the protic N—H bond by the ligand amido functionality (Scheme 9.3b) [12]. A Ti^{II}/Ti^{IV} cycle has been postulated by Manners, Lloyd-Jones, and coworkers for the titanocene-catalyzed DHC of dimethylamine borane (DMAB), which proceeds via oxidative addition of the N—H bond at the Ti^{II} metal center, concerted β-hydride transfer/B—N bond formation with an incoming DMAB molecule, and finally reductive H_2 elimination from Cp_2TiH_2 (Scheme 9.3c) [13]. This contrasts with the generic main group-catalyzed amine borane DHC mechanism, which typically involves a metal hydride intermediate undergoing dehydrogenative aminolysis, followed by β-hydride transfer and release of the aminoborane [14], as exemplified by Wright's Al^{III}-catalyzed DHC of $i Pr_2NH \cdot BH_3$ in Scheme 9.3d [15]. In many cases, however, the reaction mechanism involves a complex interplay of competitive pathways, as will become apparent in the following sections.

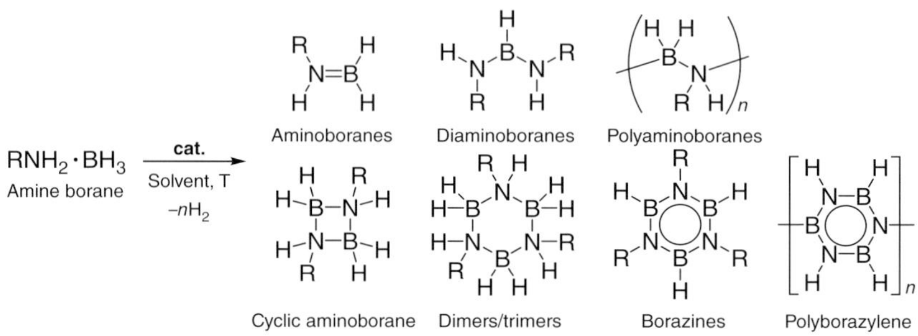

Scheme 9.4 Range of products obtained from catalytic amine borane DHC.

9.2 Early Main Group-Catalyzed Cross-DHC of Amines and Boranes

Over the past three decades, ammonia borane (AB = $H_3N \cdot BH_3$) has attracted considerable interest as a precursor for BN ceramics [16] and as a potential H_2 storage material because of its 19.6 wt% hydrogen content and ease of handling [17]. In 1926, Stock and Pohland were the first to report the uncatalyzed dehydrogenation of NH_3 and B_2H_6, into the "inorganic benzene" borazine ($B_3N_3H_6$) at 200 °C [18]. It was not until 2001 that the group of Manners published the first mild and selective TM-catalyzed DHC of amine boranes into aminoborane dimers and borazines [19]. Later, Heinekey and Manners independently reported the controlled iridium-catalyzed DHP of AB and primary amine boranes into linear polyaminoboranes [20], which are isoelectronic with polyolefins. Scheme 9.4 shows a range of B–N-coupled products that can be obtained from catalytic amine borane DHC. Although the most efficient and selective precatalysts for amine borane DHC are TM-based [21, 22], a variety of s- and p- and f-block metal complexes [21b], as well as several group 13–15 frustrated Lewis pairs (FLPs) and metal-free Lewis and Brønsted acids [14, 21], have been reported as competent catalysts for these reactions.

9.2.1 Early Stoichiometric Studies with s-Block Elements

s-Block metals have long been known to effect the stoichiometric dehydrogenation of amine boranes to generate amidoborane complexes [23]. In the past decade, these compounds have attracted renewed interest as AB substitutes for

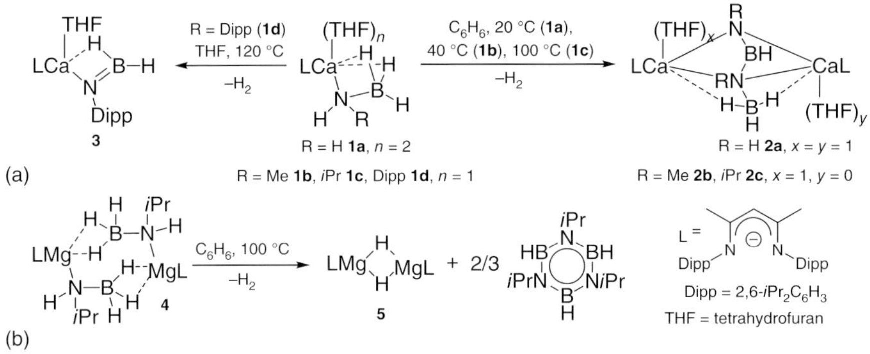

Scheme 9.5 Thermal decomposition of β-diketiminato group 2 amidoborane complexes.

hydrogen storage [24]. To model the release of H_2 from such compounds, Harder studied the thermal decomposition of β-diketiminato group 2 amidoborane complexes, LAe((N(H)R)BH$_3$)(THF)$_n$ (L = HC(C(Me)NDipp)$_2$; Dipp = 2,6-*i*Pr$_2$C$_6$H$_3$; THF = tetrahydrofuran; Ae = Mg, n = 0; Ae = Ca, n = 1; Scheme 9.5), for which the decomposition temperature drastically increased with the steric demands of the substituent R [25]. For the calcium complexes **1a–c** (R = H, Me, and *i*Pr), thermal decomposition resulted in loss of H_2 together with B–N coupling to yield dimers bridged by a $[(RN)BH(NR)BH_3]^{2-}$ dianion (**2a–c**) [25c, d], whereas dehydrogenation of the highly sterically hindered Dipp analogue **1d** yielded the monomeric amidoboryl complex **3** (Scheme 9.5a) [25c]. In contrast, thermolysis of the magnesium isopropylamidoborane dimer **4** yielded the magnesium hydride dimer **5** alongside the borazine (*i*PrNBH)$_3$, which is presumably formed via β-hydride elimination from **4**, generating *i*PrNH=BH$_2$, which then undergoes DHC to the borazine catalyzed by **5** [25b].

9.2.2 s-Block-Catalyzed Cross-dehydrogenative Synthesis of Diaminoboranes

The first example of catalytic amine borane cross-DHC with an s-block metal came from the group of Harder in 2009 [26]. In the presence of 5 mol% of the β-diketiminato magnesium amide LMgN(SiMe$_3$)$_2$ **6**, DippNH$_2$·BH$_3$ underwent twofold room temperature DHC to the dianilidoborane (Dipp(H)N)$_2$BH with loss of 1 equiv of BH_3 (Scheme 9.6). The same product was obtained from the 2 : 1 reaction of DippNH$_2$ and Me$_2$S·BH$_3$ catalyzed by commercially available di-*n*-butylmagnesium. Based on isolated intermediates and nuclear magnetic resonance (NMR) spectroscopy, the authors proposed the catalytic cycle depicted in Scheme 9.6. In a first step, precatalyst **6** is activated by aminolysis with DippNH$_2$·BH$_3$ to yield the amidoborane complex **7**, which was later isolated from the reaction of **5** with DippNH$_2$·BH$_3$ [25b]. Complex **7** undergoes dehydrogenative B—N bond formation with a second equivalent DippNH$_2$·BH$_3$ to yield an intermediate [(N(H)Dipp)BH$_2$(N(H)Dipp)BH$_3$]$^-$ complex. Subsequent β-hydride elimination results in liberation of the diaminoborane product and formation of the catalytically active magnesium tetrahydroborate dimer **8**, which

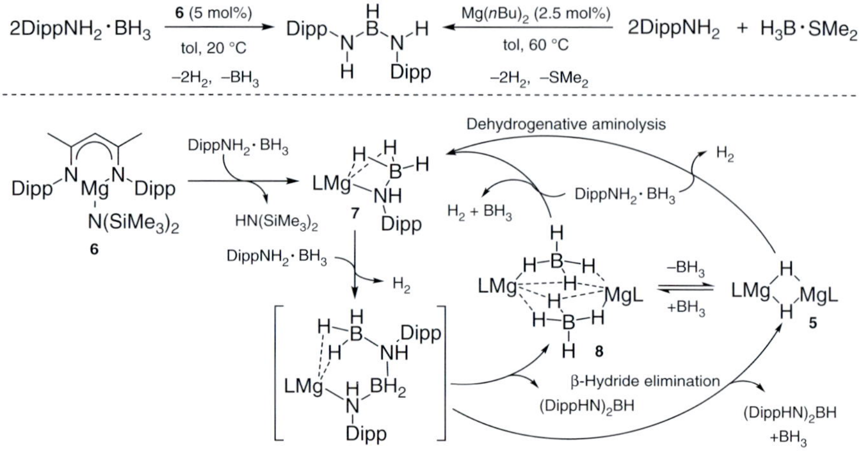

Scheme 9.6 Magnesium-catalyzed cross-DHC of $DippNH_2{\cdot}BH_3$.

was isolated from the reaction mixture. Finally, σ-bond metathesis with a further equivalent $DippNH_2{\cdot}BH_3$ induces loss of H_2 and BH_3 and regenerates **7**. The authors also proposed an alternative pathway, in which β-hydride elimination liberates both $(Dipp(H)N)_2BH$ and BH_3, thus generating the hydride complex $[LMgH]_2$ (**5**), which would be in equilibrium with **8**.

In a similar manner, Hill and coworkers employed the diamido group 2 precursors $[M\{N(SiMe_3)_2\}_2]_2$ (M = Mg **9a**, Ca **9b**) to catalyze the selective synthesis of unsymmetrical diaminoboranes by twofold DHC of $R^1R^2NH{\cdot}BH_3$ amine boranes with R^3R^4NH amines (Scheme 9.7) [27]. The magnesium catalyst **9a** was found to outperform its calcium analogue **9b**. The selectivity for the unsymmetrical product over the symmetrical one was determined by the relative basicity of the two amines. Although the DHC of DMAB and pyrrolidine produced a mixture of $(Me_2N)_2BH$ and $(Me_2N)BH(N(CH_2)_4)$, the reactions of

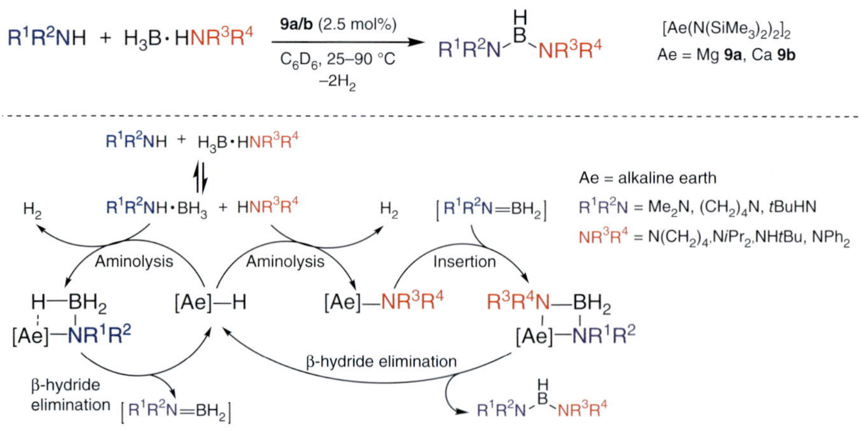

Scheme 9.7 Alkaline earth bis(amide)-catalyzed synthesis of unsymmetrical diaminoboranes.

DMAB with more Lewis acidic amines, such as $i\mathrm{Pr_2NH}$, $\mathrm{Ph_2NH}$, or $t\mathrm{BuNH_2}$, were selective for the unsymmetrical products. In the first step, the catalyst undergoes aminolysis with $\mathrm{R^1R^2NH{\cdot}BH_3}$, followed by β-hydride elimination to liberate the intermediate aminoborane $\mathrm{R^1R^2N{=}BH_2}$ and generate an alkaline earth hydride catalyst. This proposal was based on the previous observations that **9a** reacts with 4 equiv of DMAB to generate $\mathrm{Me_2N{=}BH_2}$ (see Scheme 9.10b) and also catalyzes the DHC of $i\mathrm{Pr_2NH{\cdot}BH_3}$ into $i\mathrm{Pr_2N{=}BH_2}$ [28]. In the second step, the metal hydride catalyst, rather than re-entering the first cycle, undergoes dehydrogenative aminolysis with the more Lewis acidic amine, $\mathrm{R^3R^4NH}$, to yield a metal amide intermediate, which inserts $\mathrm{R^1R^2N{=}BH_2}$. In the rather sluggish DHC of pyrrolidine borane with $t\mathrm{BuNH_2}$ catalyzed by **9b**, the resulting $[\mathrm{(N(CH_2)_4)BH_2(}t\mathrm{Bu(H)N)}]^-$ chelate was observed by ^{11}B NMR spectroscopy. Finally, β-hydride elimination liberates the unsymmetrical diaminoborane and $\mathrm{BH_3}$ and regenerates the alkaline earth hydride catalyst. If both amines are of similar Lewis acidity, the pre-equilibrium between $\{\mathrm{R^1R^2NH{\cdot}BH_3 + R^3R^4NH}\}$ and $\{\mathrm{R^3R^4NH{\cdot}BH_3 + R^1R^2NH}\}$ leads to mixtures of the unsymmetrical and symmetrical diaminoboranes.

More recently, Robertson and Mulvey reported the first alkali metal-catalyzed synthesis of cyclic diaminoboranes from the DHC of diamine boranes (Scheme 9.8) [29], a reaction thus far only catalyzed by an expensive ruthenium catalyst [30]. The lithium dihydropyridide precatalyst **10a** acts as a soluble LiH substitute and afforded 1,3,2-diazaborolidines and 1,3,2-diazaborinanes in excellent yields within two hours at 70 °C in benzene. Pyridine blocks the reaction by competing with the diamine for Lewis adduct formation with $\mathrm{BH_3}$. Furthermore, a potential catalytic intermediate in the form of the lithium (amine)amidoborane complex **11** was isolated (Scheme 9.8).

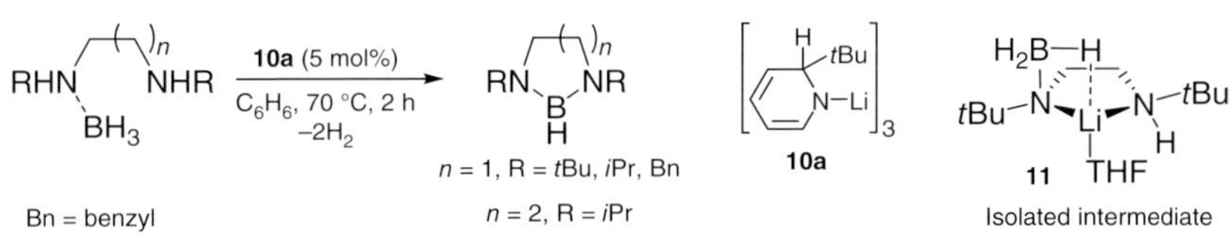

Scheme 9.8 Lithium-catalyzed synthesis of cyclic diaminoboranes.

9.2.3 s-Block-Catalyzed DHC of DMAB

Both group 1 [31] and group 2 complexes [32] have been reported as competent precatalysts for the DHC of DMAB, with catalytic activity rapidly decreasing down each group. The most active s-block precatalyst to date for this reaction is the dialkylmagnesium complex $\mathrm{Mg(CH(SiMe_3)_2)_2(THF)_2}$ (**12a**), which, at 5 mol% loading, 60 °C in $\mathrm{C_6D_6}$ and over 72 hours, catalyzes the formation of the cyclic dimer **I** with ca. 80% selectivity over the diaminoborane by-product **II** (Scheme 9.9) [32]. Although reactions involving the alkali metal 2-dihydropyridide trimers **10a–c** required heating in toluene at 80 °C for a similar performance, the selectivity for **I** over **II** could be improved to 98% when using the 4-dihydropyridide dimer **14** in cyclohexane [31a]. Mulvey and Robertson also

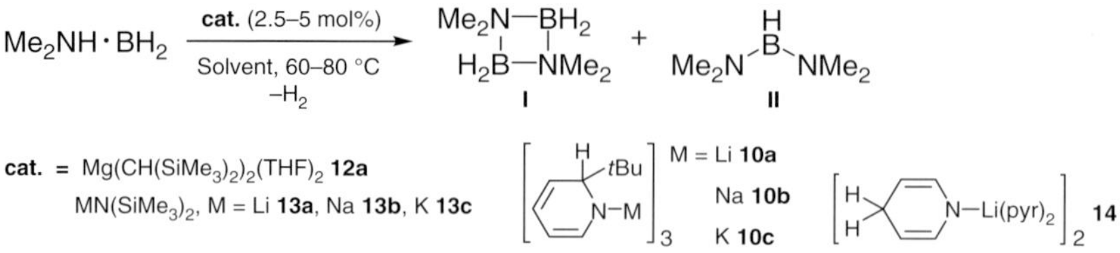

Scheme 9.9 s-block metal-catalyzed DHC of DMAB.

showed that chelating donor ligands or the use of THF as the reaction solvent strongly impede catalysis. Using pyridine as the reaction solvent did not affect catalysis rates but inverted the selectivity toward the formation of **II**, presumably because pyridine competes with Me_2NH for adduct formation with BH_3. Comparison with other non-TM catalysts for this reaction provides the following order of catalyst activity: Sc > Y [33] > Al [10] > Mg > K > Na > Li ≫ Sn [34].

The mechanism of the alkaline earth-promoted reaction has been investigated in detail by Hill and coworkers. The reaction of 4 equiv of DMAB with the dialkyl precursor **12a** led to the quantitative formation of complex **15**, in which two $[(NMe_2)BH_2(NMe_2)BH_3]^-$ ligands are bound to the metal center via the α-nitrogen and two Mg···H–B interactions of the δ-hydrides (Scheme 9.10a) [32]. Under mild heating in solution **15** slowly released the aminoborane dimer **I**, presumably with the loss of MgH_2. Although the monomeric diaminoborane $Me_2N{=}BH_2$ (**III**) was never observed under a catalytic regime, a 4 : 1 reaction between the bis(amide) precursor **9a** and DMAB resulted in near-quantitative conversion to **III** and small amounts of a heteroleptic amidoborane complex observed by NMR spectroscopy, while **9a** remained essentially intact (Scheme 9.10b), which implies that β-hydride elimination is faster than any other process in the catalytic cycle [28]. Overall, these observations led to the proposal of the mechanism presented in Scheme 9.11. In the first step, a diamidoborane intermediate undergoes β-hydride elimination, liberating the aminoborane **III**, which inserts into another amidoborane complex to yield a $[(NMe_2)BH_2(NMe_2)BH_3]^-$ complex analogous to **15**. Subsequent δ-hydride elimination liberates the cyclic dimer **I** and regenerates the group 2 hydride

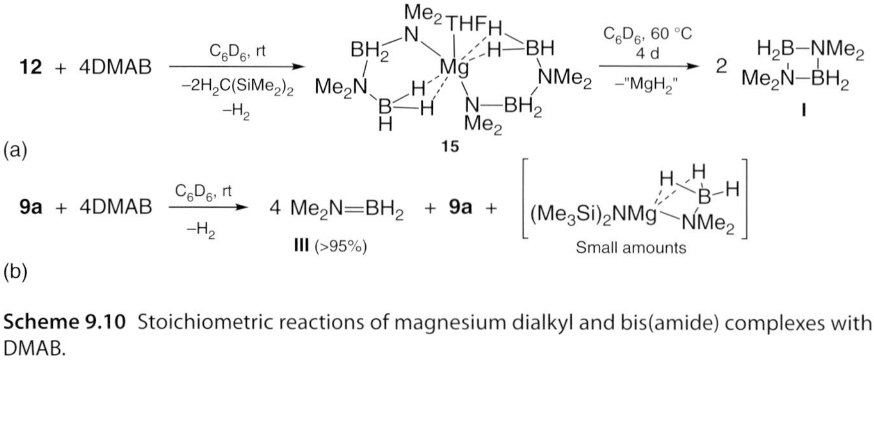

Scheme 9.10 Stoichiometric reactions of magnesium dialkyl and bis(amide) complexes with DMAB.

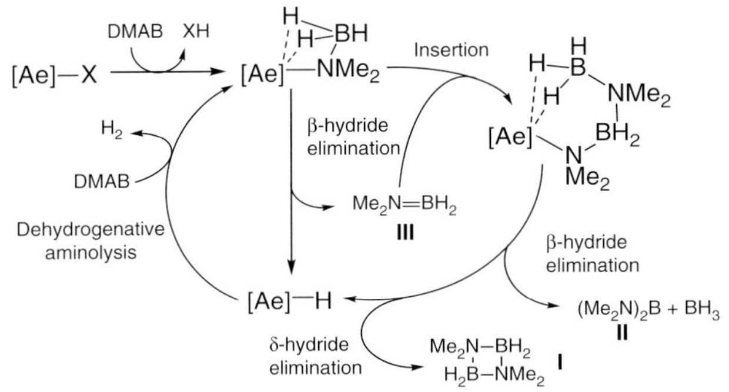

Scheme 9.11 Proposed mechanism for the alkaline earth-catalyzed DHC of DMAB.

catalyst, whereas competitive β-hydride elimination yields the diaminoborane by-product **II** and BH_3.

To probe the validity of this mechanism and determine the origin of the slower reactivity of the calcium vs. the magnesium catalyst, Sicilia and coworkers examined the reaction profiles for the model β-diketiminato complexes $LAe((NMe_2)BH_3)(THF)_n$ (Ae = Mg, $n = 0$; Ae = Ca, $n = 1$ **17**, see Scheme 9.12b) by density functional theory (DFT) calculations [35]. According to their results, the formation of **III** by β-hydride elimination and its insertion into the magnesium amidoborane intermediate are facile, whereas subsequent δ-hydride elimination to liberate **I** and regenerate LMgH presents a prohibitively high barrier of 39.6 kcal/mol. This led the authors to conclude that the cyclization of **III** to **I** must be a thermally induced, off-metal process, for which the experimentally determined barrier is only ca. 19 kcal/mol [36]. Attempts to model the same pathway with the calcium amidoborane complex **17** failed because

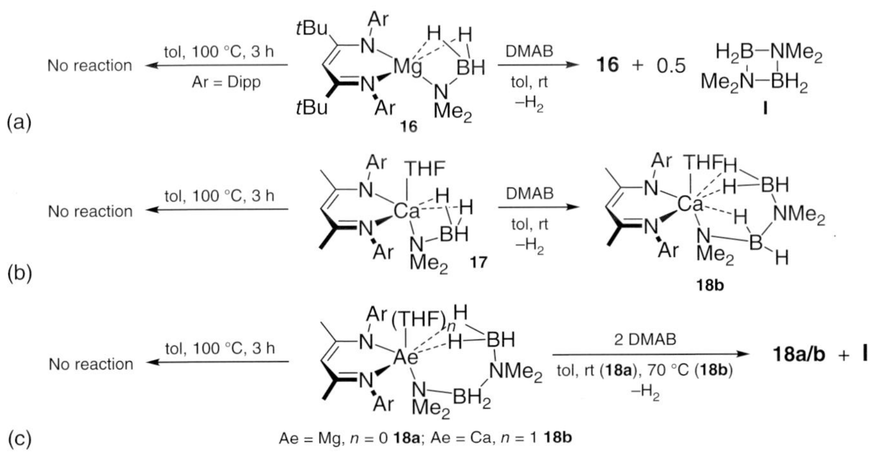

Scheme 9.12 Thermal stability and stoichiometric reactivity with DMAB of β-diketiminato magnesium and calcium $[(NMe_2)BH_3]^-$ and $[(Me_2N)BH_2NMe_2BH_3]^-$ complexes.

of the thermodynamic instability of the putative LCaH intermediate [35]. An alternative pathway was found, in which an incoming DMAB molecule displaces the THF molecule from **17** and assists with the concerted elimination of **III** and H_2 to regenerate the calcium amidoborane without the need for LCaH.

In response, Hill and coworkers examined in detail the thermal stability of isolated β-diketiminato magnesium and calcium $[(NMe_2)BH_3]^-$ and $[(Me_2N)BH_2NMe_2BH_3]^-$ complexes and showed that these remain stable toward β- and/or δ-hydride elimination at 100 °C in solution for several hours (Scheme 9.12a–c) [37]. In the presence of 1 equiv of DMAB, however, the magnesium amidoborane **16** promoted the room temperature formation of 0.5 equiv of **I** while regenerating **16** (Scheme 9.12a). The analogous reaction with the calcium amidoborane **17** led to stoichiometric H_2 elimination and the formation of the $[(NMe_2)BH_2(NMe_2)BH_3]^-$ complex **18b** (Scheme 9.12b). The 1 : 1 reaction of **17** with $Me_2ND{\cdot}BH_3$ and $Me_2NH{\cdot}BD_3$, respectively, provided small but similar kinetic isotope effects (KIEs = 1.4 and 1.6, respectively), suggesting that both N—H and B—H bond-breaking/forming processes are involved to a similar degree in the rate-determining step (RDS). Although this confirmed Sicilia and coworkers' claim against direct β-hydride elimination from **17** and direct δ-hydride elimination from **18a/b**, Hill and coworkers suggested a different, entirely DMAB-assisted mechanism, in which the $[(NMe_2)BH_3]^-$ and $[(Me_2N)BH_2NMe_2BH_3]^-$ complexes are the only intermediates, and the successive β- and δ-hydride elimination steps occur through N–H-assisted H_2 elimination without the need for metal hydride intermediates or the elimination/insertion of **III** (Scheme 9.13). This mechanism is reminiscent of the amine-assisted concerted insertion/protonolysis mechanism, which has been proposed for the group 2-catalyzed intramolecular hydroamination of aminoalkenes to account for high N–H/D KIEs and circumvent thermodynamically unstable metal alkyl intermediates [38]. In view of this new mechanism, the higher reactivity of magnesium over calcium can be ascribed to its stronger polarizing capacity, whereas the differences in stoichiometric reactivity point to the DMAB-assisted formation of the $[(NMe_2)BH_2(NMe_2)BH_3]^-$ complex being the RDS for calcium, and the DMAB-assisted elimination of **I** being the RDS for magnesium.

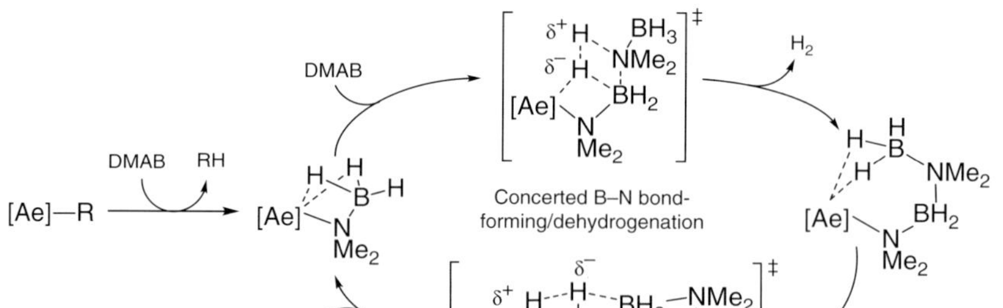

Scheme 9.13 Proposed concerted mechanism of alkaline earth-catalyzed DHC of DMAB.

9.2.4 Calcium-Catalyzed Dehydrocoupling of *tert*-Butylamine Borane

Hill and coworkers have also reported the DHC of $t\mathrm{BuNH_2{\cdot}BH_3}$ using the calcium β-diketiminate precatalyst **19** (Scheme 9.14) [39]. Although very slow, unselective, and accompanied by degradation of the catalyst to polymeric $[\mathrm{Ca(BH_4)_2}]_n$, this reaction represents the first catalytic borazine formation by an early main group metal. Given the production of the cyclic dimer $(t\mathrm{Bu(H)NBH_2})_2$ (**VII**) in the early stages of catalysis and its later consumption, the authors proposed that borazine **VIII** may be formed by calcium-mediated ring expansion of **VII** followed by catalytic dehydrogenation.

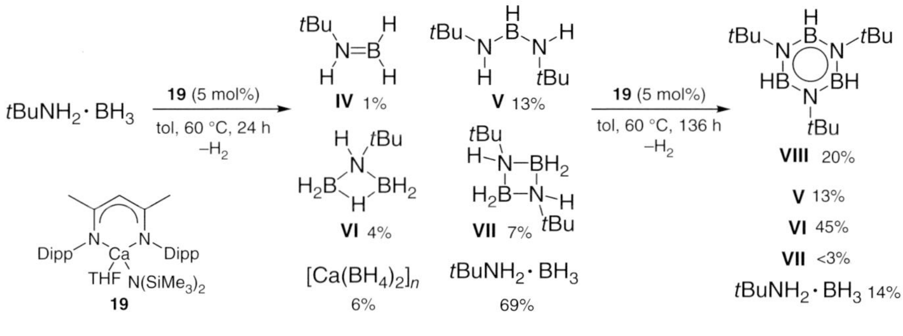

Scheme 9.14 Calcium-catalyzed DHC of $t\mathrm{BuNH_2{\cdot}BH_3}$.

9.2.5 s-Block-Catalyzed DHC of Amines and Monohydroboranes

In 2015, Hill reported the first alkaline earth-catalyzed cross-DHC of amines with pinacolborane (HBPin) and 9-borabicyclo[3.3.1]nonane (9-BBN) using the β-diketiminato magnesium and calcium precatalysts **19** and **20** (Scheme 9.15) [40]. The magnesium precursor **20** was by far superior and, at 10 mol% loading and room temperature, catalyzed the quantitative DHC of HBPin with a variety of primary and secondary amines to the corresponding aminoboranes within less than one hour, the amine substrate scope being essentially limited by sterics. The much lower reactivity of 9-BBN was at first ascribed to its lower Lewis acidity. Furthermore, **20** catalyzed the 2 : 1 DHC of HBPin with $n\mathrm{BuNH_2}$ and $t\mathrm{BuNH_2}$, respectively, to yield the corresponding bis(boryl)amines (Scheme 9.15b), whereas the same reaction with aniline stopped at the monocoupled product.

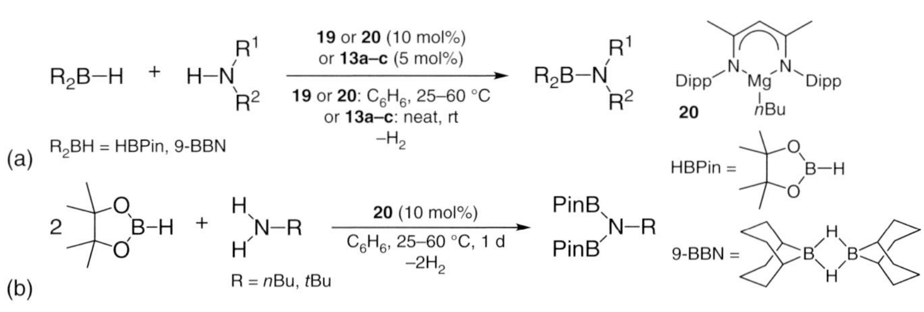

Scheme 9.15 Alkaline earth-catalyzed cross-DHC of amines and boranes.

More recently, Panda and coworkers reported that the commercially available alkali metal hexamethyldisilazides **13a–c** are also efficient precatalysts for the room temperature DHC of amines with HBPin and 9-BBN under solvent-free conditions [41]. The lithium and sodium precursors **13a** and **13b** were found to be similarly efficient and superior to the potassium analogue **13c**. The extended substrate scope employed by the authors encompassed various substituted anilines, which showed that reactivity significantly decreases with increasingly electron-withdrawing *ortho*-substituents in the order of $Me \approx H > F > NO_2$, whereas an electron-donating *para*-methoxy substituent has little effect on the reaction. Furthermore, the potassium precursor **13c** was able to catalyze the coupling of HBPin with allylamine and heteroaromatic amines, such as *o*-aminopyridine and indole.

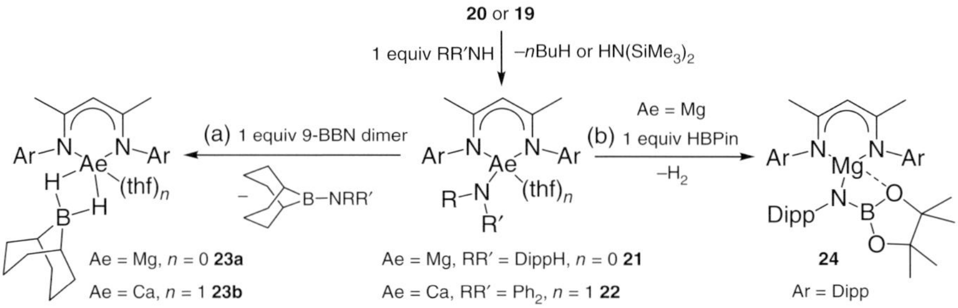

Scheme 9.16 Stoichiometric reactions of β-diketiminato magnesium and calcium anilides with (a) 9-BBN and (b) HBPin.

Stoichiometric reactions of **19**/**20** with amines and HBPin/9-BBN enabled the isolation of potential catalytic intermediates. Thus, the reaction of **20** or **19** with 1 equiv of $DippNH_2$ or Ph_2NH, respectively, followed by 9-BBN yielded the borohydride complexes **23a** and **23b**, respectively, together with the respective aminoborane by-product (Scheme 9.16a) [40, 42]. In contrast, the reaction of **20** with $DippNH_2$ followed by HBPin resulted in H_2 evolution and formation of the magnesium amidoborane complex **24** in which Mg is coordinated with the nitrogen and one of the pinacol oxygen atoms of the amidoborane ligand (Scheme 9.16b) [40]. Furthermore, entirely different rate laws were determined for the DHC of HBPin or 9-BBN with Me(Ph)NH, resulting in catalysis rates with HBPin being 2 orders of magnitude higher than with 9-BBN:

$$R_{\text{HBPin}} = -\frac{d[\text{Me(Ph)NH}]}{dt} = k\,[\mathbf{20}]^2\,[\text{Me(Ph)NH}]^0\,[\text{HBPin}]^1 \quad (9.1)$$

$$R_{\text{9-BBN}} = -\frac{d[\text{Me(Ph)NH}]}{dt} = k\,[\mathbf{20}]^{\frac{1}{2}}\,[\text{Me(Ph)NH}]^2\,[\text{9-BBN}]^{-1} \quad (9.2)$$

With HBPin, the reaction showed a second-order dependence on [**20**], a first-order dependence on [HBPin], and no dependence on [Me(Ph)NH] (Eq. (9.1)). In contrast, the use of 9-BBN led to a half-order dependence on [**20**], a second-order dependence on [Me(Ph)NH], and an inverse first-order

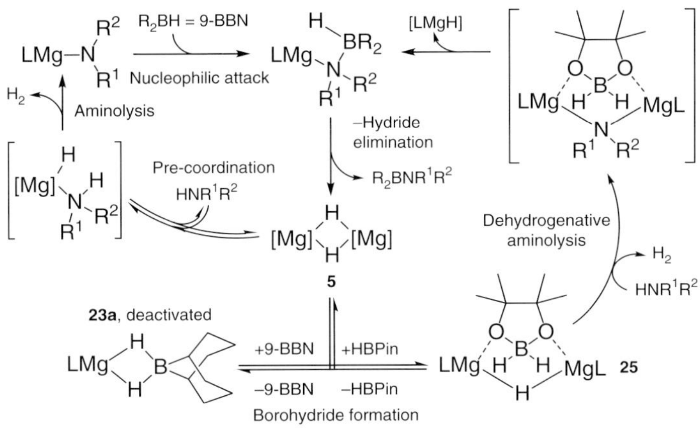

Scheme 9.17 Proposed divergent mechanisms for the magnesium-catalyzed cross-DHC of amines with HBPin and 9-BBN.

dependence on [9-BBN] (Eq. (9.2)), which was confirmed by the complete inhibition of catalytic activity when using a large excess of 9-BBN. Furthermore, although the enthalpy of activation of the 9-BBN-based reaction is three times that of the HBPin-based reaction, this is compensated by a positive activation entropy for 9-BBN, indicative of a dissociative RDS, whereas the highly negative activation entropy for HBPin indicates a very ordered transition state (TS) typical for an associative mechanism. These combined results led the authors to propose the catalytic cycle depicted in Scheme 9.17. Entry into the catalytic manifold occurs either through aminolysis of **20** with R^1R^2NH to yield the amido complex $LMgNR^1R^2$ or through σ-bond metathesis with R_2BH to yield the hydrido dimer $[LMgH]_2$ (**5**). The origin of the inverse first-order dependence in [9-BBN] in Eq. (9.2) is the reversible formation of the catalytically inactive borohydride complex **23a**. For catalytic turnover to proceed, **5** must in turn dissociate into monomers by (presumably reversible) adduct formation with 2 equiv of R^1R^2NH. It is this rate-limiting amine-mediated dimer-to-monomer step which the authors deem responsible for the half order in [catalyst], second-order dependence in [amine] and positive activation entropy. Previously, Hill and coworkers had also shown that, in the presence of excess HBPin, **5** is in equilibrium with the dinuclear hydrido borohydride complex **25** [43]. The second-order dependence in [**20**] in Eq. (9.1) and the highly negative activation entropy were proposed to stem from the rate-determining dehydrogenative aminolysis of **25**. In 2017, Xu et al. undertook DFT calculations on the mechanism of this reaction using monomeric LAeH (Ae = Be, Mg, Ca, and Sr) complexes. Their calculations, however, disregarded the formation of hydride dimers such as **5**, or borohydride intermediates such as **25**. Their results, which predict reactivity in the order of Ca > Sr > Mg ≫ Be, are to be taken with caution as they contradict experimental findings in multiple ways, yet highlighting the strong influence of Ae–H BDE in group 2 hydride catalysis.

9.3 s-Block-Catalyzed Cross-DHC of Amines and Silanes

Silylamines (or silazanes) play an important role both in organic chemistry, where they are used as mild bases, silylating agents, or protected versions of amines/silanes [44], and in materials chemistry as polysilazanes and precursors for SiN-based ceramics [45]. Although Si—N bonds are traditionally formed by chlorosilane aminolysis [46], the HCl by-product limits functional group tolerance and requires basic conditions to neutralize the reaction, which generates stoichiometric amounts of waste. Amine–silane DHC therefore offers an advantageous route to silazanes and has been reported for various TM catalysts, including heterogeneous Pd, Ru, and Au catalysts [47] and homogeneous Ti^{IV}, Cr^0, Fe^0, $Rh^{0/I}$, $Ru^{0/II}$, Cu^I [7b, e], and Zn^{II} complexes [48], as well as redox neutral d^0 rare earth, lanthanide, and actinide complexes [49], and group 13 and 15 Lewis acids [14a, 50]. As with amine borane DHC, the main issue of amine–silane DHC is selectivity, as the presence of multiple Si—H and/or N—H bonds may lead to unselective, multiple DHC, often independent of the relative stoichiometry of the substrates (Scheme 9.18). Although the reaction outcome is mainly influenced by the sterics and/or electronics of the substrates, it can be controlled by a judicious choice of catalyst.

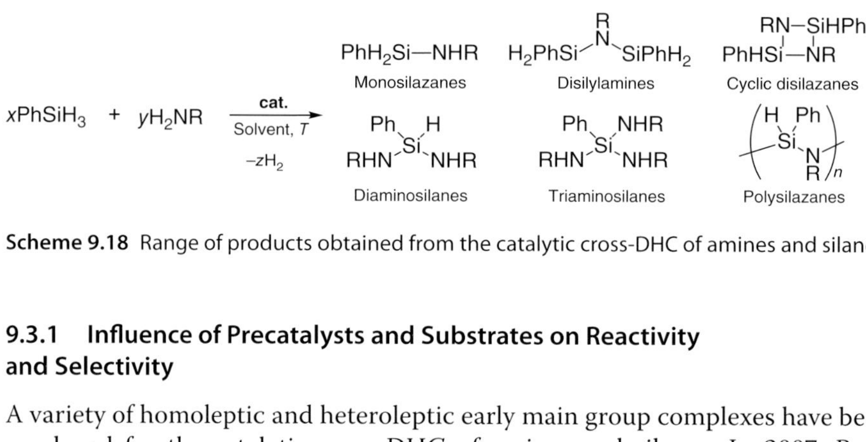

Scheme 9.18 Range of products obtained from the catalytic cross-DHC of amines and silanes.

9.3.1 Influence of Precatalysts and Substrates on Reactivity and Selectivity

A variety of homoleptic and heteroleptic early main group complexes have been employed for the catalytic cross-DHC of amines and silanes. In 2007, Buch and Harder showed that the calcium dibenzyl and azametallacyclopropane complexes **26** and **27** (Figure 9.1), respectively, provide a cheap alternative to ytterbium catalysts for the DHC of primary and secondary amines with Ph_3SiH, with similar activity and product selectivity [49]. Since then, a wide range of early main group compounds, including the alkali metal amides **13a–c**, homoleptic group 2 bis(amides) (**9a–d, 28b–d**) and dialkyls (**12b–d**), heteroleptic group 2 iminoanilides (**29a–c, 30a–c**), the magnesium tris(oxazolinyl)borate **31**, and the N-heterocyclic carbene (NHC)-supported magnesium aryl amide **32** have been found to be catalytically active (Figure 9.1). The benchmark reaction for comparing the activity of these catalysts is the DHC of Ph_3SiH and pyrrolidine (**IX**) to the corresponding silazane **X** (Scheme 9.19). Unlike in amine borane DHC, catalytic activity was typically found to increase in each group, in the

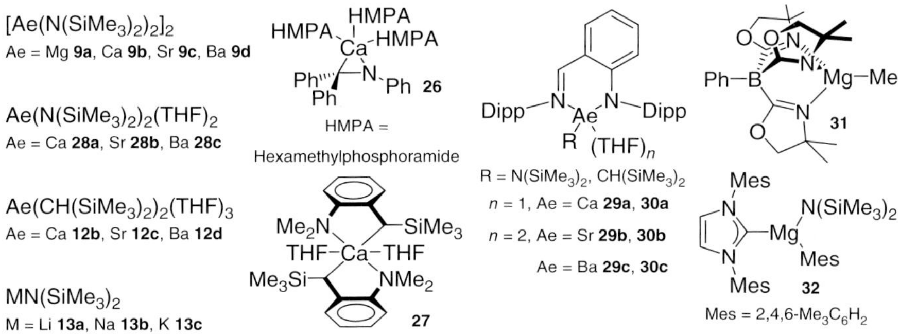

Figure 9.1 s-block precatalysts for the catalytic cross-DHC of amines and silanes.

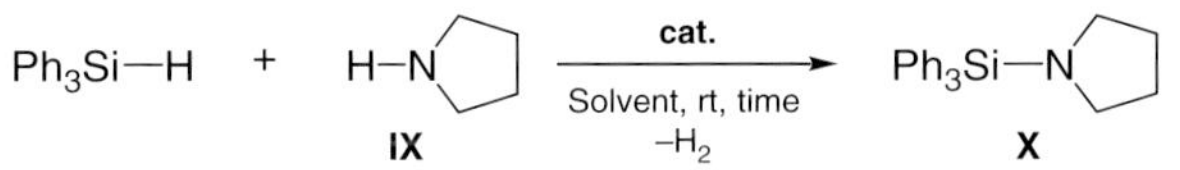

Scheme 9.19 Cross-DHC of Ph_3SiH and pyrrolidine as benchmark reaction for comparing catalyst activity.

order of Li < Na < K [51] and Mg ≪ Ca < Sr ≪ Ba [52], and group 2 catalysts outperforming group 1 catalysts within the same period. Alkyl complexes (**12b–d** and **30a–c**) were more active than their amido counterparts (**28a–c** and **29a–c**, respectively), catalyst activation by aminolysis being fast and irreversible in the former, but often slow and reversible in the latter [52]. In contrast to many other group 2-catalyzed reactions that are inhibited by THF [54], a similar catalytic activity was observed for the unsolvated group 2 bis(amides) **9b–d** and their THF-solvated counterparts **28a–c** [52]. Ligand sterics also play a major role in tuning catalyst activity: thus, the homoleptic barium precatalysts **28c** and **12d** were found to be more active than their direct heteroleptic counterparts, **29c** and **30c**, respectively [52]. Furthermore, the choice of precatalysts greatly influences the selectivity of the reaction when multiple couplings are possible (see Scheme 9.18). In 1 : 1 amine/silane reactions, the least reactive catalysts, in particular, the heteroleptic magnesium precursors **31** [55] and **32a/b** [56], were the most selective for monosilazane formation. Conversely, the most reactive precatalysts, in particular, the homoleptic barium precursors **9d** and **12d** [52], favored multiple couplings whenever possible. The selective one-step formation of cyclic disilazanes has only been observed with the potassium precursor **13c** for the 1 : 1 DHC of tBuNH$_2$ with PhSiH$_3$ and of BnNH$_2$ with Ph$_2$SiH$_2$ [51]. The most selective precatalyst to date is the magnesium tris(oxazolinyl)borate complex **31**, which is capable of catalyzing the selective monocoupling of hydrazine and ammonia with trialkylsilanes at room temperature (Scheme 9.20) [55].

For both amines and silanes, reactivity decreases with increasing substitution and substituent steric bulk, substrate scope being effectively limited by sterics [51–53, 55, 56]. In terms of electronic factors, the more nucleophilic alkylamines are the most reactive. Electron-poor silanes, such as arylsilanes and those displaying electron-withdrawing substituents, also display higher reactivity [56]. As a general rule, when multiple couplings are possible, the tendency

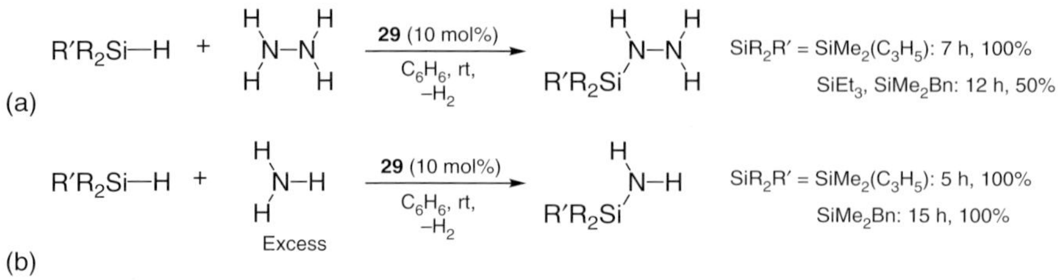

Scheme 9.20 Magnesium-catalyzed selective mono-coupling of hydrazine and ammonia with trialkylsilanes.

toward di- or triaminosilane formation (Scheme 9.18) increases for less bulky amines and at higher reaction temperatures. The selective monocoupling of Ph_2SiH_2 and tBuNH$_2$ by the magnesium complex **31**, for example, can be achieved even when using excess tBuNH$_2$, whereas the selective monocoupling of Ph_2SiH_2 and nPrNH$_2$ requires excess Ph_2SiH_2 in order to avoid the formation of $Ph_2Si(NH$n$Pr)_2$ [55]. The formation of triaminosilanes has only been observed for 1 : 3 coupling reactions between $PhSiH_3$ and the least hindered, most nucleophilic amines, nPrNH$_2$ and pyrrolidine [51–53]. The selective formation of disilylamines from 1 : 2 amine/silane reactions has only been reported for $(Ph_2SiH)_2$NBn and $(PhSiH_2)_2$NtBu and requires highly reactive precatalysts, such as **9b**, **9c** [53], or **13c** [51].

9.3.2 Mechanistic and Computational Analysis

As has now been shown for a variety of other s-block-catalyzed reactions [54], the mechanism and kinetics of alkaline earth-mediated amine–silane cross-DHC reactions are dependent not only on the nature of catalyst ligands but also on the identity of the metal center itself. The most detailed experimental and theoretical mechanistic study was carried out by Carpentier, Tobisch, and Sarazin investigating the DHC of pyrrolidine (**IX**) and Ph_3SiH (Scheme 9.19) with the heteroleptic barium precatalyst **29c**, which yields the following rate law [57].

$$R = -\frac{\mathrm{d[amine]}}{\mathrm{d}t} = k\,[\mathbf{cat}]^1\,[\mathrm{amine}]^0\,[\mathrm{silane}]^1 \quad (9.3)$$

The zero-order dependence in [amine] and first-order dependence in both [catalyst] and [silane] suggest a mononuclear TS and involvement of one silane molecule in the RDS. This was confirmed by a strong silane-dependent KIE (ca. 4.7) pointing to rate-determining Si—H bond-breaking, as well as a nonexistent amine-dependent KIE. An Eyring analysis provided a small but negative activation entropy, indicative of an associative RDS. The positive reaction constant ($\rho = 2.0$) provided by the Hammett analysis of the catalytic DHC of **IX** with $Ph_2(p\text{-R-}C_6H_4)SiH$ indicates that electron-withdrawing silane substituents lower the activation barrier, presumably by preventing a build-up of negative charge on the silicon atom. The authors undertook extensive theoretical calculations to explore alternative reaction pathways, which eventually led them to propose the catalytic cycle presented in Scheme 9.21. The active pyrrolidide catalyst is

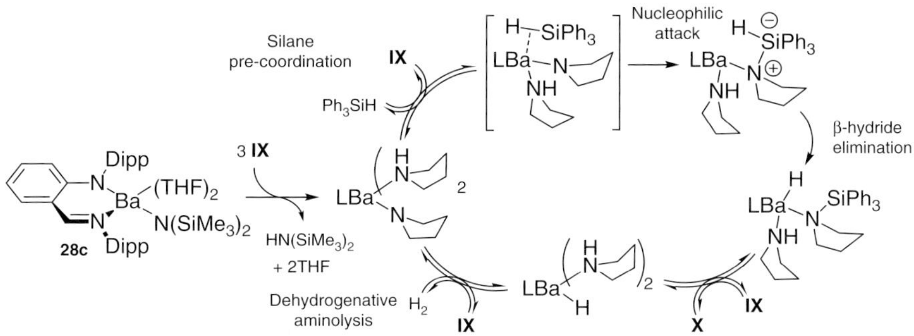

Scheme 9.21 Proposed mechanism for the barium-catalyzed cross-DHC of Ph_3SiH and pyrrolidine.

formed by aminolysis of **29c** and THF exchange with 3 equiv **IX**. The subsequent reversible side-on precoordination of Ph_3SiH is supported experimentally by the deactivated nature of catalysts with amide ligands displaying a β-Si–H functionality. Nucleophilic attack of the Ba-bound nitrogen on silicon then yields a zwitterionic silicate intermediate, which undergoes rate-determining β-hydride transfer. In the presence of **IX**, the silazane product **X** is liberated, generating a transient barium hydride, which undergoes dehydrogenative aminolysis with **IX** to regenerate the pyrrolidide catalyst. Kinetic analyses of the catalytic mono-DHC of tBuNH$_2$ and Me(Ph)SiH$_2$ with the magnesium tris(oxazolinyl)borate **31** by the group of Sadow provided the same rate law as in Eq. (9.3) and kinetics consistent with the mechanism displayed in Scheme 9.21 [55]. The absence of a silane-independent KIE points to nucleophilic attack by nitrogen on the coordinated silane rather than β-hydride transfer being the RDS, consistent with the lower nucleophilicity of the magnesium-bound nitrogen, because of strong π-donation from N to Mg, and the higher BDE of Mg–H compared to Ba–H.

For the monocoupling of Et_2NH with Ph_2SiH_2 catalyzed by the solvent-free diamido precatalysts **9a–c**, Hill and coworkers discovered two different rate laws depending on the metal center (Eqs. (9.4) and (9.5)) and differing from that of the heteroleptic complexes [53]:

$$R = -\frac{d[Et_2NH]}{dt} = k\,[\mathbf{9a/b}]^1\,[Et_2NH]^1\,[Ph_2SiH_2]^0 \qquad (9.4)$$

$$R = -\frac{d[Et_2NH]}{dt} = k\,[\mathbf{9c}]^2\,[Et_2NH]^1\,[Ph_2SiH_2]^1 \qquad (9.5)$$

Based on the order in [catalyst] in Eqs. (9.4) and (9.5), respectively, the authors proposed a mononuclear RDS for magnesium and calcium and a dimeric one for strontium. The difference presumably arises from the marked preference of strontium for higher coordination numbers and μ-bridging ligands and is at the origin of the unusual lower activity of Sr compared to Ca. The zero-order dependence in [Ph_2SiH_2] in Eq. (9.4) suggests rate-limiting aminolysis for **9a** and **9b**, whereas the first-order dependence in both substrates in Eq. (9.5) hints at a rate-limiting concerted β-hydride transfer/aminolysis for **9c**, reminiscent of the amine-assisted concerted insertion/aminolysis TS in the group 2-mediated intramolecular hydroamination/cyclization of aminoalkenes [38]. Although the rate laws in Eqs. (9.4, 9.5) do not rule out a silicate mechanism akin to that in Scheme 9.21, they are a reminder that subtle differences are bound to arise depending on the ligand environment and the identity of the catalytic s-block metal center used.

9.3.3 Application to the Synthesis of Oligo- and Polysilazanes

The highly reactive dialkylbarium complex **12d** was employed to catalyze the selective sequential DHC of diarylsilanes with amines or diamines to yield tailored oligosilazanes [58]. The most intriguing example of these is the sequential DHC of $ArPh_2SiH$ (Ar = Ph, *p*-$CF_3C_6H_4$), $BnNH_2$, Ph_2SiH_2, and $BnNH_2$ presented in Scheme 9.22: in the presence of 5 mol% **2d** at 60 °C, the resulting linear disilazane underwent intramolecular cyclization to the cyclic 1,3,2,4-diazadisiletidine $(Ph_2SiNBn)_2$ via a highly unusual N–H/Ar–Si cross-coupling reaction, which generates the arene ArH as a by-product. Although there is precedent for stoichiometric dearylative Si–N cross-coupling at molybdenum and tungsten complexes [59], this is the first example of such a catalytic reaction. The selective elimination of trifluoromethylbenzene over benzene for Ar = *p*-$CF_3C_6H_4$ led the authors to propose a silicate-based TS, in which the negative charge on silicon is stabilized by the electron-withdrawing CF_3 group (Scheme 9.22).

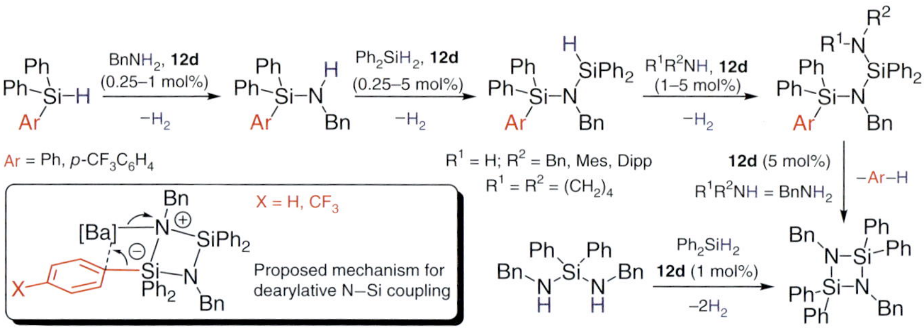

Scheme 9.22 Example of tailored oligosilazane synthesis by barium-catalyzed sequential cross-DHC of amines and silanes.

Catalyst **12d** was also employed to mediate the cross-DHP of Ph_2SiH_2 with *p*-xylylenediamine (Scheme 9.23) [60]. The reactions proceeded cleanly and rapidly to either the cyclic or linear polymers, depending on the initial monomer ratio. A strict 1 : 1 ratio of silane to diamine yielded the cyclic polymers, with up to 500 equiv of each monomer being converted within two hours at 60 °C ($M_{n,DOSY}$ = 6000 g/mol). In contrast, a slight excess of one of the comonomers systematically yielded linear polymers, the highest degree of polymerization being achieved with a $[silane]_0/[diamine]_0/[cat]_0$ = 105 : 100 : 1 ratio ($M_{n,DOSY}$ = 10 400 g/mol).

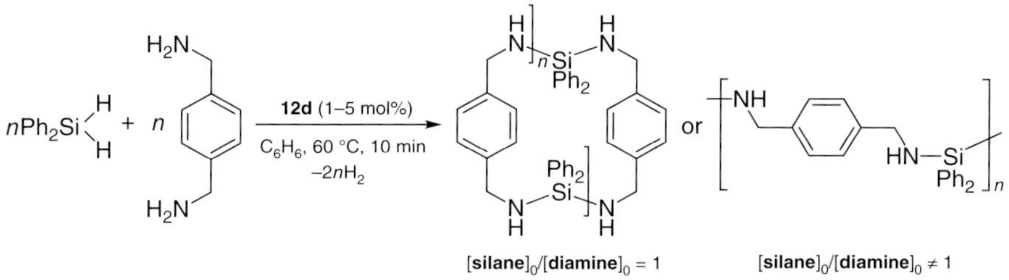

Scheme 9.23 Synthesis of cyclic and linear polymers by barium-catalyzed cross-dehydropolymerization of Ph_2SiH_2 and *p*-xylenediamine.

9.4 Other s-Block-Catalyzed Cross-DHC Reactions

9.4.1 Alkali Metal-Catalyzed DHC of Si—H and O—H Bonds

Because of the strength of the Si—O bond (Table 9.1), silyl ethers are among the most robust alcohol protecting groups [44a] and constitute the building blocks of silicones [8]. A number of commercially available alkali metal salts and complexes have been reported as efficient catalysts for the DHC of silanes and alcohols or diols (Scheme 9.24) [61]. In these reactions, all available Si—H bonds undergo Si—O coupling, except when using the very sterically hindered dihydrosilane tBu_2SiH_2, which cleanly affords mono-siloxanes [61a]. The NaOH-catalyzed reactions tolerate a surprising variety of functional groups, including ether, halide, nitro-, phthalimide, and various heteroaromatic groups, as well as functionalities that are amenable to TM-catalyzed hydrosilylation, such as alkyne, cyclic alkene, allyl, cyclopropyl, epoxide, and ester groups [61a]. Kinetic analyses for the DHC of Ph_3SiH with methanol catalyzed by the potassium amide **9c** provided a rate law showing a first-order dependence on [**9c**], [Ph_3SiH], and [MeOH], indicative of a mononuclear, silicate-based, concerted β-hydride transfer/alcoholysis RDS [61], similar to that determined for the amine–silane DHC catalyzed by the strontium bis(amide) **9c** [53].

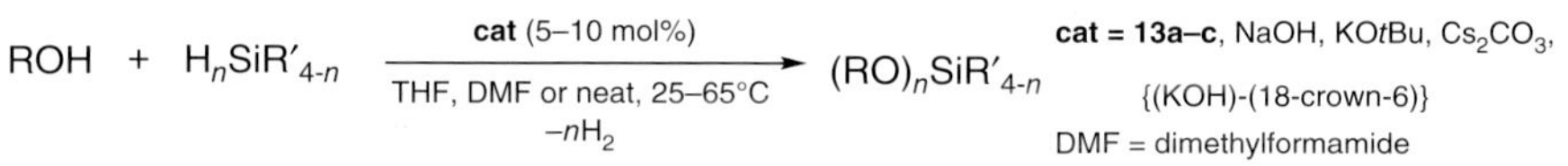

Scheme 9.24 Alkali metal-catalyzed cross-DHC of silanes and alcohols.

9.4.2 s-Block-Catalyzed DHC of Si—H and C—H Bonds

Direct C—H bond functionalization is one of the hottest topics in organic chemistry as it circumvents the need for introducing reactive C—X bonds amenable to traditional cross-coupling reactions [62]. Although the field is dominated by late TM catalysts capable of functionalizing even relatively unactivated C—H bonds, there are certain classes of substrates, such as terminal acetylenes and heteroaromatic compounds, with relatively acidic C—H bonds. In the 1990s, Itoh reported the catalytic DHC of $PhSiH_3$ and Et_3SiH with terminal acetylenes to mixtures of

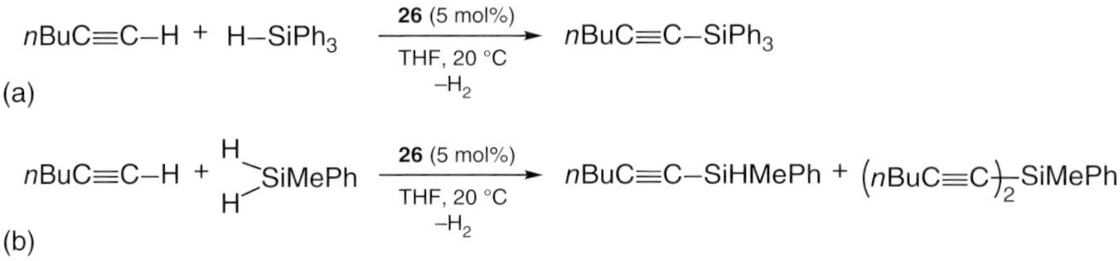

Scheme 9.25 Calcium-catalyzed cross-DHC of silanes and terminal alkynes.

mono-, bis-, and tris(alkynyl)silanes, as well as polymeric materials, using various s-block base and oxide catalysts [63]. Later Buch and Harder showed that the calcium azametallacyclopropane complex **26** catalyzes the DHC of $PhSiH_3$ and $Me(Ph)SiH_2$ with 1-hexyne under very mild conditions (Scheme 9.25) [49e]. The authors proposed a catalytic cycle based on successive σ-bond metathesis steps via calcium acetylide and hydride intermediates.

More recently, Grubbs reported that commercially available KO*t*Bu catalyzes the dehydrogenative silylation of aromatic heterocycles (Scheme 9.26), thereby offering the first TM-free direct heteroaromatic silylation methodology [64]. Scalable to several 100 g, the reactions were successfully applied to pharmaceutically relevant thiophenes. A later detailed experimental and computational mechanistic study revealed an ionic silicate-based mechanism, in which K^+ acts as a π-acid to stabilize a reactive $[Ar]^-$ intermediate [65]. The crucial role of the soft–soft K^+–arene interaction is supported by the failure of NaO*t*Bu or LiO*t*Bu to catalyze this reaction. The exact mechanism, however, remains unclear: a polar ionic route as well as a radical pathway has been proposed [65].

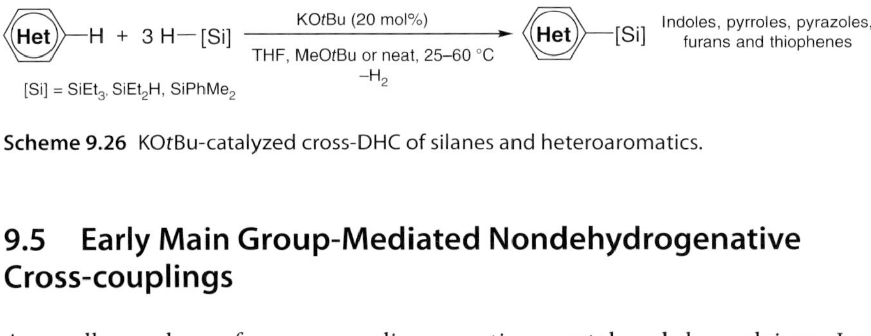

Scheme 9.26 KO*t*Bu-catalyzed cross-DHC of silanes and heteroaromatics.

9.5 Early Main Group-Mediated Nondehydrogenative Cross-couplings

A small number of cross-coupling reactions catalyzed by calcium Lewis acids have been reported by the groups of Niggemann and Yaragorla. These include the coupling of π-activated alcohols with (allyl)trimethylsilane by $Ca(NTf_2)_2/[Bu_4N]^+[BF_4]^-$ (Tf = F_3CSO_2), which generates Me_3SiOH as the by-product [66], as well as the cross-dehydrative coupling of styrenes with propargyl and benzylic alcohols by $Ca(OTf_2)_2/Bu_4NBF_4$ [67].

Furthermore, the commercially available silylborane $Me(Ph)_2SiBPin$ has become popular as a cross-coupling partner in the range of copper-catalyzed and alkoxide base-promoted silylation reactions with organic R–X electrophiles [68], these reactions being thermodynamically driven by the lability of the Si—B bond and the stability of the X-BPin by-product (Table 9.1). In 2015, the group

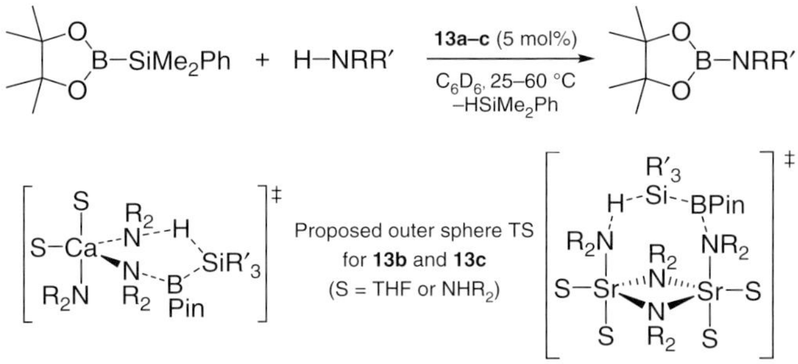

Scheme 9.27 Alkaline earth-catalyzed cross-coupling of silylboranes and amines.

of Hill reported the desilacoupling of $Me(Ph)_2SiBPin$ with a range of amines, yielding the corresponding aminoboranes and $Me(Ph)_2SiH$ as a by-product, using the homoleptic group 2 bis(amide) precatalysts **13a–c** (Scheme 9.27) [69]. The magnesium-catalyzed reactions proved extremely slow, requiring several days at 60 °C to achieve good conversions, whereas the calcium- and strontium-catalyzed reactions generally proceeded to completion over the course of one day at room temperature. Although no definite trend could be determined based on the amine substitution pattern, secondary amines reacted in general faster than primary ones. Furthermore, the strontium-catalyzed reaction with $DippNH_2$ afforded the bis-borylated product $DippN(BPin)_2$. A kinetic study on the reaction of Me(Ph)NH and $Me(Ph)_2SiBPin$ with the calcium and strontium precursors **13b** and **13c** once more provided two different rate laws, as shown in Eqs. (9.6) and (9.7), respectively:

$$R = -\frac{d[\mathrm{Me(Ph)NH}]}{dt} = k\,[\mathbf{13b}]^1\,[\mathrm{Me(Ph)NH}\,]^1\,[\mathrm{Me(Ph_2)SiBPin}\,]^1 \qquad (9.6)$$

$$R = -\frac{d[\mathrm{Me(Ph)NH}\,]}{dt} = k\,[\mathbf{13c}]^2\,[\mathrm{Me(Ph)NH}\,]^0\,[\mathrm{Me(Ph_2)SiBPin}]^1 \qquad (9.7)$$

The highly negative activation entropies determined in both cases also point to a highly ordered TS in the RDS. Given the unlikely formation of highly reactive group 2 silyl intermediates in such slow reactions, the authors proposed an outer sphere mechanism, in which the incoming silylborane undergoes concerted B—N/Si—H bond formation and N—H/B—Si bond breakage at a monomeric calcium or dimeric strontium amide complex (Scheme 9.27). The first order in [Me(Ph)NH] in Eq. (9.6) was deemed to stem from an amine–amide pre-equilibrium.

9.6 Conclusion and Outlook

The field of early main group-catalyzed cross-DHC has greatly advanced over the past decade. In the field of amine borane and amine silane DHC, the main challenge that lies ahead is the design of new systems capable of catalytic DHP to generate polyaminoboranes and polysilazanes, respectively. Furthermore, phosphine borane cross-DHC, which is less facile than its amine borane counterpart,

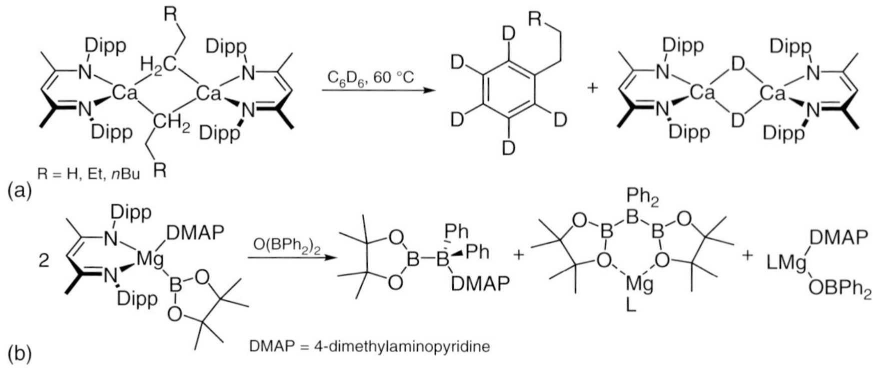

Scheme 9.28 Examples of stoichiometric alkaline earth-mediated homocoupling reactions: (a) calcium alkyl-mediated alkylation of benzene; (b) magnesium boryl-mediated boron catenation.

is yet to be achieved with group 1 or group 2 catalysts. Although there are now a handful of promising examples of s-block-catalyzed dehydrogenative C—C bond formation and nondehydrogenative cross-coupling reactions, groups 1 and 2 still have a lot of catching up to do to rival TM-based catalysts for these transformations, notably in the area of homo-DHC, of which there still is no catalytic early main group example. A couple of stoichiometric reactions recently reported by the Hill group provide a tantalizing glimpse at what might still be possible. The first of these is the calcium-mediated alkylation of benzene by σ-bond metathesis of a benzene C—H bond with dimeric β-diketiminato calcium alkyl complexes, which may hold the key to calcium-catalyzed arene-based dehydrocross-couplings (Scheme 9.28a) [70]. The second is the stoichiometric magnesium-mediated formation of an unsymmetrical diborane(**5**) and a triborate anion by σ-bond metathesis of a β-diketiminato magnesium boryl complex with $O(BPh_2)_2$, which, if made catalytic, could provide access to valuable new di- and triboranes (Scheme 9.28b) [71]. In light of these reactions, it looks like these underappreciated s-block metals were only biding their time and might very soon provide cheaper and greener alternatives for key reactions in TM catalysis.

References

1 (a) Baumgartner, T. and Jaekle, F. (eds.) (2018). *Main Group Strategies towards Functional Hybrid Materials*. Hoboken, NJ: Wiley. (b) Mark, J.E., Allcock, H.R., and West, R. (2005). *Inorganic Polymers*, 2e. New York, NY: Oxford University Press.

2 Luo, Y.R. (2007). *Comprehensive Handbook of Chemical Bond Energies*. Boca Raton, FL: CRC Press.

3 Ogier, M.J. (1880). *Ann. Chim. Phys.* 5: 37.

4 Corcoran, E.W. Jr., and Sneddon, L.G. (1985). *J. Am. Chem. Soc.* 107: 7446–7450.

5 Aitken, C., Harrod, J.F., and Samuel, E. (1985). *J. Organomet. Chem.* 279: C11–C13.
6 (a) Kim, M.H., Lee, J., Mo, S.Y. et al. (2010). *J. Chosun Nat. Sci.* 3: 61–71. (b) Corey, J.Y. (2004). *Adv. Organomet. Chem.* 51: 1–52. (c) Harrod, J.F. and Woo, H.G. (1998). *Adv. Organomet. Chem.* 42: 363–405. (d) Imori, T. and Tilley, D.T. (1994). *Polyhedron* 13: 2231–2243.
7 (a) Waterman, R. (2013). *Chem. Soc. Rev.* 42: 5629–5641. (b) Leitao, E.M., Jurca, T., and Manners, I. (2013). *Nat. Chem.* 5: 817–829. (c) Hill, M.S. (2010). Homocatenation of metal and metalloid main group elements. In: *Metal–Metal Bonding*, vol. 136 (ed. G. Parkin), 189–216. Springer-Verlag Berlin Heidelberg. (d) Waterman, R. (2008). *Curr. Org. Chem.* 12: 1322–1339. (e) Clark, T.J., Lee, K., and Manners, I. (2006). *Chem. Eur. J.* 12: 8634–8648.
8 (a) Pouget, E., Tonnar, J., Lucas, P. et al. (2010). *Chem. Rev.* 110: 1233–1277. (b) Kawakami, Y., Li, Y., Liu, Y. et al. (2004). *Macromol. Res.* 12: 156–171.
9 Johnson, H.C., Hooper, T.N., and Weller, A.S. (2015). The catalytic dehydrocoupling of amine–boranes and phosphine–boranes. In: *Synthesis and Application of Organoboron Compounds*, vol. 49 (eds. E. Fernández and A. Whiting), 153–220. Switzerland: Springer International Publishing.
10 Nako, A.E., Gates, S.J., Schädel, N. et al. (2014). *Chem. Commun.* 50: 9536–9538.
11 Huertos, M.A. and Weller, A.S. (2013). *Chem. Sci.* 4: 1881–1888.
12 Blaquiere, N., Diallo-Garcia, S., Gorelsky, S.I. et al. (2008). *J. Am. Chem. Soc.* 130: 14034–14035.
13 Sloan, M.E., Staubitz, A., Clark, T.J. et al. (2010). *J. Am. Chem. Soc.* 132: 3831–3841.
14 (a) Melen, R.L. (2016). *Chem. Soc. Rev.* 45: 775–788. (b) Liptrot, D.J. (2014). *Inorg. Chim. Acta* 422: 141–145.
15 Hansmann, M.M., Melen, R.L., and Wright, D.S. (2011). *Chem. Sci.* 2: 1554–1559.
16 Haubner, R., Wilhelm, M., Weissenbacher, R., and Lux, B. (2002). Boron nitrides — properties, synthesis and applications. In: *High Performance Non-oxide Ceramics II*, vol. 102 (ed. M. Jansen), 1–45. Berlin Heidelberg: Springer-Verlag.
17 (a) Demirci, U.B. (2016). *Int. J. Hydrogen Energy* 42: 9978–10013. (b) Smythe, N.C. and Gordon, J.C. (2010). *Eur. J. Inorg. Chem.* 4: 509–521. (c) Staubitz, A., Robertson, A.P.M., and Manners, I. (2010). *Chem. Rev.* 110: 4079–4124. (d) Hamilton, C.W., Baker, R.T., Staubitz, A., and Manners, I. (2009). *Chem. Soc. Rev.* 38: 279–293. (e) Stephens, F.H., Pons, V., and Baker, R.T. (2007). *Dalton Trans.*: 2613–2626. (f) Marder, T.B. (2007). *Angew. Chem. Int. Ed.* 46: 8116–8118.
18 Stock, A. and Pohland, E. (1926). *Ber. Dtsch. Chem. Ges.* 59: 2215–2223.
19 (a) Jaska, C.A., Temple, K., Lough, A.J., and Manners, I. (2003). *J. Am. Chem. Soc.* 125: 9424–9434. (b) Jaska, C.A., Temple, K., Lough, A.J., and Manners, I. (2001). *Chem. Commun.*: 962–963.
20 (a) Staubitz, A., Presa Soto, A., and Manners, I. (2008). *Angew. Chem. Int. Ed.* 47: 6212–6215. (b) Dietrich, B.L., Goldberg, K.I., Heinekey, D.M. et al. (2008). *Inorg. Chem.* 47: 6212–6215.

21 (a) Rossin, A. and Peruzzini, M. (2016). *Chem. Rev.* 116: 8848–8872. (b) Bhunya, S., Malakar, T., Ganguly, G., and Paul, A. (2016). *ACS Catal.* 6: 7907–7934.
22 Sewell, L.J., Lloyd-Jones, G.C., and Weller, A.S. (2012). *J. Am. Chem. Soc.* 134: 3598–3610.
23 Aftandilian, V.D., Miller, H.C., and Muetterties, E.L. (1961). *J. Am. Chem. Soc.* 83: 2471–2474.
24 Stennett, T.E. and Harder, S. (2016). *Chem. Soc. Rev.* 45: 1112–1128.
25 (a) Harder, S., Spielmann, J., and Tobey, B. (2012). *Chem. Eur. J.* 18: 1984–1991. (b) Spielmann, J., Piesik, D.F.-J., and Harder, S. (2010). *Chem. Eur. J.* 16: 8307–8318. (c) Spielmann, J. and Harder, S. (2009). *J. Am. Chem. Soc.* 131: 5064–5065. (d) Spielmann, J., Jansen, G., Bandmann, H., and Harder, S. (2008). *Angew. Chem. Int. Ed.* 47: 6290–6295.
26 Spielmann, J., Bolte, M., and Harder, S. (2009). *Chem. Commun.*: 6934–6936.
27 Bellham, P., Hill, M.S., Kociok-Köhn, G., and Liptrot, D.J. (2013). *Chem. Commun.* 49: 1960–1962.
28 Hill, M.S., Hodgson, M., Liptrot, D.J., and Mahon, M.F. (2011). *Dalton Trans.* 40: 7783–7790.
29 McLellan, R., Kennedy, A.R., Orr, S.A. et al. (2017). *Angew. Chem. Int. Ed.* 56: 1036–1041.
30 Wallis, C.J., Dyer, H., Vendier, L. et al. (2012). *Angew. Chem. Int. Ed.* 51: 3646–3648.
31 (a) McLellan, R., Kennedy, A.R., Mulvey, R.E. et al. (2017). *Chem. Eur. J.* 23: 16853–16861. (b) Bellham, P., Hill, M.S., and Kociok-Köhn, G. (2015). *Dalton Trans.* 44: 12078–12081.
32 Liptrot, D.J., Hill, M.S., Mahon, M.F., and MacDougall, D.J. (2010). *Chem. Eur. J.* 16: 8508–8515.
33 Hill, M.S., Kociok-Köhn, G., and Robinson, T.P. (2010). *Chem. Commun.* 46: 7587–7589.
34 Erickson, K.A., Wright, D.S., and Waterman, R. (2014). *J. Organomet. Chem.* 751: 541–545.
35 Butera, V., Russo, N., and Sicilia, E. (2014). *Chem. Eur. J.* 20: 5967–5976.
36 Luo, Y. and Ohno, K. (2007). *Organometallics* 26: 3597–3600.
37 Bellham, P., Anker, M.D., Hill, M.S. et al. (2016). *Dalton Trans.* 45: 13969–13978.
38 Arrowsmith, M., Crimmin, M.R., Barrett, A.G.M. et al. (2011). *Organometallics* 30: 1493–1506.
39 Bellham, P., Hill, M.S., and Kociok-Köhn, G. (2014). *Organometallics* 33: 5716–5721.
40 Liptrot, D.J., Hill, M.S., Mahon, M.F., and Wilson, A.S.S. (2015). *Angew. Chem. Int. Ed.* 54: 13362–13365.
41 Harinath, A., Anga, S., and Panda, T.K. (2016). *RSC Adv.* 6: 35648–35653.
42 Barrett, A.G.M., Crimmin, M.R., Hill, M.S. et al. (2007). *Organometallics* 26: 4076–4079.
43 Arrowsmith, M., Hill, M.S., Hadlington, T. et al. (2011). *Organometallics* 30: 5556–5559.

44 (a) Wuts, P.G.M. and Greene, T.W. (2006). *Protecting Groups in Organic Synthesis*, 4e. New York, NY: Wiley. (b) Armitage, D.A. (1991). Chapter 11: Organosilicon nitrogen compounds. In: *The Silicon–Heteroatom Bond*, 365–484. Chichester: Wiley.
45 Kroke, E., Li, Y.L., Konetschny, C. et al. (2000). *Mater. Sci. Eng. R: Rep.* 26: 97–199.
46 Fessenden, R. and Fessenden, J.S. (1961). *Chem. Rev.* 61: 361–388.
47 (a) Mitsudome, T., Urayama, T., Maeno, Z. et al. (2015). *Chem. Eur. J.* 21: 3202–3205. (b) Zahmakıran, M., Tristany, M., Philippot, K. et al. (2010). *Chem. Commun.* 46: 2938–2940. (c) Sommer, L.H. and Citron, J.D. (1967). *J. Org. Chem.* 32: 2470–2472.
48 Tsuchimoto, T., Iketani, Y., and Sekine, M. (2012). *Chem. Eur. J.* 18: 9500–9504.
49 (a) Li, N. and Guan, B.T. (2017). *Adv. Synth. Catal.* 359: 3526–3531. (b) Pindwal, A., Ellern, A., and Sadow, A.D. (2016). *Organometallics* 35: 1674–1683. (c) Nako, A.E., Chen, W., White, A.J.P., and Crimmin, M.R. (2015). *Organometallics* 34: 4369–4375. (d) Xie, W., Hu, H., and Cui, C. (2012). *Angew. Chem. Int. Ed.* 51: 11141–11144. (e) Buch, B. and Harder, S. (2007). *Organometallics* 26: 5132–5135. (f) Wang, J.X., Dash, A.K., Berthet, J.C. et al. (2000). *J. Organomet. Chem.* 610: 49–57.
50 Allen, L.K., Garcìa-Rodrìguez, R., and Wright, D.S. (2015). *Dalton Trans.* 44: 12112–12118.
51 Anga, S., Sarazin, Y., Carpentier, J.-F., and Panda, T.K. (2016). *ChemCatChem* 8: 1373–1378.
52 Bellini, C., Dorcet, V., Carpentier, J.-F. et al. (2016). *Chem. Eur. J.* 22: 4564–4583.
53 Hill, M.S., Liptrot, D.J., MacDougall, D.J. et al. (2013). *Chem. Sci.* 4: 4212–4222.
54 Hill, M.S., Liptrot, D.J., and Weetman, C. (2016). *Chem. Soc. Rev.* 45: 972–988.
55 Dunne, J.F., Neal, S.R., Engelkemier, J. et al. (2011). *J. Am. Chem. Soc.* 133: 16782–16785.
56 Baishya, A., Peddarao, T., and Nembenna, S. (2017). *Dalton Trans.* 46: 5880–5887.
57 Bellini, C., Carpentier, J.-F., Tobisch, S., and Sarazin, Y. (2015). *Angew. Chem. Int. Ed.* 54: 7679–7683.
58 Bellini, C., Roisnel, T., Carpentier, J.-F. et al. (2016). *Chem. Eur. J.* 22: 15733–15743.
59 Kanno, Y., Komuro, T., and Tobita, H. (2015). *Organometallics* 34: 3699–3705.
60 Bellini, C., Orione, C., Carpentier, J.-F., and Sarazin, Y. (2016). *Angew. Chem. Int. Ed.* 55: 3744–3748.
61 (a) Toutov, A.A., Betz, K.N., Haibach, M.C. et al. (2016). *Org. Lett.* 18: 5776–5779. (b) Harinath, A., Bhattacharjee, J., Anga, S., and Panda, T.K. (2016). *Aust. J. Chem.* 70: 724–730. (c) Grajewska, A. and Oestreich, M. (2010). *Synlett*: 2482–2484. (d) Weickgenannt, A. and Oestreich, M. (2009). *Chem. Asian J.* 4: 406–410. (e) Le Bideau, F., Coradin, T., Hénique, J., and Samuel, E. (2001). *Chem. Commun.*: 1408–1409.

62 (a) Li, J.J. (ed.) (2017). *C–H Bond Activation in Organic Synthesis*. Boca Raton, FL: CRC Press. (b) Roudesly, F., Oble, J., and Poli, G. (2017). *J. Mol. Catal. A: Chem.* 426: 275–296.

63 (a) Ishikawa, J. and Itoh, M. (1999). *J. Catal.* 185: 454–461. (b) Itoh, M., Inoue, K., Iwata, K., and Mitsuzuka, M. (1997). *Macromolecules* 30: 694–701. (c) Itoh, M., Mitsuzuka, M., Utsumi, T. et al. (1994). *J. Organomet. Chem.* 476: C30–C31.

64 Toutov, A.A., Liu, W.-B., Betz, K.N. et al. (2015). *Nature* 518: 80–84.

65 (a) Banerjee, S., Yang, Y.-F., Jenkins, I.D. et al. (2017). *J. Am. Chem. Soc.* 139: 6880–6887. (b) Liu, W.-B., Schuman, D.P., Yang, Y.-F. et al. (2017). *J. Am. Chem. Soc.* 139: 6867–6879.

66 Meyer, V.J. and Niggemann, M. (2011). *Eur. J. Org. Chem.*: 3671–3674.

67 (a) Yaragorla, S., Dada, R., and Pareek, A. (2018). *Chem. Sel.* 3: 495–499. (b) Yaragorla, S., Pareek, A., Dada, R. et al. (2016). *Tetrahedron Lett.* 57: 5841–5845.

68 Recent examples: (a)Sakaguchi, H., Ohashi, M., and Ogoshi, S. (2018). *Angew. Chem. Int. Ed.* 57: 328–332. (b) Xue, W. and Oestreich, M. (2017). *Angew. Chem. Int. Ed.* 56: 11649–11652. (c) Yamamoto, E., Ukigaia, S., and Ito, H. (2015). *Chem. Sci.* 6: 2943–2951.

69 Liptrot, D.J., Arrowsmith, M., Colebatch, A.L. et al. (2015). *Angew. Chem. Int. Ed.* 54: 15280–15283.

70 Wilson, A.S.S., Hill, M.S., Mahon, M.F. et al. (2017). *Science* 358: 1168–1171.

71 Hill, M.S., Pécharman, A.F., and Mahon, M. (2018). *Angew. Chem. Int. Ed.* 57: 10688–10691.

10

Enantioselective Catalysis with s-Block Organometallics

Philipp Stegner and Sjoerd Harder

University Erlangen-Nürnberg, Inorganic and Organometallic Chemistry, Egerlandstrasse 1, 91058 Erlangen, Germany

10.1 Introduction

Catalysis in general and particularly enantioselective catalysis is a challenging, yet rewarding research topic that still sparks vast interest across all fields of chemistry including the economically very important syntheses of fine and industrial chemicals [1]. This is constituted by the fact that the synthesis of 75% of all chemical compounds relies on the application of catalysis. The Chemistry Nobel Prize 2001 for contributions to asymmetric catalysis underscores the importance of enantiopure chiral products to society [2]. Although most chiral catalysts are based on noble transition metals, the use of early main group metals in this area is underdeveloped [3]. Transition metal catalysts are able to backdonate *d*-electrons in antibonding orbitals, thus activating the substrate [4]. In contrast, group 1 and 2 catalysts react by Lewis acidic substrate polarization followed by nucleophilic attack. Despite the different working principles, a number of attractive properties such as biocompatibility, world wide availability, and low price have encouraged research interest in catalytic applications of early main group organometallics [5]. Within the field of s-block metal catalysis, there are clearly two directions. The first group of researchers mainly exploits the Lewis acidic properties of the metal cations for numerous synthetic protocols (see Chapters 11 and 12), a method that provides an interesting alternative to transition-metal-based catalysis [5, 6]. Although these catalytic species purely rely on metal Lewis acid activation, another group of researchers is especially interested in polar organometallic catalysts that take advantage of the metal's high Lewis acidity combined with the presence of a highly reactive (strongly nucleophilic or Brønsted basic) organic rest. This chapter exclusively describes applications of these highly reactive "true" organometallics in enantioselective catalysis [7].

The chiral catalysts in the s-block can be categorized as shown in Scheme 10.1. They consist of spectator ligands (abbreviated as L), which can be negatively charged or neutral, and reactive, noninnocent, anionic groups R^-, or neutral bases B that are able to deprotonate the substrate inducing further reactivity. We

Early Main Group Metal Catalysis: Concepts and Reactions,
First Edition. Edited by Sjoerd Harder.

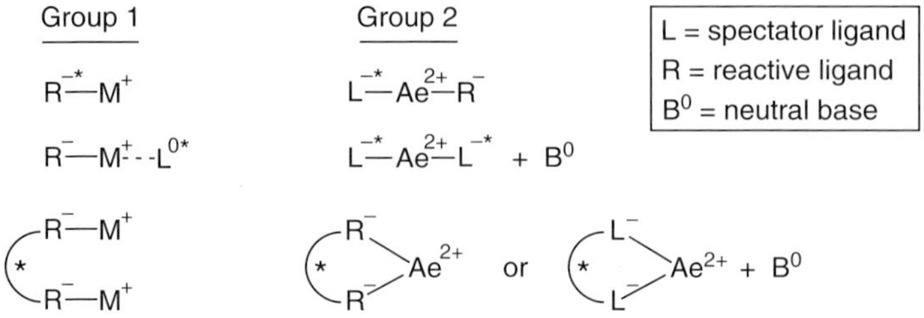

Scheme 10.1 Types of chiral catalysts in group 1 metal (M) and group 2 metal (Ae) chemistry.

systematically discuss the examples of these different types of catalysts arranged according to the metal and chiral ligand.

10.2 Lithium-Based Catalysts

Organometallic lithium compounds (e.g. *n*-butyllithium) are well-known reagents that are applied in numerous synthetic protocols, almost exclusively in a stoichiometric manner [8]. The catalytic application of organolithium reagents in comparison to transition metals is limited, particularly in the context of asymmetric catalysis [8–17]. Most of these chiral lithium catalysts are exclusively based on Lewis acid catalysis (see Chapters 11 and 12). The very few asymmetric, "true" organometallic lithium catalysts that rely on the cooperative effect of lithium-based Lewis acidity as well as on ligand-based nucleophilicity are discussed below.

10.2.1 Lithium Catalysts Based on Neutral Chiral Ligands

Lithium compounds containing the neutral chiral ligand sparteine are applied in a variety of asymmetric synthetic protocols in a stoichiometric manner [10]. Catalytic application of lithium bis(trimethylsilyl)amide and (−)-sparteine or α-isosparteine, respectively, are limited to intermolecular hydroamination of α-Me-styrene using various benzylamines [11].

The catalyst (**4**, Scheme 10.2) is formed via deprotonation of the amine **2** by $LiN(SiMe_3)_2$ (**1**) and coordination of the chiral spectator ligand **3**. Subsequently, the styrene substrate **5** is inserted into the lithium nitrogen bond generating the highly reactive alkyl lithium intermediate **6**, which is protonated to give the product **7** while regenerating the catalyst **4** (Scheme 10.2). Although use of a low catalyst loading of 2 mol% could be achieved, harsh conditions (120 °C, up to 70 hours) were necessary, yielding conversions of 48–72% and low enantiomeric excess of 14% or less [11]. The intramolecular hydroamination of unactivated amino olefins such as 2,2-dimethylpent-4-en-1-amine proceeded at 90 °C in 22 hours and resulted in a purely racemic product [12].

A more effective protocol for lithium-based asymmetric intramolecular hydroamination is based on bisoxazoline (BOX) ligands yielding enantioselectivities up to 91% ee and yields up to 99%. This protocol was utilized in the

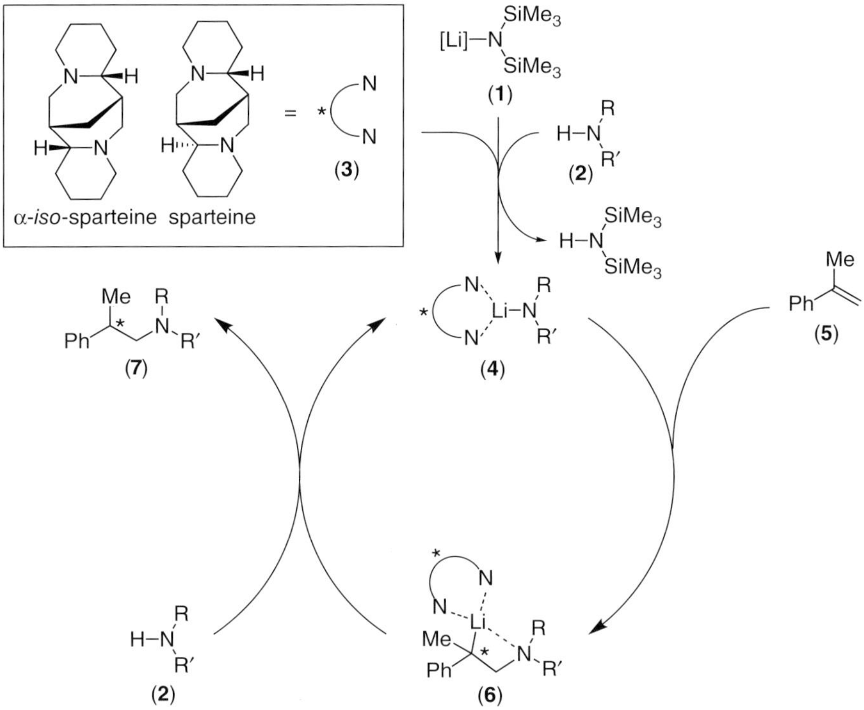

Scheme 10.2 Lithium-catalyzed hydroamination of α-Me-styrene. Source: Horrillo-Martínez et al. 2007 [11]. Reproduced with permission of John Wiley & Sons.

synthesis of pyrrolidine and piperidine derivatives bearing different substitution patterns (**8–12**, Scheme 10.3), as well as in the total synthesis of the chiral alkaloid laudanosine (**13**, Scheme 10.3) [13]. These catalytic reactions ran at −60 °C using 40 mol% BOX ligand, 20 mol% of *n*-butyllithium, and 20 mol% of diisopropylamine as a proton source. The catalyst is formed by deprotonation of *i*Pr_2NH (**14**, Scheme 10.4) with *n*BuLi and subsequent coordination of the chiral ligand **15**, resulting in the formation of the chiral lithium amide **16**. Catalysis is initiated by deprotonation of the amino olefin **17** yielding the corresponding lithium amide **18** and *i*Pr_2NH. Complex **18** will undergo cyclization by aminolithiation via the diastereomeric transition states **19** or **20** to give the alkyllithium intermediate **21**. Because of the large difference in pK_a values between compound **18** and compound **21**, the presence of an equilibrium toward **18** in this cyclization step is likely. Although the pro-(*S*) transition state **19** shows no repulsion between the aromatic moieties of the amino olefin and the BOX-ligand, the diastereomeric pro-(*R*) analog **20** is disadvantaged by steric repulsion between the aromatic moieties of the amino olefin substrate and the residues R of the BOX-ligand. This steric repulsion is the key to enantiotopic discrimination, yielding the kinetically favored (*S*)-enantiomer as the main product (Scheme 10.4). Protonation by *i*Pr_2NH regenerates the catalyst **16** and generates product **22** [13].

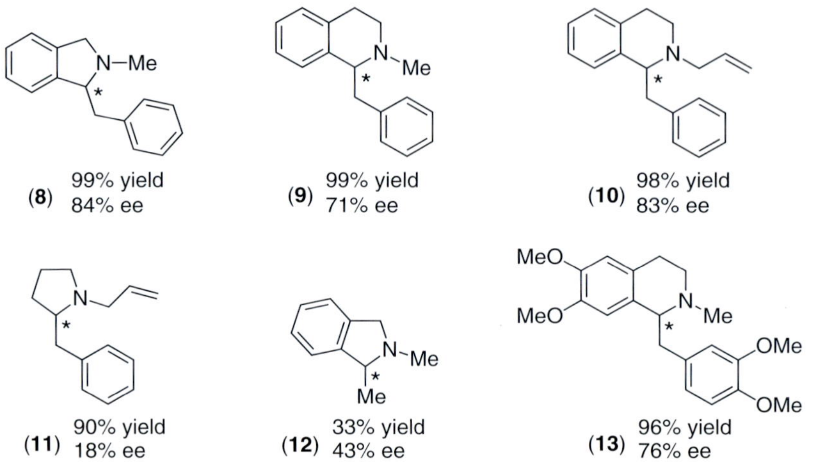

Scheme 10.3 Product scope of (BOX)LiN*i*Pr_2-catalyzed intramolecular hydroamination reactions [13]. Source: Ogata, T., et al 2012, 2007 [13].

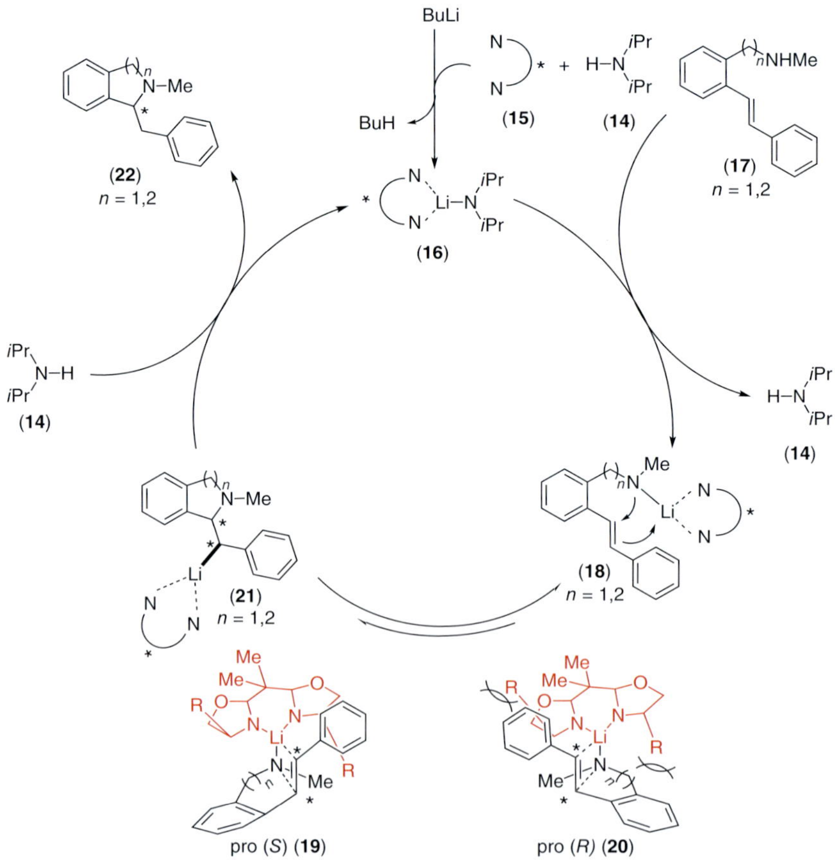

Scheme 10.4 The (BOX)LiN*i*Pr_2-catalyzed intramolecular hydroamination of aminoalkenes [13]. Source: Ogata, T., et al 2012, 2007 [13].

10.2.2 Lithium Catalysts Based on Monoanionic Chiral Ligands

Although intramolecular hydroamination catalysis generally converts unsaturated bonds to cyclic molecules, epoxide deprotonation has been used as a synthetic method to create unsaturated bonds upon ring opening. The asymmetric deprotonation of epoxides can efficiently be catalyzed by various enantiopure lithium bases (**26–31**, Scheme 10.5) [14]. Yields up to 96% and enantioselectivities up to 96% ee have been accomplished in the rearrangement of cyclohexene oxide (**32**) to 2-cyclohexan-1-olate (**33**) (Scheme 10.6) [14]. Synthetic protocols are based either on stoichiometric addition of chiral lithium organometallics or on the application of catalytic amounts in the presence of an achiral bulk base. Here, a variety of different bulk bases **23–25** and chiral bases **26–28** were applied successfully (Scheme 10.5). It is crucial that the achiral bulk bases do not react with the epoxide but that a mixed aggregate (**29–31**) of bulk and chiral base does the deprotonation. The Brønsted bases **23** and **24** are unreactive toward the epoxide, while **25** deprotonates the epoxide negligibly slow. An excellent selectivity of 96% ee and a high yield of 96% were observed applying catalytic amounts of **26** plus 1 equiv of the bulk bases **23** or **24**. Excess bulk base or stoichiometric use of **26** resulted in 93% ee. Catalytic quantities of **27** combined with 1 equiv of the bulk base gave 77% ee, while excess bulk base or stoichiometric use of **27** yielded 73% ee. The higher selectivity of **26** compared to **27** is likely related to the reduced chiral information as a decreased number of stereocenters is present in **27**. Catalytic amounts of **28** combined with 1 equiv of bulk base resulted in 76% ee. Stoichiometric use or application of excess bulk base yielded 80% ee. Using the widespread lithiumdiisopropylamide (LDA) as a bulk base quenched the enantioselectivity significantly in all cases. LDA deprotonated the epoxide, generating an unselective competing reaction yielding a racemic rearrangement product. The combination of two different

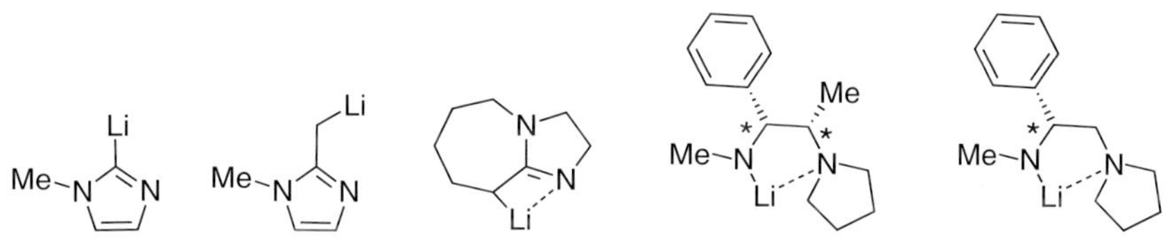

(23) (24) (25) (26) (27)

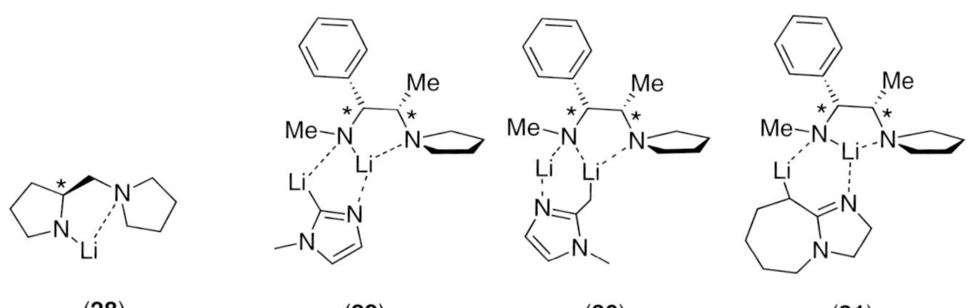

(28) (29) (30) (31)

Scheme 10.5 Organolithium reagents and catalysts used in catalytic asymmetric deprotonation. Source: Pettersen et al. 2002 [14]. Reproduced with permission of Elsevier.

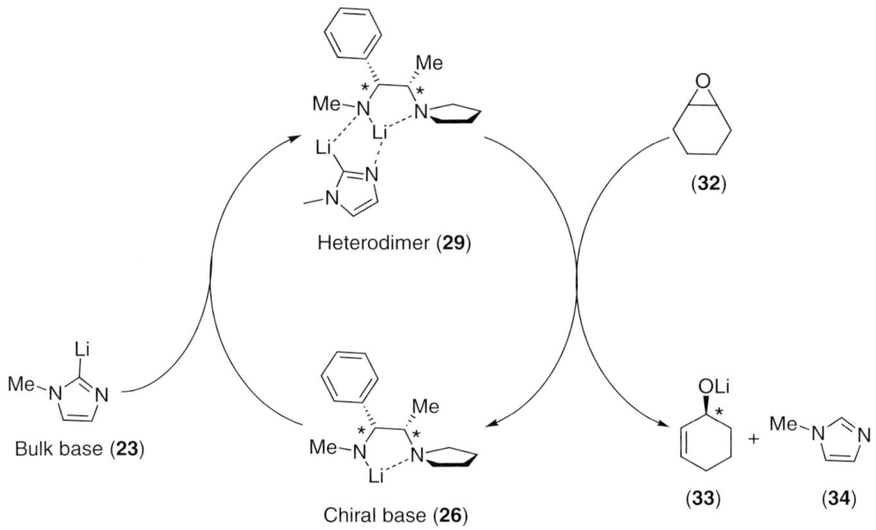

Scheme 10.6 Catalytic asymmetric deprotonation of cyclohexene oxide. Source: Pettersen et al. 2002 [14]. Reproduced with permission of Elsevier.

bases yielded a higher selectivity than the stoichiometric reaction. Therefore, it is likely that a new catalytically active species was formed. Applying various nuclear magnetic resonance (NMR) methods, the formation of the heterodimers **29**, **30**, and **31** (Scheme 10.5) was confirmed spectroscopically [14].

These heterodimers appear to be more stable than the corresponding homodimers and deprotonate cyclohexene oxide, inducing the rearrangement to the corresponding chiral allylic lithium alkoxide **33**. In addition, the 1-methylimidazol (**34**) is formed and the chiral lithium amide **26** is liberated. Subsequently, the catalytically active heterodimer **29** is regenerated by reaction of **26** with the bulk base **23** (Scheme 10.6).

A similar enantioselective cyclohexene deprotonation could also be catalyzed by a lithium organometallic based on a chiral dianionic ligand system [15]. The homochiral bis lithium amide was catalytically active but suffered from decreased activity and selectivity, resulting in inferior yield and enantiomeric excess. The substrate scope of asymmetric deprotonation reactions could be extended to 4-substituted ketones (**36**), using the same approach of combining catalytic amounts of a chiral base **35** and excess bulk base (**40**, Scheme 10.7). A variety of bases has been tested, revealing **40** as the ideal bulk base. 4-Substituted ketones (**36**) were converted into the corresponding chiral lithium enolates **37**, which were subsequently transformed into the corresponding trimethylsilyl enolates **38** by quenching with trimethylsilyl chloride [16]. In addition to various fluorinated chiral amides, several additives have been screened to enhance the level of enantioselectivity and catalytic activity. The optimum conditions are 30 mol% chiral amine **35**, 240 mol% bulk base **40**, 240 mol% HMPA (hexamethylphosphoramide), and 150 mol% of 1,4-Diazabicyclo[2.2.2]octane (DABCO) in tetrahydrofuran (THF) at −78 °C. The reaction proceeds via deprotonation of the ketone **36** by the chiral base **35** and subsequent enolisation of the ketone to the corresponding lithium enolate **37**. Subsequently, the catalytically active

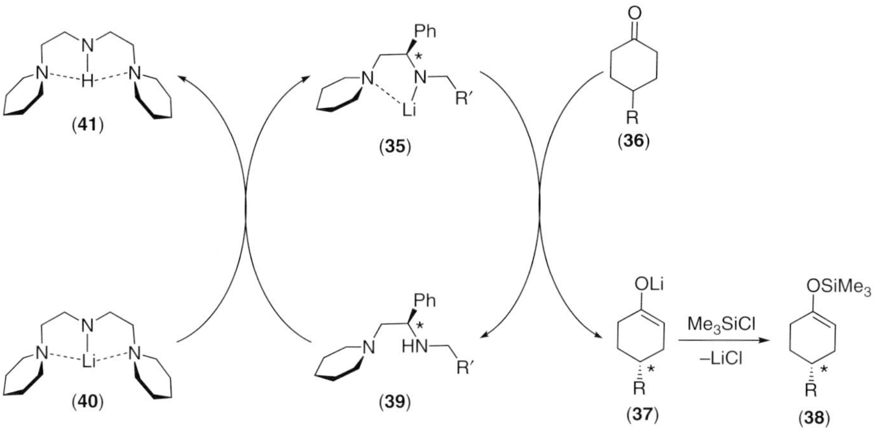

Scheme 10.7 Catalytic asymmetric deprotonation of 4-substituted cyclohexanones. Source: Yamashita et al. 1997 [16]. Reproduced with permission of Elsevier.

Table 10.1 Catalytic asymmetric deprotonation of 4-substituted cyclohexanones [16].

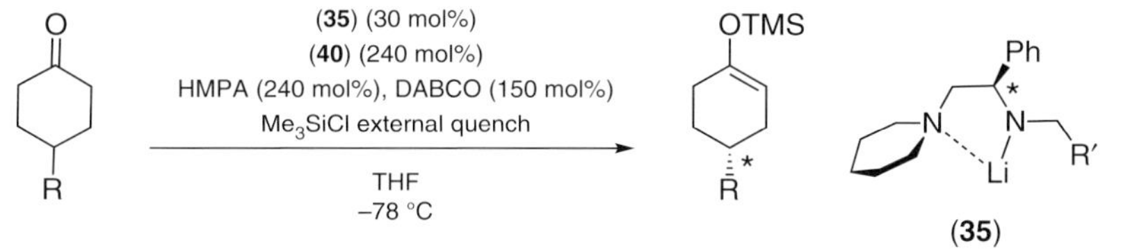

Chiral base *R′*	Substrate R	Conversion (%)	ee (%) (configuration)
CF_3	*t*Bu	83	79 (*R*)
CHF_2	*t*Bu	82	76 (*R*)
CH_2F	*t*Bu	50	12 (*R*)
CF_3	Ph	77	76 (*R*)
CF_3	*i*Pr	80	76 (*R*)
CF_3	Me	70	75 (*R*)

Source: Yamashita et al. 1997 [16]. Reproduced with permission of Elsevier.

chiral amide **35** is regenerated by lithium hydrogen exchange with the excessively present bulk base **40**. Fast deprotonation of the chiral amine **39** by bulk base **40** was found to be essential to avoid the competing racemate formation by reaction of the ketone substrate and bulk base. Therefore, the enantioselectivity is dependent on the acidity of compound **39**, which is proportional to the degree of fluorination of the residue R′ (Table 10.1) [16].

10.2.3 Lithium Catalysts Based on Dianionic Chiral Ligands

Chiral dianionic lithium organometallics derived from *R*-BINAM were applied in catalytic hydroamination of aminodienes yielding conversions of 95% or more [17]. The conversion of aminodienes to pyrrolidines ($n = 1$) proceeded at

Table 10.2 Li-catalyzed asymmetric hydroamination of aminodienes [17].

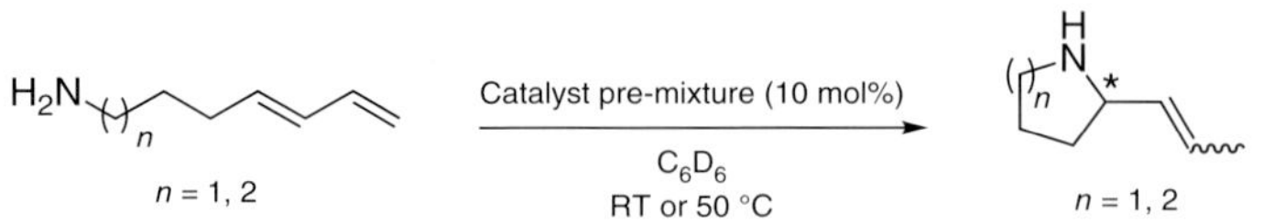

(**42**) R = CH_2Ph
(**43**) R = CH_2Mes
(**44**) R = CH_2(2-pyridyl)
(**45**) R = CH_2(2-naphthyl)
(**46**) R = cyclopentyl
(**47**) R = CH_2tBu

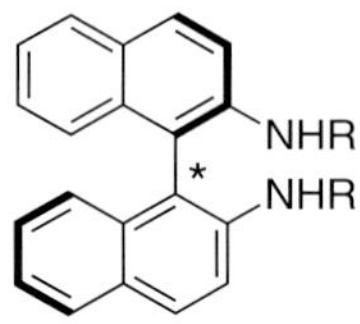

Ligand	Substrate (*n*)	Ratio *E*/*Z* (%)	Time (h)	ee (*E*) (%) (configuration)	ee (*Z*) (%) (configuration)
43	1	30/70	1	6	1
44	1	24/76	3	18 (*S*)	61 (*R*)
45	1	92/8	0.25	51 (*S*)	13 (*S*)
46	1	51/49	2	35 (*S*)	18 (*R*)
42	2[a)]	90/10	18	72 (*S*)	58 (*S*)
45	2[a)]	90/10	4	73 (*S*)	42 (*S*)
47	2[a)]	64/36	19	51 (*S*)	2

a) $T = 50\,°C$; all catalysts prepared from $LiCH(SiMe_3)_2$
Source: Deschamp et al. 2010 [17]. Reproduced with permission of John Wiley & Sons.

room temperature, while the cyclization to piperidines ($n = 2$) required longer reaction times at 50 °C [17]. The catalysts were generated *in situ* by premixing $LiCH(SiMe_3)_2$ or MeLi and the corresponding ligands **42–49** (Table 10.2 and 10.3). From several sets of experiments, it was concluded that the best lithium precursor-to-ligand ratio is 2.5 : 1 for $LiCH(SiMe_3)_2$ and 4 : 1 for MeLi in a mixture of benzene and diethyl ether as a solvent. The use of a 2 : 1 stoichiometry and subsequent isolation of the product lead in both cases to inactive catalysts, while activity was regained by addition of excess base. This observation can either be explained by the formation of aggregates incorporating excess alkyllithium or by deprotonating a small amount of substrate by the excess base resulting in increased nucleophilicity. A catalyst loading of 10 mol% catalyst displayed the best performance, while decreased amounts resulted in longer reaction times and poorer enantioselectivities. Utilizing various BINAM derivatives, the *E*/*Z* selectivity could be reversed by manipulating the bulk of the benzylic substituents (Table 10.2). Furthermore, this catalyst mixture was active in the cyclization of various unactivated aminoalkenes (Table 10.3). Although aminohexenes could, even at elevated temperatures, not be cyclized to the corresponding piperidines, aminopentenes could already at room temperature be converted into the respective pyrrolidines [17].

A significant improvement of stereoselectivity in catalytic hydroamination is achieved by replacing the hydrocarbon substituents R on the BINAM

Table 10.3 Li-catalyzed asymmetric hydroamination of aminoalkenes [17].

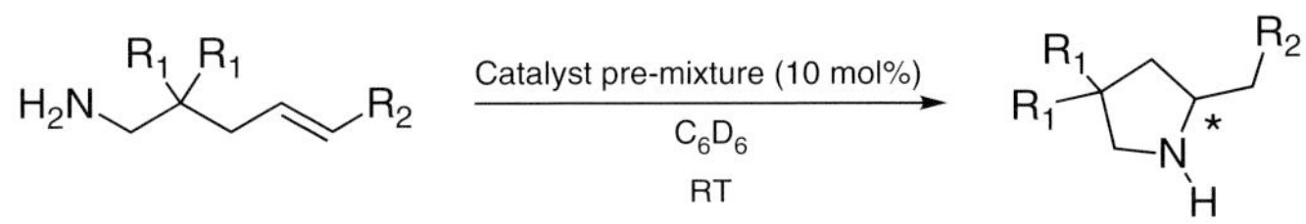

(42) R = CH_2Ph **(48)** R = cyclohexyl
(46) R = cyclopentyl **(49)** R = *i*Pr

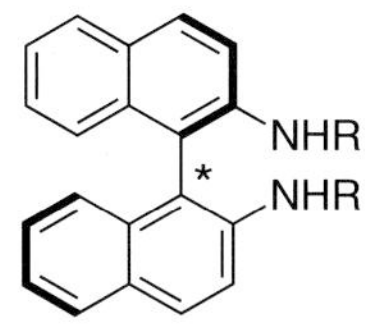

Ligand	Substrate R_1	Substrate R_2	Conversion (%)	Time (h)	ee (%)
42[a)]	$-(CH_2)_5-$	H	95	4	12
42[b)]	$-(CH_2)_5-$	Me	95	16	5
46[a)]	Ph	H	95	20	58
48[b)]	Ph	H	87	24	48
49[b)]	Ph	H	75	4	55

a) Catalyst prepared from MeLi.
b) Catalyst prepared from $LiCH(SiMe_3)_2$.
Source: Deschamp et al. 2010 [17]. Reproduced with permission of John Wiley & Sons.

moiety with amino-acid-derived *N*-methylpyrrolidine donor groups (Table 10.4). Pyrrolidines could be obtained in up to 75% enantiomeric excess. The catalytically active species exhibits a unique coordination environment for each of the Li atoms. The dimeric compound **50** was obtained by deprotonation of the (*S*,*S*,*S*)-ligand while the lithiation of the (*R*,*S*,*S*)-ligand yielded an oligomeric structure instead of a defined organometallic compound according to NMR investigations [12]. Addition of THF significantly reduced the catalytic activity, while the enantiomeric excess was barely influenced.

Control experiments with a model system that represents one half of the BINAM catalyst, i.e. the chiral Li amide **51**, revealed inferior activity and selectivity (120 °C, 2% ee) in comparison to compound **50**. Therefore, it was proposed that the close proximity of at least two lithium atoms is necessary to achieve suitable enantioselectivity and conversion. Further mechanistic investigations are needed to strengthen this theory. Until now **50**, is the most efficient lithium catalyst for asymmetric hydroamination [17]. Asymmetric catalytic protocols for the heavier homolog sodium are yet unknown; however, for potassium, two catalytic applications have been published.

10.3 Potassium-Based Catalysts

Potassium is frequently applied in various synthetic transformations either as a strongly reducing metal or in the form of highly basic reagents of which the Schlosser superbase is the most prominent example. However, these applications

Table 10.4 Hydroamination of aminoalkenes catalyzed by proline-derived Li amides [12].

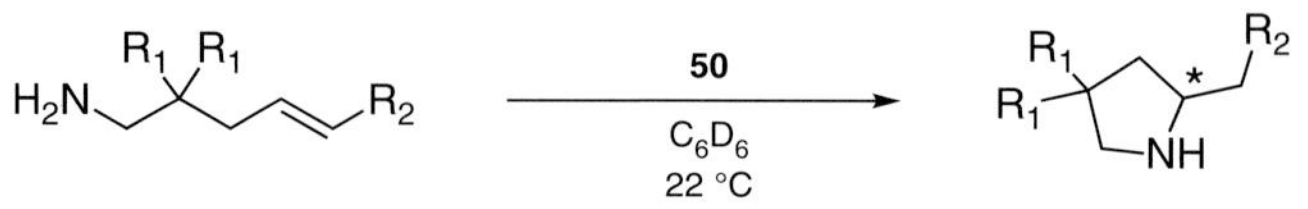

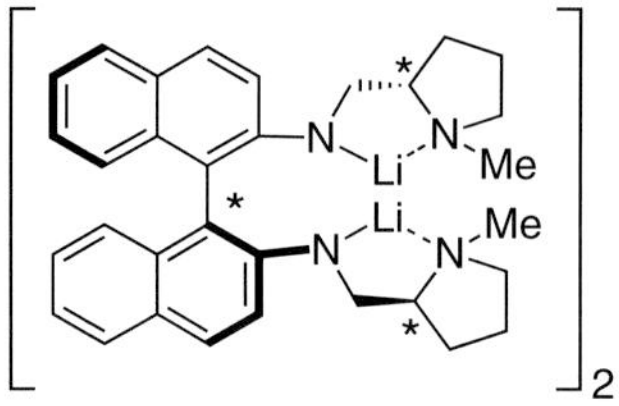

(50)

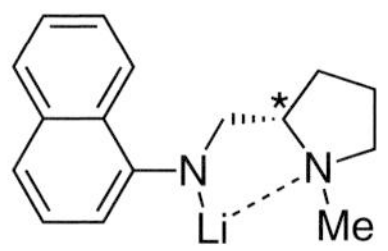

(51)

Catalyst loading (mol %)	Substrate R_1	Substrate R_2	Conversion (%)	Time	ee (%) (configuration)
2.5	$—(CH_2)_5—$	H	91	66 min	75 (*S*)
5	Ph	H	97	48 min	31 (*S*)
5	Me	H	96	42 h	68 (*S*)
5	Me	Ph	98	5 min	17 (*S*)
5	Me, $CH_2CH{=}CH_2$	H	98[a)]/79[b)]	2 h	64[a)]/72[b)]

The ratio of major to minor diastereomers is 1.2 : 1.
a) Major diastereomer.
b) Minor diastereomers.
Source: Horrillo-Martinez et al. 2006 [12]. Reproduced with permission of Royal Society of Chemistry.

require stoichiometric or excess quantities of the metal reagents [18]. In comparison, catalytic applications are in general underdeveloped and asymmetric catalytic applications are limited to two examples [19–21].

10.3.1 Potassium Catalysts Based on Monoanionic Chiral Ligands

Catalytic use of potassium complexes in asymmetric protocols based on neutral and dianionic chiral ligands is unknown. The monoanionic potassium salt **55** (Scheme 10.8) is formed via deprotonation of the chiral amine **53** with KCH_2SiMe_3 (**52**). Amide **55** catalyzed the direct addition of toluene to imines, resulting in the formation of the chiral benzylic amine **59** in 85% yield and 56% ee. This is the first and only reported example of the catalytic and enantioselective addition of toluene to an imine. A diastereoselective version of this protocol, utilizing numerous arylarenes and imines, was also reported [20]. Catalyst **55** deprotonates toluene yielding the benzyl potassium derivative **56**. Imine **57** is added across the potassium carbon bond of **56** yielding intermediate **58** (Scheme 10.8). Arguably, two different reaction pathways are possible at this point. Pathway A yields **59** via intramolecular proton transfer and regenerates catalyst **55**. Pathway B provides an alternative shortcut via direct deprotonation of toluene

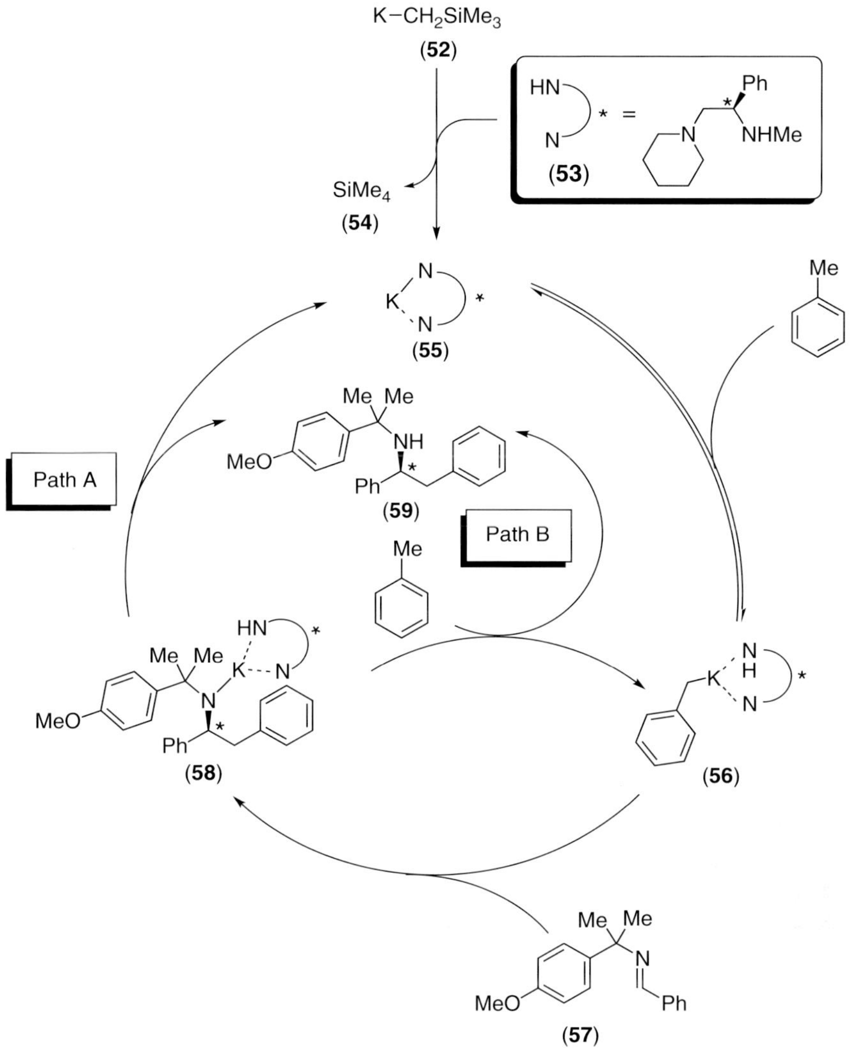

Scheme 10.8 Potassium-catalyzed asymmetric addition of toluene to an imine. Source: Yasmashita et al. 2018 [20]. Reproduced with permission of John Wiley & Sons.

by intermediate **58** liberating the product and forming intermediate **56**, which reacts with excess imine, to regenerate the catalyst. The second example of potassium catalysis is the enantioselective dehalogenation of vicinal dihalides (Table 10.5) [21]. This reaction relies on the use of a two-base system similar to the deprotonation reactions in Parts 2.2 and 2.3. The asymmetric deprotonation of **62** by the chiral alkoxide **60** yields *N*-methylephedrin (**61**), the targeted product **63**, and KBr. Subsequently, the catalyst is regenerated by reaction of **61** with MeOK, which is formed *in situ* by the reaction of KH and MeOH (Scheme 10.9). This catalytic deprotonation reaction could be conducted with excellent selectivities of up to 98% ee (Table 10.5). These high levels of enantioselectivity can only

Table 10.5 Enantioselective dehalogenation of vicinal dihalides [21].

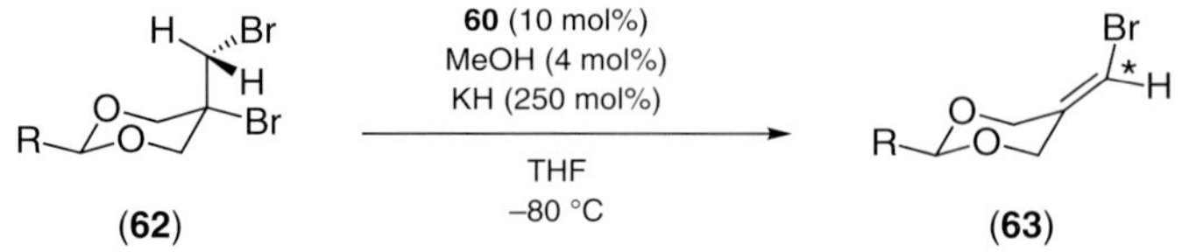

Substrate R	Conversion (%)	ee (%)
*n*Pr	78	77
*i*Pr	79	65
*t*Bu	83	90 (*R*)
4—Ph—C_6H_4	81	96 (*R*)
4—NO_2—C_6H_4	78	98
Ph	82	94
4—MeO—C_6H_4	79	98

Source: Amadji et al. 1996 [21]. Reproduced with permission of American Chemical Society.

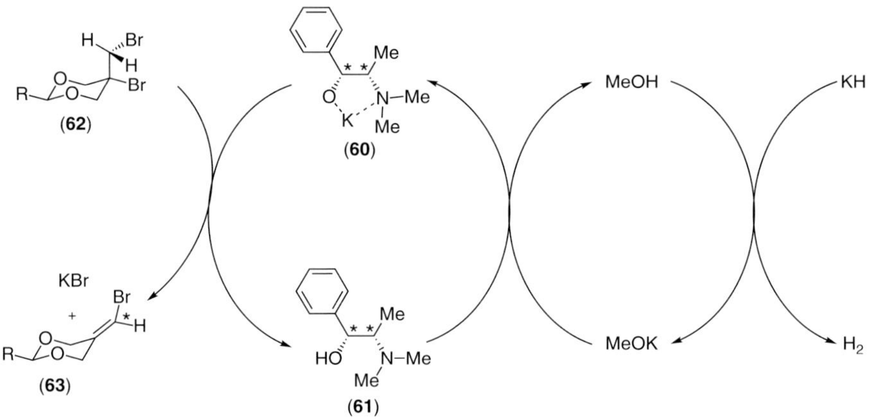

Scheme 10.9 Reaction mechanism for the catalytic enantioselective dehalogenation of vicinal dihalides. Source: Amadji et al. 1996 [21]. Reproduced with permission of American Chemical Society.

be maintained by suppressing the competing achiral reaction between MeOK and the substrate that was achieved by applying low methanol concentrations. From a series of experiments, it was concluded that 4 mol% methanol are ideal as the enantioselectivity decreases with higher alcohol concentrations because of competing reaction with potassium methoxide. For the heavier homologs rubidium and cesium, no catalytic asymmetric application is reported.

10.4 Magnesium-Based Catalysts

Since the Nobel Prize winning breakthrough of Victor Grignard more than 100 years ago, highly nucleophilic magnesium compounds known as Grignard reagents are invaluable synthetic tools for C—C bond formation [22]. In addition to their nucleophilic nature, the magnesium center also displays Lewis acidic properties [5]. Despite these dual properties, catalytic application of magnesium organometallics remains underdeveloped [23]. This is particularly true for Mg-mediated asymmetric conversions [24].

10.4.1 Magnesium Catalysts Based on Monoanionic Chiral Ligands

Magnesium compounds are plagued by a facile ligand redistribution reaction known as the Schlenk equilibrium. This leads to decay of the heteroleptic species **64** into the homoleptic counterparts **65** and **66** (Scheme 10.10) [25]. Such ligand redistribution causes the loss of chiral information in the active catalyst, resulting in diminished enantioselectivity and racemic product formation. Therefore, the key to successful enantioselective catalysis is control over the Schlenk equilibrium of the chiral organometallic magnesium catalyst. The use of bulky chiral chelating ligands to sterically hamper the formation of homoleptic complexes is the most frequently used approach [24]. This concept has successfully been applied in compound **69** using a tris(oxazolinyl)borate-based ligand system (Scheme 10.11). The magnesium heteroscorpionate **69** was synthesized by deprotonation with $Me_2Mg(dioxane)_2$ and despite the presence of strongly coordinating 1,4-dioxane in the precursor, no coordination of this ethereal solvent to the metal center was observed. The sterically demanding ligand prohibits the formation of the corresponding homoleptic complexes as coordination of two bulky heteroscorpionate ligands in a tridentate manner is sterically hampered. As a result, no ligand redistribution reaction was observed even after heating to 120 °C over a period of five days [26].

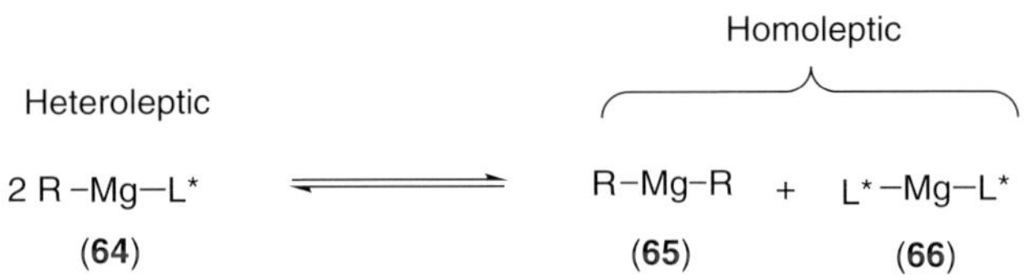

Scheme 10.10 The reactive species L*MgR may lose chiral information by the Schlenk equilibrium. Source: Schlenk and Schlenk 1929 [25]. Reproduced with permission of John Wiley & Sons.

In contrast, the heteroleptic species **67** and **68** decomposed into the corresponding homoleptic complexes during catalysis. This is due to insufficient steric bulk of the ligand and led to low enantioselectivities in catalytic hydroamination. In addition, protonation of the ligand by the substrate could liberate the chiral ligand, which may be a contributing reason for low ee's.

The bimetallic composition of complexes **67** and **68** has been proposed based on NMR data (crystallographic characterization failed because of decay of the complexes via ligand exchange) [27]. Attempted synthesis of a monometallic derivative by reacting the diprotic ligand with an equimolar ratio of $Mg(nBu)_2$ resulted in an undefined mixture of products.

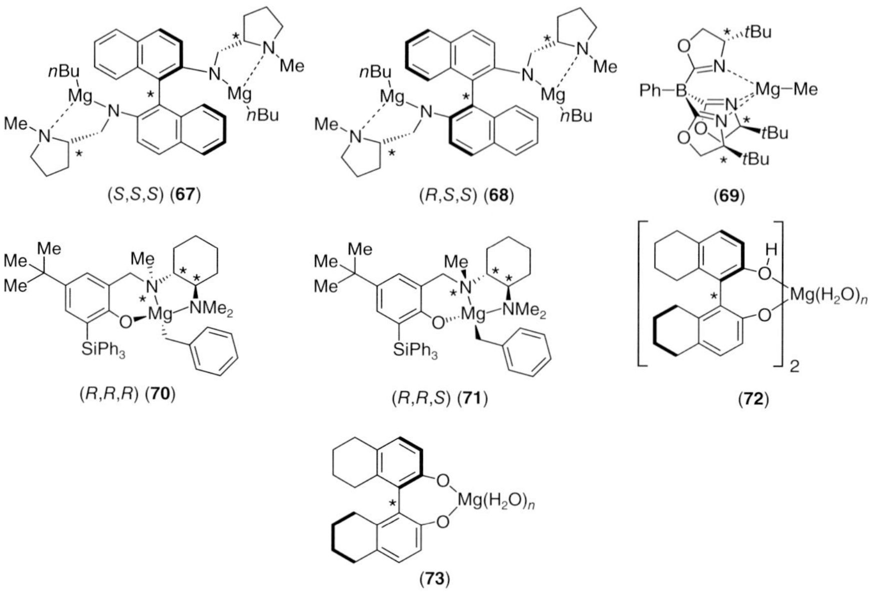

Scheme 10.11 Enantiopure heteroleptic Mg-based catalysts [26–29].

In 2012, the group of Hultzsch synthesized a heteroleptic magnesium catalyst based on a C_1-symmetric sterically demanding monoanionic *N*, *N*, O-chelating ligand (NNO) ligand (**70**, Scheme 10.11). This catalyst is inert toward ligand redistribution even during prolonged heating, resulting in excellent enantioselectivities up to 93% ee [28]. However, it consists of two diastereomers (**70**/**71**) in a ratio of circa 9/1. This mixture of diastereomers could be purified by recrystallization, but in solution, the ratio returned to 9/1 within five hours at room temperature. Heating shifted the equilibrium to 5 : 1. Because of their high activity, these catalysts were also applied in intermolecular hydroamination of styrenes. Compared to neat $Mg(nBu)_2$, all ligands have an enhancing effect in terms of selectivity and activity probably by prevention of higher aggregate formation. These complexes **67**–**70** have been applied in catalytic hydroamination, achieving yields above 95% unless noted otherwise. Their catalytic performance is summarized in Table 10.6. The catalytic cycle for hydroamination catalyzed by heteroleptic alkaline earth metal compounds is similar for Mg and Ca. Therefore, it is depicted as a generalized reaction mechanism (Scheme 10.12) [30]. Catalysis is initiated by deprotonation of amine **75** by the heteroleptic catalyst **74** (Ae = Mg). Subsequently, olefin is polarized by coordination with the Lewis acidic Ae metal center (**76**). Nucleophilic attack of the polarized double bond results in the formation of intermediate 77 via cyclization. The alkyl species **77** is immediately rearranged to the energetically more favorable amide **78**. **78** is cleaved by protonolysis via a second substrate molecule (**75**) yielding product **79** and regenerating **76**. Further details on hydroamination catalysis can be found in Chapter 3. Analog to amines in catalytic hydroamination, phosphines can be added to double bonds in an asymmetric manner resulting in chiral

Table 10.6 Magnesium-catalyzed asymmetric hydroamination of aminoalkenes [26–28].

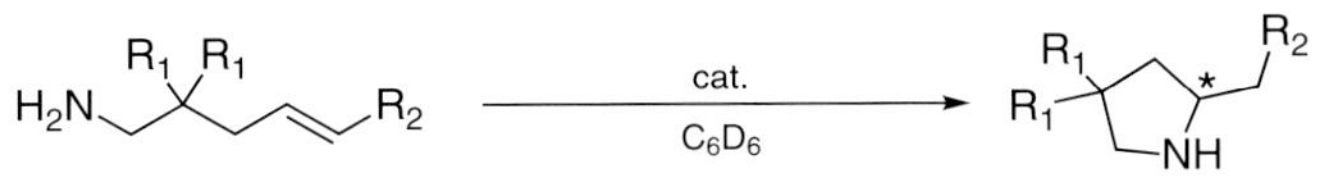

Catalyst	Temperature (°C)	Catalyst loading (mol%)	Substrate R_1	Substrate R_2	Time	ee (%) (configuration)
67[a)]	22	10	$-(CH_2)_5-$	H	3.5 h	6 (*R*)
67	100	5	Me	H	22 h	4 (*S*)
68	22	10	Ph	H	0.17 h	14 (*R*)
69	60	10	Ph	H	12 h	0
69[b)]	60	10	$-(CH_2)_5-$	H	26 h	36 (*R*)
69[a)]	80	10	Me	H	5 d	27 (*R*)
70	−20	5	$-(CH_2)_5-$	H	2 d	90
70	−20	3	Ph	H	12 h	80
70	22	5	Me	H	10 h	79
70	−20	5	Me, $CH_2CH{=}CH_2$	H	3 d	82
70[c)]	80	10	H	H	3 d	51
70	−20	3	Ph	Ph	12 h	93
70	−20	5	Me	Ph	1.5 d	92

a) Conversion 80%.
b) Conversion 93%.
c) Conversion 81%.

organophosphorus derivatives. This reactivity is explored in a wide variety of applications reaching from pharmaceuticals, such as biophosphate mimics or antiviral and antitumor agents, to ligands for transition metal catalysis [31]. A synthetic protocol for asymmetric hydrophosphinylation includes *in situ* catalyst formation by deprotonation of (*R*)-H_8-BINOL by $Mg(nBu)_2$ and subsequent partial hydrolysis by adding substoichiometric amounts of water (Table 10.7). The best results have been obtained applying a 3 : 2 : 2 ligand:metal:water ratio. A variety of potential catalytically active species can be anticipated, but the exact composition remains unknown. To shed light on the actual catalyst, a series of experiments was carried out in which the identity of **72** and **73** (Scheme 10.11) was confirmed by high-resolution electronspray ionization (ESI) mass spectrometry. Complexes **72** and **73** bear a variety of functional groups including the Lewis acidic Mg-center, the Brønsted and Lewis basic naphthoxide moiety, as well as the Brønsted acidic naphthol group. The proposed catalytically active species, however, is not in agreement with the applied 3 : 2 ligand:Mg stoichiometry, but further species could not be assigned [29]. Using this catalyst, the ketone functionality of Michael systems was hydrophosphorylated yielding a high 1,2- to 1,4-selectivity of 99 : 1. The catalysts were also active in the C=C bond functionalization of α,β-unsaturated esters, but addition of minor quantities of water as a cocatalyst was necessary in all cases. The latter hydrolysis

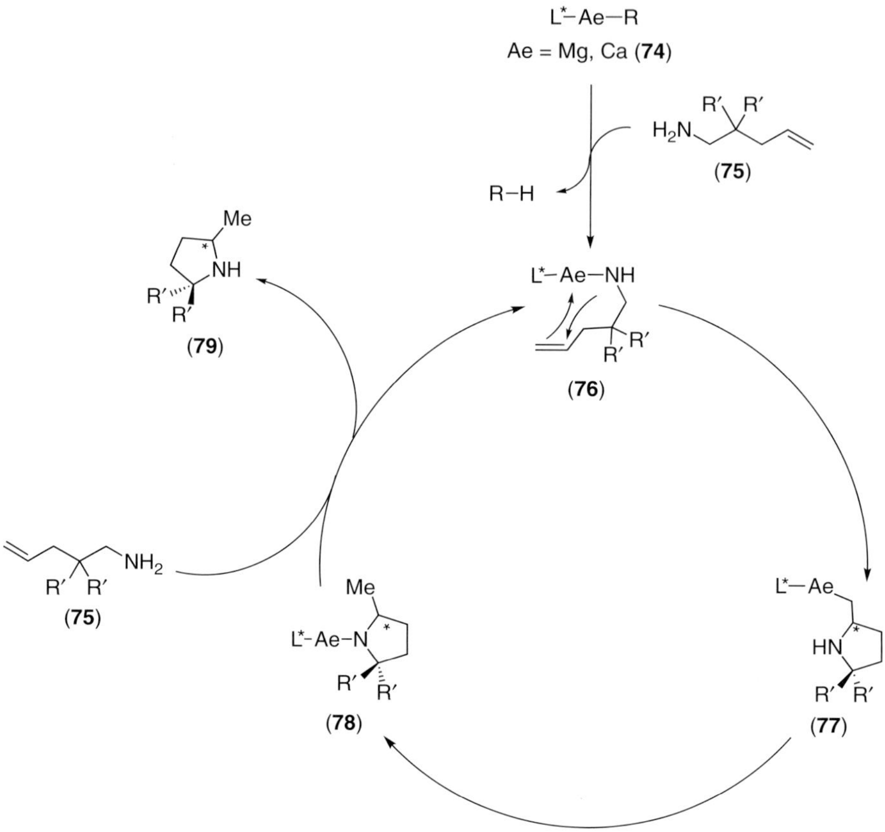

Scheme 10.12 Generalized mechanism for the intramolecular asymmetric alkene hydroamination catalyzed by heteroleptic alkaline earth metal compounds [28, 30].

likely produces a catalyst that contains a naphthol group. A mechanism is proposed in which the nucleophilic diphenylphosphine oxide is activated by the Brønsted basic naphthoxide moiety, while the electrophilic Michael system is activated by the associated Brønsted acidic naphthol moiety [29].

Therefore, the olefin functionality could be hydrophosphorylated in a high 1,4- to 1,2-selectivity of 99 : 1 (Table 10.8).

This protocol proved to be highly selective, yielding ee's of up to 96%. However, the exact composition of the actual catalytic active species is not known, as either a monoanionic or a dianionic chiral ligand could be present under reaction conditions.

10.4.2 Magnesium Catalysts Based on Dianionic Chiral Ligands

Complex **80** (Table 10.9) was synthesized by double deprotonation of an diphenylphosphine substituted (*S*,*S*)-ANDEN ligand using diphenylmagnesium. The absence of Mg–P coordination is indicated by ^{31}P NMR. Instead, the N atoms coordinate the metal forming a distorted five-membered metallacycle.

Table 10.7 Catalytic asymmetric 1,2-hydrophosphanylation of α,β-unsaturated ketones [29].

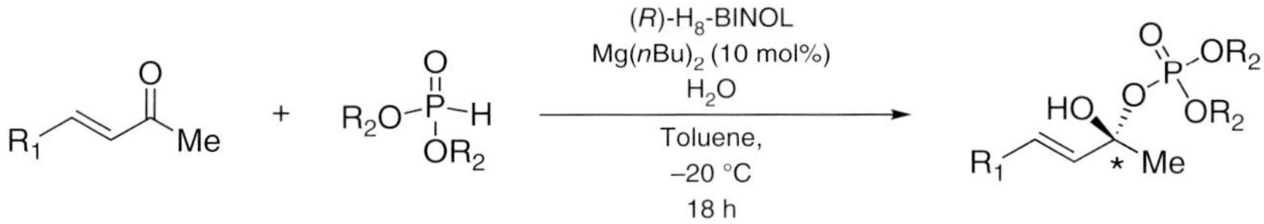

(*R*)-H_8-BINOL (mol%)	H_2O (mol%)	Substrate R_1	Substrate R_2	Conversion (%)	ee (%)
15	10	Ph	Me	89	86 (*R*)
15	10	3-Me-C_6H_4	Me	77	86 (*R*)
15	10	3,5-Cl_2–C_6H_3	Me	96	81 (*R*)
15[a)]	10	2-naphthyl	Me	74	84 (*R*)
15	10	3-thienyl	Me	63	82 (*R*)
20	—	Ph	Et	81	82 (*R*)
15[b)]	—	4-MeO-C_6H_4	Me	59	82 (*R*)

a) $T = -15\,°C$.
b) $T = -10\,°C$.
Source: Hatano et al. 2013 [29]. Reproduced with permission of John Wiley & Sons.

Table 10.8 Catalytic asymmetric 1,4-hydrophospanylation of α,β-unsaturated esters [29].

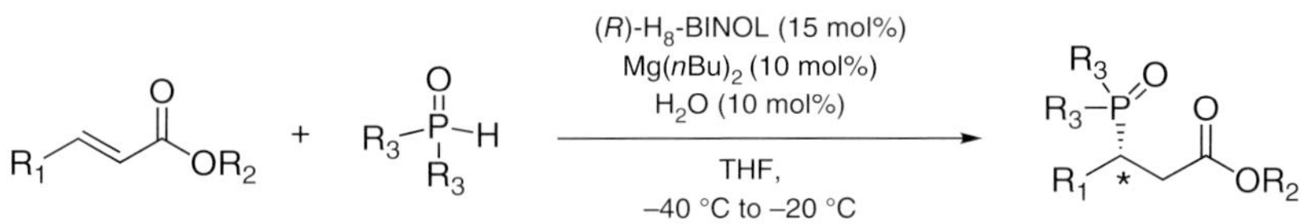

Temperature (°C)	Substrate R_1	Substrate R_2	Substrate R_3	Conversion (%)	Time (h)	ee (%)
−40	Ph	Me	Ph	91	16	95 (*R*)
−40	4-Cl-C_6H_4	Me	Ph	93	5	92 (*R*)
−20	4-MeO-C_6H_4	Me	Ph	80	4	95 (*R*)
−40	3-pyridyl	Me	Ph	89	5	85 (*R*)
−20	Cy	Me	Ph	86	18	95 (*R*)
−40	Ph	Et	Ph	78	10	91 (*R*)
−20	Ph	Me	Naphthyl	91	4	92 (*R*)
−20	Ph	Me	3,5-Xylyl	84	10	96 (*R*)

Source: Hatano et al. 2013 [29]. Reproduced with permission of John Wiley & Sons.

The backbone of the ligand system was found to be flexible [32]. This bifunctional system has been applied as a catalyst in hydroamination reactions. A generalized mechanism for this reaction is displayed in Scheme 10.13. In comparison to the catalytic cycle proposed for heteroleptic magnesium catalysts (Scheme 10.12), the reaction mechanism differs. In complexes of type **81** (Ae = Mg), the chiral dianionic ligand is noninnocent, participating actively as a Brønsted base in the

Table 10.9 Intramolecular asymmetric alkene hydroamination catalyzed by **80** [32].

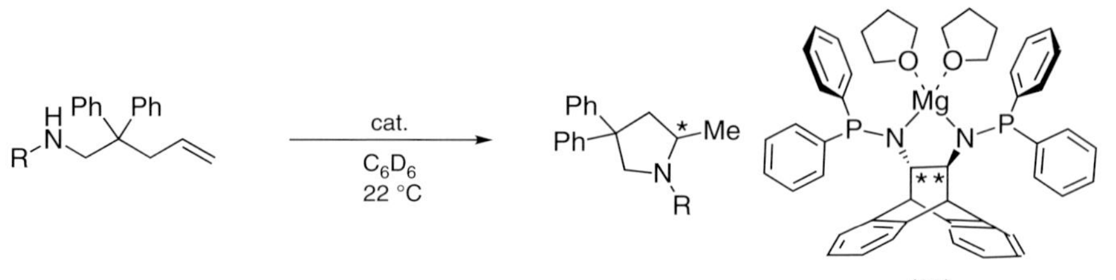

(**80**)

Catalyst loading (%)	Substrate R	Conversion (%)	Time (h)	ee (%)
5	H	98	4	7
1	H	38	72	7
5	Bn	—	72	—

Source: Schmid et al. 2016 [32]. Reproduced with permission of Royal Society of Chemistry.

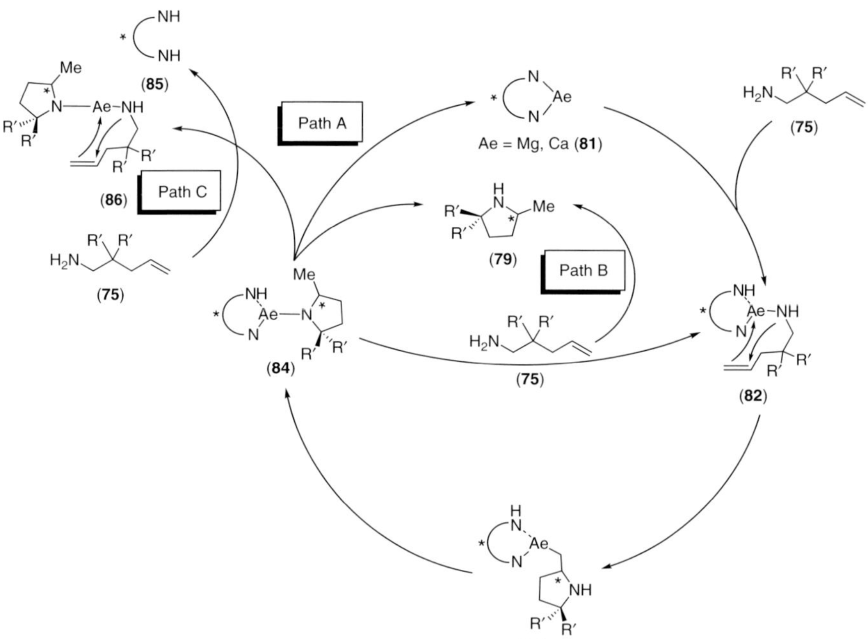

Scheme 10.13 Intramolecular asymmetric alkene hydroamination catalyzed by dianionic alkaline earth metal compounds [33].

deprotonation of the aminoalkene. Subsequently, the C=C bond is polarized by coordination to the Lewis acidic metal center (**82**) after which ring closure to **83** takes place. The latter is rearranged to the energetically more favorable amide **84**. Intramolecular proton transfer from the ligand to the pyrrolidine moiety provides product **79** and regenerates catalyst **81** (Path A). An alternative pathway produces the product **79** by intermolecular proton transfer from a second aminoalkene to cycle back straight to **84** (Path B). Apart from that, the chiral ligand in **84** can also be fully protonated by the substrate, resulting in elimination of a neutral chiral ligand (**85**) and formation of an unselective catalyst (**86**, Path C). Because ligand exchange via the Schlenk equilibrium is not observed for dianionic ligands, the latter catalyst degradation pathway may explain the very low ee's observed (Table 10.9). Similar attempts catalyzing asymmetric hydroamination reactions have been carried out using the heavier homolog calcium.

10.5 Calcium-Based Catalysts

Although Grignard reagents are widely used, first fully characterized arylcalcium analogs were reported only in 2006 [34]. The synthesis and full characterization of even more challenging alkylcalcium Grignard reagents was recently reported in 2018 [35]. Like magnesium, calcium is a nontoxic, cheap, worldwide available electropositive metal that has drawn recent interest for a variety of catalytic applications [5]. Despite the higher reactivity and more challenging handling of organocalcium compounds, they have been applied in asymmetric catalysis. Examples are the multiple consecutive syndioselective insertions in living styrene polymerization or the formation of chiral heterocycles [3, 5].

10.5.1 Calcium Catalysts Based on Monoanionic Chiral Ligands

Asymmetric calcium catalysis has been pioneered by Harder et al. utilizing bisoxazoline *S*-(BOX) and β–diketiminate-based catalysts (**87**–**92**, Scheme 10.14). Compound **87** was exclusively formed as one diastereomer, indicating chiral communication between the benzylic ligand and the chiral BOX-ligand. Despite this excellent communication between chiral centers, dissolved in benzene, complex **87** slowly epimerized already at room temperature (at 50 °C fast epimerization was observed). This process likely proceeds via a second deprotonation of the chiral ligand in benzylic position, forming intermediate **90**, which after protonation gives **87** with scrambled chiral centers. Complexes **88** and **89**, with the less basic $N(SiMe_3)_2$ amide ligand, are not prone to epimerization. However, in comparison to magnesium complexes, these calcium compounds are more sensitive to the ligand redistribution via the Schlenk equilibrium, which resulted in partial formation of the homoleptic species **91** and **92** as well as the corresponding achiral reagent $Ca[N(SiMe_3)_2]_2$ [36]. It is therefore not surprising that very low ee's (5–10%) were found for the intramolecular hydroamination or styrene hydrosilylation. The ligand redistribution equilibrium can be influenced by addition of the catalytically inactive homoleptic compounds **91** or **92**, respectively, which resulted in increased quantities of the chiral heteroleptic compounds **88** or **89**. However, no significant impact on the enantioselectivities was observed (Table 10.10). Attempts to isolate the catalytically active hydride species by reaction of **88** or **89** with phenylsilane yielded only CaH_2 and the homoleptic complexes **91** or **92**, respectively.

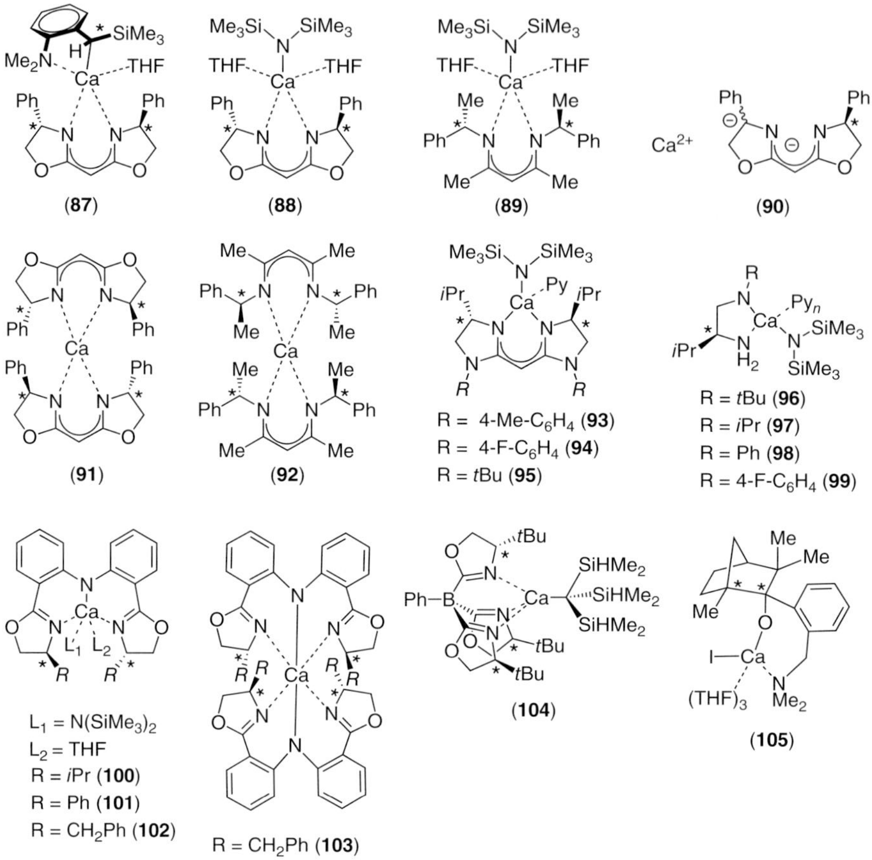

Scheme 10.14 Enantiopure heteroleptic Ca-based catalysts [36–41].

After these first attempts, bis(1-imidazolyl)methane (BIM) based calcium complexes **93–95** were tested in catalytic alkene hydroamination (Table 10.10). Although the BIM ligand system allows a better control of electronic and steric properties compared to BOXs, their complexes were found to be poorly soluble even in polar solvents such as THF. Use of pyridine was necessary for the successful synthesis of these complexes and the problem of facile ligand exchange is still present. As a result, low enantioselectivities as well as long reaction times were observed in alkene hydroamination: for catalyst **93**, only a racemic product mixture was obtained [37].

Like **93–95**, catalysts **96–99** are based on a chiral L-valine-derived ligand system in which the N-substituents have been varied (Scheme 10.14) [37, 38]. The coordination of pyridine is crucial for the stabilization of these complexes, but the exact number of coordinated pyridines is not always known. Because of facile ligand redistribution via Schlenk equilibria, the heteroleptic complexes could not be crystallized and only homoleptic complexes were isolated. Application of **96–98** in catalytic alkene hydroamination required long reaction times and resulted in

Table 10.10 Asymmetric catalytic intramolecular alkene hydroamination [36–41].

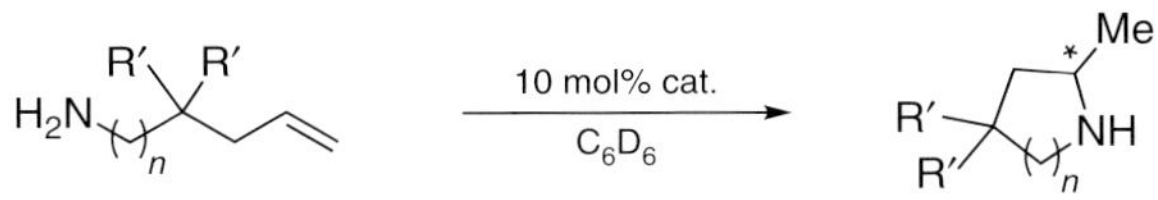

Catalyst	Temperature (°C)	Substrate *n*	Substrate R_1	Conversion (%)	Time	ee (%)
88 + **91** (1 : 1)[a]	20	1	Ph	98	2 h	5 (*R*)
89 + **92** (1 : 1)[a]	20	1	Ph	98	1 h	10 (*R*)
94	20	1	Ph	99	7 d	9
95	20	1	Ph	99	5 d	12
96	20	1	Ph	99	1 d	6
97	20	1	Me	99	5 d	12
98	20	1	Ph	80	3 d	26
100	40	1	Ph	99	24 h	22
100	80	2	Ph	80	5 d	6
101	40	1	Ph	99	24 h	40
101[b]	50	1	Ph	99	24 h	50
102	30	1	Ph	99	24 h	26
102[b]	80	2	Ph	14	5 d	16
104	20	1	Ph	99	5 min	0
104	20	1	$-(CH_2)_5-$	99	5 min	18 (*S*)
104	20	1	Me	100	5 min	18 (*S*)

a) 5 mol%.
b) 20 mol%.

low enantiomeric excess (Table 10.10). Complex **99** did not show any catalytic activity, likely because of insufficient basicity of this fluorinated compound [38].

The mechanism for intramolecular alkene hydroamination with these Ca catalysts is similar to what has been described for their Mg analogs (Ae = Ca, Scheme 10.12). Enantioselectivities up to 50% were obtained by Ward et al. utilizing catalysts **100**–**102** based on the bis(oxazolinylphenyl)amide (BOPA) ligand system (Scheme 10.14) [39]. These C_2-symmetric catalysts were prepared *in situ* and used without further purification. The ligand redistribution rates strongly depend on the substitution pattern. At room temperature, the catalysts **101** and **102** gave after one hour only 5% of the homoleptic complexes, whereas the less sterically demanding *i*Pr-substituted catalyst **100** gave complete ligand exchange within 15 minutes. Heating to 50 °C shifted the equilibrium to 80% of the homoleptic complex for **101** and 85% for **102**. Crystal structures of **100**–**102** could not be obtained, but their chemical composition was confirmed by NMR and ESI mass spectrometry. Crystallographic characterization was only possible for **103**, i.e. the homoleptic and catalytically inactive degradation product of **102**. In addition to pyrrolidine formation, the catalytic formation of piperidines was also possible, but considerably slower, as predicted by the Baldwin rules (Table 10.10) [39].

The heteroscorpionate-based complex **104** is inert toward Schlenk-like ligand redistribution and application in catalytic hydroamination resulted in full cyclization of the aminoalkene substrates to the corresponding pyrrolidines within five minutes. Unfortunately, these highly active catalysts suffered from decreased enantioselectivities of 18% ee or less.

A different hydroamination approach is based on the cooperative reactivity of a Ca complex based on an innocent chiral ligand and an external non-nucleophilic base. As a strong metal-free base, the Schwesinger phosphazene base P4 (**107**, Scheme 10.15), was applied in equimolar combination with a chiral fenchole-based calcium iodide catalyst (**105**, Scheme 10.14) [40, 41]. This system catalyzed the intramolecular hydroamination of unactivated amino olefins to pyrrolidines in selectivities of up to 36% ee (Table 10.11). A tentative mechanism (Scheme 10.15, [Ca] = Ca-I, X = NMe_2,) starts with coordination of the aminoalkene to the chiral Ca complex **106**. Formation of the Ca—N bond acidifies the NH_2 group, which now can be readily deprotonated by the neutral P4 base (**107**) to form **109**. After ring closure to intermediate **110** and subsequent proton migration to amide **111**, the pyrrolidine product is released either through protonation by P4-H^+ (**108**, Path A) or by the aminoalkene substrate (**75**) (Path B). This protocol, based on a chiral Ca complex bearing unreactive spectator ligands, and the P4 base can be extended from complexes with monoanionic fencholates to catalysts containing dianionic binaphtholates.

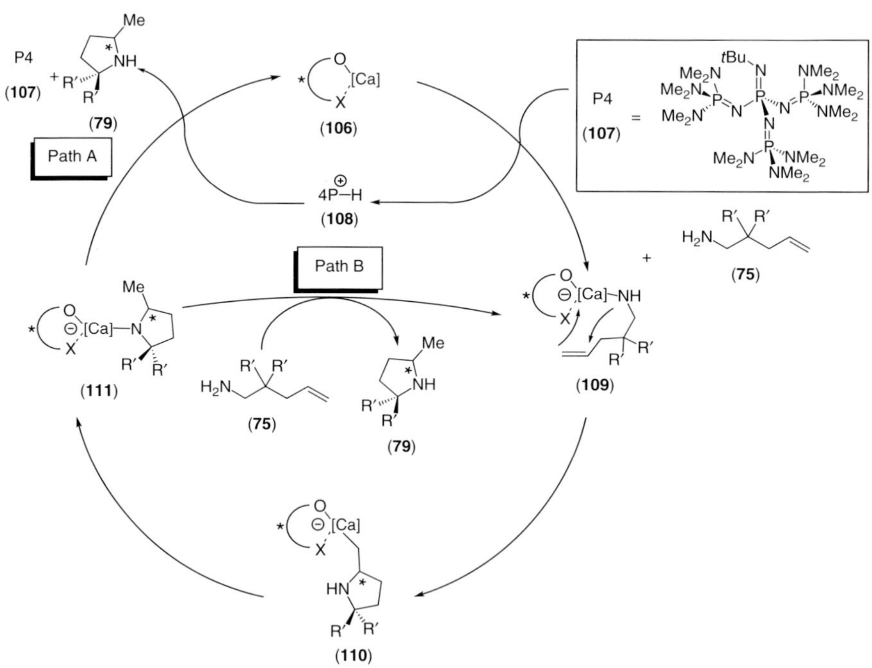

Scheme 10.15 Intramolecular asymmetric alkene hydroamination catalyzed by "innocent" calcium compounds using an external neutral base [40, 41].

Table 10.11 Asymmetric catalytic intramolecular alkene hydroamination [40, 41].

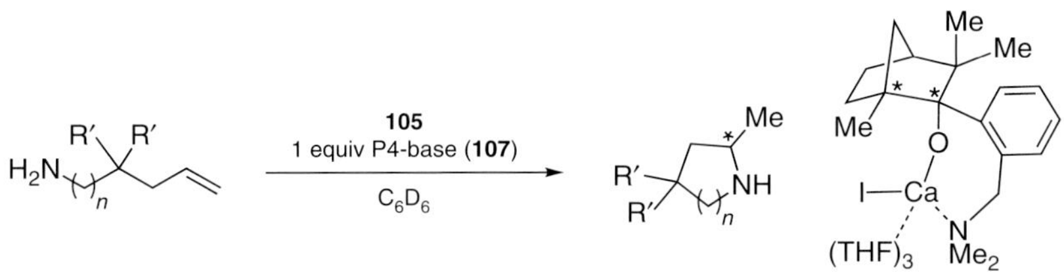
(**105**)

Substrate R	Catalyst loading (mol%)	Temperature (°C)	Conversion (%)	Time (h)	ee (%)
Ph	5	20	98	5	36
$-(CH_2)_5-$	5	20	98	8	25
Me	10	90	98	18	24

10.5.2 Calcium Catalysts Based on Dianionic Chiral Ligands

Dianionic chiral ligands can be applied in enantioselective hydroamination in two manners, either with or without the use of an external non-nucleophilic base. Using strongly Brønsted basic ligands that are capable of deprotonating the substrate molecule (e.g. BINAM derivatives), no further additives are necessary. In case of insufficiently basic ligands (e.g. BINOL), the addition of an external non-nucleophilic base, like P4 (**107**), is required to achieve catalytic activity. Catalyst **112** is based on a (*S*)-BINOL ligand with axial chirality, and in equimolar combination with P4 base (**107**) as an additive, the intramolecular alkene hydroamination reached selectivities up to 34% ee (Table 10.12) [40, 41]. The proposed mechanism is similar to catalyst **105** and shown in Scheme 10.15 ([Ca] = Ca, X = O).

Table 10.12 Asymmetric catalytic intramolecular alkene hydroamination [41].

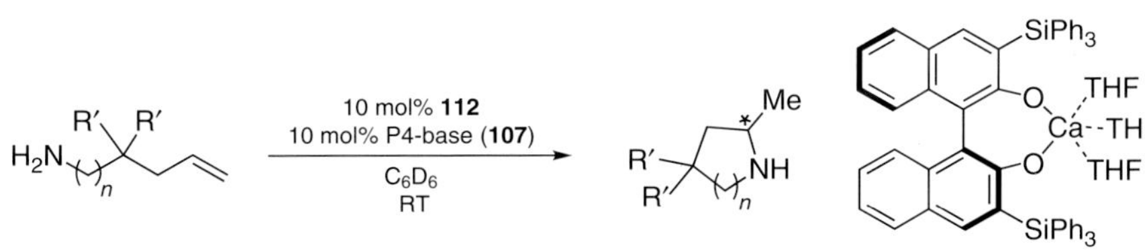
(**112**)

Substrate R	Conversion (%)	Time (h)	ee (%)
Ph	99	2	34
$-(CH_2)_5-$	90	24	30
Me	75	48	19

Table 10.13 Asymmetric intramolecular alkene hydroamination using a catalyst with a noninnocent ligand [33].

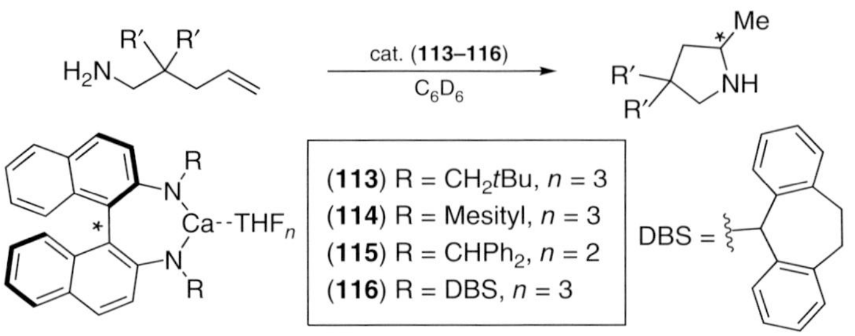

Catalyst	Catalyst loading (mol%)	Temperature (°C)	Substrate R_1	Time	ee (%)
113	5	25	Ph	2 min	16
113	5	25	$-(CH_2)_5-$	5 min	5
113	5	25	Me	75 min	8
114	10	40	Ph	10 min	27
114	10	40	$-(CH_2)_5-$	15 min	14
115	15	60	Ph	3.5 h	9
115	15	60	$-(CH_2)_5-$	4.5 h	30
116	20	25	Ph	24 h	59
116	20	40	$-(CH_2)_5-$	24 h	20
116	20	80	Me	48 h	11

This approach of using a dianionic chiral ligand for stereocontrol can be extended to cooperative metal ligand catalysis using the more Brønsted basic (*R*)-BINAM ligands in which case the application of the external P4 base is not necessary. Linking chiral unit and reactive group prevents catalyst decay via Schlenk-like ligand redistribution. The proposed catalytic cycle is similar to that discussed for analog magnesium catalysts (Scheme 10.13). A variety of calcium catalysts based on noninnocent dianionic ligands has been tested achieving conversion of at least 96% (**113–116**, Table 10.13) [33].

In contrast to chiral heteroleptic catalysts, which can easily lose chirality by the Schlenk equilibrium (Scheme 10.10), these complexes are not prone to ligand scrambling. The chiral information, however, may be lost by complete protonation of the dianionic ligand (Path C, Scheme 10.13). To overcome the aforementioned catalyst degradation by ligand liberation, it was attempted to provide increased stability by using polydentate chelating BINAM derivatives as ligands. This approach can effectively suppress ligand liberation (Path C, Scheme 10.13), but unfortunately, these catalysts show reduced activity because of the strongly reduced Lewis acidity of the calcium center, which comes as a consequence of attaching multiple donor groups to provide stability [33]. The cyclization of unactivated aminoalkenes using **113–116** yielded up to 59% ee, depending on the nature of the catalyst. This represents the highest reported selectivity in calcium-catalyzed intramolecular alkene hydroamination so far and is promising for further developments in enantioselective alkaline earth metal catalysis. Therefore, future research should be conducted toward a balanced ratio between kinetic stability and catalytic performance in order to maximize the catalyst efficiency.

10.6 Conclusion and Outlook

The horizon of early main group metal catalysis is continuously broadened by new organic transformations. A large variety of reactions that are typically catalyzed by transition metals are now also mediated by group 1 and 2 metals. Replacing transition metals for low-cost, benign, and highly abundant main group metals certainly is an attractive goal. There is, however, one particular area in catalysis in which there is hitherto no replacement for transition metals. The very high ee's obtained in reactions catalyzed by metals from the platinum group often exceed 99%. In contrast, with very few exceptions, early main group metal catalysis hardly reaches selectivities over 50% ee. There are a couple of reasons for the very tenacious development in enantioselective early main group metal catalysis. First of all, transition metal complexes have defined coordination geometries that are generally constrained by covalent bonding and orbital symmetries. In contrast, main group metal complexes are largely bound ionically and coordination geometries are determined by sterics and generally highly fluxional. This makes chiral communication between the complex and the substrates less defined, lowering the enantioselectivity in catalysis. A second problem in stereocontrolled catalysis is the propensity of early main group metal complexes to exchange ligands, often resulting in a mixture of catalysts among which are active species that do not carry chiral information anymore. Also, chiral ligands may be completely lost by protonation, either by water impurities or by the substrates itself. Although we are slowly starting to understand the problems related to enantioselective early main group metal catalysis, finding creative solutions is still a challenging process and progress is hitherto slow. The future likely will see breakthroughs in ligand development but also in broadening the scope to new reactions that may be better suited for a stereoselective protocol.

List of Abbreviations

ANDEN	9,10-dihydro-9,10-ethanoanthracene-11,12-diamine
BIM	bis(1-imidazolyl)methane
BINAM	[1,1′-binaphthalene]-2,2′-diamine
BINOL	[1,1′-binaphthalene]-2,2′-diol
Bn	benzyl
BOPA	bis(oxazolinylphenyl)amide
BOX	bisoxazoline
cat.	catalyst
Cy	cyclohexyl
d	day(s)

DABCO	1,4-Diazabicyclo[2.2.2]octane
ee	enantiomeric excess
ESI	electronspray ionization
HMPA	hexamethylphosphoramide
H_8-BINOL	5,5′,6,6′,7,7′,8,8′-octahydro-[1,1′-binaphthalene]-2,2′-diol
*i*Pr	isopropyl
LDA	lithiumdiisopropylamide
Me	methyl
Mes	mesityl
NMR	nuclear magnetic resonance
*n*BuLi	*n*-butyllithium
*n*Bu	*n*-butyl
*n*Pr	*n*-propyl
Ph	phenyl
*t*Bu	*tert*-butyl
THF	tetrahydrofuran

References

1 Lindström, B. and Petterson, L. (2003). *Cattech* 7: 130–138.

2 (a) Hagen, J. (2006). *Industrial Catalysis: A Practical Approach*, vol. 2, 2. Weinheim, Germany: Wiley-VCH. (b) Frängsmyr, T. (ed.) (2002). *The Nobel Prizes 2001*. Stockholm: Nobel Foundation.

3 Wilkins, L. and Melen, R. (2016). *Coord. Chem. Rev.* 324: 123–139.

4 (a)Chatt, J. and Duncanson, L.A. (1953). *J. Chem. Soc.*: 2939–2947. (b)Elschenbroich, C. (2011). *Organometallics*, 3e, 635. Weinheim: Wiley-VCH.

5 Harder, S. (2010). *Chem. Rev.* 110: 3852–3876.

6 Tsubogo, T., Yamashita, Y., and Kobayashi, S. (2009). *Angew. Chem. Int. Ed.* 48: 9117–9120.

7 Hill, M., Liptrot, D., and Weetman, C. (2016). *Chem. Soc. Rev.* 45: 972–988.

8 Reich, H. (2012). *J. Org. Chem.* 77: 5471–5491.

9 Halasa, A.F., Schulz, D.N., Tate, D.P., and Mochel, V.D. (1980). *Adv. Organomet. Chem.* 18: 55–97.

10 Clark, N., Garcıa-Alvarez, P., Kennedy, A., and O'Hara, C. (2009). *Chem. Commun.*: 5835–5837.

11 Horrillo-Martínez, P., Hultzsch, K., Gil, A., and Branchadell, V. (2007). *Eur. J. Org. Chem.*: 3311–3325.

12 Horrillo-Martinez, P., Hultzsch, K., and Hampel, F. (2006). *Chem. Commun.*: 2221–2223.

13 (a) Ogata, T., Kimachi, T., Yamada, K. et al. (2012). *Heterocycles* 86: 469–485. (b) Ogata, T., Ujihara, A., Tsuchida, S. et al. (2007). *Tetrahedron Lett.* 48: 6648–6650.

14 Pettersen, D., Amedjkouh, M., and Ahlberg, P. (2002). *Tetrahedron* 58: 4669–4673.

15 Tiereney, J., Alexakis, A., and Mangeney, P. (1997). *Tetrahedron Asym.* 8: 1019–1022.

16 Yamashita, T., Sato, D., Kiyoto, T. et al. (1997). *Tetrahedron* 53: 16987–16998.

17 (a) Deschamp, J., Olier, C., Schulz, E. et al. (2010). *Adv. Synth. Catal.* 352: 2171–2176. (b) Deschamp, J., Collin, Q., Hannedouche, J., and Schulz, E. (2011). *Eur. J. Org. Chem.* (18): 3329–3338.

18 (a) Carraro, M., Pisano, L., and Azzena, U. (2017). *Synthesis* 49: 1931–1937. (b) Mulvey, R. (2009). *Acc. Chem. Res.* 42: 743–755.

19 Buch, F., Brettar, J., and Harder, S. (2006). *Angew. Chem.* 118: 2807–2811.

20 Yasmashita, Y., Suzuki, H., Io, S. et al. (2018). *Angew. Chem. Int. Ed.* 57: 6896–6900.

21 Amadji, M., Vadecard, J., Plaquevent, J. et al. (1996). *J. Am. Chem. Soc.* 118: 12483–12484.

22 (a) Grignard, V. (1901). *Ann. Chim.* 24: 433. (b) Seyferth, D. (2009). *Organometallics* 28: 1598–1605.

23 Anker, M.D. and Hill, M.S. (2017). *"Encyclopedia of Inorganic and Bioinorganic Chemistry" Section "Alkaline Earth Catalysis"*, 1. Wiley.

24 Harder, S. (2013). *Alkaline-Earth Metal Compounds: Oddities and Applications*, 2. Berlin-Heidelberg: Springer-Verlag.

25 Schlenk, W. and Schlenk, W. Jr., (1929). *Chem. Ber.* 62: 920–924.

26 Neal, S., Ellern, A., and Sadow, A. (2011). *J. Organom. Chem.* 696: 228–234.

27 Horrillo-Martinez, P. and Hultzsch, K. (2009). *Tetrahedron Lett.* 50: 2054–2056.

28 Zhang, X., Emge, T., and Hultzsch, K. (2012). *Angew. Chem. Int. Ed.* 51: 394–398.

29 Hatano, M., Horibe, T., and Ishihara, K. (2013). *Angew. Chem. Int. Ed.* 52: 4549–4553.

30 Crimmin, R., Arrowsmith, M., Barrett, A. et al. (2009). *J. Am. Chem. Soc.* 131: 9670–9685.

31 (a) Engel, R. (1977). *Chem. Rev.* 77: 349–367. (b) Kafarski, P. and Lecjak, B. (1991). *Phosphorus, Sulfur Silicon Relat. Elem.* 63: 193–215. (c) Kolodiazhnyi, O. (2005). *Tetrahedron Asym.* 16: 3295–3340. (d) Guiry, P. and Saunders, C. (2004). *Adv. Synth. Catal.* 346: 497–537.

32 Schmid, B., Frieß, S., Herrera, A. et al. (2016). *Dalton Trans.* 45: 12028–12040.

33 Stegner, P., Ballmann G., Harder, S., manuscript in preparation.

34 Fischer, R., Görls, H., and Westerhausen, M. (2005). *Inorg. Chem. Commun.* 8: 1159–1161.

35 Wolf, B., Stuhl, C., Maichle-Mössmer, C., and Anwander, R. (2018). *J. Am. Chem. Soc.* 140: 2373–2383.

36 Buch, F. and Harder, S. (2008). *Z. Naturfosch* 63b: 169–177.

37 Wixey, J. and Ward, B. (2011). *Dalton Trans.* 40: 7693–7696.

38 Wixey, J. and Ward, B. (2011). *Chem. Commun.* 47: 5449–5451.

39 Nixon, D. and Ward, B. (2012). *Chem. Commun.* 48: 11790–11792.

40 Penafiel, J. (2016). Alkaline earth organometallic compounds in homogeneous catalysis. PhD thesis. University of Groningen, p. 172.

41 Stegner, P., Fischer, C., Nguyen, T., Harder, S., manuscript in preparation.

11

Early Main Group Metal Lewis Acid Catalysis

Marian Rauser, Sebastian Schröder, and Meike Niggemann

RWTH Aachen University, Institute of Organic Chemistry, Landoltweg 1, 52074 Aachen, Germany

11.1 Introduction

According to Pearson's definition, any species with an inherent ability to accept electron pairs displays Lewis acidic behavior. As this comprises a vast amount of chemical entities, numerous types of classifications for Lewis acids have been devised in order to facilitate the prediction of the reactivity of a given Lewis acidic compound. Nevertheless, the synthetic chemist's fundamental question – which Lewis acid is going to be the most effective in my transformation? – is still not easily answered, and the screening of a range of different Lewis acids remains the typical starting point for the development of a new reaction in this area of chemistry. As a thorough discussion of the different types of classifications and theoretical analyses of Lewis acidity is beyond the scope of this review, this introduction shall be focused on explaining why the prediction of Lewis acidic behavior remains challenging, in addition to pointing out a few general trends.

Generally, six different types of Lewis acids have been recognized (see Figure 11.1). Among these, the s-block metal cation Lewis acids, the protagonists of this book, have undoubtedly not yet received the attention they deserve. This is surprising as they come with an inherently low toxicity (with the exception of Be^{2+} and the radioactive elements) and are readily available at low cost. The electropositive character of these elements and the resultant polarization of the M–Base bonding provide systems displaying a high degree of charge separation (M^+/$Base^-$), which, as a result, are extremely nucleophilic sources of the base [1]. Therefore, they have been traditionally used as counter cations for many of the stronger bases, such as alkoxides, amides and organometallics, and hence appear in every reaction that requires the stoichiometric or catalytic deprotonation of a compound with low acidity. More recently, they are also found to take part in an increasing amount of enantioselective reactions (see Chapter 12). The dominating reactivity in all these reactions is attributed to the anion, and the role of the s-block metal has long been considered to be merely that of an innocent spectator. That this is not the case, and the nature of the s-block metal certainly plays a distinct role, is obvious as the switch from one s-block metal to another often has a huge impact on the reaction outcome. This

Early Main Group Metal Catalysis: Concepts and Reactions,
First Edition. Edited by Sjoerd Harder.

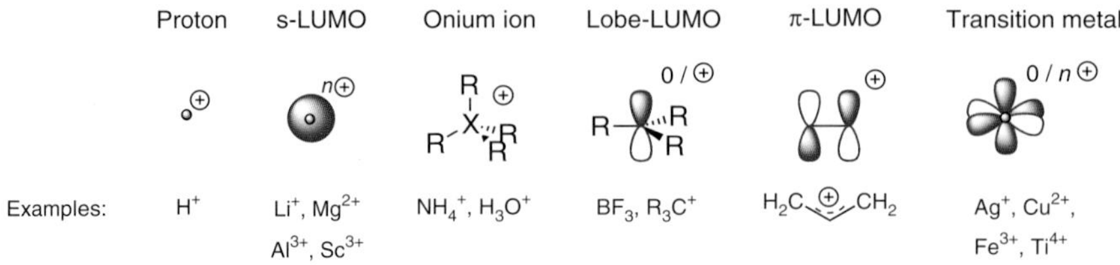

Figure 11.1 Recognized types of Lewis acids, classification according to frontier molecular orbital theory (FMO).

is strongly supported by the increasing evidence that the reactivity of early main group metal organics relies on the interplay of both nucleophilicity and metal Lewis acidity [2]. To shed light on the Lewis acidity, as the major contribution to their *distinctive, individual* reactivity, the present review shall summarize the state of the art in reactions that are induced by s-block metal cation reactivity with *spectator anions.*

11.1.1 Lewis Acidity of s-Block Metal Cations

Because of their position in the periodic table, s-block metals are characterized by an extremely stable +1 (alkali metals = $M^{(1)}$) or +2 (alkaline earth metals = $M^{(2)}$) oxidation state. Therefore, reactivity that is based on the catalyst's redox activity, typical for many transition metal (TM)-catalyzed processes, is precluded. For their application as Lewis acids, this might be even beneficial, as alternative reaction pathways cannot interfere. A frontier molecular orbital analysis of their electronic structure leaves them with a shell-like LUMO (2s, 3s, 4s… AO), which is superimposed upon a varying amount of closed, inaccessible electron shells. This defines their interaction with Lewis bases, including ligands, counter anions, substrate, solvent, and water molecules. As inner shells are closed and inaccessible, s-block metals do not back-donate, or accept, any electron density to, or from, the Lewis base with these orbitals, which markedly distinguishes them from most TM Lewis acids. In TM cation Lewis acid–Lewis base interactions, even though the primary contribution to bonding is still the σ-donation from the Lewis base to the metal, the ability of the base to accept or donate electrons in π-orbitals of the TM substantially modulates the character of the bonding.

The strength of the σ-donation from the Lewis base to the metal are generally defined by the following: (i) The charge of the metal: the higher the charge, the stronger the interaction. This generally makes the $M^{(2)}$ metals stronger Lewis acids as the $M^{(1)}$ series. (ii) The ionic radius: the larger the ionic radius, the weaker the interaction. Hence, Na^+ is a less active Lewis acid than Li^+ and the activity of K^+, Rb^+, and Cs^+ is even weaker. It might be for this reason that our search revealed no reaction that was solely based on the Lewis activity of these heavier alkaline metals. The same applies for the $M^{(2)}$ series, reactions are confined to Mg^{2+} and Ca^{2+}, with the smaller Mg^{2+} being the stronger Lewis acid. Therefore, this chapter will henceforth be focused on these four metals. The interplay between charge and ionic radius alone, as Lewis base back-bonding is precluded, hence sets the principal element for the relative Lewis acid activity

Table 11.1 Ionic radii of and ionic potential (z/r_{ion}) of selected metal ions.

	Na^+	Li^+	Ca^{2+}	Mg^{2+}	Sc^{3+}	Ce^{3+}
r_{ion} (Å)	1.16	0.90	1.14	0.86	0.885	1.15
z/r_{ion} (charge/Å)	0.86	1.11	1.75	2.33	3.33	2.61

Source: I. Persson 2010 [3].

of the metals in this review (Table 11.1) and also allows for a classification according to the hard-soft-acid-base (HSAB) principles: All s-block metal cations are considered hard Lewis acids, with Mg^{2+} being the hardest and Na^+ the softest in this chapter. For further details on the relative Lewis acidity of s-block metals, see Section 11.1.7.

A class of established and widely used Lewis acid catalysts that also bears high similarities to the reactivity of the s-block metal cations is the rare earth metal-based ones [4]. In analogy to the s-block metals, the rare earth metals also have (relatively) stable oxidation states, albeit of +3, and only inaccessible electron shells beneath a shell-like LUMO (4s, 6s AO). Their electron configuration thus also precludes participation in back-bonding and their cation ligand interaction is based solely on σ-donation from the ligand. As all of them have a +3 charge, Sc^{3+}, which has the smallest ionic radius, is the most active Lewis acid and has been used in most organic reactions. It may come with no surprise that Ca^{2+}, which is isoelectronic to Sc^{3+}, dominates s-block metal Lewis acid catalysis.

Based on these considerations, a prediction of the Lewis acid behavior of s-block metals is seemingly quite straightforward – if chemical reactions were run in the gas phase, with only one metal cation and one molecule of the substrate. In reality, the situation is much more complicated and a few of the most influential factors shall be discussed in the following.

11.1.2 Interactions with More than One Lewis Base

Depending on the size of the cation, s-block metals are able to accommodate four to eight Lewis bases in their first coordination sphere, with an average coordination number of CN = 4 for Li^+ to CN = 8 for Ca^{2+}. For main group metal cations in general, also for s-block metal cations, the Lewis base–metal interaction becomes increasingly weak with every new Lewis base being coordinated [5]. This is a consequence of the raising electron density at the metal and enhanced steric and electronic repulsion between the coordinated Lewis bases. For main group metals, this continues as a very linear trend until the first solvent shell is completed. In TM, this trend is more often than not very nonlinear because of back-bonding-modulated electronic variations and induced rehybridizations. For example, the bond dissociation energy for a second ligand may even exceed that of the first, when the first pays the cost for the induced rehybridization. Hence, although a proficient choice of a residual ligand might even enhance the activity of a TM Lewis acid, the s-block metal's Lewis acidity effectively decreases as more coordination sites are occupied.

Because of its importance for understanding all of the complexity of Lewis acid-catalyzed reactions, it shall just be mentioned, as it is beyond the scope of this review to be discussed in greater detail that the influence of the metal cation is by no means restricted to the first solvent shell. Even though the interaction is certainly strongest with ligands bound directly at the metal, it has been found that, e.g. for Ca^{2+}, the bond dissociation energy is markedly changed for up to 20 water molecules arranged in up to 3 solvent spheres [6]. This effect is most pronounced in the presence of polar protic molecules, such as water, alcohol, or nontertiary amines, as metal ions polarize and strengthen hydrogen bond networks in their immediate vicinity (see also Section 11.3.3). In the same regard, it should be emphasized that different ligands can of course interact among each other, particularly well so when spatially confined by their respective binding to the metal cation [7]. A favorable interaction, such as a hydrogen bridge, between an incoming and a residual ligand therefore might significantly increase its binding to the metal, and if the incoming ligand is the substrate, it might even be the interaction with the other ligand and not with the metal itself that is responsible for the observed catalytic activity.

Another aspect that is elemental to understanding mechanisms in Lewis acid-catalyzed reactions when more than one coordinating species is present is the exchange of one ligand for another [8]. A vacant coordination site in a cationic complex, in solution, is a very unlikely occurrence, as such a truly vacant site would be rapidly filled. Hence, for an efficient activation of the substrate, it must outcompete a range of other coordinating species, such as the counter anion, solvent molecules, by-products, and the product. Generally, when the ligand exchange rates become faster, the interaction with the metal cation becomes weaker. As no back-bonding further strengthens their interaction, s-block metals typically exchange first shell ligands within the picosecond scale, which is significantly faster than the rates observed for most TMs. This also means that a σ-donation, and thus the very activation of the substrate, of the same order of magnitude may coincide with a much faster exchange rate for an s-block metal than for a TM involved in back-bonding. This makes an s-block metal a much more efficient Lewis acid catalyst in the presence of coordinating species that compete with the substrate, as even if a heteroatom containing molecule (for example) coordinates to the s-block metal catalyst, the fast exchange for the desired interaction with the substrate restores the catalytic activity. This is particularly important in the presence of water molecules and will therefore be discussed in further detail below (see Section 11.1.6).

11.1.3 Counter Anions

Metal cations obviously always come along with a counter anion, which has a significant influence on the properties of the cation. The Lewis acidity of the cation directly correlates with the strength of the interactions between cation and counter anion. As in s-block metals, a strong interaction implies a strong σ-donation, the electron density at the metal is significantly reduced. Hence, the weaker the interaction between the cation and counter anion, the higher is the Lewis acidity of the cation. As a result, common salts with a strong interaction like NaCl do not provide catalytic activity. To enhance the activity, weakly or noncoordinating counter anions have been established as a widely used tool in Lewis acid catalysis [9]. In such a weakly coordinating anion, the negative charge is delocalized over a large part of the molecule so that the actual point charge at the binding site to the cation remains low. This effect is further enhanced, when steric hindrance spatially separates the anionic charge from the cation, a concept that has been taken to an extreme in frustrated Lewis pair (FLP) catalysis [10, 11]. Since the first examples of Li^+, Na^+, or K^+ perchlorates (ClO_4^-) in the late 1920s, different types of weakly coordinating counter anions have been developed such as perfluorinated phosphates (PF_6^-), and antimonates (SbF_6^-), triflates (TfO^-), triflimides (Tf_2N^-), or borates (BR_4^-).

A second aspect that can be achieved with the typically very large weakly coordinating counter anions is that the salts display enhanced solubility in solvents with lower dielectric constants [12], as lattice energies vary linearly with the inverse cube root of the formula unit volume. This allows for the solvation of the intact cation anion pair, and thus its accessibility to the substrate, without the saturation of the metal cation with strongly coordinating solvents.

Because of the poor solubility of their salts in organic solvents, and their less pronounced Lewis acidity, s-block metal catalysis would not be possible without these minimally interacting spectator anions. In many of the reactions described below, not only one but even two different ones are used to achieve optimal catalytic activity.

11.1.4 Solvation

Evidently, when a chemical reaction is run in solution, the interaction of the catalyst with the solvent molecules must be taken into account. Several of the commonly used solvents contain Lewis basic heteroatoms. Dimethylformamide (DMF), dimethyl sulfoxide (DMSO), CH_3CN, tetrahydrofuran (THF), and other ethers are therefore often referred to as "coordinating solvents." As stated above, the Lewis acidity of an s-block metal decreases with every coordination site being occupied; therefore, a saturation with coordinating solvents significantly reduces their activity, often to the point that none is left. It should also be noted that solvent molecules compete with other ligands. Therefore, if the solvent is a stronger Lewis base compared with the counter anion, the latter is replaced, because of the overwhelming ratio of solvent molecules to counter anions present. This effectively annihilates the benefit of weakly coordinating counter anions (see Section 11.1.3). Thus, noncoordinating solvents such as aromatic, aliphatic, or chlorinated hydrocarbons are used to maintain a high Lewis acidity, with the chlorinated hydrocarbons typically providing higher efficiency as their higher polarity increases the solubility of the metal salts (see also Section 11.1.5).

11.1.5 Solubility and Aggregation

Dissolving metal cation salts in organic solvents is generally achieved via one or a combination of the following approaches. A coordinating solvent is used to saturate the metal, thereby creating a hydrophobic shell around the metal and dissolving the salt as a solvent-separated ion pair (SSIP). Analogously, specifically tailored ligands can be coordinated with the metal to create the hydrophobic shell, a concept prevailing in TM catalysis. As ligands are not tightly bound in most cases, and saturation with strongly coordinating solvent molecules is undesired for maintaining a high Lewis acidity, the third possibility proved most efficient for s-block metal cations. As mentioned above (see Section 11.1.3), weakly coordinating counter anions can be used for the solvation of an intact cation anion pair, i.e. a contact ion pair (CIP). In this scenario, the salt dissolves as a polar solute instead of two charged species, and the solvent molecules are arranged in a hydrophobic shell around the entire polar solute. It should be noted that this type of solvation tends to be imperfect, and larger aggregates of oligomeric salts certainly prevail.

11.1.6 Water Tolerance

Traditional Lewis acidic catalysts, such as $AlCl_3$, $TiCl_4$, and BCl_3, promptly react with even trace amounts of water in the reaction medium, resulting in their decomposition to hydroxides or oxides and the immediate or gradual loss of their Lewis acidity. The formation of hydroxides, which always precedes further reaction to oxides, is initiated via the coordination of water to the metal. In that coordinated water molecule, the Lewis acidity of the metal enhances the polarization of one of the O—H bonds so that the water molecule is essentially turned into an acid. The Brønsted acidity of this metal–water complex is directly linked to the metal oxygen interaction, so that a stronger Lewis acid results in a more acidic complex. The metal hydroxide remains as the conjugated base, after the proton has been donated (see Eq. (11.1), Table 11.2). This process can be quantified and is reflected by the hydrolysis constant $\mathrm{p}K_\mathrm{h}$ [13]. The $\mathrm{p}K_\mathrm{h}$ is to be seen in relation to the $\mathrm{p}K_\mathrm{W} = 14$ of pure water, as it is, in principle, the corresponding value for water with the analyzed metal dissolved in it. Hence, the $\mathrm{p}K_\mathrm{h}$ is a good indicator for the stability of metal cations in water toward hydrolysis.

Apart from being stable, the Lewis acid must also be able to maintain catalytic activity in a strongly coordinating solvent such as water. This means that the substrate must be able to compete for a coordination site, on a Lewis acid that

Table 11.2 $\mathrm{p}K_\mathrm{h}$ values and water exchange rate constant (WERC) of selected metal ions.

$$x\ \mathrm{M}^{z+} + y\ \mathrm{H_2O} \rightleftharpoons \mathrm{M}_x(\mathrm{OH})_y^{(xz-y)+} + y\ \mathrm{H}^+ \qquad (11.1)$$

$$\mathrm{p}K_\mathrm{h} = -\log K_\mathrm{h}$$

$$\mathrm{M(H_2O)}_x^{\ y+} + \mathbf{H_2O} \underset{}{\overset{k\,=\,\mathrm{WERC}}{\rightleftharpoons}} \mathrm{M(H_2O)}_{x-1}(\mathbf{H_2O})^{y+} + \mathrm{H_2O} \qquad (11.2)$$

	Li^+	Na^+	Mg^{2+}	Ca^{2+}	Ni^{2+}	Sc^{3+}	Al^{3+}
$\mathrm{p}K_\mathrm{h}$	13.64	14.18	11.44	12.85	9.89	4.3	1.14
WERC ($M^{-1}\,s^{-1}$)	4.7×10^7	1.9×10^8	5.3×10^5	5.0×10^7	2.7×10^4	4.8×10^7	1.6×10^0

is saturated with water molecules. As already mentioned above, this is indeed possible if the exchange of inner sphere ligands is fast. A measurable value for the continuous exchange of water molecules is the water exchange rate constant WERC (see Eq. (11.2), Table 11.2) [13]. The first class of Lewis acid catalysts that was found to perform well in the presence of water, even as a solvent, are rare earth metal salts. Hence, rare earth metals are the protagonists of the only systematic study on water compatibility of Lewis acids in organic synthesis [14, 15]. In these studies and as major contributors to the development of rare earth metal catalysis in water, the *Kobayashi* group correlated the pK_h and WERC of group 1–15 metal chlorides, perchlorates, and triflates, with the outcome of an aldol reaction with acid-sensitive enol silyl ethers in aqueous medium. The threshold for the WERC was found to be at c. $3.2 \times 10^6\ M^{-1}\ s^{-1}$. Metals with lower exchange rates performed poorly, regardless of their Lewis acidity. The study also revealed that at pK_h values below 4, the Lewis acid decomposes and/or the high amount of acid generated initiates side reactions, whereas, at pK_h values above 10, even though the Lewis acid is stable, the authors of the study claim that its Lewis acidity is too weak to activate the aldehyde. Unfortunately, for pK_h values above 10, including the four alkaline and alkaline earth metals in this chapter, only metal chlorides were tested. Nevertheless, according to their pK_h and WERC, water compatibility should be even higher than for the rare earth metals, and all the reactions discussed below provide ample proof for their Lewis acid activity, when they are paired with a more suitable counter anion.

11.1.7 Relative Lewis Acid Activity of Alkaline and Alkaline Earth Metals

Given that the Brønsted acidity of a metal–water complex is directly linked to the metal oxygen interaction, and thus the Lewis acidity of the metal, it should be possible to quantitatively predict the precise activity of any given metal salt based on its pK_h value [13]. Even though, more often than not, experimental facts contradict such a theoretically deduced Lewis acid activity, because of the complex interplay of other influences in (nonaqueous) solution, a correlation of the pK_h with charge and ionic radius shall be used to give an estimation of the relative s-block metal's Lewis acidity compared to other Lewis acids. As shown in Figure 11.2a, the pK_h values indeed correlate quite well with the ionic potential (z/r_{ion}), for metals that are not involved in significant back-bonding [3], placing the small, highly Lewis acidic Al^{3+} in the upper right corner and the big, almost neutral K^+ in the lower left.

A second way to obtain quantitative data for the interaction of a Lewis acid is to look at the bond distance between the metal and a coordinated heteroatom [3]. The values for metal cations that are saturated with water molecules in Figure 11.2b have been measured by extended X-ray absorption fine structure (EXAFS), large-angle X-ray scattering (LAXS), and/or large-angle neutron scattering (LANS), which allows for the direct assignment of interatomic distances in solution. These data clearly show a much more pronounced interaction of alkali metal cations with coordinated water molecules than would be expected from their pK_h values.

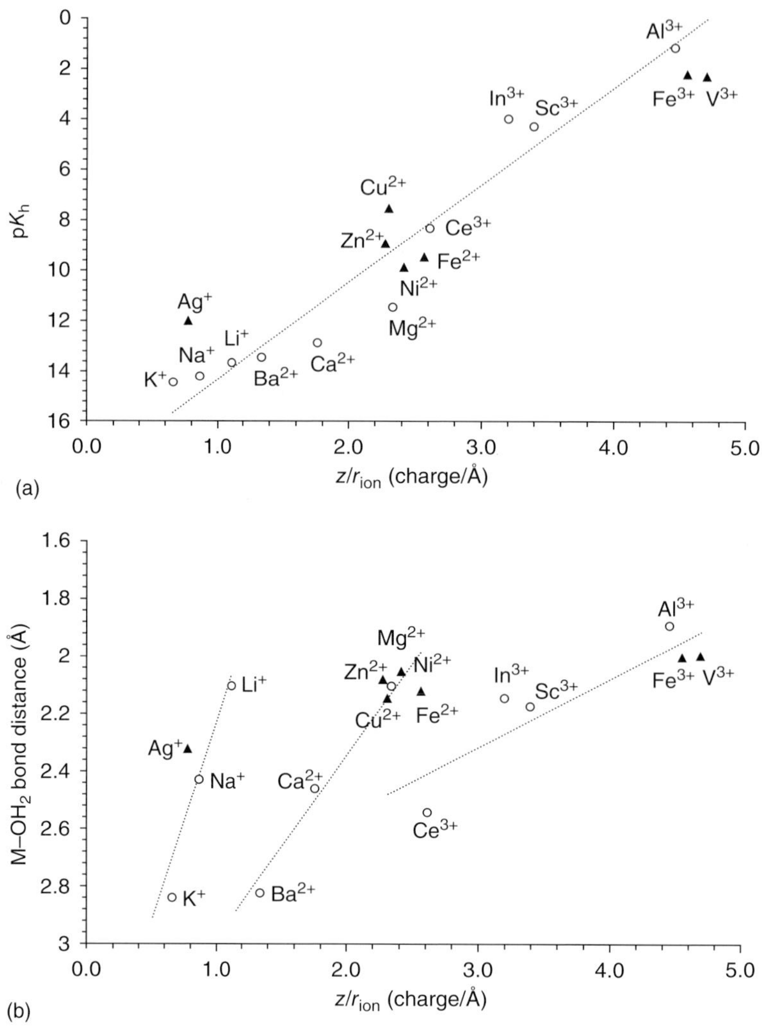

Figure 11.2 pK_h (a) and **M–OH**$_2$ distances (b) to ionic potential (z/r_{ion}) for selected metals. Source: I. Persson 2010 [3].

Where, on the spectrum of Lewis acidity, does this now leave our s-block metals? Na^+ and Li^+ with their +1 charge should be *very mild* Lewis acids, according to their pK_h. Nevertheless, the analysis of their bond distances clearly shows that under certain circumstances, their interaction with a Lewis base can be pushed into a medium range that equals that of a divalent Lewis acid such as Zn^{2+}. They might be used most efficiently for reactions that require just a little encouragement and are probably excellent at achieving selectivity for the activation of one Lewis base among others. Also, functional group tolerance is expected to be at its prime. Mg^{2+} and Ca^{2+} are in a medium range with Zn^{2+} or Ni^{2+}, according to both analyses. This should result in less sensitivity to the reaction conditions and a much broader substrate scope than for the $M^{(1)}$ metals, with good selectivities and functional group tolerance still to be expected. Their reactivity might be best referred to as *balanced*.

11.1.8 Hidden Brønsted Acid

The data discussed in the previous paragraphs also brings to the fore that as soon as both, any metal cation and water molecules, even in adventitious amounts, are present in a reaction mixture, the likelihood that protons are generated is eminent, a tendency that increases with increased Lewis acidity. This certainly is not even limited to water molecules, but may happen with any molecule containing a heteroatom that bears a hydrogen atom. Being a type of Lewis acid themselves, according to the definition above, protons are able to provide the same catalytic reactivity as metal cation Lewis acids. Hence, the question often arises whether they and not the metal cation are the true catalytic species of a reaction, a phenomenon generally referred to as hidden Brønsted acid catalysis [16]. As the generation of protons is inherent to Lewis acid catalysis, it is overwhelmingly difficult to distinguish one from the other, and some ambiguity remains for every single Lewis acid catalyzed reaction. As an attempt at differentiation, the addition of 2,6-di-*tert*-butyl-4-methylpyridine (DTBMP) to the reaction has been promoted [17]. DTBMP is said to be able to scavenge protons but presumably does not coordinate with metals, for the steric repulsion by the two *tert*-butyl groups would prevent the approach of the metal to the Lewis basic nitrogen atom. Nevertheless, results are often inconclusive, as in merely slowing down the conversion, which may be due to the DTBMP's altering of some part in the catalytic system, such as a change in counter anion composition or degree of coordination. Alternatively, and more likely, the generated protons and the Lewis acid cooperate to achieve optimal catalytic performance and the, thus far elusive, true catalytic species is a more complex entity altogether. Obviously, the only way to prevent hidden Brønsted acids is to rigorously exclude water from the reaction.

As little is known about any "true catalytic species" in solution, one might argue that if a *catalytic system as a whole* provides a reactivity or selectivity in a reaction that cannot be achieved via the simple addition of an acid, it is a valuable addition to the organic chemist's toolbox. If the "new" catalyst is simply a replacement for a previously known acid, with no apparent benefit also regarding functional group compatibility, catalyst handling, and cost efficiency, it might not be worthwhile.

11.2 Polarized Carbon–Heteroatom Double Bonds

Polarized carbon–heteroatom double bonds such as carbonyl, imines, or, rarely, thiocarbonyl groups are classical substrates for Lewis acid-catalyzed reactions. Typically, electron density is withdrawn by the coordination of one of the heteroatom's lone pairs to the Lewis acid. Therefore, the polarity of the double bond is enhanced and its natural reactivity is amplified. Thus, nucleophile addition to the carbon atom is facilitated. Alternatively, in the presence of α-acidic H-atoms, enolization and subsequent addition to electrophiles is more efficient. Early main group metal-based Lewis acids have been successfully used to activate polarized carbon–heteroatom double bonds.

11.2.1 Carboxylates: Anhydrides and Carbonates

Anhydrides are often used as relatively mild transfer reagents for carboxyl groups. This is often warranted for in the realm of protection group chemistry. For the introduction of a carboxyl protection group to an alcohol or amine, the anhydride is typically combined with a base to enhance the alcohol's or amine's nucleophilicity. If basic conditions are incompatible or undesired, a Lewis acid activation of the anhydride might present a viable alternative. It is of particular importance that the Lewis acid chosen for this task is at the same time mild enough to leave the newly formed carboxylated product intact, while still efficiently enhancing the anhydrides reactivity, an ideal playground for the mildly Lewis acidic early main group metals.

It was indeed shown that in the presence of 20–30 mol% LiOTf, a variety of primary, allylic, benzylic, hindered, and unhindered secondary and even sterically congested tertiary alcohols were acetylated with acetic anhydride in good to excellent yields under essentially neutral conditions, at room temperature (Scheme 11.1a) [18]. The same procedure also proved highly efficient for the protection of various aldehydes as 1,1-diacetates, a product that is even more sensitive under Lewis acidic conditions (Scheme 11.1b) [18].

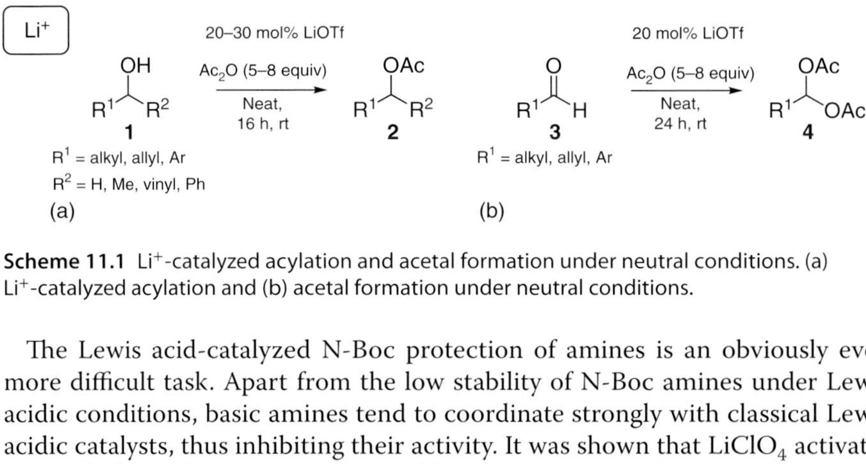

Scheme 11.1 Li^+-catalyzed acylation and acetal formation under neutral conditions. (a) Li^+-catalyzed acylation and (b) acetal formation under neutral conditions.

The Lewis acid-catalyzed N-Boc protection of amines is an obviously even more difficult task. Apart from the low stability of N-Boc amines under Lewis acidic conditions, basic amines tend to coordinate strongly with classical Lewis acidic catalysts, thus inhibiting their activity. It was shown that $LiClO_4$ activates Boc-anhydride for the protection of less reactive secondary or aryl amines, which cannot be protected in the absence of a catalyst (Scheme 11.2) [19].

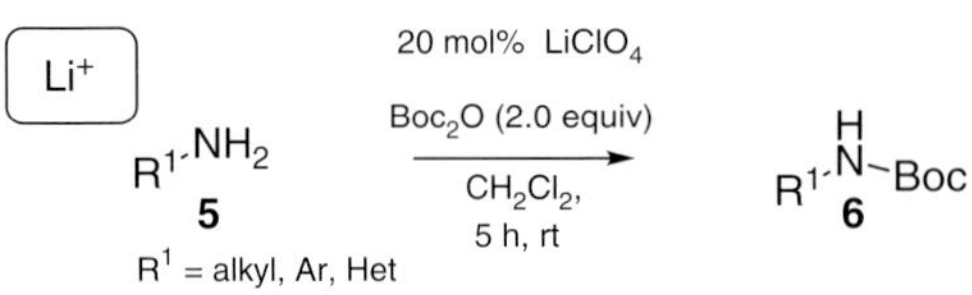

Scheme 11.2 Li^+-catalyzed Boc-protection under neutral conditions.

A closely related reaction is the formation of mixed aryl alkyl carbonates. These compounds are usually made by carbonate interchange. This equilibrium reaction proceeds to displace less nucleophilic alcohols with more nucleophilic ones. Consequently, the formation of aryl alkyl carbonates from inexpensive alkyl carbonates, such as diethyl carbonate, is unlikely to occur.

As a solution to this problem and as a viable alternative to carbonate exchange, a $Mg(ClO_4)_2$-catalyzed procedure was devised, reacting phenols with diethyl dicarbonate under base-free conditions (Scheme 11.3) [20].

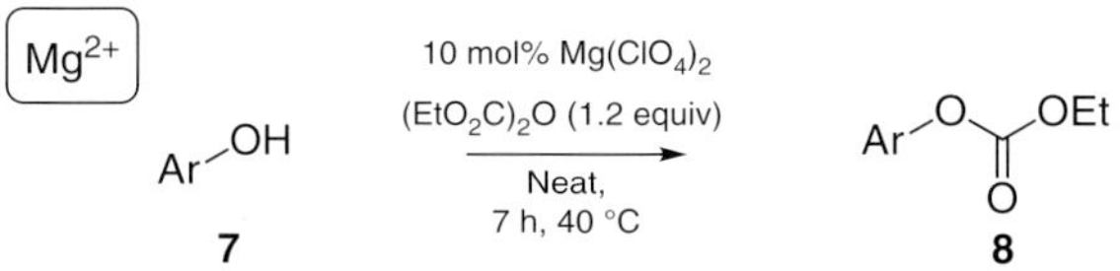

Scheme 11.3 Mg^{2+}-catalyzed formation of mixed aryl alkyl carbonates.

11.2.2 Aldehydes, Ketones, and Formates

In analogy to the reactions presented in the previous paragraph, the formation of acetals from aldehydes or ketones requires a Lewis acidic catalyst with a well-balanced reactivity, as the product is again highly sensitive under the reaction conditions. The range of potential catalysts is obviously even more restricted, when the synthesis of a thioacetal is envisioned, as the catalyst must retain its activity in the presence of thiols. The proficiency of a LiOTf catalyst was demonstrated for this task, as typical for this type of catalyst, under essentially neutral reaction conditions (Scheme 11.4) [21, 22].

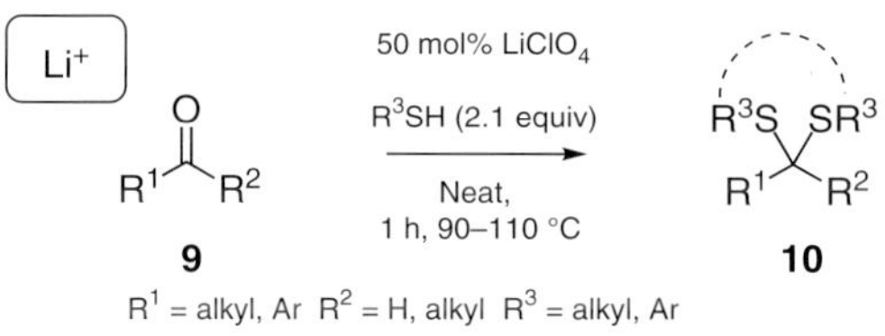

Scheme 11.4 Li^{+}-catalyzed formation of thioacetals.

The equilibrium of this type of reaction was shifted to the opposite side by an aqueous medium, in the presence of a $NaBAr^F{}_4$ catalyst (Scheme 11.5) [23]. The reaction tolerated various functional groups such as amines, alcohols, or even a TBDMS (*tert*butyldimethylsilyl) ether.

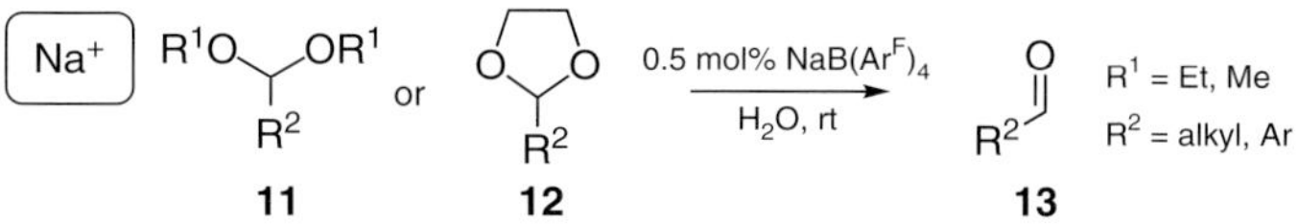

Scheme 11.5 Na^{+}-catalyzed deprotection of acetals.

An interesting example showcasing the similarity of the s-block metal Lewis acids to the more prominent lanthanide-based catalysts is the Friedel–Crafts reaction of aldehydes with indols in water (Scheme 11.6) [24]. With an appropriate counter anion, a range of different alkali and alkaline earth metal Lewis acids, among which $NaBAr^F_4$ was most competent, efficiently activated the aldehyde **15** for the addition of indole nucleophiles **14**. The activity of the catalyst was not only retained in the aqueous medium but the presence of water and the formation of colloidal particles were even found crucial for the reaction outcome. The authors claim that the BAr^F_4-anion plays two roles in this reaction: as a counter anion and as a surfactant.

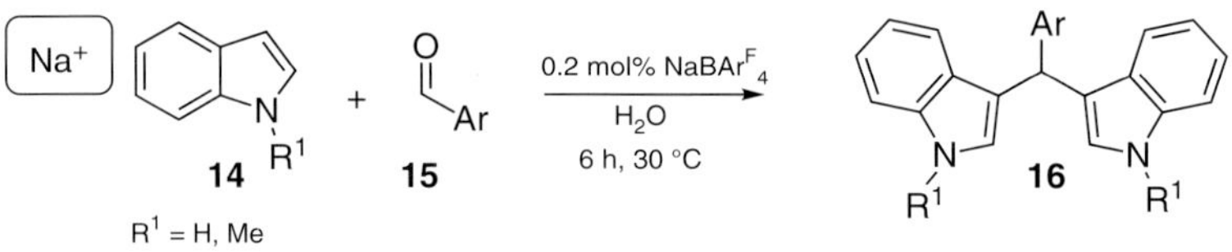

Scheme 11.6 Na^+-promoted Friedel–Crafts reaction.

That the activity of early main group metal Lewis acids is not only mild but can also indeed be quite powerful was testified to in the following two examples of carbonyl activation. Predictably, the more active alkaline earth metal Lewis acids were chosen for these reactions. The difficult Knoevenagel condensation of poorly reactive β-diketones **17** with aldehydes **18** was enabled at room temperature by a $Mg(ClO_4)_2$ catalyst (Scheme 11.7) [25].

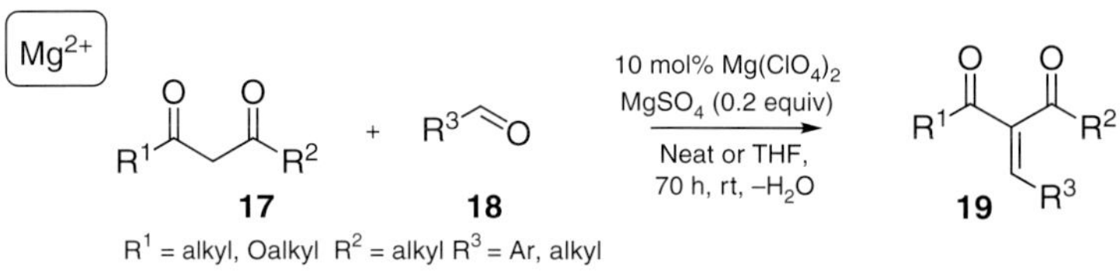

Scheme 11.7 Mg^{2+}-catalyzed Knoevenagel condensation of poorly reactive β-diketones **17**.

In this reaction, the catalyst enhances both the enolization of β-diketones **17** as well as the electrophilicity of the aldehyde.

Alkyl formates are an inexpensive and widely available feedstock, and thus an ideal, nontoxic, and noncorrosive substitute for DMF or formic acid. The formylation of poorly nucleophilic anilines is particularly difficult and was found to be highly efficient in the presence of a $Ca(NTf_2)_2$ catalyst (Scheme 11.8) [26].

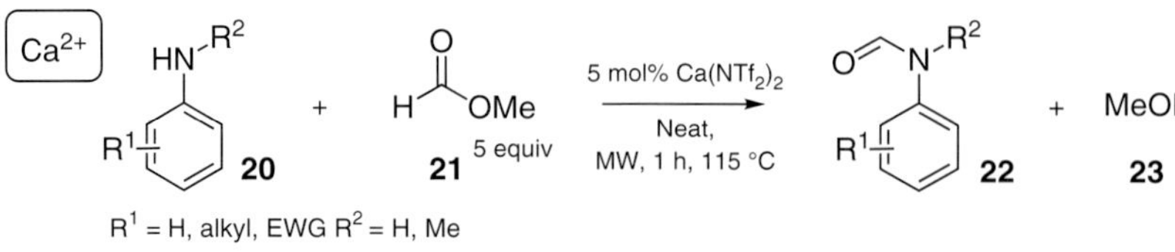

Scheme 11.8 Ca^{2+}-catalyzed formylation of anilines with methyl formate.

11.2.3 α,β-Unsaturated Carbonyl Compounds

Not only the reaction of simple C=O double bonds but also the reactivity of α,β-unsaturated carbonyl compounds can be efficiently amplified by Lewis acidic catalysts. As nucleophiles can react via 1,2- as well as 1,4-addition, the catalyst must selectively guide the reaction toward one or the other.

A selective 1,4-addition of α,β-unsaturated diesters **25** to indoles **24** has been achieved with a range of Ca^{2+} or Mg^{2+} catalysts among which MgI_2 performed best (Scheme 11.9) [27].

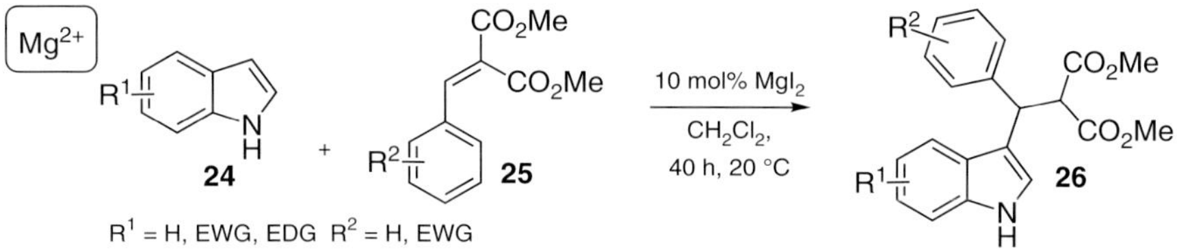

Scheme 11.9 Mg^{2+}-promoted selective 1,4-addition of indoles **24** to α,β-unsaturated esters **25**.

A second example for a selective 1,4-addition in alkaline earth metal catalysis is the synthesis of pyranocoumarins **27** and tetrahydrochromenones **31** (Scheme 11.10) [28].

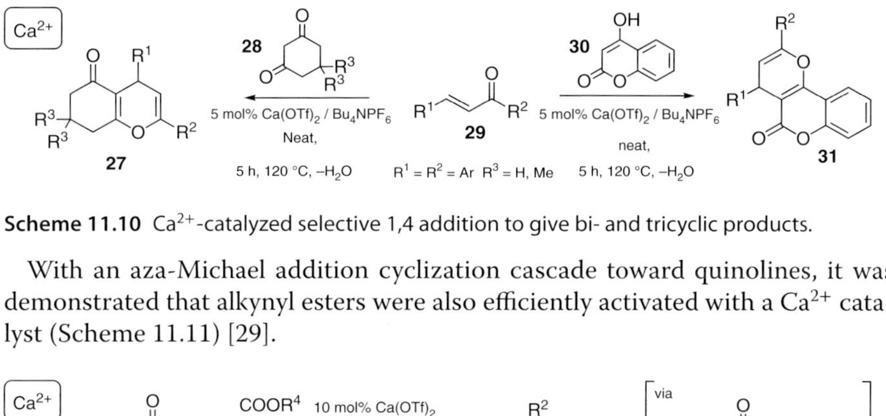

Scheme 11.10 Ca^{2+}-catalyzed selective 1,4 addition to give bi- and tricyclic products.

With an aza-Michael addition cyclization cascade toward quinolines, it was demonstrated that alkynyl esters were also efficiently activated with a Ca^{2+} catalyst (Scheme 11.11) [29].

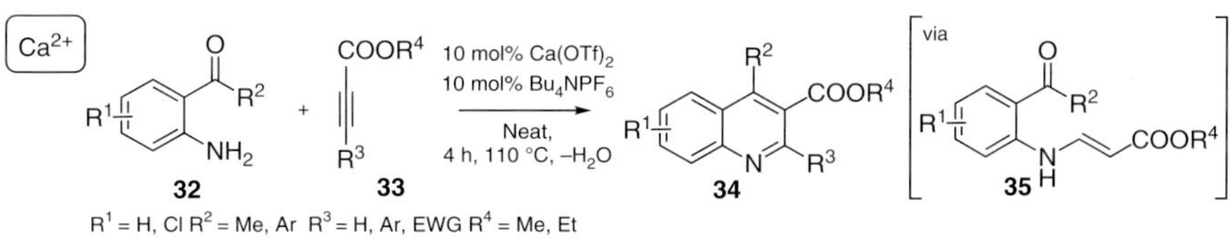

Scheme 11.11 Ca^{2+}-catalyzed Michael addition to alkenyl esters in a quinoline synthesis.

Besides a Sc^{3+} catalyst, Mg^{2+} was found most competent for the activation of α-nitroacrylate esters **36** for an intramolecular 1,5-H-shift, followed by a Mannich cyclization. Remarkably, the reaction was carried out in CH_3CN, a coordinating solvent. As hidden acid formation is unlikely – protic acids did not catalyze the reaction – the high temperature (80 °C) might be crucial for a fast ligand exchange at the Mg^{2+}, restoring catalytic activity (Scheme 11.12) [30].

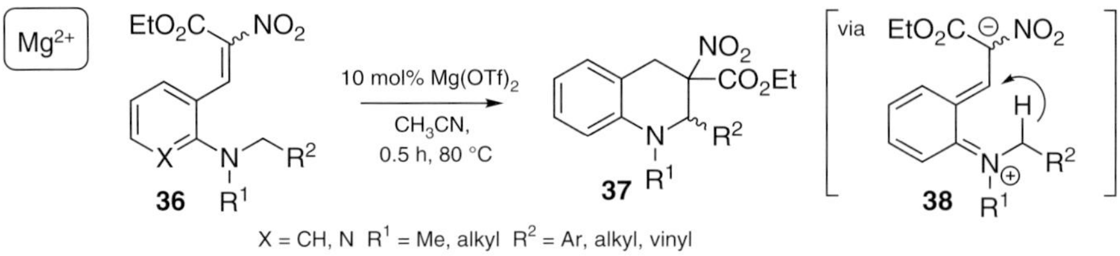

Scheme 11.12 Mg^{2+}-catalyzed 1,5-H-shift – Mannich cyclization cascade.

Even though the 1,4-addition selectivity is governed by the organocatalyst, it is noteworthy that a $LiClO_4$ catalyst performs well in the presence of an amine base, water, and a phosphonium ylide, which is remarkable for a Lewis acid (Scheme 11.13) [31]. The authors state that $LiClO_4$ is used to enhance both the addition and release of the organocatalyst to the substrate **40** and the subsequent addition of the phosphonium ylide to the aldehyde in **43**.

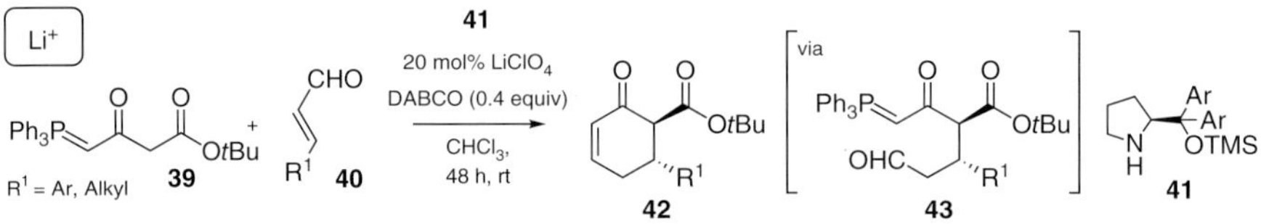

Scheme 11.13 Li^{+}-accelerated, organocatalyzed 1,4 addition – cyclization cascade.

Based on the hypothesis that alkaline earth metal Lewis acid catalysts are highly efficient binders to 1,3 dicarbonyl compounds, a Nazarov cyclization was efficiently catalyzed by a calcium-based catalyst system (Scheme 11.14) [32].

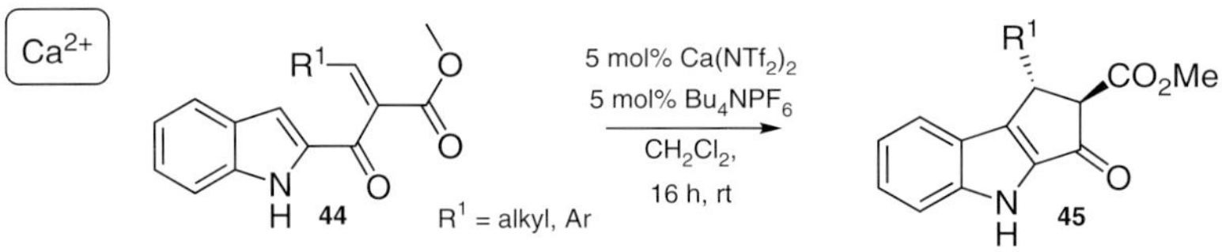

Scheme 11.14 Ca^{2+}-catalyzed Nazarov cyclization.

This study, which is also one of the very few mechanistic studies on s-block metal catalytic systems revealed that the substrate is bound to not only one but two calcium cations for an efficient activation.

11.2.4 Imines and Enamines

Because of their inherent instable nature under Lewis acidic conditions, imines are, as much as the acetals, carbonates and esters covered above, a class of molecules that demand for activation by a balanced Lewis acid catalyst.

Predictably, a Li^{+} catalyst is ideally suited for a Strecker synthesis (Scheme 11.15) [33].

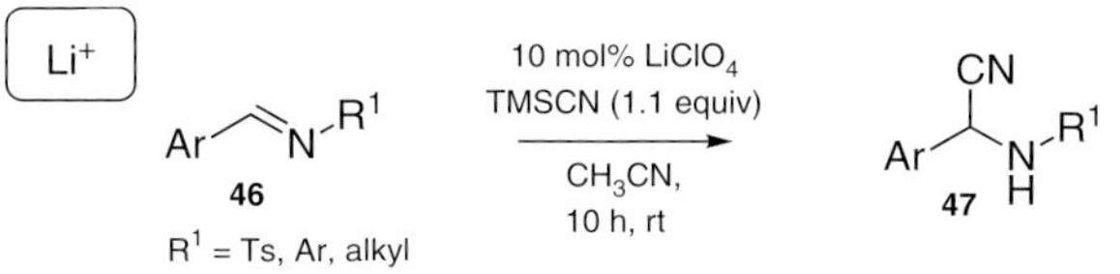

Scheme 11.15 Li^+-catalyzed Strecker synthesis.

In situ imine formation is by far the better option, not only for the step economy but also because the isolation of imines, especially aliphatic ones can be cumbersome. As the condensation of a carbonyl compound with an amine comes along with the formation of water molecules, such a process calls for a water-tolerant Lewis acid catalyst.

Calcium hexafluoroisopropoxide (i.e. the deprotonated HFIP) proved an excellent choice, and Pictet–Spengler reactions directly provide heterocycles **50** from ketones **49** and amines **48** (Scheme 11.16) [34].

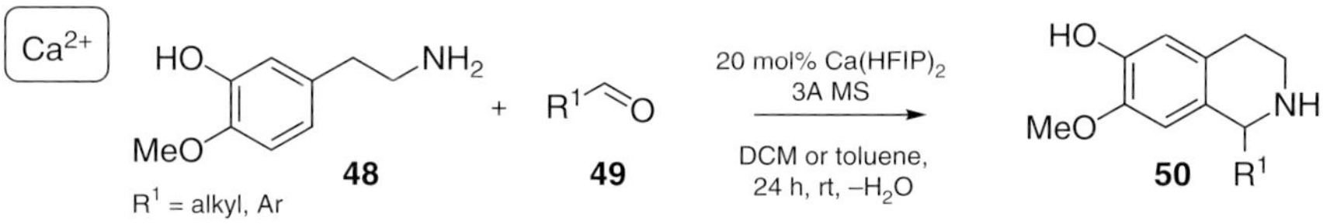

Scheme 11.16 Ca^{2+}-induced Pictet–Spengler reaction.

Analogously, a $Mg(ClO_4)_2$ catalyst was competent for the *in situ* condensation of β-keto esters **51** with amines **52** to the enamines **55**, as well as the subsequent double Michael addition to the alkynyl ester **53** and the ring closure to dihydropyridines **54** (Scheme 11.17) [35].

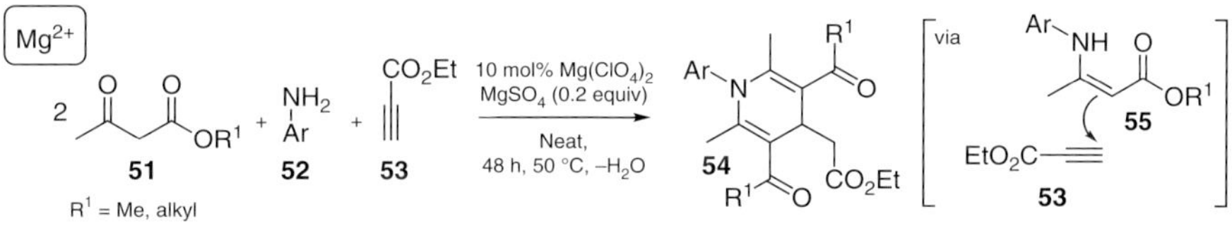

Scheme 11.17 Mg^{2+}-promoted amine-condensation – double Michael addition.

Further extending this concept, a $Ca(OTf)_2$-catalyzed condensation to an analogous enamine **60** initiates a Michael addition – aldol condensation cascade under oxidative conditions, yielding anilines **59** in a single synthetic operation (Scheme 11.18) [36].

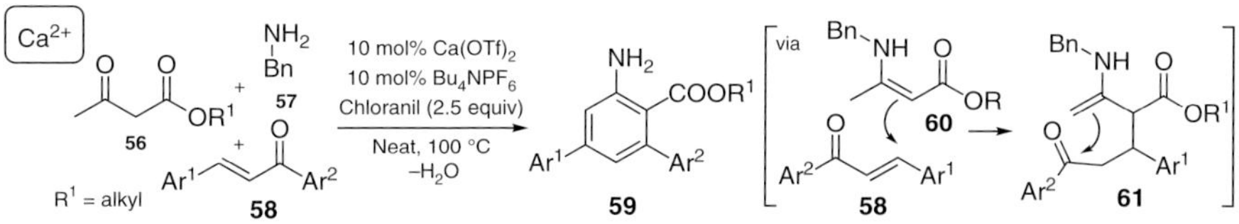

Scheme 11.18 Ca^{2+}-catalyzed Michael addition – aldol cascade under oxidative conditions.

11.2.5 Mannich Reactions

The first ever reported reaction of a water-tolerant s-block metal Lewis acidic catalyst is also a condensation of an amine to an aldehyde as part of a classical Mannich reaction. For this reaction, run in water as a reaction medium, the very mild Lewis acidity of NaOTf proved highly beneficial, as it did not initialize hydrolysis of the silyl enol ether **64**. Kinetic measurements revealed that the activity of the sodium catalyst is enhanced in the presence of the β-amino ketone product **65** (Scheme 11.19a) [37].

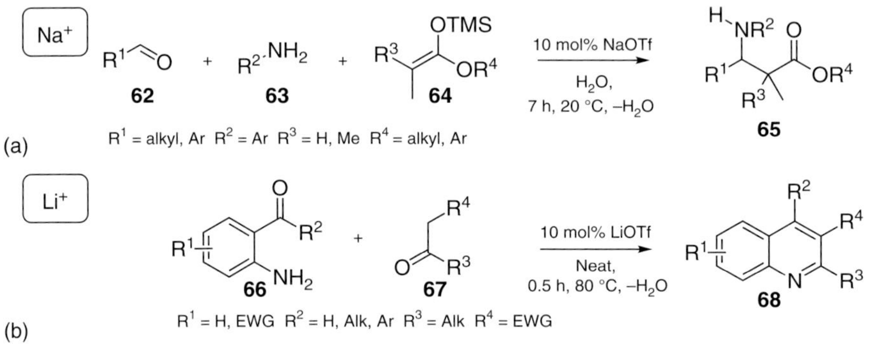

Scheme 11.19 (a) Na^+-catalyzed Mannich reaction in water and (b) Li^+-catalyzed Friedlander reaction.

The reaction was extended to also include an *in situ* enolization of the imine's reaction partner with a $NaBAr^F{}_4$ catalyst (not shown) [38].

As an intramolecular variation, a Li^+-catalyzed Friedlander reaction is a straightforward approach toward highly functionalized quinoline derivatives **68** (Scheme 11.19b) [39].

11.2.6 Oxidation and Reduction

The use of (aqueous) H_2O_2 as a more benign oxidant, replacing the traditionally used percarboxylic acids in Bayer–Villinger oxidations is highly desirable. This obviously calls for a Lewis acidic catalyst, which at the same time is water tolerant, and mild enough to leave the lactone products of this reaction intact. Acting presumably as a Lewis acid-activated acid, a combination of Li^+ or Ca^{2+} with oxalic acid was up to the challenge (Scheme 11.20a) [40].

The Luche reduction of α,β-unsaturated carbonyl compounds requires the guidance of a Lewis acid to discriminate a 1,4- in favor of a 1,2-addition of a hydride donor. As a replacement of the traditionally used Ce^{3+} salts, Ca^{2+} proved highly effective (Scheme 11.20b) [41].

The same reaction conditions were also applied to the selective reduction of α,β-epoxy ketones [42].

11.2.7 Donor–Acceptor Cyclopropanes

A further class of molecules that largely benefits from a Lewis acid enhancement of their reactivity are donor–acceptor (DA) cyclopropanes. For the ring opening

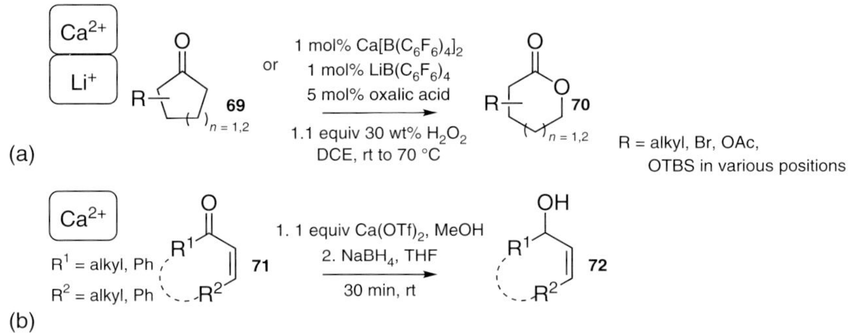

Scheme 11.20 (a) Ca^{2+}- or Li^{+}-catalyzed oxidation and (b) Ca^{2+}-mediated Luche reduction.

of these molecules, a powerful Lewis acid catalyst is required if the donor ability of the donor moiety is low. Also, as the stabilization of the cation in the CIP, generated upon the fission of the cyclopropane, is much lower than for the heteroatom-based traditional DA cyclopropanes, too strong a Lewis acid easily promotes oligomerization side reactions.

Predictably, the more active alkaline earth metal-based Lewis acidic catalysts and $Ca(NTf_2)_2$ in particular were found efficient in this regard. Thus, an aryl-substituted DA cyclopropane **73** was added to indoles **74** in a Friedel–Crafts-type reaction. Remarkably, this reaction is carried out in 2-MeTHF, a coordinating solvent, with no additive (Scheme 11.21a) [43].

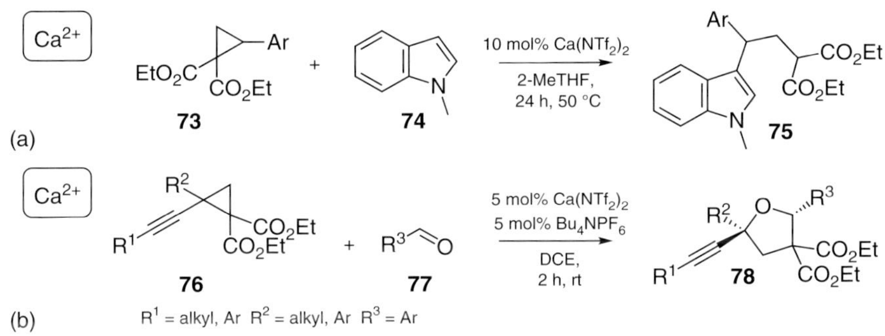

Scheme 11.21 Ca^{2+}-catalyzed reaction of DA cyclopropanes with indoles (a) or with aldehydes (b).

The original development of this type of activation capitalizes of the more classical $Ca(NTf_2)_2/Bu_4NPF_6$ combination and allows for the activation of DA cyclopropanes with even weaker alkynyl donors (Scheme 11.21b) [44].

11.2.8 Diels–Alder Reaction and Cycloaddition

Another important C—C bond forming reaction, which is efficiently promoted by a Lewis acidic catalyst, is the Diels–Alder reaction.

In the only example of thiocarbonyl activation, $LiClO_4$ was used to promote the cycloaddition of thiobenzophenones **79** with acetylenedicarboxylates **80**

(Scheme 11.22a) [45], based on a previous report that Li^+ catalysts perform well for this type of reaction (not shown) [46].

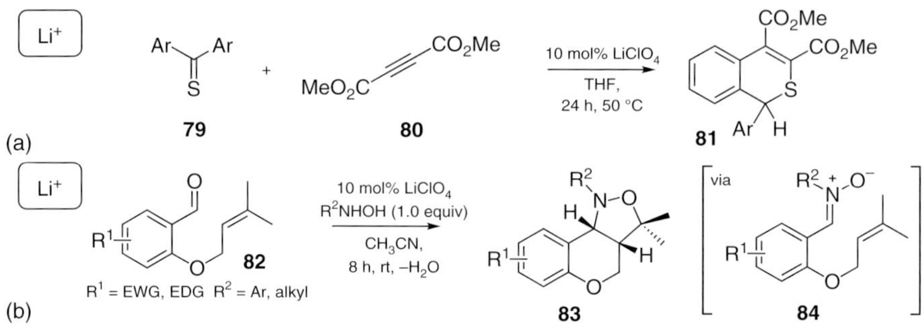

Scheme 11.22 Li^+-promoted Diels–Alder reaction (a) and 1,3-dipolar cycloaddition (b).

In line with these results, a range of other alkaline metals also catalyzed a hetero-Diels–Alder reaction of Danishefsky dienes with imines in water (not shown) [47].

Finally, $LiClO_4$ was found an excellent catalyst for an *in situ* condensation 1,3-dipolar cycloaddition sequence via nitrones (Scheme 11.22b) [48].

11.3 Activation of Polarized Single Bonds

Because of the difference in electronegativity, carbon heteroatom bonds are typically polarized. The coordination of a Lewis acid with the heteroatom increases the polarization and enhances its natural reactivity. This is typically called for if a cleavage of the bond is desired in the envisioned synthetic operation.

11.3.1 Opening of Three-Membered Heterocycles

In small heterocyclic ring systems, the formation of a Lewis acid base pair between the heteroatom and the catalyst further weakens the strained heteroatom–carbon bond and accelerates the ring cleavage process. The nucleophilic attack at a thus activated epoxide or aziridine provides interesting 1,2-difunctionalized compounds. Most nucleophiles preferably attack at the site of lesser steric congestion, though electronics, which may be modulated by a Lewis acidic catalyst, are able to invert this preference.

Stoichiometric s-block metal Lewis acid mediated opening reactions of epoxides have been known for centuries. However, catalytic applications just arose with the development of weakly coordinating anions. Remarkably, even though it is known that aminolysis selectively occurs in the benzylic position with most stronger Lewis acids, it was shown that this preference is partly (LiOTf) [49] and almost completely reversed ($LiNTf_2$) [50] in the presence of a mildly Lewis acidic Li^+ catalyst (Scheme 11.23). The same catalyst also reversed the regioselectivity of aziridine aminolysis, albeit less efficiently.

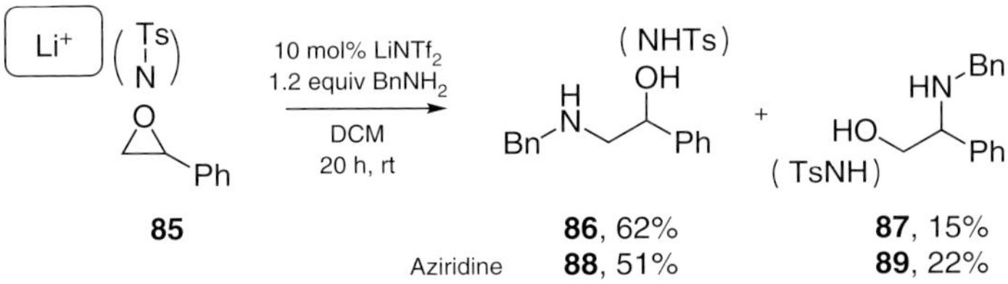

Scheme 11.23 Regioselectivity of $LiNTf_2$-catalyzed aminolysis, against the electronic bias.

For the epoxide opening with an isocyanide, as part of a Passerini-type multicomponent reaction (MCR), the selectivity for a LiOTf catalyst, in PEG-400, a polyethylene glycol, was back to the traditionally observed one, presumably because of the higher reaction temperature of 60 °C. Notably, the LiOTf catalyst outperformed $Sc(OTf)_3$, $Zn(OTf)_2$, and even $BF_3{\cdot}OEt_2$ in this unusual reaction medium, while acids gave no conversion (Scheme 11.24) [51].

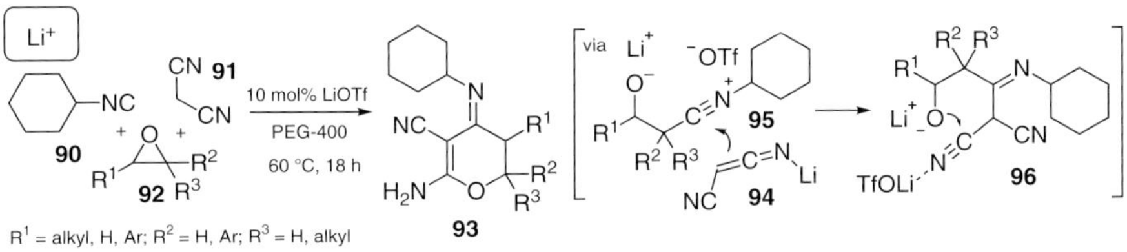

Scheme 11.24 Li^+-catalyzed MCR of epoxides, isocyanides, and nitriles in PEG-400.

Analogously run at higher temperatures of 80 °C, a ring-opening cyclization that was initiated by a Friedel–Crafts-type addition of a 2-vinylindole to an aziridine under LiOTf catalysis also occurred with the traditional selectivity for the benzylic position (not shown) [52].

11.3.2 Leaving Groups

Nucleophilic substitutions at sp^3-carbon atoms are among the most fundamental transformations in organic synthesis. In many cases, Lewis acids are used to promote these reactions' efficiency or selectivity via the activation of the leaving group (LG) (Scheme 11.25).

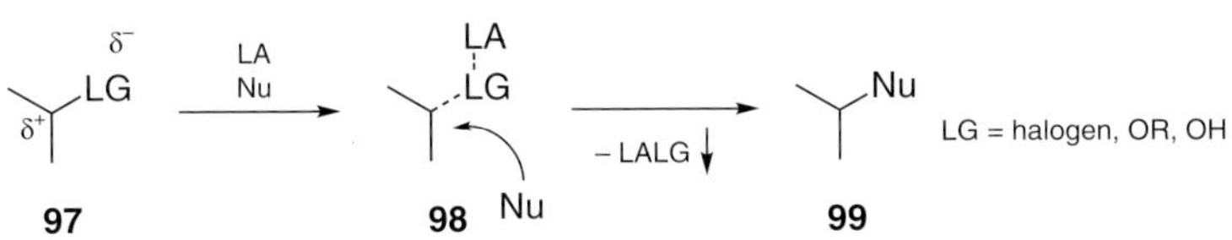

Scheme 11.25 Lewis acid-promoted substitution reaction.

The activation of the LG by the catalyst in this reaction type is twofold. The coordination with the LG weakens the LG—C bond. Equally important is the contribution of the nature of the departed LA–LG adduct. If this species is poorly soluble in the reaction medium, which is generally the case for s-block metal LG adducts, the equilibrium of the reaction is shifted by the removal of this by-product. In the reactions of this chapter, in which the s-block metal is used only in catalytic amounts, the challenge lies with the regeneration of the catalyst for turnover.

The Friedel–Crafts alkylation with benzyl halides or mesylates commonly requires strong Lewis acids such as Al^{3+} or Fe^{3+} catalysts. However, the combination of one of the most weakly coordinating anions, $B(C_6F_5)_4^-$, and the absence of a coordinating solvent boosted the activity of a Li^+ catalyst sufficiently to allow for the conversion of the more reactive mesylates **102** at room temperature in three hours. For benzyl chlorides, **100** a higher temperature and the addition of MgO as an HCl scavenger were found necessary (Scheme 11.26) [53].

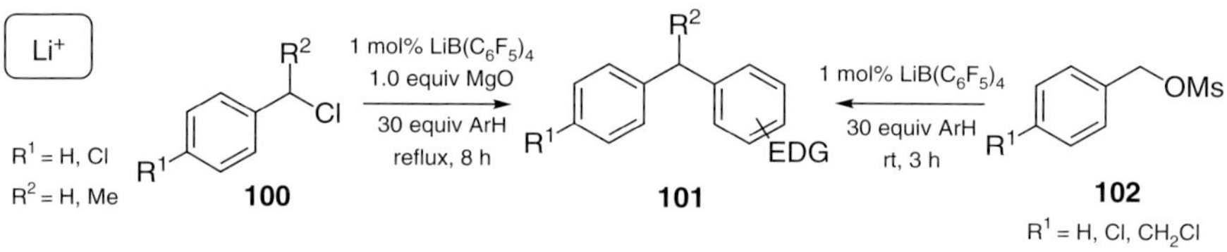

Scheme 11.26 Li^+-catalyzed Friedel–Crafts alkylation with benzyl halides and mesylates.

Even though the Friedel–Crafts sulfonylation with Tos-Cl can be promoted by a range of different Lewis and protic acids, the essentially neutral reaction conditions allowed for by a $NaClO_4$ catalyst might prove beneficial in the presence of acid-sensitive groups [54]. The bis-arylation of sulfonylchloride was achieved under analogous conditions (Scheme 11.27) [55].

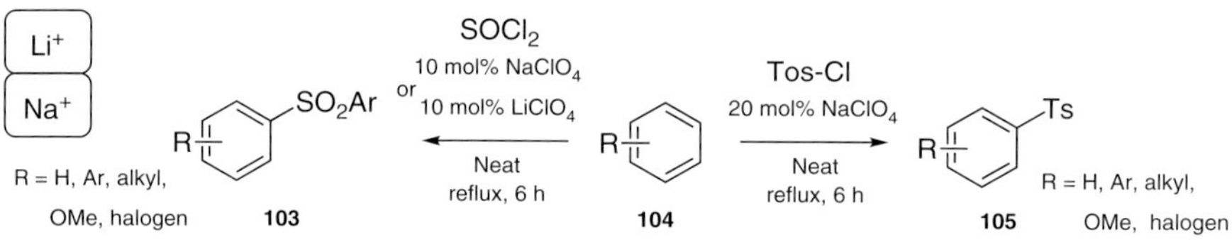

Scheme 11.27 Na^+- or Li^+-catalyzed Friedel–Crafts sulfonylation under neutral conditions.

Hexadimethylsilazane (HMDS), a readily available and inexpensive reagent, is an ideal source of trimethylsilyl (TMS) groups for the protection of alcohols. However, HMDS as a weak TMS-donor demands for further activation. Lewis acidic complexation of the amine moiety weakens the nitrogen–silicon bond for a transfer of both silyl groups to the alcohols oxygen moiety.

Early main group metal catalysts such as $LiClO_4$ [56] or $Mg(OTf)_2$ [57] proved to be able to activate HMDS **107**, or more likely the alcohol **110** for this transformation (Scheme 11.28). Comparison of the reaction conditions of both protocols nicely showcases the difference in activity between those two catalyst systems.

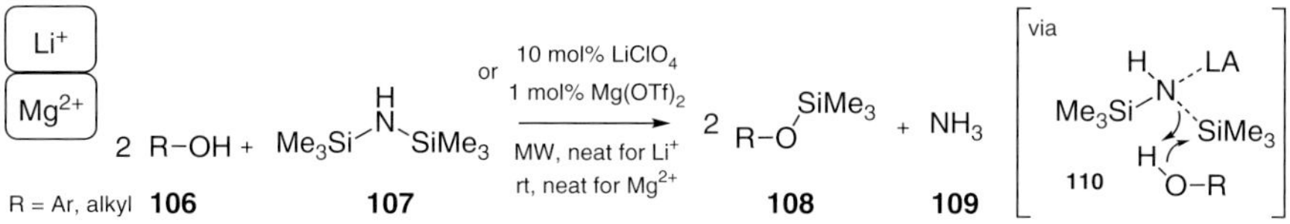

Scheme 11.28 Rate difference in Li^{+}- or Mg^{2+}-catalyzed activations of HMDS.

11.3.3 Ca^{2+}-Catalyzed Dehydroxylation as a Special Case

Alcohols are readily available and their conversion into other functional groups highly desirable. Because of the poor LG ability of the hydroxyl anion, transformation into halides, carboxylates, carbonates, or phosphates is a commonly necessary deviation. Even though the direct conversion of alcohols was successful with a still severely limited number of (Lewis) acid and TM catalysts, reactions often demand for harsh reaction conditions [58, 59].

As a unique feature of an early main group metal Lewis acid, that is unparalleled by any other catalyst thus far known, Ca^{2+} proved a high proficiency for a dehydroxylation of alcohols at room temperature, under almost neutral reaction conditions, in many instances. Consequently, this body of work constitutes the largest and most systematically investigated part of all s-block metal-catalyzed reactions.

A potential reason for this outstanding reactivity may be found in the combination of the following features:

1. A well-balanced Lewis acidity, which is at the same time strong enough for an efficient polarization of the C—O bond, but not too strong to preclude a rapid exchange of alcohol molecules, so that the catalyst activity is retained.
2. The poor solubility of calcium hydroxides (LA–LG) in organic solvents that enhances hydroxyl group cleavage by shifting the equilibrium according to Le Chatelier's principle.
3. The relatively high hydrolysis constant of $pK_h = 12.85$ that allows for a straightforward protonation of hydroxides bound to calcium, thus restoring the catalytic activity.

Our current understanding of the catalytic cycle is summarized in Scheme 11.29. The coordination of an alcohol **111** results in the polarization of the C—O bond. The cleavage of that bond is certainly further assisted by a hydrogen bridge from a fellow coordinated alcohol or water molecule, as the exclusive coordination of only a single oxygen with a Ca^{2+} cation in a reaction medium that contains an increasing amount of water, is very unlikely. This hydrogen bridge is much stronger than in bulk water/alcohol as the coordination with the calcium enhances the acceptor ability of the proton by several orders of magnitude. Upon C–O cleavage, the cation **113** is formed alongside the poorly soluble calcium hydroxide species **114**. The reaction of the cation with the nucleophile generates protons, which slowly react with the crashed out calcium hydroxides to regenerate the catalyst for the next turnover.

As the immediate products generated by the catalyst are carbocations, the restrictions inherent to carbocation formation dictate the substrate

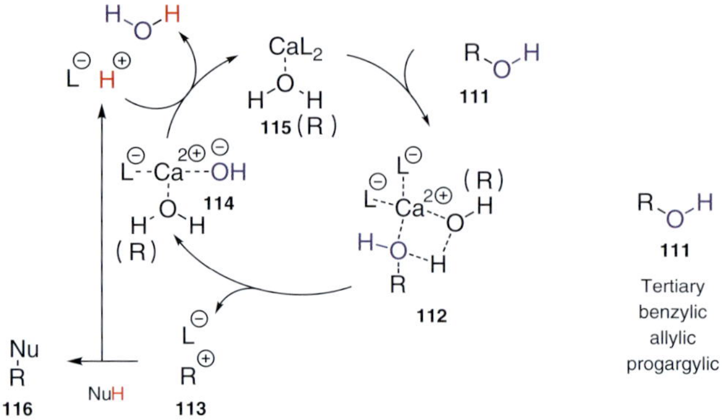

Scheme 11.29 Proposed mechanism for the Ca^{2+}-catalyzed dehydroxylation of alcohols.

scope of this reaction. It applies well, with decreasing reactivity to benzylic < allylic < propargylic = tertiary alkyl alcohols. The calcium salt generally used consists of $Ca(NTf_2)$ as the calcium source and an additive of a solubility enhancing cation (e.g. Bu_4N^+) with a second weakly coordinating anion. This second anion considerably varies from reaction to reaction, with anions such as PF_6^-, BF_4^-, SbF_6^-, and $BAr^F{}_4^-$ as the typical candidates, and is thought to improve the overall solubility of the polar solute (cf. Section 11.1.5) in the polar, noncoordinating solvents that are commonly used. The enhanced solubility is presumably a result of an anion metathesis, which has been supported by nuclear magnetic resonance (NMR) studies [32, 60].

Starting from the original discovery of the calcium-catalyzed dehydroxylation reaction, which was described for Friedel–Crafts alkylation of electron-rich arenes with alcohols [61], a range of substitutions with different nucleophiles were reported (Scheme 11.30). The amination of alcohols was found to proceed with anilines, carbamates, and sulfonamides [60]. Nitriles have been employed yielding amides in a *Ritter*-type reaction [62]. C—C bond formation was achieved with unsaturated organosilicon or -boron compounds as carbanion surrogates [63–65]. Alternatively, the addition of the cation to styrene derivatives also gave rise to new C—C bonds [66]. The long-standing challenge of the deoxygenation of tertiary propargylic alcohols has been solved with triethyl silanes as a hydride source [67].

As an extension of the Ca^{2+}-catalyzed Friedel–Crafts alkylation, the natural tendency of carbocations to induce Wagner–Meerwein hydride shifts was capitalized on for the diastereoselective formation of highly substituted indanes and tetralines [68] from diastereomeric mixtures of alcohols **127** (Scheme 11.31) [68]. Although without a preceding stereo-redistributing equilibration, a high

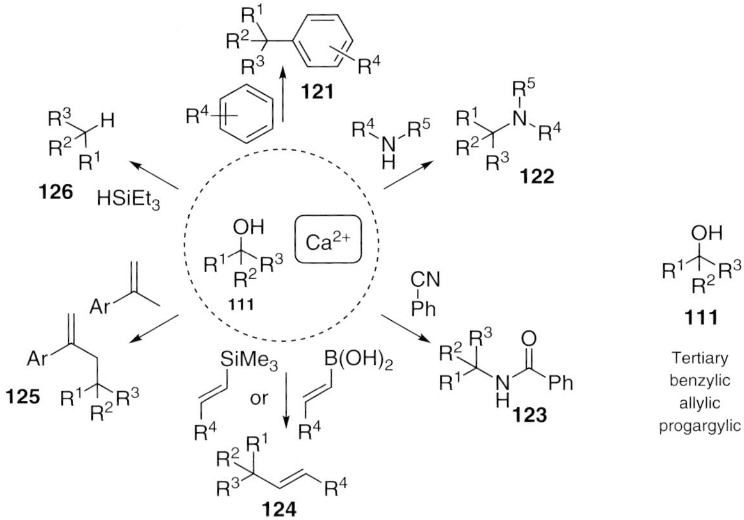

Scheme 11.30 Ca^{2+}-catalyzed dehydroxylation – substitution with nucleophiles.

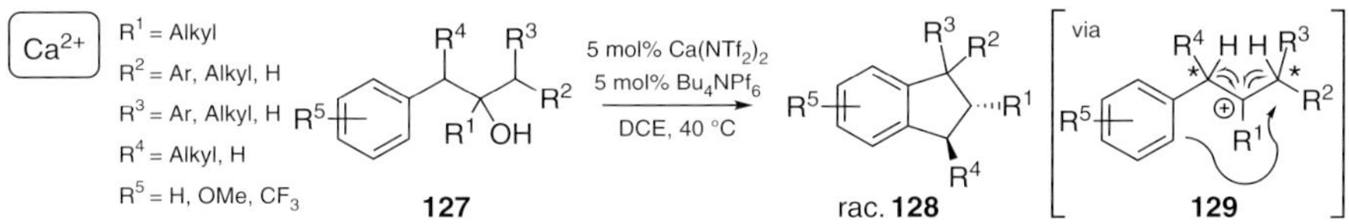

Scheme 11.31 Ca^{2+}-catalyzed, stereo-redistributing diastereoselective synthesis of indanes.

diastereoselectivity was also observed in a Friedel–Crafts cyclization of tethered aminoalcohols toward benzazepines (not shown) [69].

As a highly efficient approach to vinyl cation intermediates **135**, the dehydroxylated alcohols **130** were added to alkynes **132**. In an intramolecular reaction, the vinyl cation was added to a tethered arene **131** [70]. In one of the few intermolecular reactions of these intermediates, the water generated in the dehydroxylation step was used as the nucleophilic reaction partner of the vinyl cation, followed by the tautomerization to ketone **131** [71]. Notably, this reaction was enabled by the addition of an equivalent of cyclopentanone as an electron–pair donor, presumed to stabilize the vinyl cation (Scheme 11.32).

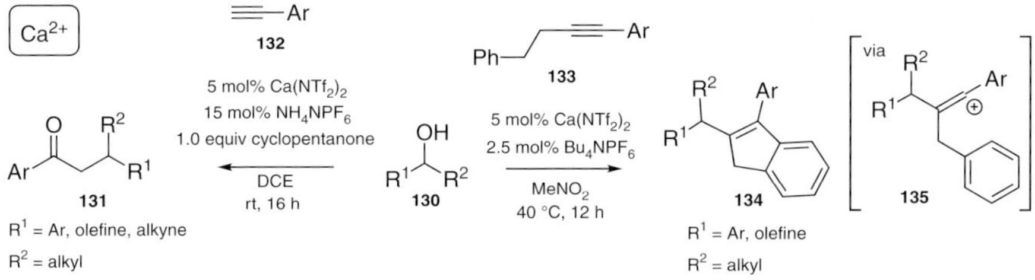

Scheme 11.32 Ca^{2+}-catalyzed inter- and intramolecular alkylation of alkynes.

The addition of an analogously generated vinyl cation to a tethered olefin **140** led to the diastereoselective formation of bicyclic amines **139** via a dynamic MCR. Here, the high Lewis acidity of the Ca^{2+}-based catalyst system was key, as it allowed for a reversible formation of covalent bonds and thus the amplification of the thermodynamically most stable amine product from a range of different intermediates (Scheme 11.33) [72].

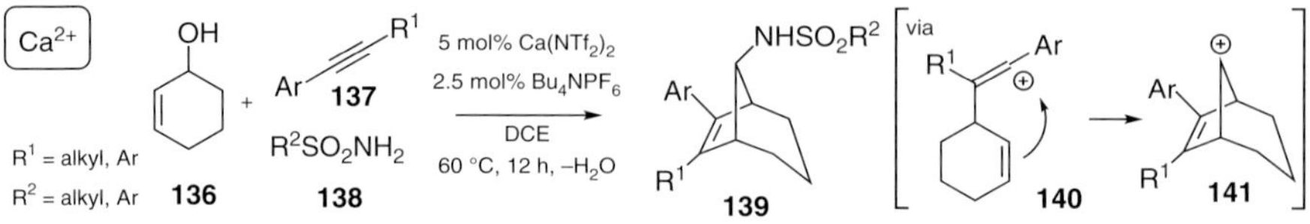

Scheme 11.33 Ca^{2+}-catalyzed dynamic MCR of alcohols, alkynes, and amines.

The addition of an enolized cyclic diketone **142** to a dehydroxylated propargylic alcohol **143** was shown to initiate a subsequent Michael cyclization toward highly substituted furans **144** (Scheme 11.34) [73]. A similar addition–cyclization sequence was achieved via the addition of phenols, naphthols, and hydroxycoumarins that cyclize via a series of electrocyclic rearrangements (not shown) [74, 75].

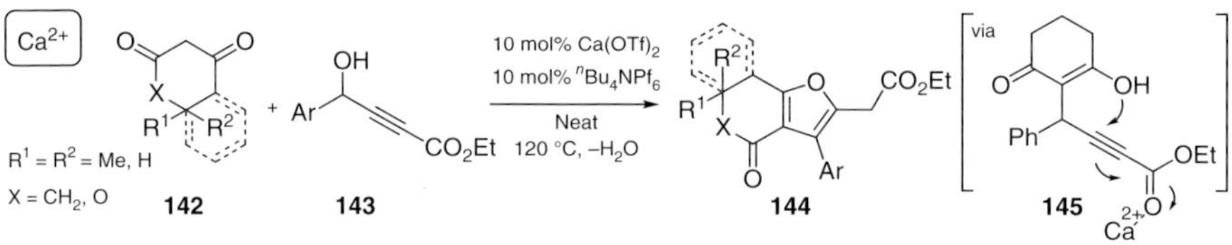

Scheme 11.34 Ca^{2+}-catalyzed regioselective synthesis of polyfunctionalized furans.

As already shown in Scheme 11.31, the initially formed carbocation can undergo rearrangements before its addition to the nucleophile. This gives rise to alternative reactive intermediates that are difficult to access via other methods, including direct dehydroxylations.

The rearrangement of a furyl benzyl alcohol **146** to the oxonium ion **149** was used for an aza-Piancatelli reaction with electron-poor and electron-rich anilines **147** (Scheme 11.35). A higher reaction temperature was found necessary to accelerate the exchange rate of ligands in the inner sphere of the calcium cation as electron-rich anilines are otherwise inhibiting the catalyst (cf. Sections 11.1.2 and 11.1.6) [76].

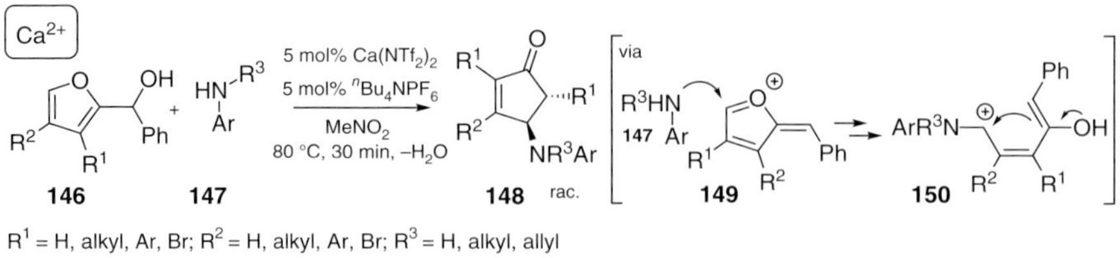

Scheme 11.35 Ca^{2+}-catalyzed aza-Piancatelli reaction of alcohols **146** with anilines **147**.

As an improvement of the reaction conditions, it was later found that changing the solvent to hexafluoropropanol (HFIP) restored reactivity at room temperature and also allowed for an extension of the substrate scope, including more nucleophilic secondary anilines [76]. Supported by density functional theory (DFT) calculations and ^{1}H-NMR-based Lewis acidity analysis (Childs method), the authors claim that in HFIP, the alcohol is no longer activated by Ca^{2+} directly. Instead, the hydrogen bridges of the two HFIP molecules in **151**, the acceptor ability of which is enhanced by Ca^{2+} complexation, promote a cooperative activation of the C—O bond cleavage [77–80].

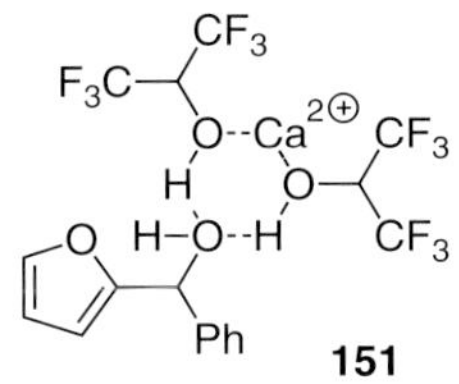

151

In a sequential reaction, extended by a copper-catalyzed hydroamination, the aza-Piancatelli reaction of propargylic amines also gave access to cyclopentapyrols in a one-pot protocol (not shown) [81, 82].

The rearrangement of a cyclopropyl-substituted alcohol **152** gave rise to 1,3-dienes **153** with a catalyst loading of just 1 mol% in a formal homo-Nazarov cyclization (Scheme 11.36) [83].

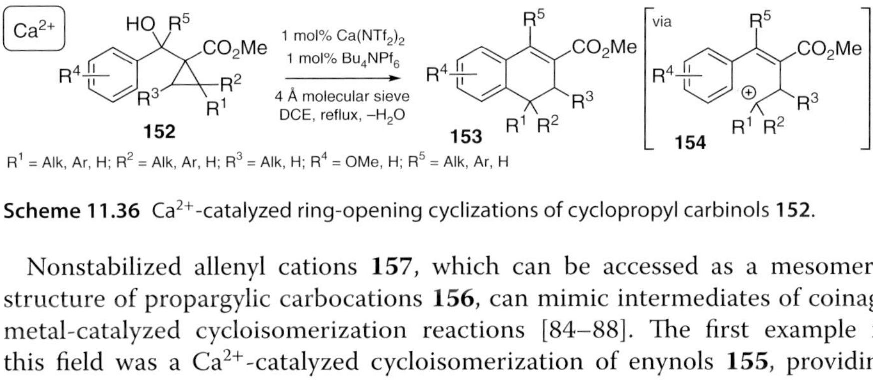

R^1 = Alk, Ar, H; R^2 = Alk, Ar, H; R^3 = Alk, H; R^4 = OMe, H; R^5 = Alk, Ar, H

Scheme 11.36 Ca^{2+}-catalyzed ring-opening cyclizations of cyclopropyl carbinols **152**.

Nonstabilized allenyl cations **157**, which can be accessed as a mesomeric structure of propargylic carbocations **156**, can mimic intermediates of coinage metal-catalyzed cycloisomerization reactions [84–88]. The first example in this field was a Ca^{2+}-catalyzed cycloisomerization of enynols **155**, providing densely substituted cyclopropanes **159** (Scheme 11.37) [89]. Supported by a computational analysis, the high diastereoselectivity of the reaction originates

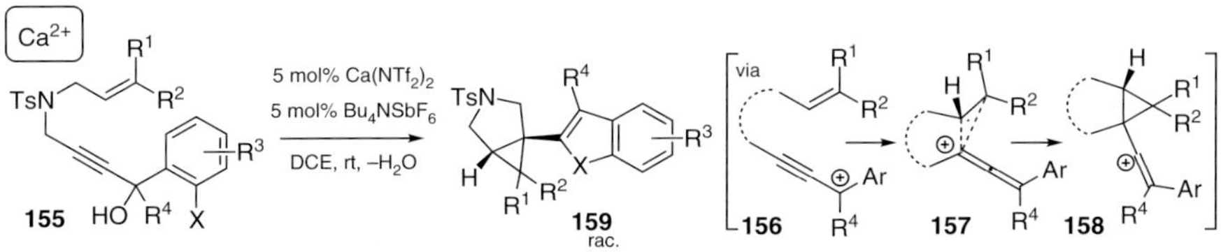

R^1 = alkyl, Ar, H; R^2 = alkyl, H; R^3 = H, Cl, Me; R^4 = alkyl; X = OH, NHTs

Scheme 11.37 Ca^{2+}-catalyzed cyclopropanation of enynols.

in a biomimetic concerted asynchronous carbocation cascade, which starts with the dehydroxylation of the propargylic alcohol in **155**.

The allenyl–cyclopropyl cation rearrangement, allowing for the key step of the carbocation cascade, was also capitalized on for a formal [2+2+2] cycloaddition of enynols **160** with aldehydes **161** (Scheme 11.38) [90]. Here, the allenyl cation intermediate was intercepted by the aldehyde oxygen **161**, followed by a *Prins*-type cyclization.

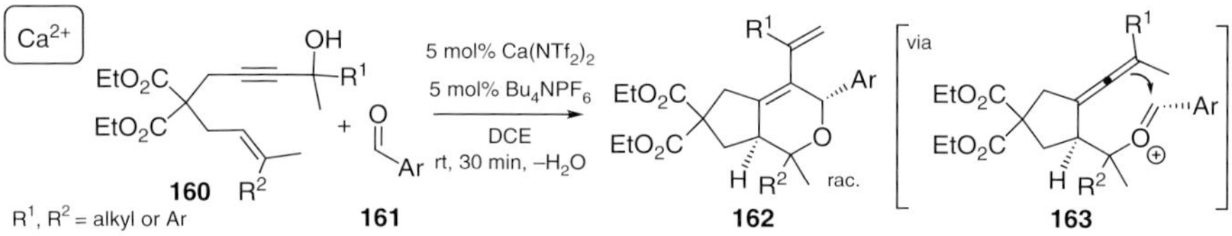

Scheme 11.38 Ca^{2+}-catalyzed formal [2+2+2]-cycloaddition of enynols with aldehydes.

The outcome of a Ru-catalyzed cycloisomerization of diynols **164** [91] was efficiently reproduced by a cooperative catalyst system of $Ca(NTf_2)_2$ and camphor sulfonic acid (CSA) (Scheme 11.39) [92]. The CSA that was found to markedly increase the reactivity of the catalyst system either acts as a Ca^{2+}-assisted Brønsted acid or allows for a more efficient catalyst recovery from LCa(OH) species (cf. Scheme 11.29). The reaction further benefits from the addition of benzaldehyde as a vinyl cation stabilizing electron-pair donor (cf. Scheme 11.32).

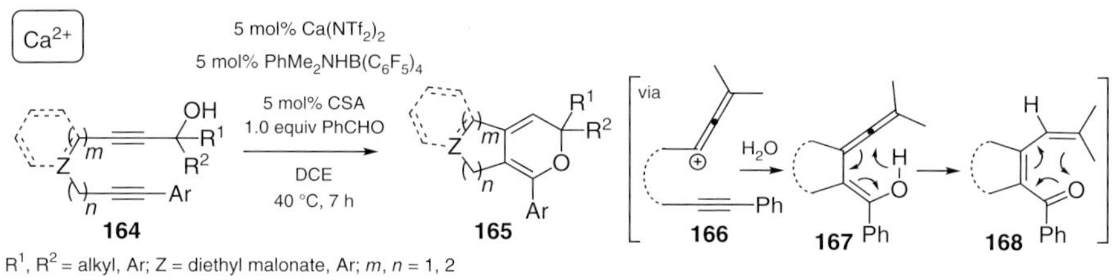

Scheme 11.39 Cooperative Ca^{2+}- and CSA-catalyzed cycloisomerization of diynols **164**.

The simple addition of water to the allenyl cation **172** results in the formation of α,β-unsaturated enones **174**, a reaction generally referred to as Meyer–Schuster rearrangement (Scheme 11.40) [93]. The Michael addition of a tethered β-keto ester yields seven-membered ring systems **170**. The sequence was found to halt at the Meyer–Schuster product in the absence of 5 equiv of ^{i}PrOH. The role of the additive was explained in analogy to the results obtained in the presence of HFIP (see **151** and Scheme 11.40). Notably, the reaction is promoted by a potassium salt (KPF_6) instead of the typically used ammonium-based additives (e.g. $^nBu_4NPF_6$).

In the presence of molecular sieves, the reaction took a different course. As the addition of water for the initial Meyer–Schuster rearrangement is efficiently prevented, a carbonyl oxygen of the tethered β-keto ester directly reacts with the allenyl cation **172** to give furans **171** after olefin isomerization [94].

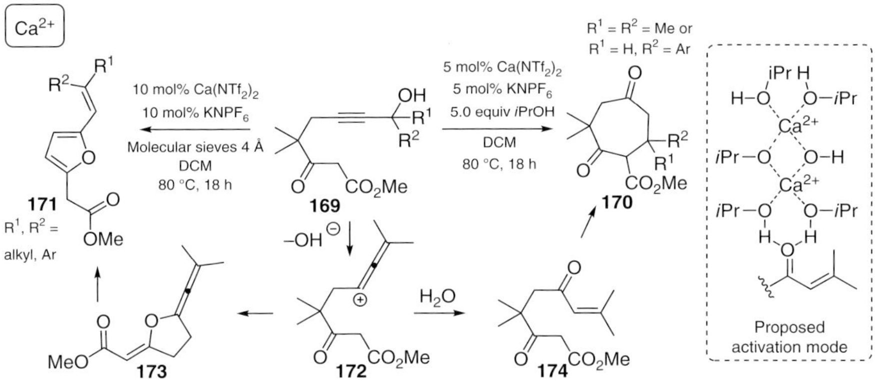

Scheme 11.40 Divergent pathways in Ca^{2+}-promoted dehydroxylations of alcohols **169**.

11.4 Activation of Unpolarized Double Bonds

Bearing no heteroatom with a free electron pair, the activation of unpolarized double bonds in olefins is undoubtedly the most challenging among the substrates in this chapter, for Lewis acids in general. Thus, only a handful of reactions have been shown to proceed with the assistance of the milder s-block metal catalysts. Also, it comes with no surprise that all of these reactions are catalyzed by the highly active Ca^{2+}-based system ($Ca(NTf_2)_2$ with a second weakly coordinating anion pioneered by the *Niggemann* group. As the intermediary formation of alcohols has been observed in none of these reactions, the activation of the double bond is presumably proceeding via one of the two options shown in Scheme 11.41.

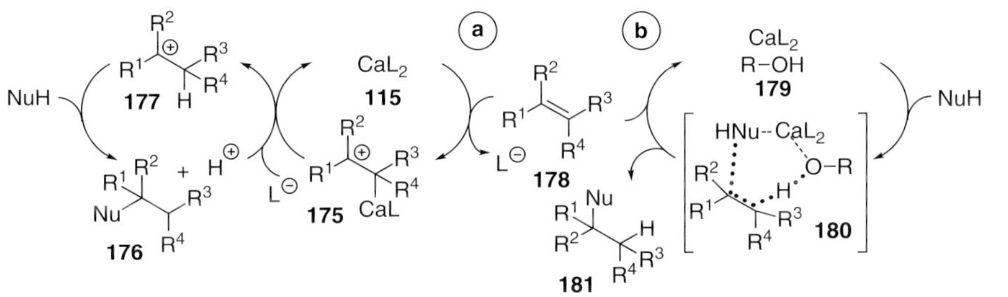

Scheme 11.41 Proposed mechanisms for activation of olefins by Ca^{2+}.

Either the Ca^{2+} coordinates with the double bond, resulting in a Lewis acid-bound carbocation **175**, which is subjected to immediate protodemetalation (Scheme 11.41**a**). The thus formed carbocation **177** is then intercepted by the nucleophile, thereby simultaneously producing a proton. Alternatively, as supported by the computational analysis (DFT) of a reaction in HFIP by the groups of *Gandon* and *Leboeuf*, the self-assembly of a Ca^{2+}-bridged complex **180** of the substrate, the nucleophile and HFIP, a concerted protonation–nucleophile addition gives rise to the product (Scheme 11.41**b**).

The first example of an olefin activation by a calcium-based Lewis acid is the hydroarylation of styrenes and dienes **182** with electron-rich arenes **183** in Scheme 11.42 [95]. This reaction is more likely to proceed via mechanism **a**, as protic molecules are not present in the reaction medium and a complex formation with the nucleophile is unlikely.

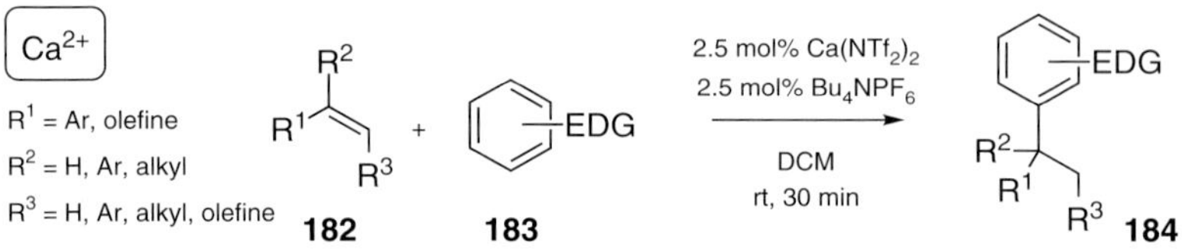

Scheme 11.42 Ca^{2+}-catalyzed hydroarylation of olefins **183**.

Based on this initial result, two intramolecular hydrofunctionalization reactions have been developed: a hydroalkoxylation at room temperature in dichloromethane (DCM) as a more classical solvent (Scheme 11.43a) [96] and a hydroamination at 80 °C in HFIP (Scheme 11.43b) [97].

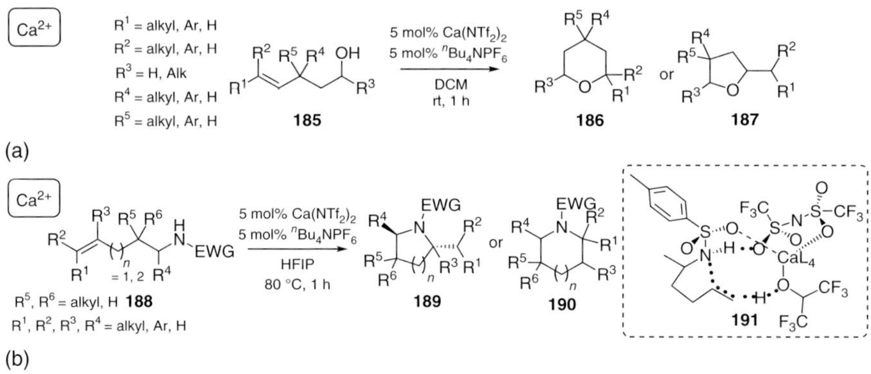

Scheme 11.43 Ca^{2+}-catalyzed intramolecular hydroalkoxylation (a) and hydroamination (b).

Notably, on account of both, the higher temperature and the solvent change, the substrate scope was more extensive in the hydroamination reaction, allowing for terminal olefins and also the intermolecular version of the reaction.

As it is supported by the abovementioned DFT-based computational analysis for the hydroamination, and also quite likely to occur in the hydroalkoxylation, these reactions proceed via mechanism **b**, as shown in Scheme 11.41.

Finally, in analogy to some of the reactions above (Schemes 11.42 and 11.43), the activation of olefins can be used for the generation of vinyl cations **195** (Scheme 11.44) [98].

Here, the carbocation, which is more likely to be generated via mechanism **a**, attacks a tethered alkyne. To provide defined amounts of water for the subsequent interception of the vinyl cation, $MgSO_4 \times H_2O$ was used as an easily handled source.

Scheme 11.44 Ca^{2+}-catalyzed cycloisomerization of enynes via vinyl cation formation.

11.5 Summary and Conclusions

All of the above reactions clearly prove that alkaline and alkaline earth metal cations are much more than the innocent bystanders they are traditionally considered to be. Even though their Lewis acidity is on the milder end of the spectrum, a proficient choice of counter anions and reaction medium can unleash their Lewis acidity, and reactions that are generally considered to require strong Lewis acids efficiently catalyzed. To conclude this chapter, the following summary of the properties of s-block metal cations may be given as a guideline for the decision, whether it may be worthwhile to include them in a catalyst screening for a new Lewis acid-catalyzed reaction:

1. They provide mild – Na^+ – or balanced – Mg^{2+} and Ca^{2+} – Lewis acidity, with Li^+ belonging to either category, depending on the catalyst system as a whole.
2. Selective activation of one Lewis base among others has been shown in many examples.
3. Tolerance toward substrates containing coordinating heteroatoms, and water can be generally expected, and they clearly outperform TM in this regard.
4. Their tendency to generate "hidden Brønsted acids" is the smallest of all Lewis acids.
5. Acid-sensitive groups are better tolerated, as with any other Lewis acid.
6. Ca^{2+} is a highly efficient catalyst for deoxygenation reactions.
7. Background reactions based on changes in oxidation state of the metal are precluded.
8. Their features can be capitalized on best, if they are combined with a weakly coordinating anion and used in a noncoordinating yet polar solvent.

References

1 Schlosser, M. (2002). *Organometallics in Synthesis: A Manual*, 2e (ed. M. Schlosser), 8–10. Chichester: Wiley.
2 Westerhausen, M., Langer, J., Krieck, S. et al. (2013). *Top. Organomet. Chem.* 45: 29–72.
3 Persson, I. (2010). *Pure Appl. Chem.* 82: 1901–1917.
4 Kobayashi, S. (1999). *Eur. J. Org. Chem.*: 15–27.
5 Rodgers, M.T. and Armentrout, P.B. (2016). *Chem. Rev.* 116: 5642–5687.
6 Bruzzi, E. and Stace, A.J. (2017). *R. Soc. Open Sci.* 4: 160671.
7 Beyer, M.K. (2007). *Mass Spectrom. Rev.* 26: 517–541.

8 Rode, B.M., Schwenk, C.F., Hofer, T.S., and Randolf, B.R. (2005). *Coord. Chem. Rev.* 249: 2993–3006.
9 Strauss, S.H. (1993). *Chem. Rev.* 93: 927–942.
10 Stephan, D.W. (2016). *Science* 354: aaf7229-1–aaf7229-8.
11 Stephan, D.W. (2015). *Acc. Chem. Res.* 48: 306–316.
12 Krossing, I. and Rabe, I. (2004). *Angew. Chem. Int. Ed.* 43: 2066–2090.
13 Ohtaki, H. and Radnai, T. (1993). *Chem. Rev.* 93: 1157–1204.
14 Kobayashi, S., Nagayama, S., and Busujima, T. (1998). *J. Am. Chem. Soc.* 120: 8287–8288.
15 Kobayashi, S. and Ogawa, C. (2006). *Chem. Eur. J.* 12: 5954–5960.
16 Dang, T.T., Boeck, F., and Hintermann, L. (2011). *J. Org. Chem.* 76: 9353–9361.
17 Brown, H.C. and Kanner, B. (1966). *J. Am. Chem. Soc.* 88: 986–992.
18 Karimi, B. and Maleki, J. (2003). *J. Org. Chem.* 68: 4951–4954.
19 Heydari, A. and Hosseini, S.E. (2005). *Adv. Synth. Catal.* 347: 1929–1932.
20 Bartoli, G., Bosco, M., Carlone, A. et al. (2006). *Eur. J. Org. Chem.* 2006: 4429–4434.
21 Firouzabadi, H., Karimi, B., and Eslami, S. (1999). *Tetrahedron Lett.* 40: 4055–4058.
22 Firouzabadi, H., Eslami, S., and Karimi, B. (2001). *Bull. Chem. Soc. Jpn.* 74: 2401–2406.
23 Chang, C.-C., Liao, B.-S., and Liu, S.-T. (2007). *Synlett*: 0283–0287.
24 Liu, S.-T., Liao, B.-S., and Chen, J.-T. (2007). *Synthesis*: 3125–3128.
25 Bartoli, G., Bosco, M., Carlone, A. et al. (2008). *Tetrahedron Lett.* 49: 2555–2557.
26 Tan, E., Ung, S., and Corbet, M. (2016). *Eur. J. Org. Chem.* 2016: 1836–1840.
27 Beletskaya, I., Feofanov, M., Anokhin, M., and Averin, A. (2017). *Synthesis* 49: 5045–5058.
28 Yaragorla, S., Saini, P.L., and Singh, G. (2015). *Tetrahedron Lett.* 56: 1649–1653.
29 Yaragorla, S. and Pareek, A. (2018). *Tetrahedron Lett.* 59: 909–913.
30 Briones, J.F. and Basarab, G.S. (2016). *Chem. Commun.* 52: 8541–8544.
31 Liu, Y.K., Ma, C., Jiang, K. et al. (2009). *Org. Lett.* 11: 2848–2851.
32 Davies, J. and Leonori, D. (2014). *Chem. Commun.* 50: 15171–15174.
33 Prasad, B.A.B., Bisai, A., and Singh, V.K. (2004). *Tetrahedron Lett.* 45: 9565–9567.
34 Vanden Eynden, M.J., Kunchithapatham, K., and Stambuli, J.P. (2010). *J. Org. Chem.* 75: 8542–8549.
35 Bartoli, G., Bosco, M., Galzerano, P. et al. (2008). *Eur. J. Org. Chem.*: 3970–3975.
36 Yaragorla, S. and Dada, R. (2017). *ACS Omega* 2: 4859–4869.
37 Loncaric, C., Manabe, K., and Kobayashi, S. (2003). *Adv. Synth. Catal.* 345: 1187–1189.
38 Chang, C.-T., Liao, B.-S., and Liu, S.-T. (2006). *Tetrahedron Lett.* 47: 9257–9259.
39 Atar, A.B., Dindulkar, S.D., and Jeong, Y.T. (2013). *Monatsh. Chem.* 144: 695–701.

40 Uyanik, M., Nakashima, D., and Ishihara, K. (2012). *Angew. Chem. Int. Ed.* 51: 9093–9096.
41 Forkel, N.V., Henderson, D.A., and Fuchter, M.J. (2012). *Green Chem.* 14: 2129.
42 Forkel, N.V., Henderson, D.A., and Fuchter, M.J. (2014). *Tetrahedron Lett.* 55: 5511–5514.
43 Dulin, C.C., Murphy, K.L., and Nolin, K.A. (2014). *Tetrahedron Lett.* 55: 5280–5282.
44 Haubenreisser, S., Hensenne, P., Schroeder, S., and Niggemann, M. (2013). *Org. Lett.* 15: 2262–2265.
45 Mloston, G., Grzelak, P., Mikina, M. et al. (2015). *Beilstein J. Org. Chem.* 11: 576–582.
46 Augé, J., Gil, R., Kalsey, S., and Lubin-Germain, N. (2000). *Synlett*: 0877–0879.
47 Loncaric, C., Manabe, K., and Kobayashi, S. (2003). *Chem. Commun.*: 574–575.
48 Yadav, J.S., Reddy, B.V.S., Narsimhaswamy, D. et al. (2003). *Tetrahedron Lett.* 44: 3697–3700.
49 Auge, J. and Leroy, F. (1996). *Tetrahedron Lett.* 37: 7715–7716.
50 Cossy, J., Bellosta, W., Hamior, C., and Desmurs, J.R. (2002). *Tetrahedron Lett.* 43: 7083–7086.
51 Samzadeh-Kermani, A. and Ataeifar, S. (2017). *Monatsh. Chem.* 149: 47–56.
52 Sayyad, M., Wani, I.A., Babu, R. et al. (2017). *J. Org. Chem.* 82: 2364–2374.
53 Mukaiyama, T., Nakano, M., Kikuchi, W., and Matsuo, J. (2000). *Chem. Lett.*: 1010–1011.
54 Bandgar, B.P., Kamble, V.T., Fulse, D.B., and Deshmukh, M.V. (2002). *New J. Chem.* 26: 1105–1107.
55 Bandgar, B.P. and Makone, S.S. (2004). *Synth. Commun.* 34: 743–750.
56 Bandgar, B.P. and Kasture, S.P. (2001). *Monatsh. Chem.* 132: 1101–1104.
57 Firouzabadi, H., Iranpoor, N., Sobhani, S., and Ghassamipour, S. (2004). *J. Organomet. Chem.* 689: 3197–3202.
58 Emer, E., Sinisi, R., Capdevila Montse, G. et al. (2010). *Eur. J. Org. Chem.* 2011: 647–666.
59 Kumar, R. and Van der Eycken, E.V. (2013). *Chem. Soc. Rev.* 42: 1121–1146.
60 Haubenreisser, S. and Niggemann, M. (2011). *Adv. Synth. Catal.* 353: 469–474.
61 Niggemann, M. and Meel, M.J. (2010). *Angew. Chem. Int. Ed.* 49: 3684–3687.
62 Yaragorla, S., Singh, G., Lal Saini, P., and Reddy, M.K. (2014). *Tetrahedron Lett.* 55: 4657–4660.
63 Meyer Vera, J. and Niggemann, M. (2011). *Eur. J. Org. Chem.* 2011: 3671–3674.
64 Lebœuf, D., Presset, M., Michelet, B. et al. (2015). *Chem. Eur. J.* 21: 11001–11005.
65 Qi, C., Gandon, V., and Lebœuf, D. (2017). *Adv. Synth. Catal.* 359: 2671–2675.
66 Yaragorla, S., Pareek, A., and Dada, R. (2017). *Adv. Synth. Catal.* 359: 3068–3075.
67 Meyer, V.J. and Niggemann, M. (2012). *Chem. Eur. J.* 18: 4687–4691.

68 Begouin, J.-M., Capitta, F., Wu, X., and Niggemann, M. (2013). *Org. Lett.* 15: 1370–1373.
69 Damsen, H. and Niggemann, M. (2015). *Eur. J. Org. Chem.*: 7880–7883.
70 Fu, L. and Niggemann, M. (2015). *Chem. Eur. J.* 21: 6367–6370.
71 Stopka, T. and Niggemann, M. (2015). *Org. Lett.* 17: 1437–1440.
72 Gao, S., Stopka, T., and Niggemann, M. (2015). *Org. Lett.* 17: 5080–5083.
73 Yaragorla, S., Dada, R., Pareek, A., and Singh, G. (2016). *RSC Adv.* 6: 28865–28870.
74 Yaragorla, S., Pareek, A., and Dada, R. (2017). *Tetrahedron Lett.* 58: 4642–4647.
75 Yaragorla, S., Dada, R., Singh, G. et al. (2016). *ChemistrySelect* 1: 6902–6906.
76 Lebœuf, D., Marin, L., Michelet, B. et al. (2016). *Chem. Eur. J.* 22: 16165–16171.
77 Berkessel, A. and Adrio Jens, A. (2004). *Adv. Synth. Catal.* 346: 275–280.
78 Berkessel, A. and Adrio, J.A. (2006). *J. Am. Chem. Soc.* 128: 13412–13420.
79 Berkessel, A., Adrio, J.A., Hüttenhain, D., and Neudörfl, J.M. (2006). *J. Am. Chem. Soc.* 128: 8421–8426.
80 Berkessel, A., Krämer, J., Mummy, F. et al. (2012). *Angew. Chem. Int. Ed.* 52: 739–743.
81 Marin, L., Gandon, V., Schulz, E., and Lebœuf, D. (2017). *Adv. Synth. Catal.* 359: 1157–1163.
82 Marin, L., Guillot, R., Gandon, V. et al. (2018). *Org. Chem. Front.* 5: 640–647.
83 Cavitt, M.A., Phun, L.H., and France, S. (2014). *Chem. Soc. Rev.* 43: 804–818.
84 Hashmi, A.S.K. (2007). *Chem. Rev.* 107: 3180–3211.
85 Gorin, D.J., Sherry, B.D., and Toste, F.D. (2008). *Chem. Rev.* 108: 3351–3378.
86 Jiménez-Núñez, E. and Echavarren, A.M. (2008). *Chem. Rev.* 108: 3326–3350.
87 Patil, N.T. and Yamamoto, Y. (2008). *Chem. Rev.* 108: 3395–3442.
88 Furstner, A. (2009). *Chem. Soc. Rev.* 38: 3208–3221.
89 Haven, T., Kubik, G., Haubenreisser, S., and Niggemann, M. (2013). *Angew. Chem. Int. Ed.* 52: 4016–4019.
90 Meyer, V.J., Ascheberg, C., and Niggemann, M. (2015). *Chem. Eur. J.* 21: 6371–6374.
91 Trost, B.M. and Rudd, M.T. (2002). *J. Am. Chem. Soc.* 124: 4178–4179.
92 Rauser, M., Schroeder, S., and Niggemann, M. (2015). *Chem. Eur. J.* 21: 15929–15933.
93 Presset, M., Michelet, B., Guillot, R. et al. (2015). *Chem. Commun.* 51: 5318–5321.
94 Morcillo, S.P., Presset, M., Floquet, S. et al. (2016). *Eur. J. Org. Chem.* 2016: 2688–2694.
95 Niggemann, M. and Bisek, N. (2010). *Chem. Eur. J.* 16: 11246–11249.
96 Kena Diba, A., Begouin, J.-M., and Niggemann, M. (2012). *Tetrahedron Lett.* 53: 6629–6632.
97 Qi, C., Hasenmaile, F., Gandon, V., and Lebœuf, D. (2018). *ACS Catal.* 8: 1734–1739.
98 Meyer, V.J., Fu, L., Marquardt, F., and Niggemann, M. (2013). *Adv. Synth. Catal.* 355: 1943–1947.

12

Enantioselective Group 2 Metal Lewis Acid Catalysis

Yasuhiro Yamashita[1], *Tetsu Tsubogo*[2], *and Shū Kobayashi*[1]

[1] *The University of Tokyo, Department of Chemistry, School of Science, Hongo 7-3-1, Bunkyo-ku, Tokyo, 113-0033, Japan*
[2] *Tokyo University of Science, Faculty of Pharmaceutical Sciences, 2641 Yamazaki, Noda-Shi, Chiba, 278-8510, Japan*

12.1 Introduction

Asymmetric synthesis is an important methodology for efficient preparation of optically active molecules. Especially, catalytic enantioselective synthesis using a chiral catalyst is one of the most promising approaches for supplying a large amount of optically active molecule because only a small amount of chiral source is required [1]. Chiral Lewis acid-promoted enantioselective reactions are traditional and reliable methods in asymmetric catalysis. To date, several kinds of chiral metal Lewis acids have been developed and applied for asymmetric reactions [2]. Among them, transition metal Lewis acid catalysts have been well developed; however, most of them are sometimes toxic and harmful on not only our health but also the earth environment. Therefore, the use of safe and harmless metal catalysts in organic synthesis is now strongly preferred from a viewpoint of green sustainable chemistry (GSC) [3]. Compared with typical transition metals, group 2 metals, especially magnesium (Mg), calcium (Ca), strontium (Sr), and barium (Ba), are readily available metals in the earth's crust. Although they are known to be economical metals and often employed in our life, their use as catalysts, especially chiral Lewis acid catalysts, has not been established yet.

Group 2 metal compounds have interesting characters among group 1–3 metal compounds. For examples, group 1 metal phenoxides (lithium, sodium, and potassium phenoxides) show strong Brønsted basicity, whereas group 3 metal phenoxides (scandium, yttrium, and lanthanide phenoxides) show significant Lewis acidity, although the stable valences of the ions are different. However, group 2 metal phenoxides show both characters. Among group 2 metal compounds, Mg compounds have relatively strong Lewis acidity compared to typical metal Lewis acids including transition metals and have been used for strong activation of substrates via coordination with their Lewis basic functionalities to promote Lewis acid catalysis. In the presence of basic species, such as tertiary amines, proton transfer Lewis acid/Brønsted base cooperative

Early Main Group Metal Catalysis: Concepts and Reactions,
First Edition. Edited by Sjoerd Harder.

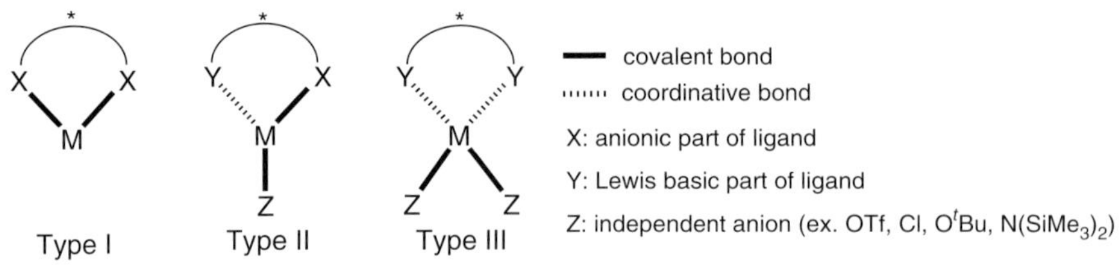

Figure 12.1 Categories of chiral alkaline earth metal complexes.

catalysis, which can realize very efficient chemical processes from an atom economical perspective, is also possible. On the other hand, Lewis acidity of calcium, strontium, and barium compounds is not so strong, and they have not been well investigated as Lewis acid catalysts. The most interesting function of these metal compounds is activation of substrates by both a mild but significantly Lewis acidic metal part and a strongly Brønsted basic counter anion part on the same molecule in a pseudo-intramolecular manner for proton transfer Lewis acid/Brønsted base cooperative catalysis. Therefore, the development of chiral group 2 metal catalysts is one of the hottest topics in catalytic enantioselective synthesis [4].

Chiral group 2 metal complexes developed until now can be classified into three types, type I, type II, and type III because of their divalent stable oxidation state (Figure 12.1). The type I complex is bearing two covalent bonds between metal and ligand, which can control an asymmetric environment strictly by tight interaction. However, fine control of Lewis acidity and Brønsted basicity of a complex is sometimes difficult because well-designed asymmetric environment of the complex is easily affected by the structure of the anionic part of the ligand. The type II complex is bearing one covalent bond and one coordinative bond between the metal and the ligand. When a chiral ligand is introduced in type II complex, an asymmetric environment is constructed by a combination of one covalent bond and other coordinative bonds in a more than a bidentate manner, which leads to strict control of the environment. Moreover, the remaining counter anion can be optimized to realize appropriate Lewis acidity or Brønsted basicity of the complex. Therefore, total design of the catalyst, both its reactivity and selectivity, is possible. The type III complex is bearing no covalent bond between metal and ligand, and the ligand interacts with the metal via only coordinative bonding. Because of significant Lewis acidity of group 2 metals, this type III complex is sometimes the most promising among the type I–III complexes because both Lewis acidic salts and Brønsted basic metal salts and many well-known coordinative chiral ligands are available. Moreover, the type III complexes prepared from Lewis acidic or neutral salts are expected to be more robust under ambient conditions compared with type I and II complexes, which are sometimes air and moisture sensitive. By using these types of complexes, many kinds of useful bond-forming reactions have been achieved. In this chapter, we introduce catalytic enantioselective reactions using type I–III chiral group 2 metal Lewis acid (/Brønsted base) catalysts, especially chiral magnesium, calcium, strontium, and barium catalysts.

12.2 Catalytic Enantioselective Reactions Using Chiral Magnesium Complexes

Magnesium compounds can show strong Lewis acidity among group 2 metal compounds; therefore, their use in Lewis acid catalysis is a major way compared to pseudo-intramolecular Lewis acid/Brønsted base cooperative catalysis. Moreover, a smaller ionic radius among group 2 metals can realize a precise asymmetric environment. In enantioselective reactions, chiral magnesium Lewis acid catalysts have been well developed, and excellent enantioselectivities were obtained. Because of its strong Lewis acidity, all types of chiral complexes (type I–III) have been prepared and employed in the reactions. Although the first discovery of an asymmetric reaction using a chiral magnesium complex was in 1990s, here we mainly discuss crucial articles reported in 2000 and later.

12.2.1 Chiral Magnesium-Catalyzed Diels–Alder and 1,3-Dipolar Cycloaddition Reactions

Enantioselective Diels–Alder and 1,3-dipolar cycloaddition reactions are powerful tools for the synthesis of contiguous chiral carbon centers. These reactions have a long history dating back to chiral magnesium-catalyzed Diels–Alder reactions in the early 1990s. In this section, we describe typical examples of Diels–Alder reactions and 1,3-dipolar cycloaddition reactions.

The enantioselective Diels–Alder reaction of 3-acryloyl-1,3-oxazolidine-2-one with cyclopentadiene using a $Mg(SbF_6)_2$-bisoxazoline (Box) ligand complex (type III) was reported by Corey et al. in 1992 (Scheme 12.1) [5]. The catalyst was

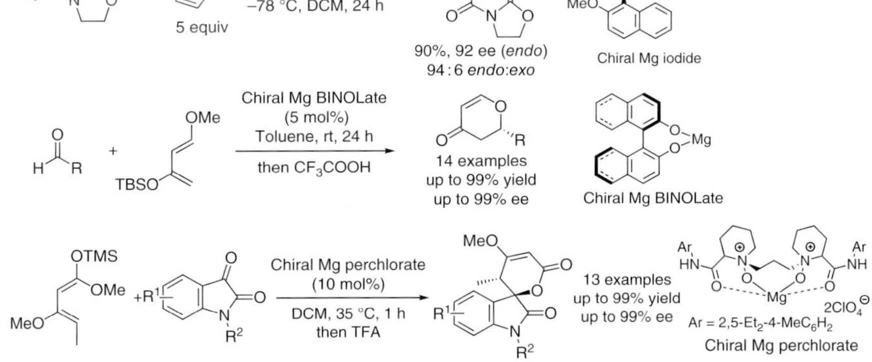

Scheme 12.1 Enantioselective Diels–Alder reactions catalyzed by chiral magnesium complexes.

prepared from magnesium iodide, silver hexafluoroantimonate, and Box ligand. The desired reaction proceeded in good yield with high diastereo- and enantioselectivities. Furthermore, they showed that MgI_2-Box and $Mg(Ph_4B)_2$-Box catalyzed the same reaction. The same Diels–Alder reaction catalyzed by a MgI_2–oxazoline-sulfoxide ligand complex (type III) was also reported by Hiroi et al. in 2001 [6]. They assumed that a tetrahedral magnesium complex coordinated with the nitrogen of the oxazoline ring, the sulfinyl oxygen, and the two carbonyl groups of the substrate. Cyclohexadiene attacks the opposite side of the bulky 2-methoxy-1-naphthyl group to give the 2*S*-product. Enantioselective hetero-Diels–Alder reactions of Danishefsky's diene with aldehydes were reported by Ding et al. using a magnesium BINOLate complex (type I) in 2008 [7]. A variety of 2-substituted 2,3-dihydro-4*H*-pyran-4-ones were obtained in high yields with excellent ee values. They assumed that this reaction proceeded via a concerted cycloaddition mechanism. Enantioselective hetero-Diels–Alder reactions of Brassard's dienes with isatins were reported using a chiral *N*,*N*′-dioxide-magnesium complex (type III) by Feng et al. in 2014 [8]. The desired chiral spirolactones bearing tetrasubstituted centers were obtained in high yields with excellent diastereo- and enantioselectivities. This reaction predominantly occurred through not Diels–Alder reaction but stepwise Mukaiyama aldol-cyclization pathway.

Enantioselective [3+2] cycloadditions of racemic cyclopropanes and aldehydes via dynamic kinetic resolution using a magnesium-pyridinebisoxazoline (Pybox) ligand complex (type III) were reported by Johnson et al. in 2009 (Scheme 12.2) [9]. The desired tetrahydrofuran derivatives were obtained in high yields with excellent diastereo- and enantioselectivities. When (*S*)-cyclopropane substrate was used, the optically pure product was also obtained in excellent yield. Enantioselective construction of pyrroloindolines using a quinine–magnesium complex (type II) was reported by Wang et al. in 2014 [10]. Addition of achiral methyl picolinate is important to improve the enantioselectivities. The products were obtained in moderate to good yields with high diastereo- and enantioselectivities through aziridine ring-opening by indoles, followed by intramolecular Mannich-type reactions, that is, formal [3+2] cycloaddition reactions. A variety of optically active pyrroloindolines could be obtained by this method.

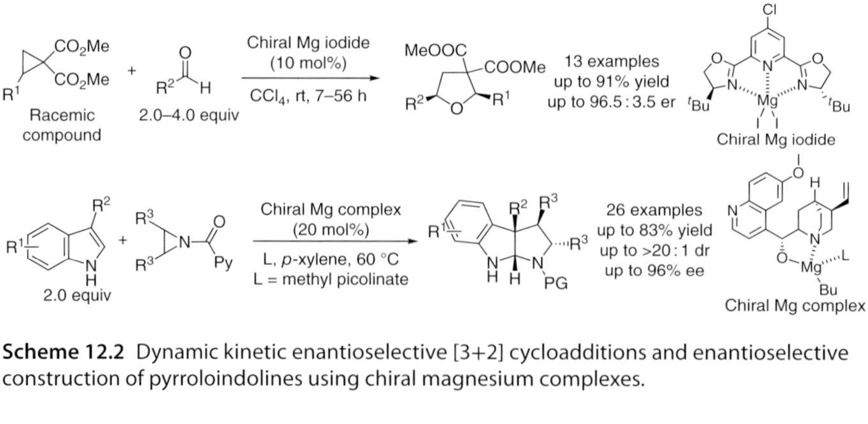

Scheme 12.2 Dynamic kinetic enantioselective [3+2] cycloadditions and enantioselective construction of pyrroloindolines using chiral magnesium complexes.

12.2.2 Chiral Magnesium-Catalyzed 1,4-Addition Reactions

Chiral magnesium-catalyzed enantioselective 1,4-addition reactions were described in this section. Many types of 1,4-addition reactions, including radical, Friedel–Crafts type, alkenylboron, phosphorous nucleophile additions to α,β-unsaturated carbonyl compounds, were described. In addition, cascade-type reactions including 1,4-addition were shown.

A radical addition to 3-acryloyl-1,3-oxazolidin-2-one derivatives using a Box–magnesium complex (type III) was reported by Sibi and Porter et al. in 1996 (Scheme 12.3) [11]. In this report, catalytic reactions using the magnesium complex were described to obtain the products in moderate enantioselectivities; however, a stoichiometric amount of the chiral Lewis acid catalyst showed higher enantioselectivity.

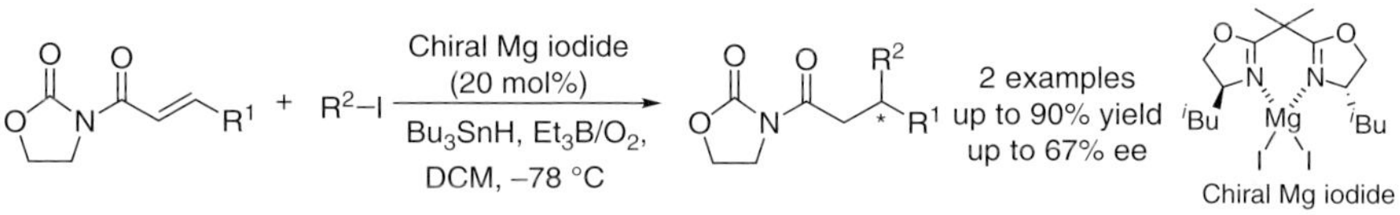

Scheme 12.3 A Box-magnesium catalyst for radical addition.

Enantioselective Friedel–Crafts reactions of β,γ-unsaturated α-ketoesters using a chiral phosphoric acid and magnesium fluoride (other type) were reported by Luo et al. in 2011 (Scheme 12.4) [12]. Phenols and indoles were applicable to these Friedel–Crafts reactions. In this reaction, the use of both chiral phosphoric acid and magnesium fluoride is a key to promote the reactions in high yields with excellent enantioselectivities. Because some acid salts could not promote the reaction, a free acid was required for effective catalysis. An enantioselective 1,4-addition of alkenylboronic acid to indole-appended enones using a BINOL derivative and a magnesium alkoxide (other type) was reported by May et al. in 2011 [13]. The reaction proceeded with high enantioselectivity. A range of α-branched indole derivatives were synthesized. The actual catalyst

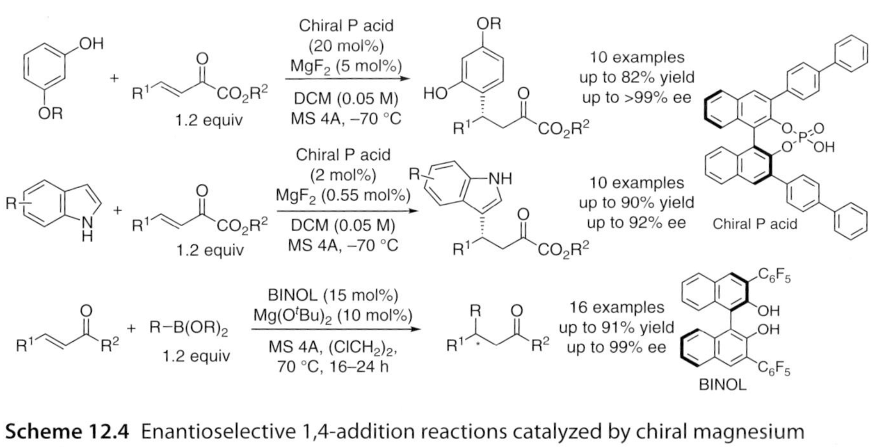

Scheme 12.4 Enantioselective 1,4-addition reactions catalyzed by chiral magnesium complexes.

structure was unclear; whether deprotonation of BINOL derivatives occurred or not.

Direct site-specific and highly enantioselective γ-functionalizations of linear α,β-unsaturated ketones using a Salen–magnesium complex (type I) were reported by Wang et al. in 2013 (Scheme 12.5) [14]. The catalyst was prepared from dibutyl magnesium with Salen ligand. This reaction proceeds via γ-deprotonation, 1,4-addition, and cyclization reactions continuously. The optically active cyclohexane frameworks were obtained in high yields with excellent diastereo- and enantioselectivities.

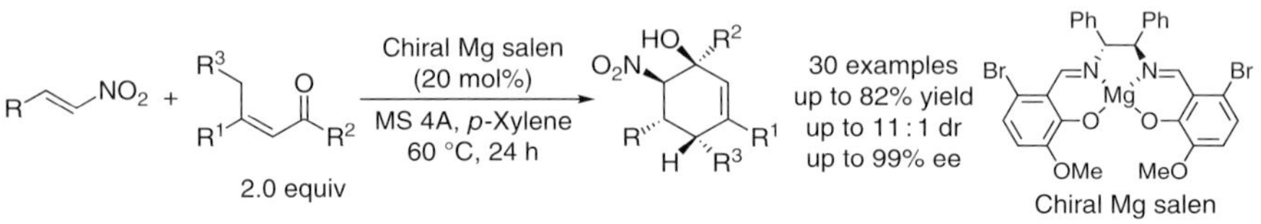

Scheme 12.5 Direct site-specific and highly enantioselective γ-functionalization reactions of linear α,β-unsaturated ketones using a chiral Salen–magnesium complex.

An enantioselective dearomatization of indoles through a 1,4-addition/Friedel–Crafts-type cascade reaction to construct polycyclic spiroindoles using a chiral *N*,*N*′-dioxide–magnesium complex (type III) was reported by Liu and Feng et al. in 2015 (Scheme 12.6) [15]. The fused polycyclic indoles containing three stereocenters were obtained in good yields with excellent diastereo- and enantioselectivities. A 1,4-addition in enantioselective dearomatization of β-naphthols using a phenol-oxazoline-magnesium complex (type II) was reported by Wang et al. in 2015 [16]. The catalyst was prepared from dibutyl magnesium and phenol–oxazoline ligand. The desired 1,4-addition products were obtained in good yields with high *Z*/*E* ratios and enantioselectivities, which were investigated by a computational study.

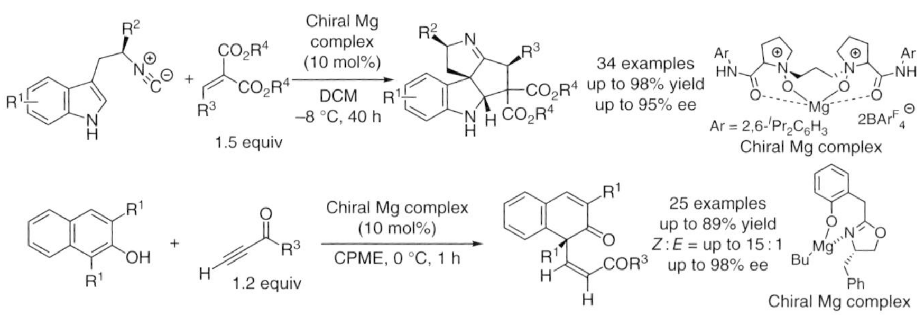

Scheme 12.6 Enantioselective dearomatization reactions by chiral magnesium complexes.

A highly enantioselective addition of phosphorous nucleophiles to α,β-unsaturated esters using a chiral magnesium (II) binaphtholates (type I or II) as a cooperative Brønsted/Lewis acid-base catalyst was reported by Ishihara et al. (Scheme 12.7) [17]. The catalyst was prepared from BINOL, dibutyl magnesium, and water. In this system, water was essential to induce

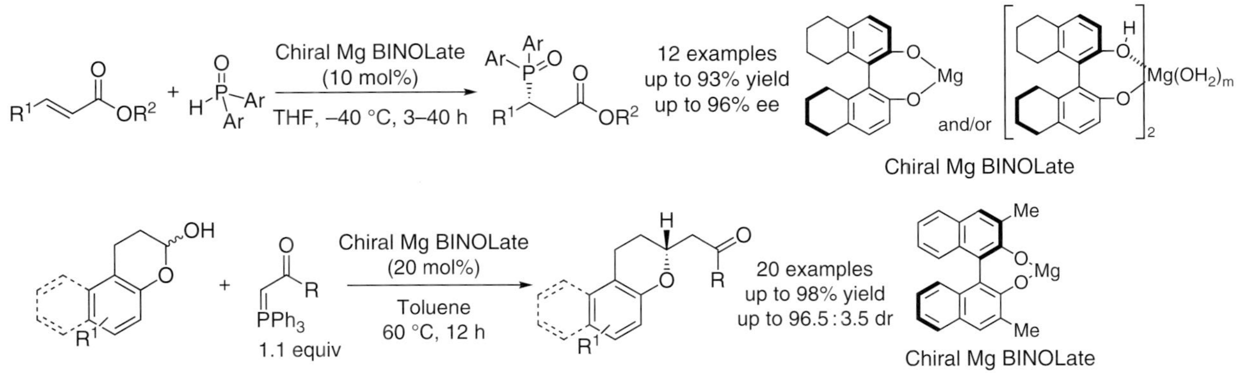

Scheme 12.7 Enantioselective addition reactions by chiral magnesium BINOLate complexes.

the catalytic activity. They proposed a catalyst bearing two covalent or covalent and coordinative Mg—O bonds. This is the first examples of enantioselective 1,4-hydrophosphinylation of α,β-unsaturated esters with diaryl phoshine oxides. Enantioselective reactions of hemiacetals and phosphorous ylides using a magnesium BINOLate complex (type I) were reported by Yang and Wang et al. in 2018 [18]. This reaction is a tandem Wittig-*oxa* Michael reaction. Interestingly, the by-product, triphenylphosphine oxide, was an essential additive for this enantioselective reaction. Chromans were obtained in high yields with excellent enantioselectivities and could be transformed into (−)-erythrococcamide B.

12.2.3 Chiral Magnesium-Catalyzed Addition Reactions to Carbonyl Compounds

In this section, enantioselective addition reactions with carbonyl groups were focused. Here, cascade reactions including aldol reactions, ene-reactions, 1,2-hydrophosphonylations of ketones, and 1,2-additions of ethyl diazoacetate to aldehydes were described.

Enantioselective aldol reactions of α-isothiocyanato esters with ketones following cyclization reaction using a Salen–magnesium complex (type I) were reported by Matsunaga and Shibasaki et al. in 2009 (Scheme 12.8) [19]. The desired protected α-amino-β-hydroxy esters with contiguous tetrasubstituted chiral carbon stereocenters were obtained in high yields with excellent diastereo- and enantioselectivities and could be transformed into the corresponding amino alcohols. Enantioselective addition reactions of alkyl enol ethers to 1,2-dicarbonyl compounds using a chiral *N*,*N*′-dioxide– magnesium complex (type III) were reported by Feng et al. in 2011 [20]. The products of substituted 3-alkyl-3-hydroxyoxindoles were obtained in high yields with excellent enantioselectivities. One of the products could be converted into (*R*)-convolutamydine A

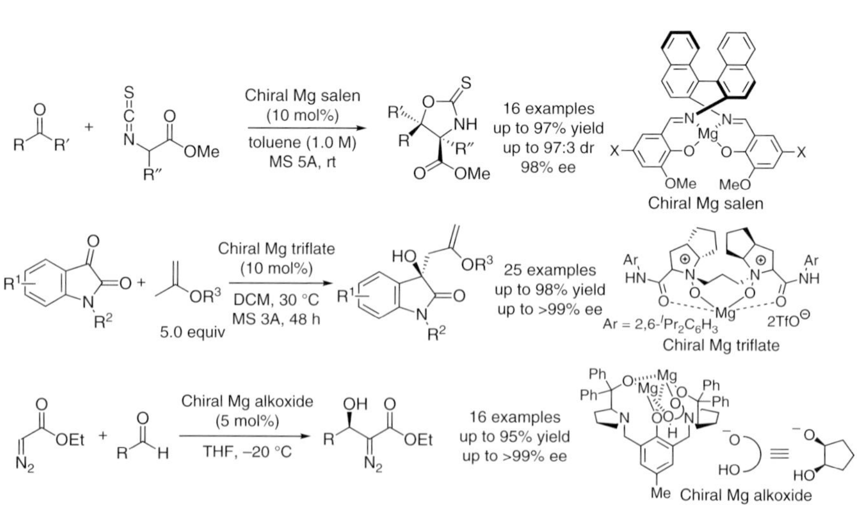

Scheme 12.8 Enantioselective addition reactions to carbonyl compounds catalyzed by chiral magnesium complexes.

in high yield. Moreover, in this catalyst system, α-ketoesters were applicable, and the products were obtained in high enantioselectivities. Enantioselective aldol reactions of ethyl diazoacetate with aldehydes using a chiral magnesium phenoxide (type I) were reported by Trost et al. in 2012 [21]. In this reaction, addition of *meso-cis*-1,2-cyclopentandiol is a key to increase the enantioselectivity. The obtained hydroxy diazoacetates could be transformed into chiral 1,2-diols.

An enantioselective 1,2-hydrophosphonylation of ketones using a magnesium BINOLate complex (type I or II) was reported by Ishihara et al. in 2013 (Scheme 12.9) [17]. The products were obtained in high yields with excellent regioselectivities and moderate to good enantioselectivities; however, the products were highly crystalline, and they were recrystallized from ethanol to remove opposite enantiomers without any serious loss of yields.

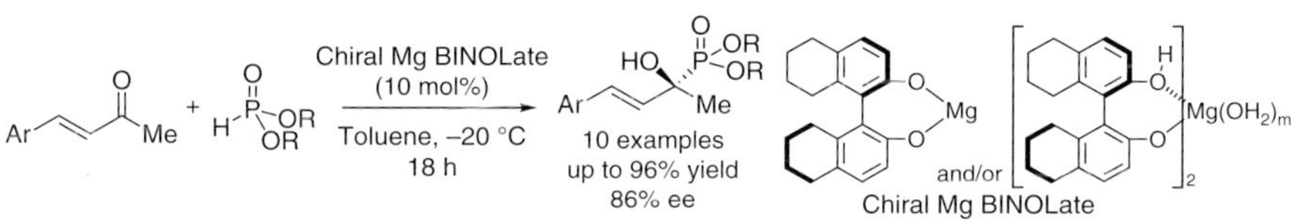

Scheme 12.9 Enantioselective 1,2-hydrophosphonylation reactions catalyzed by a chiral magnesium BINOLate complex.

12.2.4 Chiral Magnesium-Catalyzed Addition Reactions with Imines

Cascade reactions including Mannich reactions of 1,3-diesters with imines, phosphination of imines, and alkoxylation of imines are described.

Enantioselective domino Mannich/cyclization reactions using a DBFox–magnesium complex (type III) were reported by Willis et al. in 2007 (Scheme 12.10) [22]. Mannich reaction followed by cyclization gives α,β-dimamino acid derivatives in high yields and high diastereo- and enantioselectivities. The obtained product could be transformed to a *syn*- diamino acid derivative. Enantioselective direct Mannich-type reactions with malonates using a magnesium-binaphtholate (type I) were reported by Ishihara et al. in 2010 [23]. The desired optically active β-amino esters and α-halo-β-amino ester were obtained in high yields with excellent enantioselectivities. Without any loss of the enantioselectivity, β-phenyl-substituted β-lactam was obtained in good yield. Enantioselective *aza*-Darzens reactions using a chiral magnesium VAPOL phosphate (type I) were reported by Antilla et al. in 2011 [24]. A variety of aziridines were prepared from imines and α-chlorodiketones through Mannich-type reactions followed by cyclization reactions, that is Darzens reaction, the products were obtained in good yields with high enantioselectivities. This is the first example of magnesium phosphate salt-catalyzed enantioselective Mannich-type reactions.

A phosphination of imines using a chiral magnesium BINOL phosphate (type I) was reported by Antilla et al. in 2011 (Scheme 12.11) [25]. The enantio-enriched α-amino phosphine oxides were obtained in high yields. The dibenzocycloheptene-protected imines afforded improved enantioselectivities.

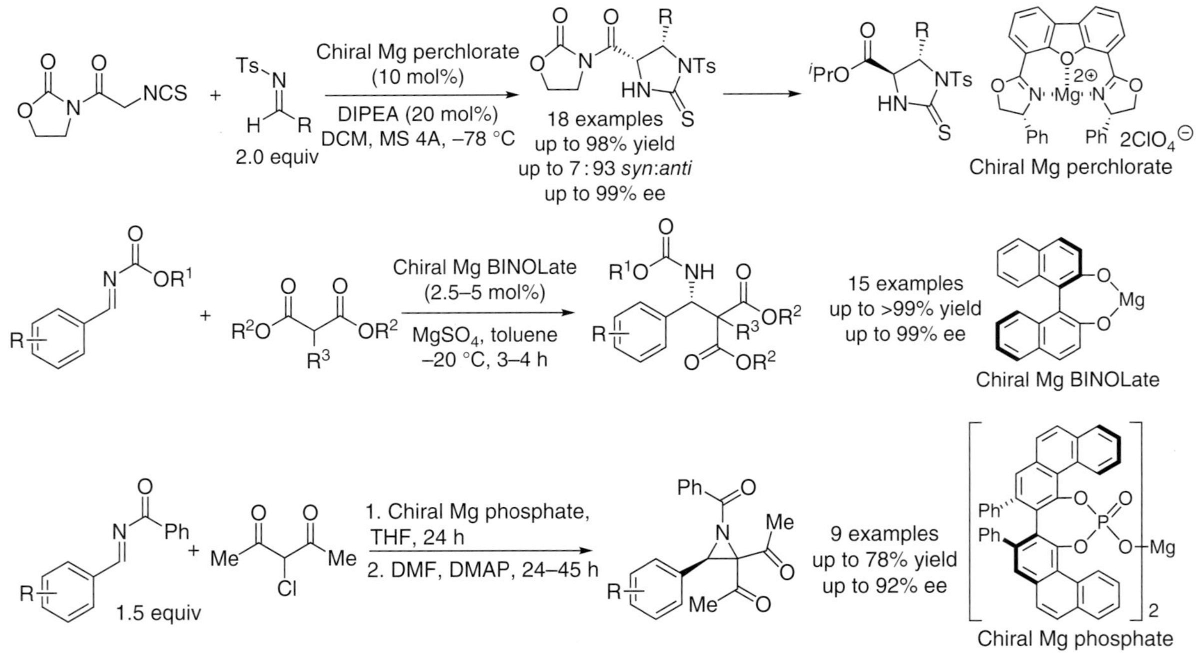

Scheme 12.10 Enantioselective addition reactions to imines catalyzed by chiral magnesium complexes.

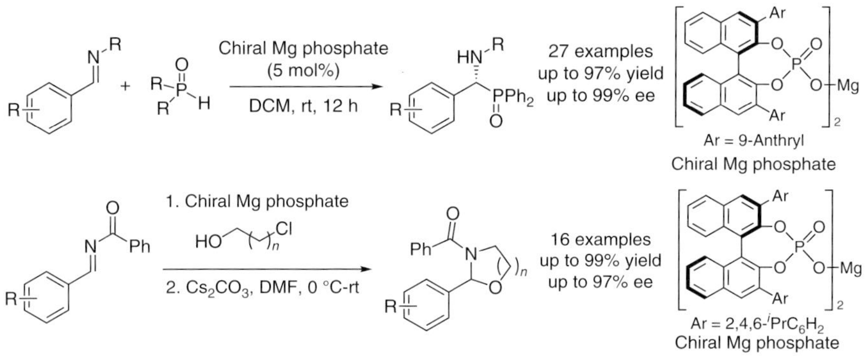

Scheme 12.11 Enantioselective phosphination and alkoxylation by a chiral BINOL phosphate magnesium complex.

The reactions of substituted diphenylphosphine oxide nucleophiles also showed good results. An enantioselective one-pot synthesis of 1,3-oxazolidines and 1,3-oxazinanes using a chiral BINOL phosphate magnesium salt (type I) was reported by Antilla et al. in 2014 [26]. The intermediates of *N*,*O*-acetals were obtained in high yields with high enantioselectivities. The products, 1,3-oxazolidines and 1,3-oxazinanes, that were formed after basic work-up were obtained in a one-pot procedure.

12.2.5 Chiral Magnesium-Catalyzed Ring-Opening Reactions of Epoxide and Aziridine

Enantioselective ring-opening reactions of *meso*-epoxides with aromatic and aliphatic amines using magnesium complexes of BINOL derivatives (type I) were reported by Ding et al. in 2010 (Scheme 12.12) [27]. In this reaction, the products were obtained in high yields with moderate to high enantioselectivities. Using

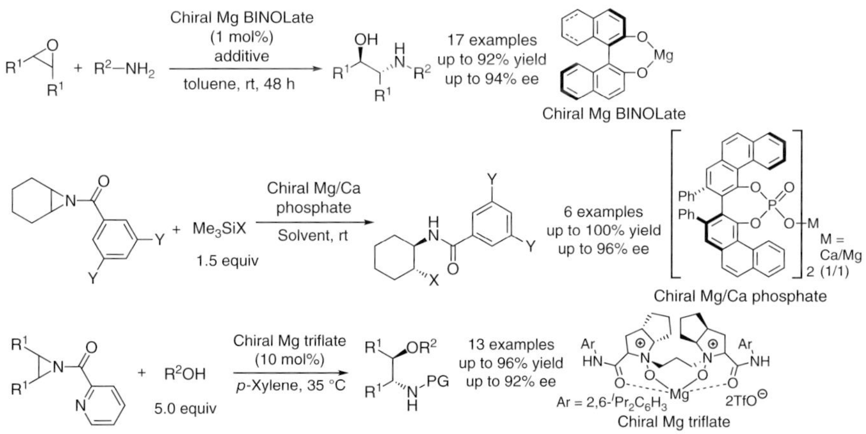

Scheme 12.12 Enantioselective ring-opening reactions catalyzed by chiral magnesium complexes.

aromatic amines, addition of methyl lithium affected the enantioselectivities, but the actual catalyst structure is unclear. In 2013, Sala investigated enantioselective desymmetrization reactions of *meso*-aziridines with silylated nucleophiles using a magnesium–phosphate complex (type I) [28]. The detailed re-examination of previously reported catalyst structures was conducted, and it was found that a 1 : 1 mixture of calcium and magnesium VAPOL phosphates worked well for the ring-opening reactions of (benzylthio)trimethylsilane (Me_3SiSBn), (methylthio)trimethylsilane (Me_3SiSMe), and (trimethylsilyl)isothiocyanate (Me_3SiNCS) with epoxides. The desired products were obtained in high yields with high enantioselectivities. Enantioselective ring-opening reactions of *meso*-aziridines with primary alcohols using a chiral *N,N′*-dioxide–magnesium complex (type III) was reported by Feng et al. in 2014 [29]. The desired *trans*-amino ethers were obtained in high yields with good to high enantioselectivities. In this reaction system, water could be applicable as a nucleophile to form *trans*-amino alcohols in moderate yields with high enantioselectivities. They also investigated an intermediate of the catalyst and aziridine complex by HRMS and found that magnesium ion, aziridine, ligand, and triflate ion were included in the complex.

An enantioselective synthesis of indole homo-Michael adducts via kinetic Friedel–Crafts alkylation with cyclopropanes using a Pybox-magnesium iodide complex (type III) was reported by Johnson et al. in 2013 (Scheme 12.13) [30]. The products were obtained in good to high yields with good to high enantioselectivities. *Iso*-propyl ester showed the best enantioselectivities. The presence of a 3-methyl substituent induced cyclization of the putative intermediate to afford the pentannulation product in good enantioselectivities, albeit in modest yields with *endo* diastereoselectivities. Intermolecular enantioselective dearomatization reactions of β-naphthols with *meso*-aziridines using a phenol–oxazoline magnesium complex (type II) were reported by Wang et al. in 2015 [31]. The desired products were obtained in high yields with excellent diastere- and enantioselectivities. The product was converted into [6-6-5] tricyclic core skeleton, which often appears in many natural products.

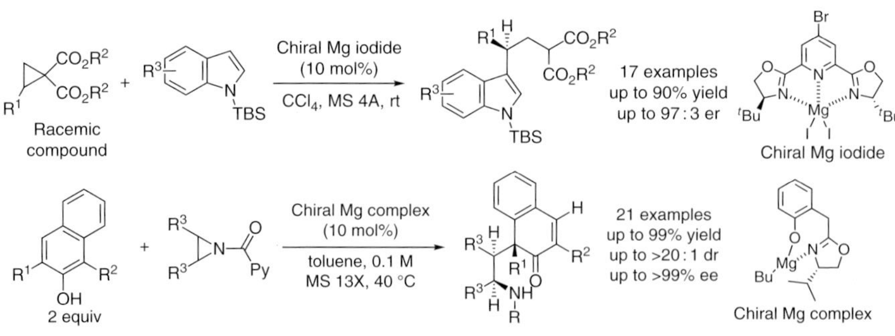

Scheme 12.13 Enantioselective ring-opening reactions through C—C bond formation by chiral magnesium complexes.

12.2.6 Chiral Magnesium-Catalyzed α-Functionalization Reactions of Carbonyl Compounds

In this section, addition reactions of azodicarboxylates with nitrosocarbonyl compounds and hydroxylation reactions using *tert*-butyl hydroperoxide were described. An enantioselective amination of *N*-acyloxazolidinones using a chiral magnesium bis(sulfonamide) complex (type I) was reported by Evans et al. in 1997 (Scheme 12.14) [32]. The desired arylglycine derivatives were obtained in high yields with high enantioselectivities, and these compounds were recrystallized to improve the enantioselectivities. Addition of *N*-methyl-*p*-toluenesulfonamide increased the reaction rate, but the catalytic process was not completely elucidated. Enantioselective α-hydroxylations of β-ketoesters and β-ketoamides using a chiral *N*,*N*′-dioxide–magnesium complex (type III) were reported by Feng et al. in 2013 [33]. The desired α-hydroxy carbonyl compounds were obtained in high yields with high enantioselectivities. Addition of tetramethylethylenediamine (TMEDA) and water were important to improve the enantioselectivities. They estimated that TMEDA was effective for deprotonation of the starting material and that water was effective for protonation of the product. Furthermore, the precursor of daunomycin was prepared. Direct aminations of β-ketoesters using *in situ* generated nitrosocarbonyl compounds as nitrogen sources and a chiral *N*,*N*′-dioxide-magnesium complex (type III) were reported by Maji and Yamamoto et al. in 2014 [34]. In these types of amination reactions, azodicarboxylates had often been employed. However, the cleavage of the N—N bond required harsh conditions. To address the issue, they investigated addition reactions of β-ketoesters with *in situ* generated nitrosocarbonyl compounds. The desired products were obtained in excellent yields with high enantioselectivities. Furthermore, the undesired *O*-NA (aminooxylation) products were not detected in any of these cases. The obtained hydroxyamination products could be converted into the corresponding amines.

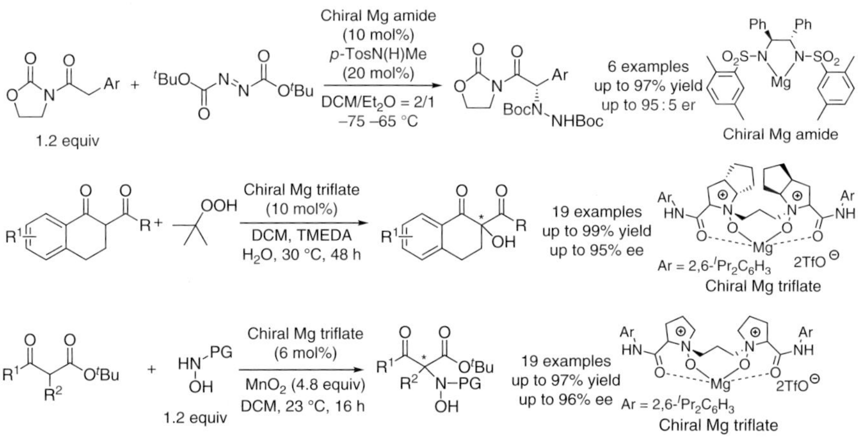

Scheme 12.14 α-Functionalization reactions of carbonyl compounds by chiral magnesium complexes.

12.2.7 Various Chiral Magnesium-Catalyzed Reactions

Other enantioselective reactions using chiral magnesium complexes have also been reported. An enantioselective epoxidation of chalcones with *tert*-butyl hydroperoxide using a diethyl tartrate magnesium complex (type I) was reported by Jackson et al. in 1997 (Scheme 12.15) [35]. The desired products were obtained in moderate yields with high enantioselectivities. An enantioselective hydroamination of aminoalkens using a chiral phenoxyamine magnesium complex (type II) was also reported by Hultzsch et al. in 2012 [36]. Enantioselective intramolecular hydroamination proceeded to form enantio-enriched substituted pyrrolidines. In some substrates, the highest enantioselectivities were detected among other published results. An enantioselective double C(sp^3)—H bond functionalization based on a sequential hydride shift/cyclization process using a chiral magnesium bisphosphate complex (type I) was reported by Mori and Akiyama et al. in 2018 [37]. This reaction consists of stereoselective domino C(sp^3)—H bond functionalization. The first [1, 5]-hydride shift reaction was performed in high diastereo- and enantioselectivities by chiral magnesium bisphosphate. The second reaction was conducted by using an achiral catalyst. Totally, the desired products were obtained in high yields with excellent diastereo- and enantioselectivities.

12.3 Catalytic Enantioselective Reactions Using Chiral Calcium Complexes

Calcium compounds have a relatively strong Lewis acidity among group 2 metals. Compared to the corresponding magnesium compounds, however, their Lewis acidity is lower. In enantioselective reactions, calcium compounds were employed in Lewis acid catalysis and in pseudo-intramolecular Lewis acid/Brønsted base cooperative catalysis. The relatively small ionic radius among group 2 metals also leads to a precise asymmetric environment, and excellent enantioselectivities were achieved. Because of its relatively strong Lewis acidity, all types of chiral complexes (type I–III) have been prepared and employed in the reactions.

12.3.1 Chiral Calcium-Catalyzed Addition Reactions to Carbonyl Compounds

In 2000, a catalytic enantioselective Morita–Baylis–Hillman reaction using tributyl phosphine and a chiral calcium catalyst prepared from calcium isopropoxide ($Ca(O^iPr)_2$) and (*R*)-BINOL (type I) was reported by Ikegami et al. (Scheme 12.16) [38]. The calcium complex worked as a Lewis acid and promoted the reaction of cyclopentenone with 3-phenylpropanal to afford the desired adduct in 62% yield with 56% ee.

Catalytic direct-type enantioselective aldol reactions using a chiral calcium–diolate complex *in situ* formed from calcium bis(trimethylsilyl)amide tetrahydrofuran complex ($Ca(HMDS)_2(thf)_2$) and a chiral diol (type I) were

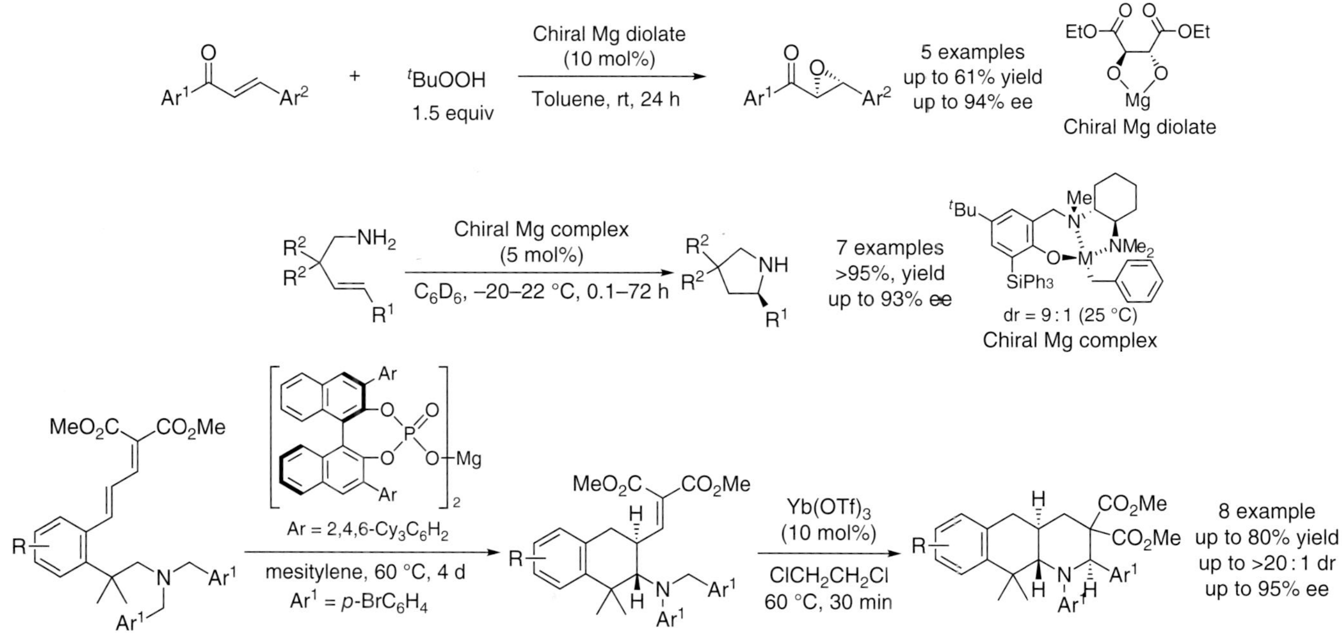

Scheme 12.15 Enantioselective epoxidations, hydroaminations, and cyclization reactions promoted by chiral magnesium complexes.

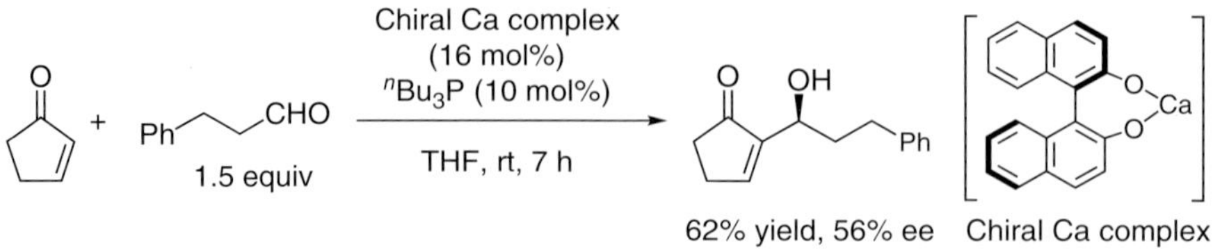

Scheme 12.16 Enantioselective Morita–Baylis–Hillman reaction catalyzed by a chiral calcium BINOLate with tributyl phosphine.

reported by Shibasaki and Noyori et al. in 2001 (Scheme 12.17) [39]. Potassium thiocyanate (KSCN) was found to be a good additive, and the enantioselectivity was dramatically improved. Analysis of the catalyst using cold spray ionization mass spectrometry (CSI-MS) revealed that it exists in a highly aggregated form. A ketone enolate might form kinetically in the presence of the chiral calcium complex.

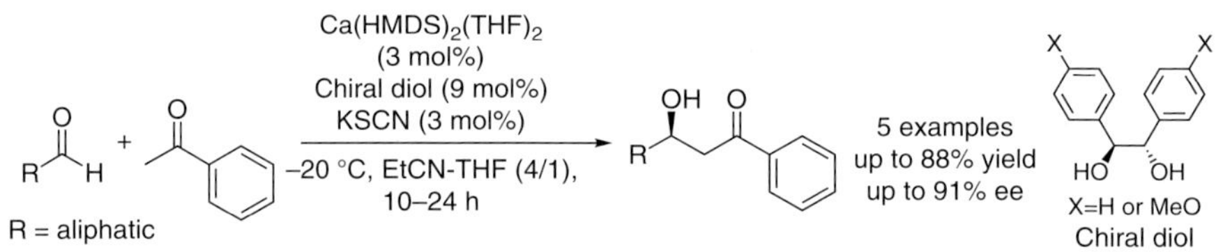

Scheme 12.17 Enantioselective aldol reactions catalyzed by chiral calcium alkoxide.

Catalytic enantioselective carbonyl-ene reactions and Friedel–Crafts reactions of indole derivatives were also reported by Rueping [40] and Guo [41] in 2012 and 2017, respectively (Scheme 12.18). A BINOL phosphoric acid calcium salt (type I) promoted the reaction, and the desired adducts were obtained in high yields with high enantioselectivities.

12.3.2 Chiral Calcium-Catalyzed 1,4-Addition Reactions

Chiral calcium complexes are also good catalysts for 1,4-addition reactions. In 2001, Kumaraswamy et al. reported that a chiral calcium–BINOL complex (type I) promoted enantioselective 1,4-addition reactions of malonates or β-ketoesters with α,β-unsaturated carbonyl compounds (Scheme 12.19) [42]. The desired reactions proceeded in good to high yields with moderate to good enantioselectivities. The phenoxide part of the calcium complexes worked well as good Brønsted basic sites.

In 2007, Kobayashi et al. reported that catalytic enantioselective 1,4-addition reactions and [3+2] cycloaddition reactions of Schiff bases of α-aminoesters with α,β-unsaturated carbonyl compounds using chiral calcium complexes prepared from calcium alkoxides and chiral Box ligands bearing methylene moieties with an acidic hydrogen atom (type II) (Scheme 12.20) [43]. The Box ligand coordinated with the calcium atom in a covalent manner via deprotonation and formed an excellent asymmetric environment. Box ligands can be prepared from several

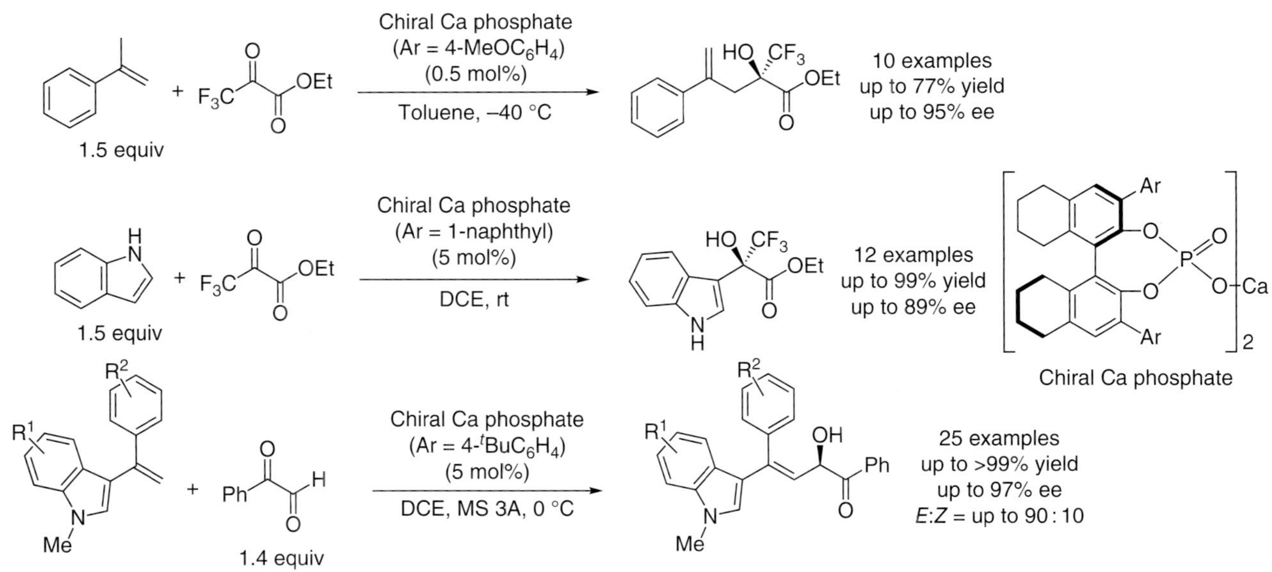

Scheme 12.18 Enantioselective ene-type reactions and Friedel–Crafts reactions catalyzed by chiral calcium phosphate complexes.

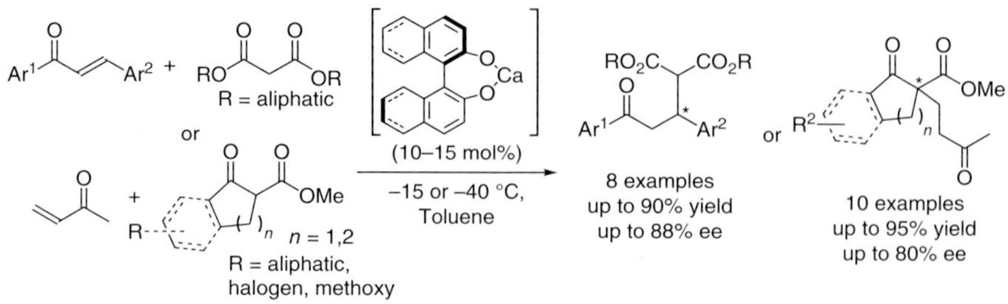

Scheme 12.19 Enantioselective reactions catalyzed by chiral calcium BINOLates.

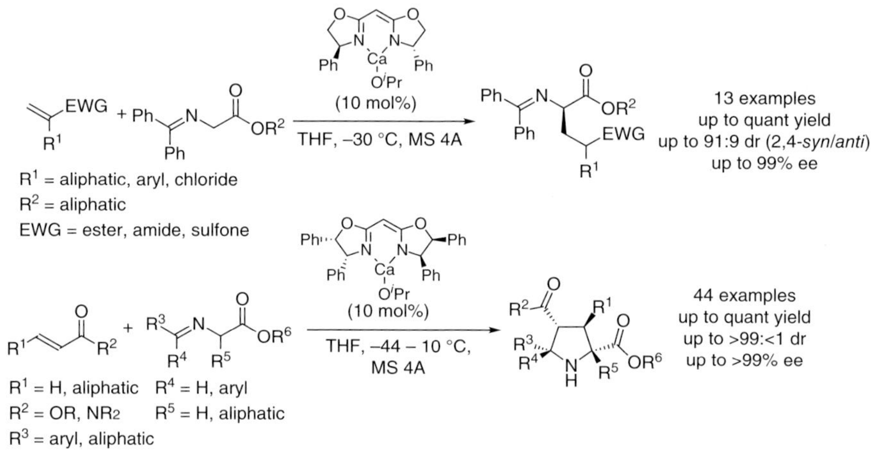

Scheme 12.20 Enantioselective 1,4-addition and [3+2] cycloaddition reactions.

chiral α-amino acids and 1,2-amino alcohols giving for a high degree of variability. The calcium complex still has an alkoxide counterion showing strong Brønsted basicity to promote the desired reactions efficiently with high chemo- and enantioselectivities. The chiral calcium–Box complexes were also applied in highly diastereoselective reactions with β-substituted-α,β-unsaturated carbonyl compounds [44] or in 1,4-addition reactions in the presence of an external amine [45]. It should be noted that calcium chloride with a chiral ligand could be used for asymmetric catalysis; indeed, this is the first example of chiral type II calcium chloride-catalyzed enantioselective reactions.

Also, in the enantioselective reactions using chiral calcium–Box catalysts, the other types of type III were applied. Kobayashi et al. showed in 2009 (Scheme 12.21) that calcium aryloxide–Pybox complexes catalyzed enantioselective 1,4-addition reactions of malonates with nitroalkenes, and the desired 1,4-adducts were obtained in high yields with high enantioselectivities [46]. The catalyst loading could be decreased to 1 mol%. The Lewis acidic calcium metal accepted a chiral ligand via only coordinative bonds constructing a highly asymmetric environment. Pyboxes are easily available chiral ligands and can be prepared from various chiral α-amino acids and 1,2-amino alcohols. It was found that enantioselective 1,4-addition reactions of malonates with

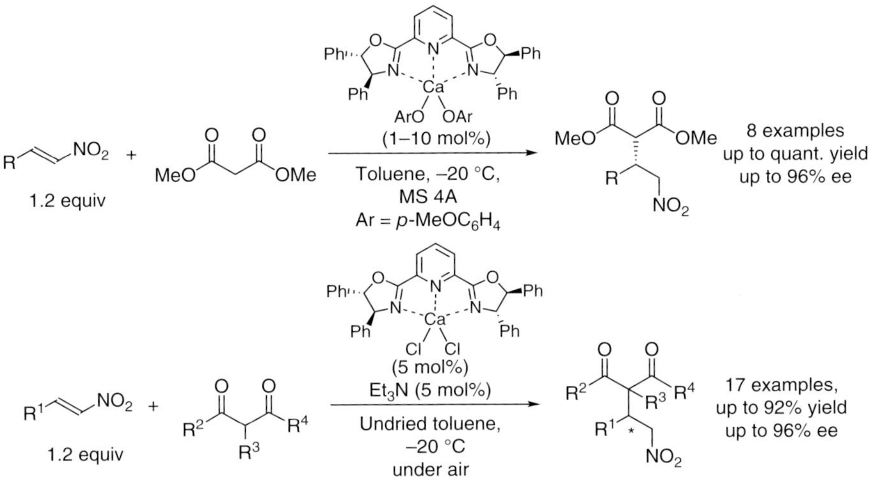

Scheme 12.21 Enantioselective 1,4-addition to nitroalkenes.

nitroalkenes also proceed in good yields with high enantioselectivities using a chiral calcium chloride–Pybox complex and triethylamine as an additional amine base (Scheme 12.21) [47]. This is the first example of chiral calcium chloride-catalyzed asymmetric reactions.

The 1,4-addition reactions were also conducted using a chiral calcium chloride–polymer-supported Pybox (PS-Pybox) complex (type III) under continuous-flow conditions by Kobayashi et al. (Scheme 12.22). Continuous-flow synthesis is currently a hot research field for the environmentally benign, efficient, and safe preparation of fine chemicals. Especially continuous-flow reactions using an immobilized chiral catalyst is a powerful method to supply optically active compounds with high performance using an expensive chiral source. However, it is known that immobilization of chiral catalysts is often problematic because of (i) a decrease in the enantioselectivity of the reaction upon immobilization, (ii) a decrease in catalyst activity upon use over longer periods of time, and (iii) leaching of precious metal catalysts, which is a serious issue that can lead to a contamination of the product by toxic material. On the other hand, calcium chloride is ubiquitous, air-stable, less toxic, less harmful, and a more economical metal salt, which is often used in chemical procedures

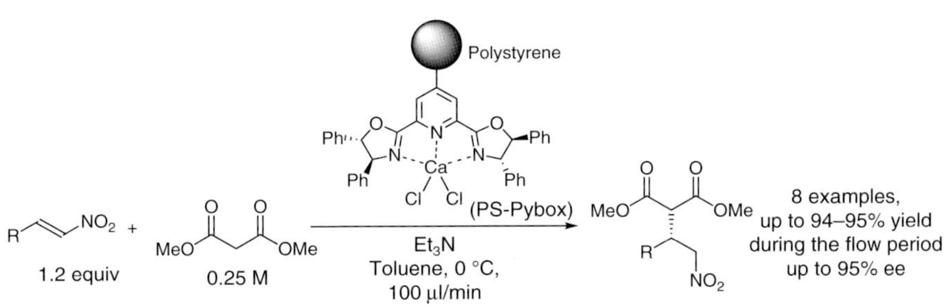

Scheme 12.22 Enantioselective reactions catalyzed by a Pybox–calcium chloride complex.

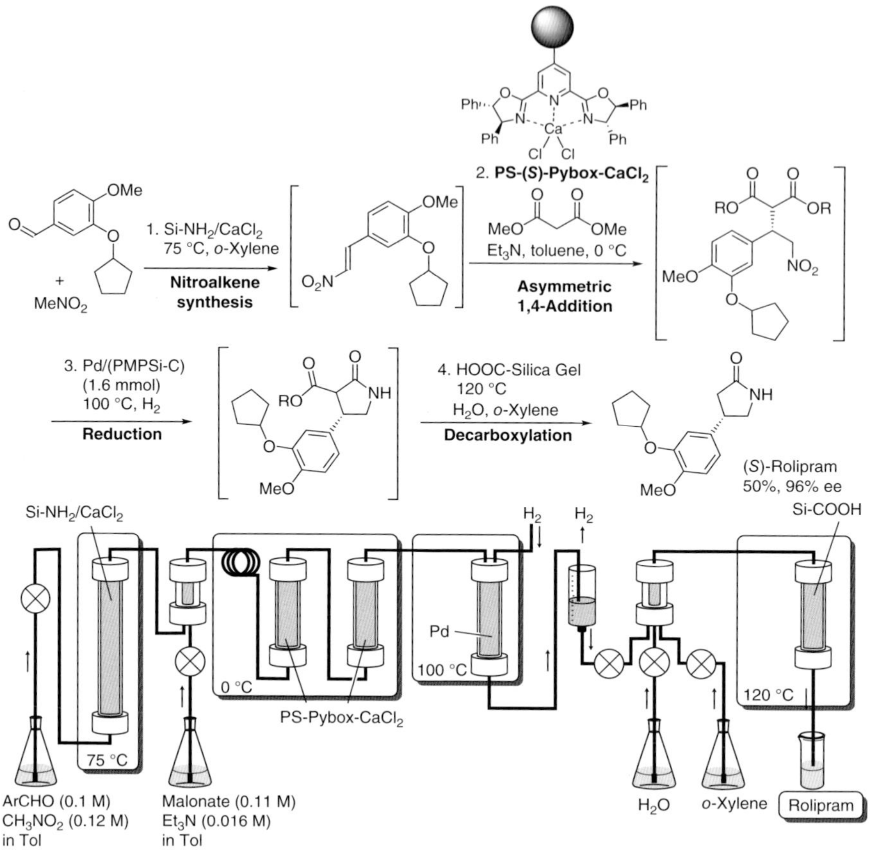

Figure 12.2 Continuous-flow synthesis of chiral Rolipram.

for drying chemicals under vacuum but also in our daily life as a desiccant, food additive, snow melting agent, etc., and, consequently, it is one of the most attractive metal compounds for continuous-flow reactions. Prepared calcium chloride–PS-Pybox is a stable catalyst even after immobilization, and the desired 1,4-addition reaction of a malonate with a nitroalkene proceeded even under air to afford the desired product with high enantioselectivity.

Furthermore, the enantioselective 1,4-addition reaction of a malonate with a nitroalkene using the chiral calcium chloride–PS-Pybox complex was successfully applied as a key step for the multistep continuous-flow synthesis of an optically active anti-inflammatory drug, Rolipram (Figure 12.2) [48]. By connecting reaction columns with the flow reactor, (*R*)-Rolipram was obtained in high yield with excellent enantioselectivity. Advantages of the chiral calcium chloride catalyst were clearly shown in the continuous-flow system. This is the first example of the synthesis of an optically active drug under continuous-flow conditions by using only heterogeneous catalysts.

Catalytic 1,4-addition and enantioselective protonation reactions were also reported by Kobayashi et al. in 2010 (Scheme 12.23) [49]. Enantioselective

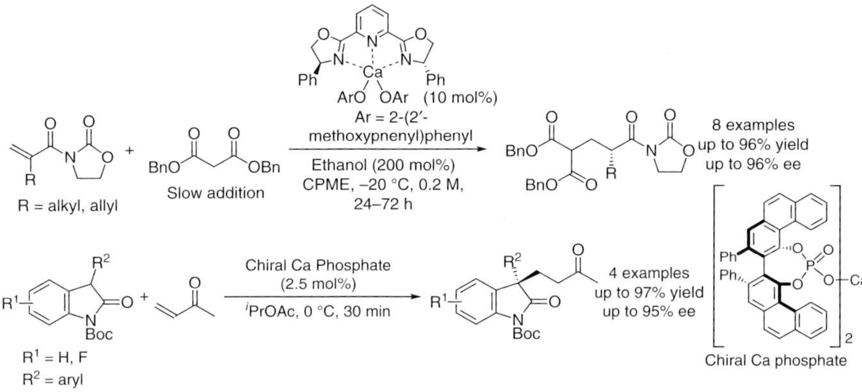

Scheme 12.23 Catalytic 1,4-addition and enantioselective protonation reactions and 1,4-addition reactions of 3-substituted oxindoles catalyzed by chiral calcium catalysts.

protonation at α-position of a carbonyl compound *via* enolate formation is a good method to introduce a chiral center at the α-position. In this reaction, a catalytically formed calcium enolate after the 1,4-addition reaction of dibenzyl malonate was protonated enantioselectively. A chiral calcium aryloxide–Pybox complex (type III) worked well in the presence of an excess amount of ethanol. Enantioselective 1,4-addition reactions of 3-substituted oxindoles with methyl vinyl ketone using a chiral calcium phosphate (type I) were reported by Antilla et al. in 2011. The desired reactions proceeded in high yields with high enantioselectivities [50]. As other enantioselective 1,4-addition reactions using chiral calcium phosphates, the 1,4-addition reactions of azlactones [51], indole [52], and alkylamines [53] were also reported.

12.3.3 Chiral Calcium-Catalyzed Addition Reactions with Imines

Chiral calcium-catalyzed addition reactions with imines were developed. In 2010, a chiral calcium alkoxide–Pybox complex (type III) was employed in enantioselective Mannich-type reactions of malonates with *N*-Boc imines by Kobayashi, et al. (Scheme 12.24) [54]. The reactions proceeded in high yields, but the enantioselectivities observed were not so high. Not only aromatic but also aliphatic Boc imines reacted with malonates, giving high yields and enantioselectivities. In some cases, however, free calcium aryloxide or alkoxide without Pybox might promote the desired reaction, leading to a decrease in the enantioselectivity. This was caused by incomplete complexation of the calcium alkoxide and the chiral ligand because of the lower Lewis acidity of calcium. Later, a chiral calcium iodide–Pybox complexes was successfully applied for the Mannich-type reactions of malonates [55]. Enantioselective Mannich-type reactions of β-ketothioesters with Boc imines using a chiral calcium complex prepared from calcium alkoxide and a phosphoric acid bearing a modified BINOL were reported by Ishihara et al. [56] This complex exists in an oligomeric structure in the absence of substrates, but it forms a calcium-ligand 1 : 2 complex by deaggregation for the actual catalysis. Other similar enantioselective

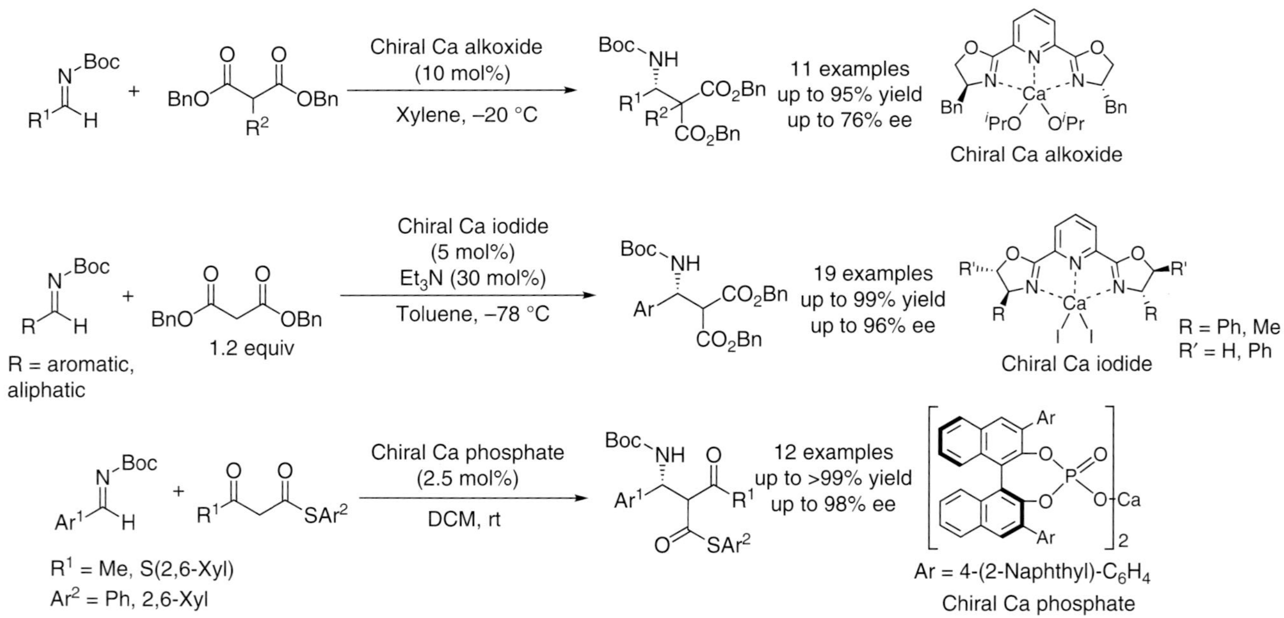

Scheme 12.24 Enantioselective Mannich-type reactions of 1,3-dicarbonyl compounds using chiral calcium–Pybox and phosphate complexes.

Mannich-type reactions were also promoted by a chiral calcium phosphate complex [57].

It was also reported that oxyindole derivatives reacted with Boc imines in high enantioselectivities by Kobayashi et al. in 2006 (Scheme 12.25) [58]. The calcium complex could control stereochemistry of the product strictly, and adjacent quaternary and tertiary carbon centers were constructed efficiently.

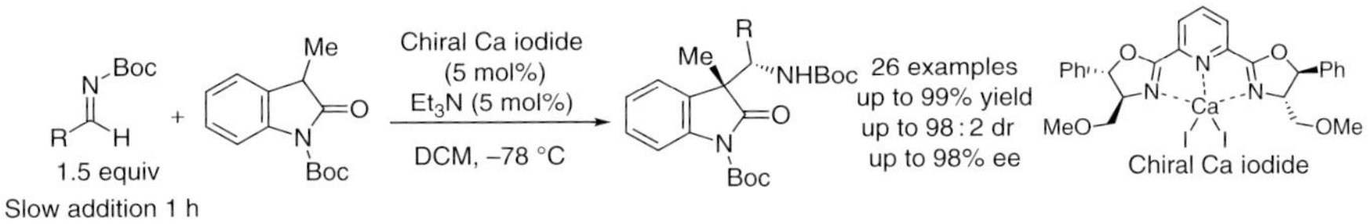

Scheme 12.25 Enantioselective Mannich-type reactions catalyzed by a chiral calcium iodide.

Additionally, catalytic enantioselective phosphination (Scheme 12.26) [24a], aziridine formation [24b], and alcohol addition reactions [25] were reported using chiral calcium–phosphate complexes.

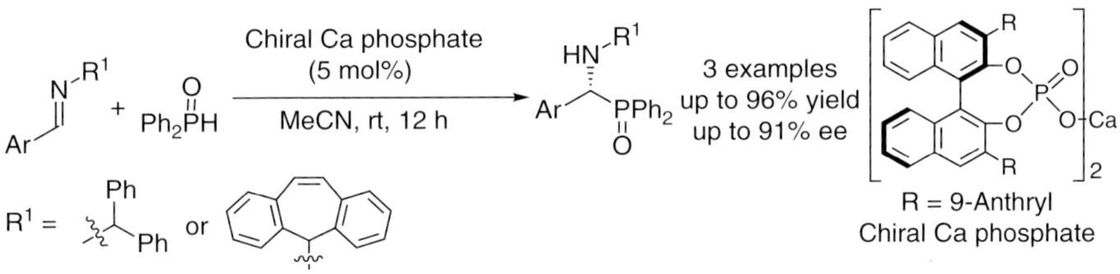

Scheme 12.26 Enantioselective phosphination reactions with imines catalyzed by chiral calcium phosphates.

12.3.4 Chiral Calcium-Catalyzed α-Functionalization Reactions with Carbonyl Compounds

Introduction of heteroatoms to α-position of carbonyl groups is an important method to obtain highly functionalized carbonyl compounds. Until now, enantioselective introduction of nitrogen, oxygen, or halogen atoms has been investigated, and chiral calcium–phosphate complexes (type I) were found to be good catalysts. Chiral calcium-catalyzed enantioselective amination reactions of enamides with an azodicarboxylate were reported by Masson et al. in 2011 (Scheme 12.27) [59]. The products were obtained as 2-hydrazinoketones or 1,2-diamines after hydrolysis or diastereoselective reduction, and the enantioselectivities obtained were very high. It was also reported that a similar catalyst promoted the enantioselective bromination of enecarbamates using *N*-bromosuccinimide (NBS) [60]. Instead of NBS, *N*-chlorosuccinimide (NCS) also worked in the halogenation reactions [61].

Enantioselective oxidation and chlorination reactions with 3-substituted oxindoles were also developed by Antilla et al. in 2011 using a chiral calcium–phosphate bearing VAPOL (type I), and high enantioselectivities were achieved (Scheme 12.28) [62].

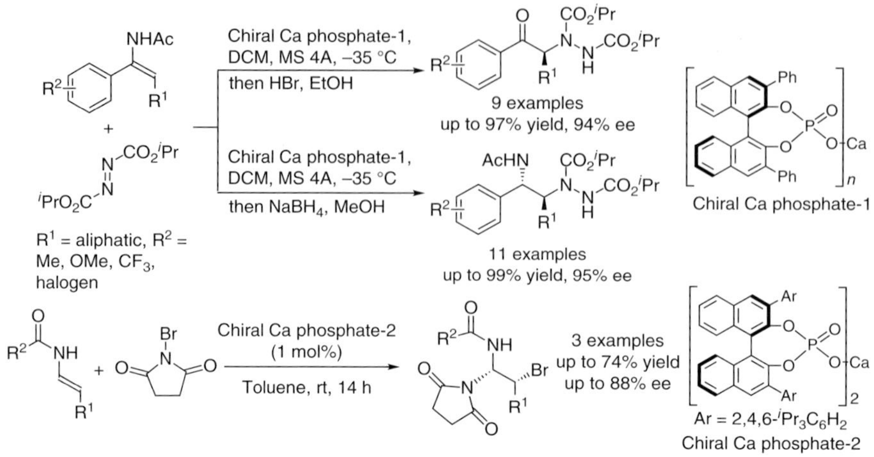

Scheme 12.27 Enantioselective amination and bromination reactions catalyzed by chiral calcium phosphates.

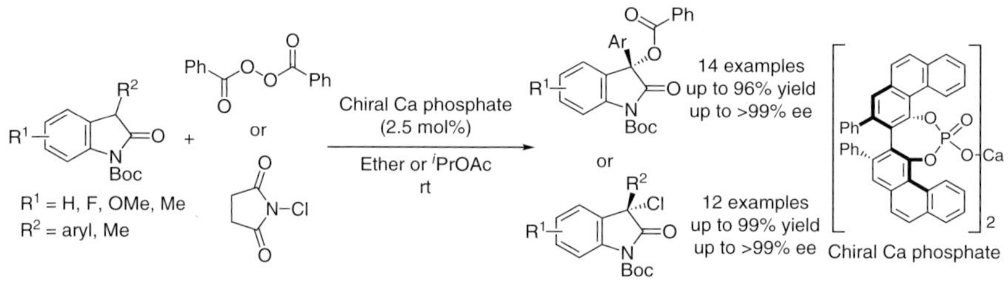

Scheme 12.28 Enantioselective oxidation and chlorination reactions catalyzed by a chiral calcium phosphate bearing VAPOL.

12.3.5 Chiral Calcium-Catalyzed Cycloaddition Reactions

Catalytic enantioselective [4+2] cycloaddition reactions were reported by Antilla et al. in 2013 using chiral calcium–phosphate complexes (type I) (Scheme 12.29). Enantioselective Diels–Alder reactions of 3-siloxydiene with an activated alkene were investigated and the desired product was obtained in moderate yield and enantioselectivity [63]. Hetero-Diels–Alder reactions of siloxydienes with α-ketoesters were also found to be promoted to afford the desired cycloadducts in high yields with high enantioselectivities [64]. Zhu and Cheng et al. reported that enantioselective *oxo*-hetero-Diels–Alder reactions of heterodienes with vinyl ethers proceeded with high enantioselectivities [65].

12.3.6 Chiral Calcium-Catalyzed Hydroamination Reactions

Catalytic enantioselective hydroamination and hydrosilylation reactions by using chiral calcium catalysts prepared from $Ca(HMDS)_2(thf)_2$ and a chiral diimine or

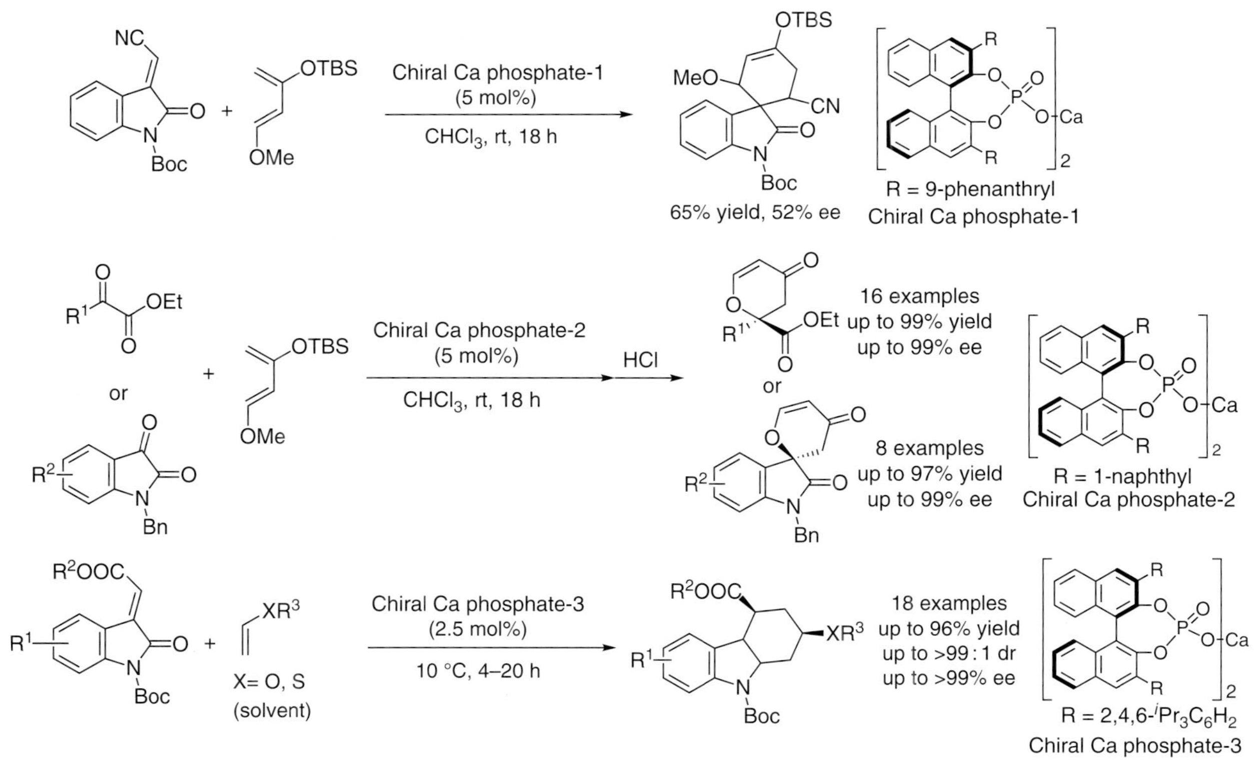

Scheme 12.29 Enantioselective [4+2] cycloaddition reactions using chiral calcium phosphate complexes.

Box-ligand (type II) were investigated by Harder et al. in 2008 (Scheme 12.30) [66]. Although the yields were high in both reactions, the enantioselectivities were very low. This was explained by a ligand exchange equilibrium between the chiral catalysts (Box)Ca(HMDS) and $(Box)_2Ca$ ad $Ca(HMDS)_2$. The latter Ca amide species might mediate undesired achiral background reactions.

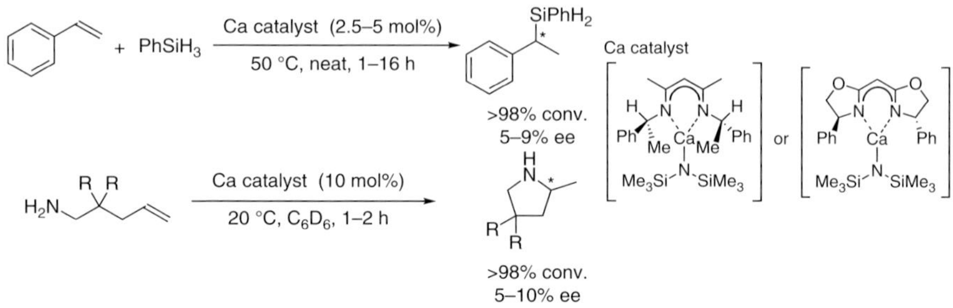

Scheme 12.30 Enantioselective alkene hydrosilylation and hydroamination reactions catalyzed by chiral calcium amide complexes.

12.3.7 Chiral Calcium-Catalyzed Epoxidation Reactions

Enantioselective epoxidation of alkenes is one of the most commonly used reactions for the formation of optically active epoxides. Although many enantioselective epoxidations using chiral transition metal catalysts have been developed, chiral calcium catalysts were also successfully applied (Scheme 12.31). A chiral calcium–BINOL complex of type I was employed by Kumaraswamy et al. for epoxidation of chalcone derivatives using *tert*-butyl hydrogen peroxide as an oxidant [42]. The reaction proceeds in high yields with good enantioselectivities. Aqueous hydrogen peroxide solution was also used for the epoxidation reaction. A water-tolerant chiral calcium catalyst prepared from calcium bistrifluoromethanesulfonimide ($Ca(NTf_2)_2$) and Pybox ligand (type III) was employed by Kobayashi et al. in 2018 [67]. The desired epoxides were obtained in good enantioselectivities. The use of hydrogen peroxide in the reactions is ideal from an atom-economical perspective because only water forms as a coproduct during the reaction.

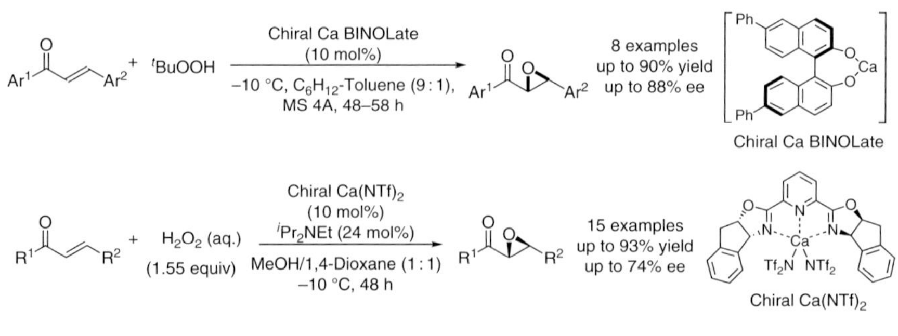

Scheme 12.31 Enantioselective epoxidation reactions using chiral calcium catalysts.

12.3.8 Chiral Calcium-Catalyzed Aziridine Ring-Opening Reaction

Enantioselective ring-opening reactions of aziridines with Me_3SiNCS were reported by Nakamura et al. in 2014 using a calcium BINOL-based chiral imidazoline-phosphonate complex (type II) (Scheme 12.32) [68]. The desired reactions proceeded with good to high enantioselectivities. The 2-pyridylsulfonyl group of the aziridines might coordinate with the Lewis acidic calcium metal to facilitate an efficient enantioselective transformation.

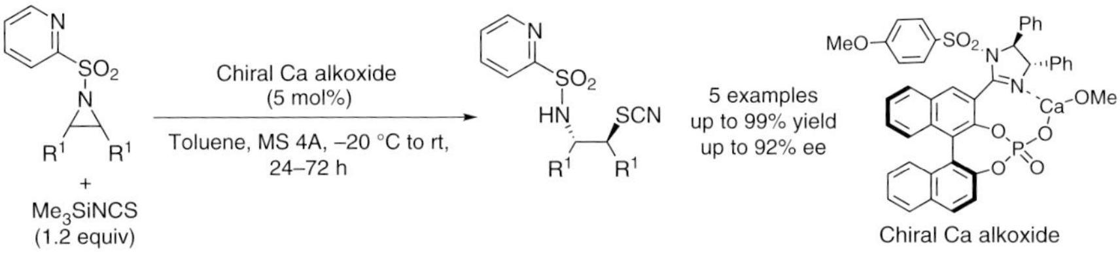

Scheme 12.32 Enantioselective aziridine ring-opening reactions catalyzed by a calcium BINOL-based chiral imidazoline–phosphonate complex.

12.4 Catalytic Enantioselective Reactions Using Chiral Strontium Complexes

Strontium compounds possess a mild Lewis acidity. Compared to calcium complexes, however, their Lewis acidity is a little lower, which is due to the larger ionic radius of strontium relative to that of calcium. Strontium compounds do not possess a significantly different characteristic nature as metal catalysts; therefore, development of enantioselective reactions using strontium compounds as catalysts have not been well investigated. Because of its low Lewis acidity, chiral ligands are mostly introduced by covalent bonding (type I).

12.4.1 Chiral Strontium-Catalyzed 1,4-Addition Reactions

In 2008, the first chiral strontium-catalyzed enantioselective reaction, a 1,4-addition reaction of chalcone with malonate, was reported by Kobayashi et al. (Scheme 12.33) [69]. A chiral strontium complex prepared from strontium *iso*propoxide and bis(sulfonamide) derived from 1,2-diphenylethylenediamine (type I) promoted the 1,4-addition reactions effectively. Among other group 2 metal compounds, strontium compounds showed the best enantioselectivity. Strontium hexamethyldisilazide ($Sr(HMDS)_2$) was found to be a good precursor for the strontium catalyst, forming a more active catalyst [70].

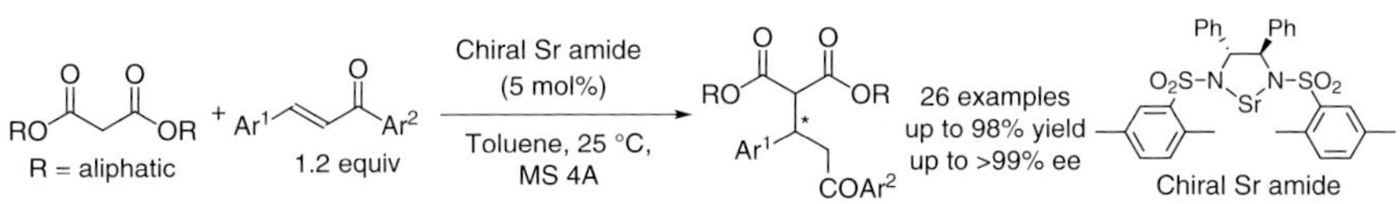

Scheme 12.33 Enantioselective 1,4-addition reactions catalyzed by a chiral strontium complex prepared from bis(sulfonamide) ligand.

Enantioselective 1,4-addition reactions of silyl cyanide with β,β-disubstituted α,β-unsaturated carbonyl compounds were also developed by Shibasaki and Kanai et al. in 2010 using a chiral strontium complex prepared from a chiral diol ligand (type I) (Scheme 12.34) [71]. ESI-MS analysis suggested that the strontium catalyst exists in an oligomeric form and that two strontium centers worked cooperatively to activate both the substrate and cyanide. In this reaction, strontium showed better results than similar lanthanoid metal complexes, and chiral quaternary carbon stereocenters were formed with high enantioselectivities.

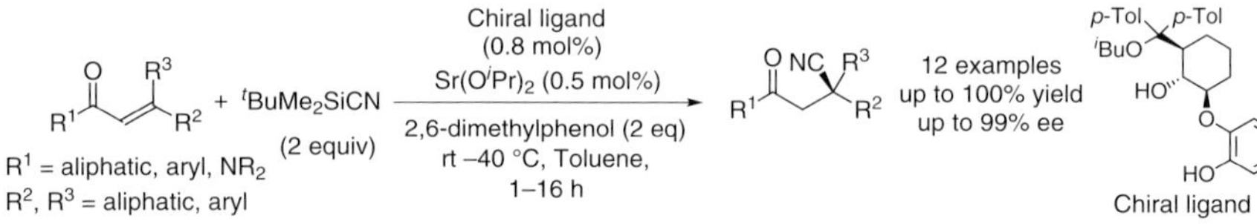

Scheme 12.34 Enantioselective 1,4-addition reactions of cyanide.

12.4.2 Chiral Strontium-Catalyzed Addition Reactions with Imines

The strontium-catalyzed enantioselective additions to imines have also been developed. In 2009, Kobayashi et al. reported that an enantioselective Mannich-type reaction of sulfonylimidate, a less acidic ester equivalent, with a Boc imine proceeded smoothly, and the desired product was obtained in good yield with moderate enantioselectivity in the presence of additional tertiary amine (Scheme 12.35) [72]. In 2011, Shibasaki and Matsunaga et al. reported that a strontium–salen complex (type I) promoted enantioselective Mannich-type reactions of α-isothiocyanate esters with ketoimines effectively, and the desired products were obtained in high yields with good to high anti-selectivities and high enantioselectivities [73]. When the same reaction was conducted by using a magnesium–salen complex, opposite syn selectivity was observed. Switching of the diastereoselectivity might be due to the difference in the dihedral angles of the binaphthyl unit in the metal complexes.

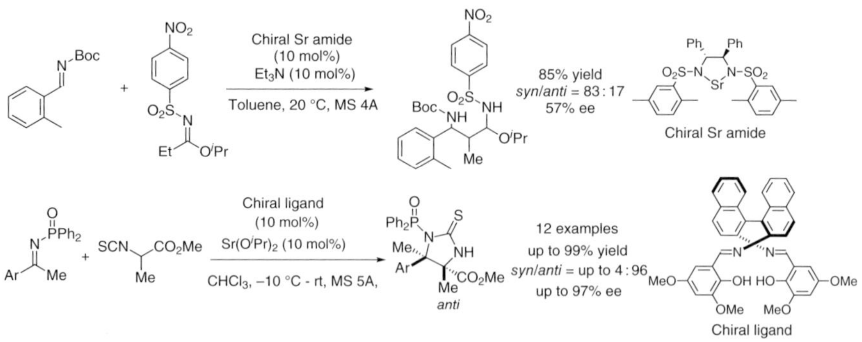

Scheme 12.35 Enantioselective Mannich-type reactions catalyzed by chiral strontium complexes.

12.4.3 Chiral Strontium-Catalyzed Oxime Formation

Catalytic enantioselective synthesis of novel axially chiral 4-substituted cyclohexylidene oximes was reported by Antilla et al. in 2017 using a chiral strontium–phosphate complex (type I) (Scheme 12.36). The reactions of 4-substituted cyclohexanones with *O*-arylhydroxylamines afforded the desired oximes with high enantioselectivities [74]. The unique chirality was caused by the restricted rotation of C=N double bond. The enantioselectivity was better than that in reactions using the chiral calcium– and magnesium–phosphate complexes.

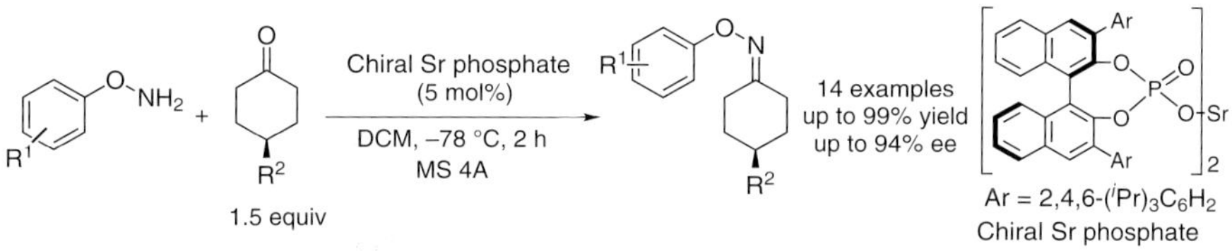

Scheme 12.36 Enantioselective synthesis of novel axially chiral 4-substituted cyclohexylidene oximes using a strontium catalyst.

12.5 Catalytic Enantioselective Reactions Using Chiral Barium Complexes

Barium compounds have the weakest Lewis acidity among group 2 metal compounds. Therefore, barium compounds were employed in pseudo-intramolecular Lewis acid/Brønsted base cooperative catalysis. Their stronger basicity can realize smooth deprotonation of less acidic substrates. The large ionic radius among group 2 metals normally makes construction of a precise asymmetric environment difficult; however, well-designed chiral barium complexes can realize high enantioselection. Similar to strontium complexes, chiral ligands were introduced by covalent bonding (type I) because of its low Lewis acidity.

12.5.1 Chiral Barium-Catalyzed Addition Reactions to Carbonyl Compounds and Imines

In 1998, Shibasaki et al. reported that a chiral barium catalyst can promote enantioselective direct aldol reactions of a ketone (Scheme 12.37). The stronger Brønsted basicity of barium phenoxide gave smooth deprotonation of acetophenone, and addition to an aliphatic aldehyde occurred enantioselectively [75]. The desired products were obtained with good enantioselectivities. Two monomethylated BINOLs coordinate with the barium atom in a covalent manner (type I). Enantioselective direct aldol reactions using activated esters were also reported by Kobayashi et al. in 2006. The aldol reaction of Boc protected-*N*-arylacetamide gave the desired product with moderate enantioselectivity using a chiral barium catalyst based on 3,3′-disilyl-substituted BINOL (type I) [76].

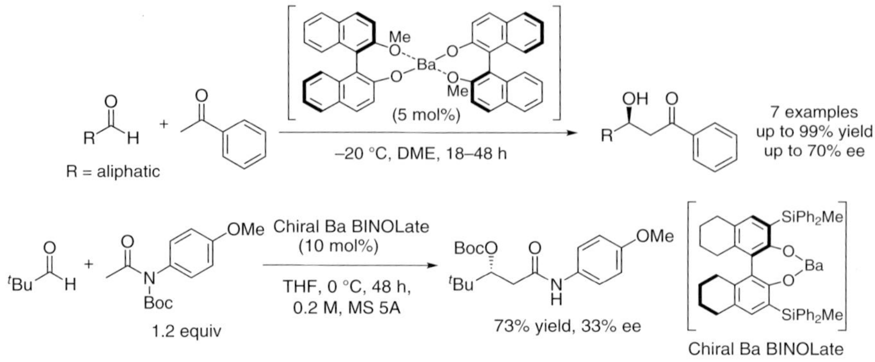

Scheme 12.37 Enantioselective aldol reactions catalyzed by chiral barium BINOLates.

Shibasaki and Matsunaga et al. reported that enantioselective direct aldol-type reactions of a β,γ-unsaturated ester were also promoted by a chiral barium complex in 2009 (Scheme 12.38) [77]. By using the chiral barium-BINOL catalyst (type I), the desired Morita–Baylis–Hillman-type products were similarly obtained in good yields with high enantioselectivities after isomerization. The chiral barium–biphenol complex (type I)-catalyzed addition reaction of the β,γ-unsaturated ester was expanded to the addition reaction of an imine by the same research group. Mannich and *aza*-Morita–Baylis–Hillman-type products were obtained in good yields with good enantioselectivities [78].

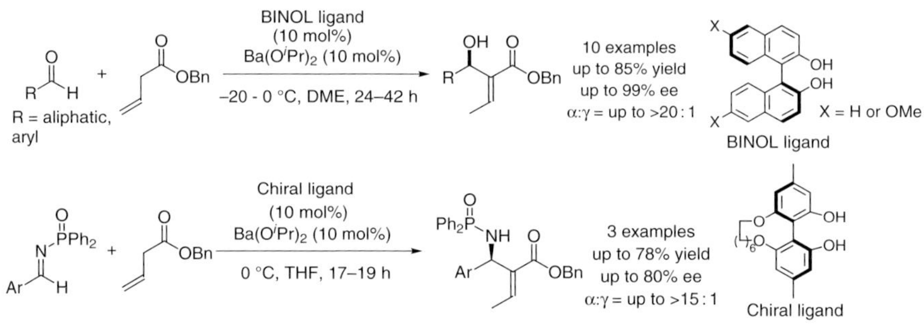

Scheme 12.38 Enantioselective aldol-type reactions of a β,γ-unsaturated ester catalyzed by chiral barium BINOLates.

12.5.2 Chiral Barium-Catalyzed 1,4-Addition Reactions

Chiral barium-catalyzed 1,4-adition reactions of chalcone derivatives with indoles were also reported by Kobayashi et al. in 2010 (Scheme 12.39). A chiral barium complex prepared from barium bis(trimethylsilyl)amide ($Ba(HMDS)_2$) and H_8-3,3′-$(SiPh_3)_2$-BINOL (type I) worked well, and the desired adducts were obtained with high enantioselectivities [52]. In this reaction, the most acidic proton of the indole might be deprotonated to form a chiral barium–indole species.

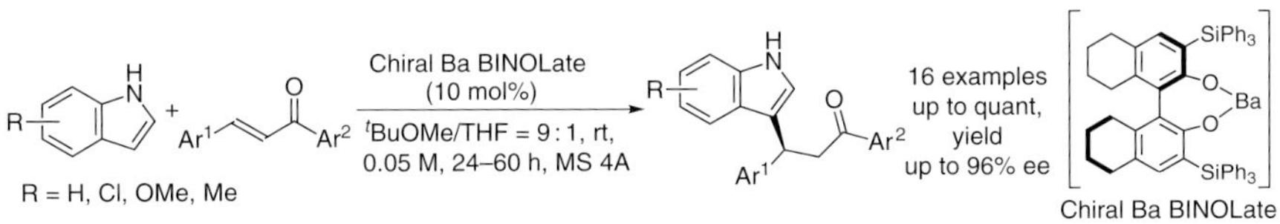

Scheme 12.39 Enantioselective 1,4-addition reactions of indole derivatives catalyzed by chiral barium H_8-BINOLate.

12.5.3 Chiral Barium-Catalyzed Diels–Alder Reactions

An enantioselective Diels–Alder reaction also proceeded by using a chiral barium catalyst (Scheme 12.40). A chiral barium complex prepared from barium alkoxide and a chiral diol bearing a phosphine oxide moiety (type I) was successfully used by Shibasaki and Kanai et al. in 2009, and the desired reaction of a siloxydiene with fumarate gave a substituted cyclohexene with high enantioselectivity [79]. The product is a precursor for the synthesis of optically active Tamiflu®. A chiral barium-activated diene formed by transmetalation of silicon with barium in the presence of fluoride anion (F^-), then the reaction proceeded efficiently.

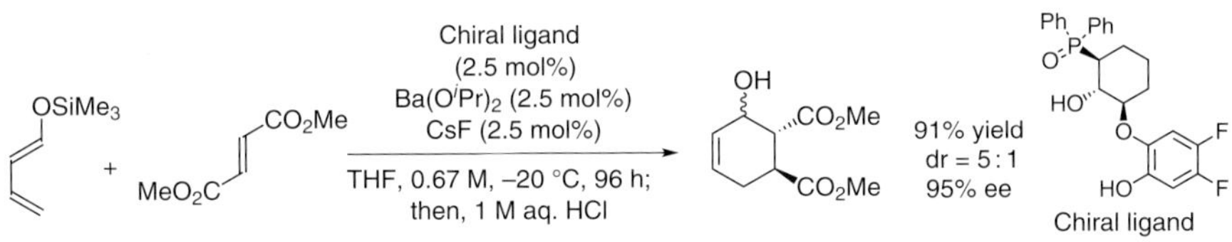

Scheme 12.40 Enantioselective Diels–Alder reaction catalyzed by a chiral barium alkoxide complex.

12.6 Summary and Outlook

This chapter describes enantioselective Lewis acid catalysis by chiral group 2 metal compounds. Similar to typical transition metal Lewis acids, magnesium, calcium, strontium, and barium compounds have a high potential as chiral catalysts, promoting various enantioselective bond-forming reactions. Because of their mild but significant Lewis acidity, the complexes can activate substrates effectively, and Lewis acid catalysis or Lewis acid/Brønsted base cooperative catalysis promotes the desired reactions. Among group 2 metal compounds, magnesium compounds showed the strongest Lewis acidity and promoted Lewis acid catalysis in a major manner. On the other hand, calcium compounds were employed in both Lewis acid catalysis and pseudo-intramolecular Lewis acid/Brønsted base cooperative catalysis because of their relatively strong Lewis acidity and significant Brønsted basicity. The Lewis acidity of strontium and barium compounds is not so strong, and these compounds were employed in pseudo-intramolecular Lewis acid/Brønsted base cooperative catalysis. As a catalyst structure, type I–III chiral complexes were prepared and employed in magnesium and calcium catalysis, and type I chiral complexes were employed in strontium and barium catalysis.

The most striking advantages of group 2 metal catalysts are their significant Lewis acidity and Brønsted basicity, their ubiquitous and harmless nature, and both essential ingredients for environmentally benign catalysis. Moreover, considering these aspects of group 2 metal catalysts, their application in continuous-flow fine synthesis, especially in catalytic proton transfer, is very promising. Further development of heterogeneous catalysts based on group 2 elements under continuous-flow conditions is to be expected.

References

1 (a) Jacobsen, E.N., Pfaltz, A., and Yamamoto, H. (eds.) (1999). *Comprehensive Asymmetric Catalysis*. Heidelberg: Springer. (b) Ojima, I. (ed.) (2010). *Catalytic Asymmetric Synthesis*, 3e. Hoboken, NJ: Wiley.
2 (a) Yamamoto, H. (ed.) (2000). *Lewis acid in Organic Synthesis*, vol. 1 and 2. Weinheim: Wiley-VCH. (b) Yamamoto, H. and Ishihara, K. (eds.) (2008). *Acid Catalysis in Modern Organic Synthesis*, vol. 1 and 2. Weinheim: Wiley-VCH.
3 Anastas, P.T. and Warner, J.C. (1998). *Green Chemistry: Theory and Practice*. New York: Oxford University Press.
4 (a) Pellissier, H. (2017). *Org. Biomol. Chem.* 15: 4750–4782. (b) Yamashita, Y., Tsubogo, T., and Kobayashi, S. (2018). *Topics in Organometallic Chemistry*, vol. 62 (ed. K. Mikami), 121–145. (c) Yamashita, Y., Tsubogo, T., and Kobayashi, S. (2012). *Chem. Sci.* 3: 967–975. (d) North, M. and Clark, J.H. (eds.) (2016). *Sustainable Catalysis: With Non-endangered Metals, Part 1*, 49–76. Cambridge, UK: Royal Society of Chemistry.
5 Corey, E.J. and Ishihara, K. (1992). *Tetrahedron Lett.* 33: 6807–6810.
6 Hiroi, K., Watanabe, K., Abe, I., and Koseki, M. (2001). *Tetrahedron Lett.* 42: 7617–7619.
7 Du, H., Zhang, X., Wang, Z. et al. (2008). *Eur. J. Org. Chem.*: 2248–2254.
8 Zheng, J.J., Lin, L., Kuang, Y. et al. (2014). *Chem. Commun.* 50: 994–996.
9 Parsons, A.T. and Johnson, J.S. (2009). *J. Am. Chem. Soc.* 131: 3122–3123.
10 Wang, L., Yang, D., Han, F. et al. (2015). *Org. Lett.* 17: 176–179.
11 Sibi, M.P., Ji, J., Wu, J.H. et al. (1996). *J. Am. Chem. Soc.* 118: 9200–9201.
12 Lv, J., Li, X., Zhong, L. et al. (2010). *Org. Lett.* 12: 1096–1099.
13 Lundy, B.J., Jansone-Popova, S., and May, J.A. (2011). *Org. Lett.* 13: 4958–4961.
14 Yang, D., Wang, L., Han, F. et al. (2013). *Angew. Chem. Int. Ed.* 52: 6739–6742.
15 Zhao, X., Liu, X., Mei, H. et al. (2015). *Angew. Chem. Int. Ed.* 54: 4032–4035.
16 Yang, D., Wang, L., Kai, M. et al. (2015). *Angew. Chem. Int. Ed.* 54: 9523–9527.
17 Hatano, M., Horibe, T., and Ishihara, K. (2013). *Angew. Chem. Int. Ed.* 52: 4549–4553.
18 Wang, L., Yang, D., Li, D. et al. (2018). *Angew. Chem. Int. Ed.* 57: 1–6.
19 Yoshino, T., Morimoto, H., Lu, G. et al. (2009). *J. Am. Chem. Soc.* 131: 17082–17083.
20 Zheng, K., Yin, C., Liu, X. et al. (2011). *Angew. Chem. Int. Ed.* 50: 2573–2577.

21 Trost, B.M., Malhotra, S., Koschker, P., and Ellerbrock, P. (2012). *J. Am. Chem. Soc.* 134: 2075–2084.
22 Cutting, G.A., Stainforth, N.E., John, M.P. et al. (2007). *J. Am. Chem. Soc.* 129: 10632–10633.
23 Hatano, M., Horibe, T., and Ishihara, K. (2010). *Org. Lett.* 12: 3502–3505.
24 Larson, S.E., Li, G., Rowland, G.B. et al. (2011). *Org. Lett.* 13: 2188–2191.
25 Ingle, G.K., Liang, Y., Mormino, M.G. et al. (2011). *Org. Lett.* 13: 2054–2057.
26 Nimmagadda, S.K., Zhang, Z., and Antilla, J.C. (2014). *Org. Lett.* 16: 4098–4101.
27 Bao, H., Wu, J., Li, H. et al. (2010). *Eur. J. Org. Chem.*: 6722–6726.
28 Sala, G.D. (2013). *Tetrahedron* 69: 50–56.
29 Li, J., Liao, Y., Zhang, Y. et al. (2014). *Chem. Commun.* 50: 6672–6674.
30 Wales, S.M., Walker, M.M., and Johnson, J.S. (2013). *Org. Lett.* 15: 2558–2561.
31 Yang, D., Wang, L., Han, F. et al. (2015). *Angew. Chem. Int. Ed.* 54: 2185–2189.
32 Evans, D.A. and Nelson, S.G. (1997). *J. Am. Chem. Soc.* 119: 6452–6453.
33 Yin, C., Cao, W., Lin, L. et al. (2013). *Adv. Synth. Catal.* 355: 1924–1930.
34 Maji, B., Baidya, M., and Yamamoto, H. (2014). *Chem. Sci.* 5: 3941–3945.
35 Elston, C.L., Jackson, R.F.W., MacDonald, S.J.F., and Murray, P.J. (1997). *Angew. Chem. Int. Ed. Engl.* 36: 410–412.
36 Zhang, X., Emge, T.J., and Hultzsch, K.C. (2012). *Angew. Chem. Int. Ed.* 51: 394–398.
37 Mori, K., Isogai, R., Kamei, Y. et al. (2018). *J. Am. Chem. Soc.* 140: 6203–6207.
38 Yamada, Y.M.A. and Ikegami, S. (2000). *Tetrahedron Lett.* 41: 2165–2169.
39 Suzuki, T., Yamagiwa, N., Matsuo, Y. et al. (2001). *Tetrahedron Lett.* 42: 4669–4671.
40 Rueping, M., Bootwicha, T., Kambutong, S., and Sugiono, E. (2012). *Chem. Asian J.* 7: 1195–1198.
41 Li, X.-Y., Yuan, W.-Y., Tang, S. et al. (2017). *Org Lett.* 19: 1120–1123.
42 (a) Kumaraswamy, G., Sastry, M.N.V., and Jena, N. (2001). *Tetrahedron Lett.* 42: 8515–8517. (b) Kumaraswamy, G., Jena, N., Sastry, M.N.V. et al. (2005). *Adv. Synth. Catal.* 347: 867–871. (c) Kumaraswamy, G., Sastry, M.N.V., Jena, N. et al. (2003). *Tetrahedron: Asymmetry* 14: 3797–3803. (d) Kumaraswamy, G., Jena, N., Sastry, M.N.V., and Ramakrishna, G. (2005). *ARKIVOC*: 53–58.
43 (a) Saito, S., Tsubogo, T., and Kobayashi, S. (2007). *J. Am. Chem. Soc.* 129: 5364–5365. (b) Tsubogo, T., Saito, S., Seki, K. et al. (2008). *J. Am. Chem. Soc.* 130: 13321–13332.
44 Kobayashi, S., Tsubogo, T., Saito, S., and Yamashita, Y. (2008). *Org. Lett.* 10: 807–809.
45 (a) Hut'ka, M., Tsubogo, T., and Kobayashi, S. (2013). *Adv. Synth. Catal.* 355: 1561–1569. (b) Lippur, K., Kaabel, S., Järving, I. et al. (2015). *J. Org. Chem.* 80: 6336–6341.
46 Tsubogo, T., Yamashita, Y., and Kobayashi, S. (2009). *Angew. Chem. Int. Ed.* 48: 9117–9120.
47 (a) Tsubogo, T., Yamashita, Y., and Kobayashi, S. (2012). *Chem. Eur. J.* 18: 13624–13628. (b) Tsubogo, T., Yamashita, Y., and Kobayashi, S. (2014). *Top. Catal.* 57: 935–939.

48 Tsubogo, T., Oyamada, H., and Kobayashi, S. (2015). *Nature* 520: 329–332.
49 Poisson, T., Yamashita, Y., and Kobayashi, S. (2010). *J. Am. Chem. Soc.* 32: 7890–7892.
50 Zheng, W., Zhang, Z., Kaplan, M.J., and Antilla, J.C. (2011). *J. Am. Chem. Soc.* 133: 3339–3341.
51 Tsubogo, T., Kano, Y., Ikemoto, K. et al. (2010). *Tetrahedron: Asymmetry* 21: 1221–1225.
52 Tsubogo, T., Kano, Y., Yamashita, Y., and Kobayashi, S. (2010). *Chem. Asian. J.* 5: 1974–1977.
53 Uno, B.E., Dicken, R.D., Redfern, L.R. et al. (2018). *Chem. Sci.* 9: 1634–1639.
54 Poisson, T., Tsubogo, T., Yamashita, Y., and Kobayashi, S. (2010). *J. Org. Chem.* 75: 963–965.
55 Tsubogo, T., Shimizu, S., and Kobayashi, S. (2013). *Chem. Asian. J.* 8: 872–876.
56 (a) Hatano, M., Moriyama, K., Maki, T., Ishihara, K., Angew. Chem. Int. Ed. 2010, 49, 3823-3826 (b) Hatano, M. and Ishihara, K. (2010). *Synthesis*: 3785–3801.
57 Rueping, M., Bootwicha, T., and Sugiono, E. (2011). *Synlett*: 323–326.
58 Shimizu, S., Tsubogo, T., Xu, P., and Kobayashi, S. (2015). *Org. Lett.* 17: 2006–2009.
59 (a) Drouet, F., Lalli, C., Liu, H. et al. (2011). *Org. Lett* 13: 94–97. (b) Lalli, C., Dumoulin, A., Lebée, C. et al. (2015). *Chem. Eur. J.* 21: 1704–1712.
60 Alix, A., Lalli, C., Retailleau, P., and Masson, G. (2012). *J. Am. Chem. Soc.* 134: 10389–10392.
61 Lebée, C., Blanchard, F., and Masson, G. (2016). *Synlett* 27: 559–563.
62 Zhang, Z., Zheng, W., and Antilla, J.C. (2011). *Angew. Chem. Int. Ed.* 50: 1135–1138.
63 Li, G., Liang, T., Wojtas, L., and Antilla, J.C. (2013). *Angew. Chem. Int. Ed.* 52: 4628–4632.
64 Liang, T., Li, G., Wojtas, L., and Antilla, J.C. (2014). *Chem. Commun.* 50: 14187–14190.
65 Mao, Z., Li, W., Shi, Y. et al. (2013). *Chem. Eur. J.* 19: 9754–9759.
66 Buch, F. and Harder, S. (2008). *Naturforsch Z.* 63b: 169–177.
67 Yamashita, Y., Macor, A.J., Fushimi, S. et al. (2018). *Chem. Pharm. Bull.* 66: 847–850.
68 Nakamura, S., Ohara, M., Koyari, M. et al. (2014). *Org. Lett.* 16: 4452–4455.
69 Agostinho, M. and Kobayashi, S. (2008). *J. Am. Chem. Soc.* 130: 2430–2431.
70 Kobayashi, S., Yamaguchi, M., Agostinho, M., and Schneider, U. (2009). *Chem. Lett.* 38: 299–300.
71 Tanaka, Y., Kanai, M., and Shibasaki, M. (2010). *J. Am. Chem. Soc.* 132: 8862–8863.
72 (a) Nguyen, H.V., Matsubara, R., and Kobayashi, S. (2009). *Angew. Chem. Int. Ed.* 48: 5927–5929. (b) Matsubara, R., Florian, B., Nguyen, H.V., and Kobayashi, S. (2009). *Bull. Chem. Soc. Jpn.* 82: 1083–1102.
73 Lu, G., Yoshino, T., Morimoto, H. et al. (2011). *Angew. Chem. Int. Ed.* 50: 4382–4385.

74 Nimmagadda, S.K., Mallojjala, S.C., Woztas, L. et al. (2017). *Angew. Chem. Int. Ed.* 56: 2454–2458.

75 Yamada, Y.M.A. and Shibasaki, M. (1998). *Tetrahedron Lett.* 39: 5561–5564.

76 Saito, S. and Kobayashi, S. (2006). *J. Am. Chem. Soc.* 128: 8704–8705.

77 Yamaguchi, A., Matsunaga, S., and Shibasaki, M. (2009). *J. Am. Chem. Soc.* 131: 10842–10843.

78 Yamaguchi, A., Aoyama, N., Matsunaga, S., and Shibasaki, M. (2007). *Org. Lett.* 9: 3387–3390.

79 Yamatsugu, K., Yin, L., Kamijo, S. et al. (2009). *Angew. Chem. Int. Ed.* 48: 1070–1076.

13

Miscellaneous Reactions

Michael S. Hill

University of Bath, Department of Chemistry, Claverton Down, Bath, BA2 7AY, UK

13.1 Introduction

The primary intent of this contribution is to perform a "mopping up" of s-block-centered catalytic transformations, which are not encompassed by the specific mandates of the earlier chapters of this monograph. Although the absence of an overarching theme risks a consequent lack of overall coherence, the identification of a smorgasbord of reactions that are insufficiently developed to merit more focused coverage also provides the author with an opportunity to reflect upon the challenges and opportunities presented to the future development of the area. Reactivity will necessarily be introduced from the perspective provided by the more in-depth consideration of individual reaction types provided by previous chapters and in earlier more focused reviews [1–10]. Any catalytic cycle is also necessarily the summation of a sequence of discrete molecular reaction steps. With no attempt to provide exhaustive coverage, therefore, this chapter will also highlight some of the more unusual or promising stoichiometric reactivity that may plausibly be exploited in the construction of future catalytic systems.

13.2 Privileged Substrates and s-Block Reactivity

A vast majority of the catalytic reactions described in the earlier chapters of this book are predicated on the Lewis acidity of the s-block elements in their positive oxidation states and/or the significantly polar character of their bonding to p-block substituents. The elements of both groups 1 and 2 are highly electropositive and their respective +1 and +2 ions display the expected periodic variations arising from their relative positions in the periodic table. Although a case may be made for the initiation of reactivity by single electron transfer (SET) processes (*vide infra*), the primarily ionic binding of typical p-block centered ligands ensures that the reaction chemistry of s-block complexes is most commonly restricted to two-electron processes. Scheme 13.1 depicts a simple conceptual framework that allows a generalized picture of the reactivity

Early Main Group Metal Catalysis: Concepts and Reactions,
First Edition. Edited by Sjoerd Harder.

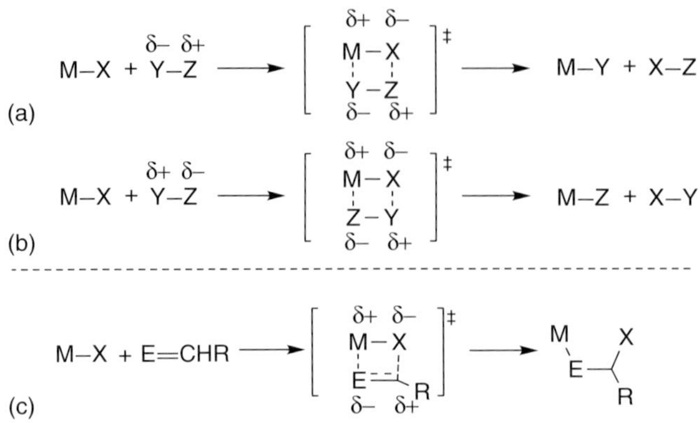

Scheme 13.1 Generalized reactivity of group 1 and group 2 M—X bonds. (a, b) σ-bond metathesis; (c) polarized insertion.

of s-block complexes to be presented. Within this model, the reactivity of an M—X-bonded unit (where M represents any group 1 or group 2 element center and X is either hydrogen or a σ-bonded p-block element) is rationalized as a consequence of both the Lewis acidity of the electropositive metal cation and the basicity and nucleophilic nature of the bound substituent, X. Irrespective of the identity of M, the difference in electronegativity ensures that the electron density across the formally two-electron M—X interaction will always be polarized significantly in a δ- sense toward the more electronegative X substituent. Although the charge distribution will necessarily vary with the precise identity of the M—X unit, the polarized nature of these interactions ensures that the reactivity of an M—X bond with any σ-bonded substrate, Y—Z, is effectively confined to the 2σ–2σ exchange of electron density between the bonds being broken and those being formed (Scheme 13.1a,b). This reactivity may be broadly classified as σ-bond metathesis [11], envisioned to ensue through the assembly of a classical polarized four-center transition state. From this perspective, certain Y—Z substrates may be defined as "privileged" in possessing a predisposition toward their engagement in the assembly of the metathesis transition state. When the Y and Z atoms are dissimilar, the regiochemical outcome of the reaction will be dictated by the initial polarization of the M—X bond, whereupon the Y—Z substrate adopts an orientation that places the more electronegative component (e.g. the nitrogen center of a secondary amine or the hydride ligand of a tertiary silane) in an α-position with respect to the s-block center (Scheme 13.1a,b). Although the thermodynamic viability of the reaction will clearly depend on the relative strength of the bonds being broken and those being formed, the mode of assembly and the consequent distribution of charge across such a transition state will dictate the kinetic facility of the metathesis.

The reaction chemistry of M—X bonds is also characterized by an ability to effect polarized insertion of multiply bonded substrates. Again, an analogous element of "entitlement" is conferred to more polarized substrates. Precoordination of an unsaturated C=E bonded molecule to the s-block center dictates the orientation of the insertion transition state when E is, for example, the oxygen center of a ketone or aldehyde, or the nitrogen atom of an imine (Scheme 13.1c). In contrast, polarization across the C—C multiple bonds of alkene and alkyne substrates must necessarily be induced during encounters of their polarizable π electrons with the highly polar M—X bond [12]. The kinetic facility toward insertion will then be dictated by a substrate's ability to stabilize the high degree of charge separation induced during assembly of the insertion transition state. For this reason, albeit with the exception of several notable recent examples [13, 14], a majority of known intermolecular C=C insertion reactions are limited to dienes or vinyl-substituted arenes [15–21] due to their conjugative (allylic or benzylic) stabilization of negative charge accumulation during the course of the insertive process [12, 20]. Ensembles of secondary polarization effects have also been shown to be crucial for reactions to progress through polarized transition states reminiscent of that shown in Scheme 13.1c, where the rates of reaction reflect the ability of the constituent atoms to stabilize the consequent distribution of partial charges [22, 23].

The sequential combination of "protic" and "hydridic" σ-bond metathesis steps represented in Scheme 13.1a,b into catalytic cycles has allowed the development of a range of dehydrocoupling processes for the construction of Si—N, B—N, and Si—C containing molecules (Scheme 13.2a) [24]. This chemistry, however, naturally highlights that a majority of the σ-bond metathesis reactivity derived from s-block reagents is currently limited to the activation of prepolarized, either hydridic (Y—H) or protic (Z—H), element to hydrogen bonds.

The wide variety of multiple bond heterofunctionalization catalyses described in the earlier chapters of this book have also been achieved through sequences of σ-bond metathesis reactions and the introduction of polarized insertion steps (Scheme 13.2b).

In common with dehydrocoupling reactivity, however, reported s-block-catalyzed C=E heterofunctionalization processes almost exclusively encompass σ-bonded substrates comprising protic (N—H [25], P—H [21], C≡C—H), or hydridic (B—H [26, 27], Si—H [28]) hydrogen atoms as a component of the Y—Z substrate.

Catalytic reactivity that necessitates the activation of σ bonds to elements other than hydrogen is only starting to emerge (Section 13.5), while the only current examples of s-element-based catalysis to require the rupture of a nonpolarized σ-bonded substrate are limited to the activation of the H—H bond of dihydrogen. The reactivity of this latter type has been invoked during the reaction of H_2 with the potassium silyl [K(Me_6TREN)$SiPh_3$] [**1**, Me_6TREN = (tris{2-(dimethylamino)ethyl}amine)], which occurs with cleavage of silicon–phenyl bonds to give the corresponding hydridosilyl compound and benzene (Scheme 13.3) [29], and in various calcium-based imine [30] and alkene hydrogenation catalyses [13, 20]. Notably, computational analysis of this latter reactivity implicated a very high degree (effectively ionic) of charge separation during the requisite Ca—C/H—H transition states [31].

In keeping with this view of the current "state of the art" in s-block catalysis, a majority of the "miscellaneous" reactivity presented in this chapter will emphasize polarized or similarly "privileged" σ- and π-bonded substrates. Recent advances, for example, in the catalytic functionalization of the C(sp^2)—H bonds

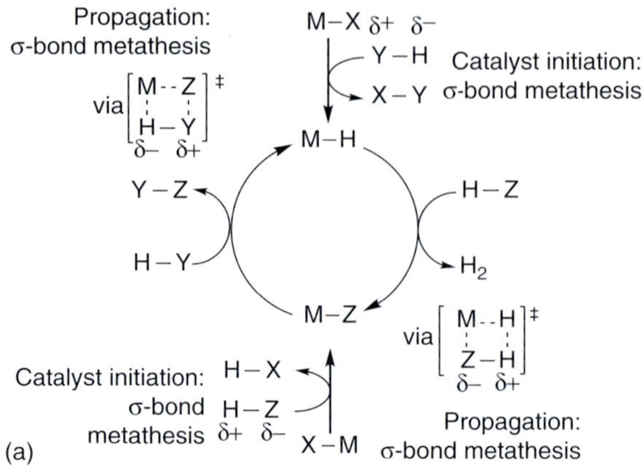

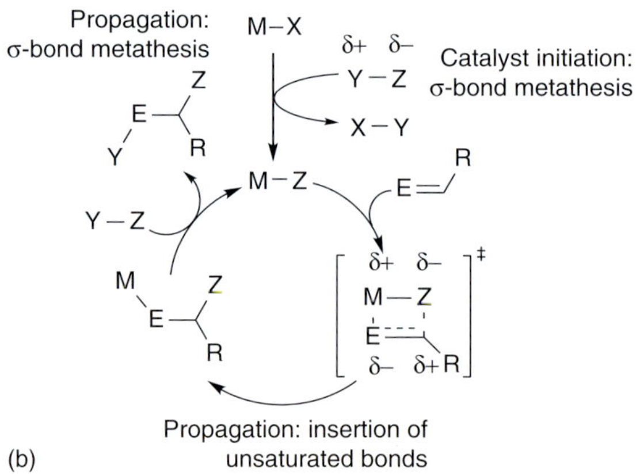

Scheme 13.2 Prototypical mechanisms for (a) s-block-catalyzed dehydrogenative cross metathesis and (b) multiple bond heterofunctionalization.

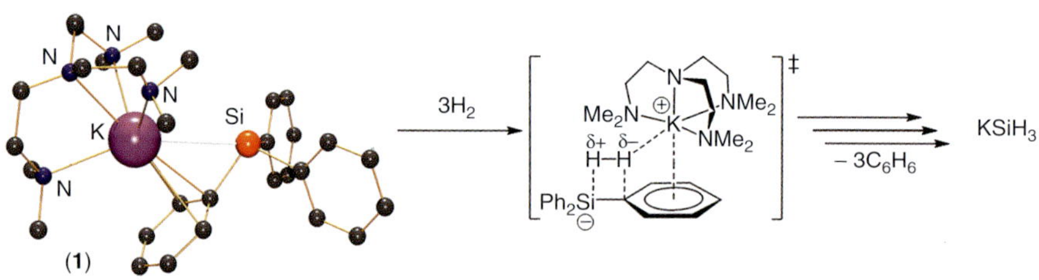

Scheme 13.3 Stepwise hydrogenolysis of the potassium silyl [K(Me$_6$TREN)SiPh$_3$] (**1**) via the four-center transition state proposed for the silicon–phenyl bond cleavage by dihydrogen.

of heterocycles (Section 13.4) [32] and the catalytic activation of nonhydrogenous Y—Z substrates (Section 13.5), however, clearly indicate that significantly more diverse and unique reactivity remains to be discovered. The chapter will conclude, therefore, with some attempted (and possibly specious) futurology and a few reflections on the opportunities and possibilities that may arise from the achievement of increasing ambitious and challenging s-block-derived bond activation processes.

13.3 Reactivity with Multiply Bonded Substrates

13.3.1 Tishchenko Dimerization of Aldehydes

Group 1 and group 2 species are active catalysts for the disproportionation of aldehydes to the corresponding carboxylic esters (the Tishchenko or Claisen–Tishchenko reaction). Several simple, and even off the shelf, alkali metal reagents including sodium and potassium hydride [33], a dilithium 2-aminopyrrolyl derivative (**2**, Scheme 13.4) [34], lithium bromide [35], and potassium *tert*-butoxide [36] have been reported to provide efficient dimerization of both aliphatic and aromatic aldehydes (Scheme 13.5a), while the transformation may even be achieved under solvent-free conditions through ball milling of the neat carbonyl compound with NaH [37]. Although competitive aldol condensation has been observed for aldehydes possessing an enolisable methylene group at the α-position to the aldehyde functionality, in general, high yields of the ester products may be obtained with catalyst loadings as low as 1 mol%. Similar observations have been reported for a variety of group 2 precatalysts, including the heavier alkaline earth amides, $[M\{N(SiMe_3)_2\}_2(THF)_2]$ (M = Ca, Sr, and Ba) [38], homo- and heteroleptic magnesium and calcium guanidinates [39, 40], magnesium amidinates [41], and a series of bromomagnesium thiolates that were generated by addition of the mercaptan to phenylmagnesium bromide [42]. Although this latter system also enables aldehyde–ketone coupling, a consistent mechanistic interpretation of this reactivity has emerged, which encompasses both the group 1 and group 2-centered catalysis and is similar to what was previously described for related lanthanide-catalyzed reactions (Scheme 13.5b) [43–46].

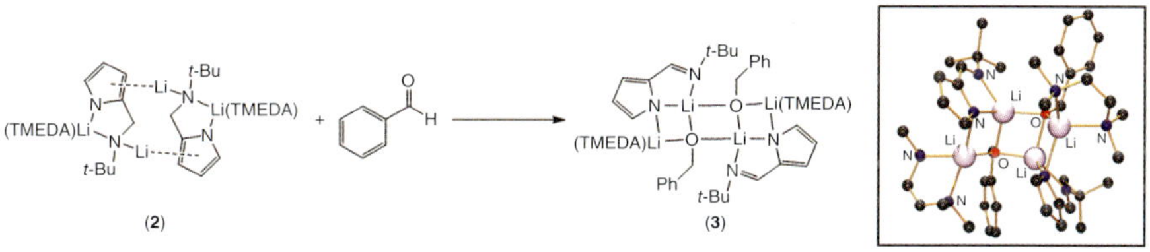

Scheme 13.4 The conversion of compound **2** to compound **3** and the X-ray structure of **3**.

The key step of the reaction requires conversion of the precatalyst to a metal alkoxide originating from a Meerwein–Ponndorf–Verley-type reduction of the aldehyde. Subsequent addition to a second equivalent of aldehyde followed by a further similar hydride transfer is suggested to result in the formation of the ester products with regeneration of the catalytic metal alkoxide. Direct evidence for the operation of such a redox process has been obtained through the reaction of compound **2** with benzaldehyde, during which the secondary lithium amide was oxidized to an imino unit with concomitant reduction of the aldehyde to a benzyloxide (Scheme 13.4).

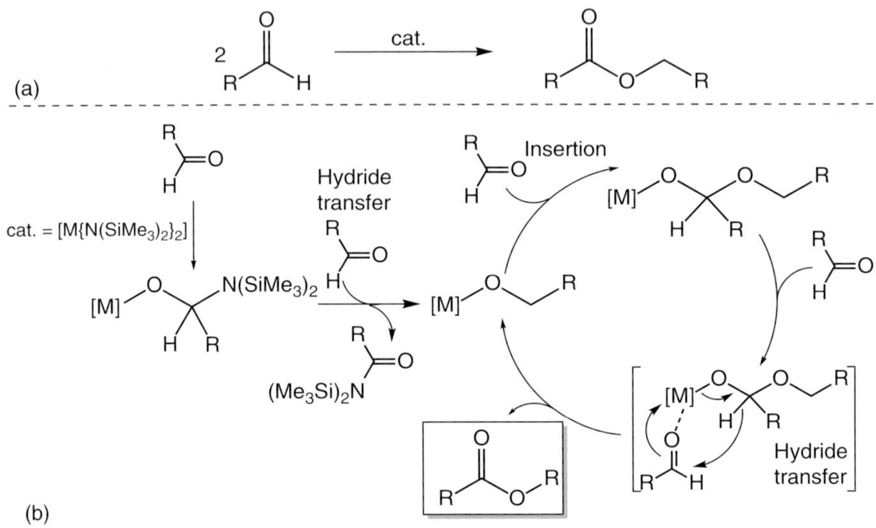

Scheme 13.5 (a) Tishchenko coupling of aldehydes and (b) the suggested mechanism involving polarized insertion and formal hydride transfer.

13.3.2 Trimerization of Organic Isocyanates

A number of alkali metal-derived systems for the cyclotrimerization of isocyanates (RNCO) to isocyanurates have been recently reported, including several lithium amide and pyrrolide derivatives [47–50]. Hevia and coworkers have also observed that the mixed metal reagents $[NaMg(CH_2SiMe_3)_3]$ (**4**) and $[(THF)NaMg(NPh_2)_3(THF)]$ (**5**) can act as precatalysts to selectively promote either the hydroamination or the analogous isocyanate trimerization of (RNCO) depending on the identity of the R substituent [51]. A catalytic cycle (Scheme 13.6) was proposed in which **4** or **5** reacts with amine to form a bimetallic tris(amide), which can itself undergo isocyanate insertion to yield

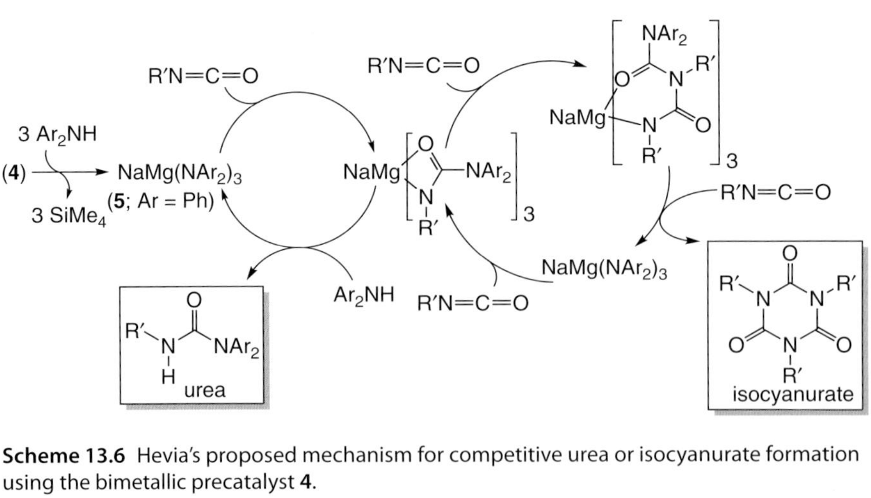

Scheme 13.6 Hevia's proposed mechanism for competitive urea or isocyanurate formation using the bimetallic precatalyst **4**.

tris(ureido) species. Although protonolysis by excess amine generates the urea hydroamination product with relatively bulky isocyanate N-substitution, use of the less sterically demanding and more electrophilic *p*-tolNCO initiated a cascade of insertion reactions and production of the isocyanurate.

In a similar manner, Harder has confirmed that reactions of the calcium methandiide (**6**) [52] with cyclohexyl isocyanate provides the double-insertion product (**7**), which was identified as an intermediate in the catalytic trimerization reaction (Scheme 13.7) [53]. A 5 mol% solution of **7** catalyzed the trimerization of CyNCO with 80% conversion after one week at 50 °C, whereas reaction with the more electrophilic PhNCO provided smooth conversion to the corresponding isocyanurate at 20 °C in three hours with only a 1 mol% loading.

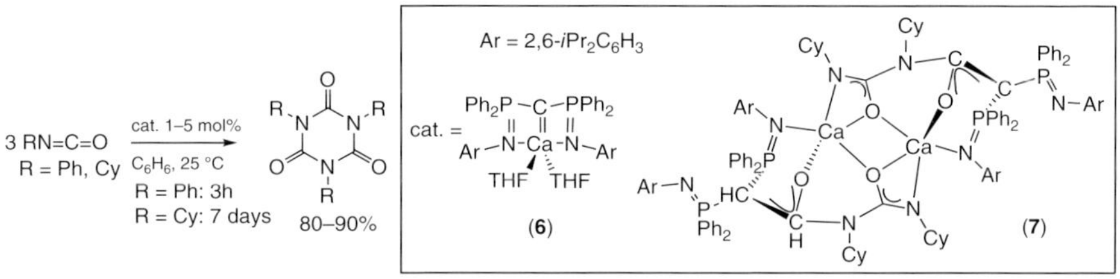

Scheme 13.7 Isocyanate trimerization catalyzed by the calcium methanediide **6** and the X-ray structure of the product of the double insertion of cyclohexylisocyanate, compound **7**.

13.3.3 Hydroalkoxylation of Alkynyl Alcohols

As an advance from their earlier demonstration of intramolecular hydroamination catalysis [22], Hill and coworkers have shown that the heavier alkaline earth bis(trimethylsilyl)amides homoleptic heavier alkaline earth amides, $[M\{N(SiMe_3)_2\}_2]_2$ (M = Ca, Sr, and Ba), are effective precatalysts for the regioselective intramolecular hydroalkoxylation/cyclization of a wide range of alkynyl and allenyl alcohols (Scheme 13.8) [54]. Cyclization of alkynyl alcohols produces

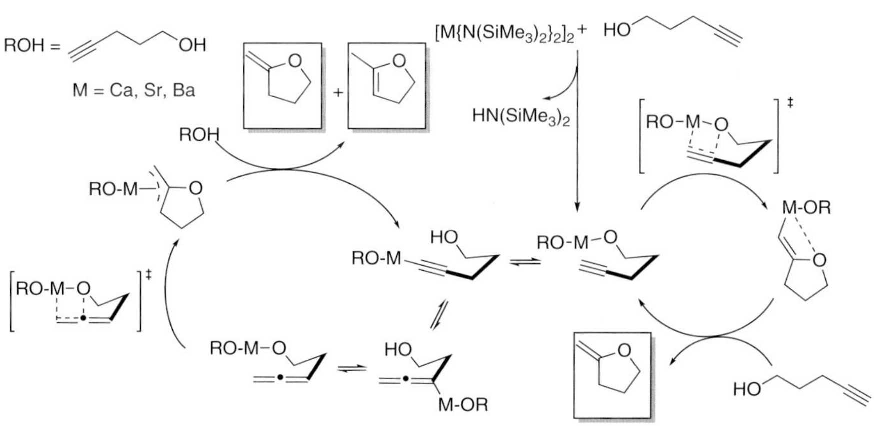

Scheme 13.8 Proposed mechanism for the alkaline earth-catalyzed cyclohydroxylation of alkynylalcohols.

mixtures of the possible endo- and exocyclic enol ether products, which was rationalized to be a result of isomerization of the initially formed alkynylalkoxide intermediates to the corresponding allene derivatives. Cyclization rates for terminal alkynyl alcohols were found to be significantly higher than for substrates bearing internal alkynyl substituents. Emphasizing the periodic variations associated with the study of the series of congeneric precatalysts, kinetic studies highlighted a significant dependence of the reaction rate on the identity of the alkaline earth center and demonstrated a pronounced inhibition of the catalysis with increasing (substrate). This latter feature was interpreted to indicate that rate-determining C≡C insertion in the M—O bond requires the dissociation of Lewis basic substrate molecules away from the catalytic metal center.

Potassium *tert*-butoxide has also been found to be an effective catalyst for the cyclization of aromatic alkynols [55]. In this case, however, and highlighting the greater propensity of the dicationic group 2 elements toward isomerization of polarizable carbon–carbon triple bonds (*vide infra*), alkynols were converted to the corresponding exo-cyclic enol ethers as pure *Z*-stereoisomers with 100% selectivity and moderate to excellent yields.

13.3.4 Catalytic Isomerization and C–C Coupling with Terminal Alkynes

Underlining the above noted ability of heavier alkaline earth centers to polarize even nonpolar C—C multiple bonds, Barrett et al reported the head–head dimerization of donor-functionalized terminal alkynes (propargyl ethers and amines) to a mixture of *E* and *Z* butatriene products under the action of Ca and Sr reagents [56, 57]. Echoing an earlier report that a “side-on” interaction between the polarizable C≡C unit and the calcium center in dimeric β-diketiminato calcium alkynides results in a dissipation of the negative charge over both alkynide carbon centers [58], the reactions were proposed to be promoted by polarization of the electron density within the alkynide C≡C bonds, an effect that is amplified by the secondary coordination of the alkynide through its donor substituent in molecules such as **8** (Scheme 13.9). The resultant dissipation of charge away from the coordinated α-carbon was suggested to reduce the mutual repulsion of the carbon centers and promote the formation of a transition state in which the C4 bridging bis(alkynide) fragment has significant butatrienyl diyl character during protonolysis. Similar, and reversible, μ-(*Z*)-butatriene

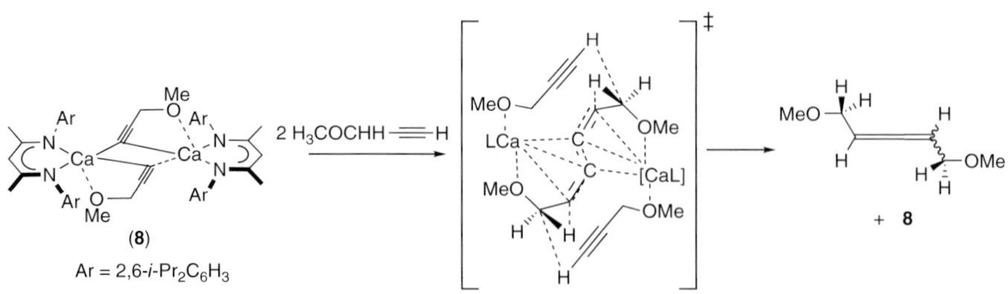

Scheme 13.9 Calcium-mediated coupling of terminal alkynes via compound **8**.

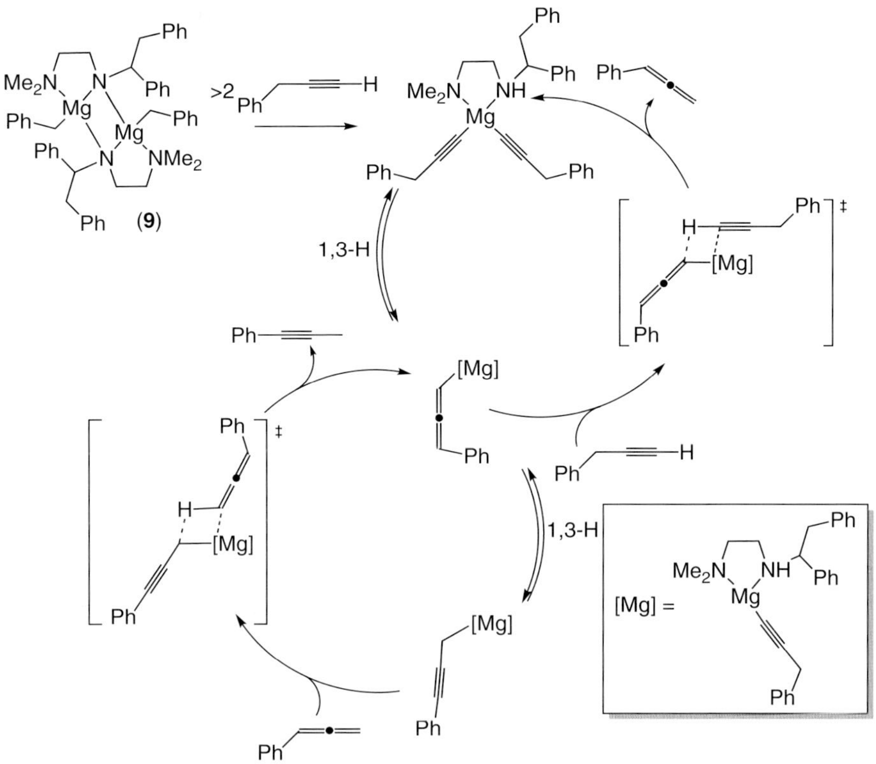

Scheme 13.10 Isomerization of terminal to internal alkynes catalyzed by compound **9**.

diyl bridging units had previously been shown to exist with group 3 or the 4f elements [59, 60]. Although reactions promoted by homoleptic silylamides of Ca and Sr were restricted to a single turnover because of precipitation of insoluble alkynide intermediates, catalytic dimerization could be achieved when the group 2 metal center was supported by stabilizing triazenide or dearomatized 1,2-bis(imino)acenaphthene spectator ligands.

Emphasizing the facility of group 2 species to promote selective transformations of alkynes, Rochat et al. have reported that benzyl magnesium species such as **9** (Scheme 13.10) can effect the transformation of terminal alkynes into allenes and further to internal alkynes under mild conditions [61]. The reactions were deduced to proceed through temporally separated autotandem catalysis, thus allowing the isolation of the allene or internal alkyne species in good yields. In observations reminiscent of the proposed intermediates illustrated in Scheme 13.8 [54], mechanistic experiments suggested that the catalytically active tetraalkynyl complexes consist of a tautomeric mixture of alkynyl-, allenyl-, and propargylmagnesium species.

The facile formation of alkaline earth alkynides has also been exploited in C—C bond forming catalysis. In 2006, Richeson and coworkers reported that the commercially available lithium precursor, [Li{N(SiMe$_3$)$_2$}], was an efficient catalyst for the hydroacetylenation of carbodiimides [62]. In all cases, catalysis was shown to proceed via the insertion of a carbodiimide molecule into the metal-acetylide bond of the catalyst to form an intermediate ethynylamidinate complex. Subsequent protonolysis with alkyne releases the ethynylamidine and regenerates the active metal acetylide species as shown in Scheme 13.11.

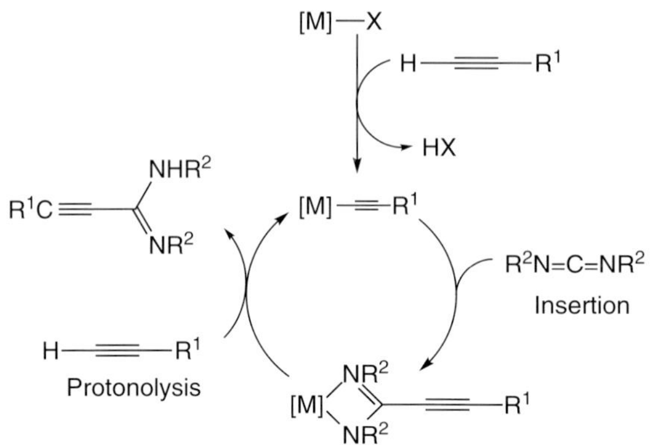

Scheme 13.11 Mechanism of C—C coupling through the s-block-catalyzed reaction of terminal alkynes and carbodiimides.

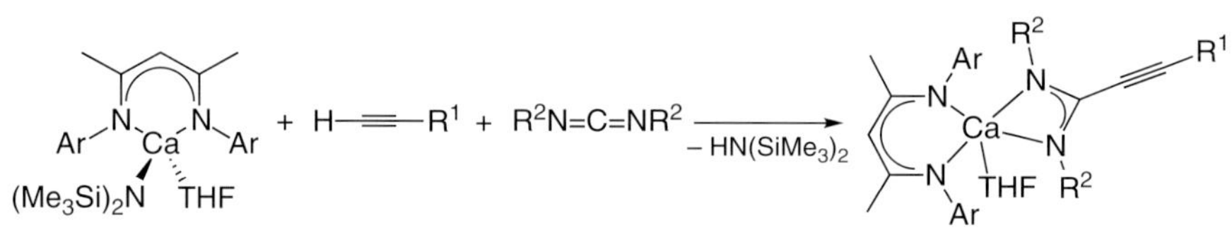

(**10**; Ar = 2,6-*i*-$Pr_2C_6H_3$))

Scheme 13.12 Synthesis of calcium propargyl amidinates.

Reactions of 1,3-diorganocarbodiimides with alkynides generated through protonolysis of the β-diketiminato calcium amide (**10**) with terminal alkynes have also been shown to provide the corresponding heteroleptic calcium C-propargyl amidinate complexes (Scheme 13.12) [63]. Although an initial extension of this reactivity to catalysis demonstrated the application of **10** (5 mol%) to the catalytic hydroacetylenation of 1,3-di-isopropylcarbodiimide with phenylacetylene, yielding the corresponding propargyl amidine, this process also resulted in competitive protonolysis of the β-diketiminate supporting ligand. Subsequent investigations by the groups of Coles and Hill, therefore, were concentrated on the alkaline earth bis-hexamethyldislazides, $[M\{N(SiMe_3)_2\}_2(THF)_2]$ (M = Mg; Ca, Sr), and a wide range of organo- and amidomagnesium species [64, 65]. Although all the compounds studied were shown to act as efficient precatalysts for the hydroacetylenation of organic carbodiimides with alkyl- and arylacetylenes, catalytic activity was observed to increase with the size of the group 2 metal center employed. Proposed intermediate dimeric calcium and strontium bis(amidinate) complexes, $[\{PhC{\equiv}CC(N\mathit{i}Pr)_2\}_2M]_2$ (M = Ca, Sr), were isolated and crystallographically characterized, while kinetic analyses of the catalysis revealed that carbodiimide insertion provides the rate-limiting step for reactions that are self-inhibiting with increasing [product].

Although the formation of substituted ureas by the hydroamination of organic isocyanates has also been shown to proceed with considerable efficacy [51, 66], attempted extension of the hydroacetylenation chemistry of Scheme 13.11 to RNCO substrates provided notably divergent behavior [67]. Addition of Bu_2Mg to a 4 : 1 mixture of a range of RNCO substrates and PhC≡CH was found, after hydrolysis, to provide structurally complex bis(imidazolin-2,4-dione) molecules. Although not catalytic, this reactivity was found to provide the bicyclic compounds with 100% atom efficiency and with complete diastereoselectivity to provide the *R,R/S,S* pair of diastereomers. The reactions were proposed to ensue through the hydroacetylenation, isocyanate insertion, intramolecular hydroamination, and protonolysis steps shown in Scheme 13.13, with the initially formed heterocycle culminating from an intramolecular ring closure strongly reminiscent of the processes implicated in the hydroamination of aminoalkenes and -alkynes [25].

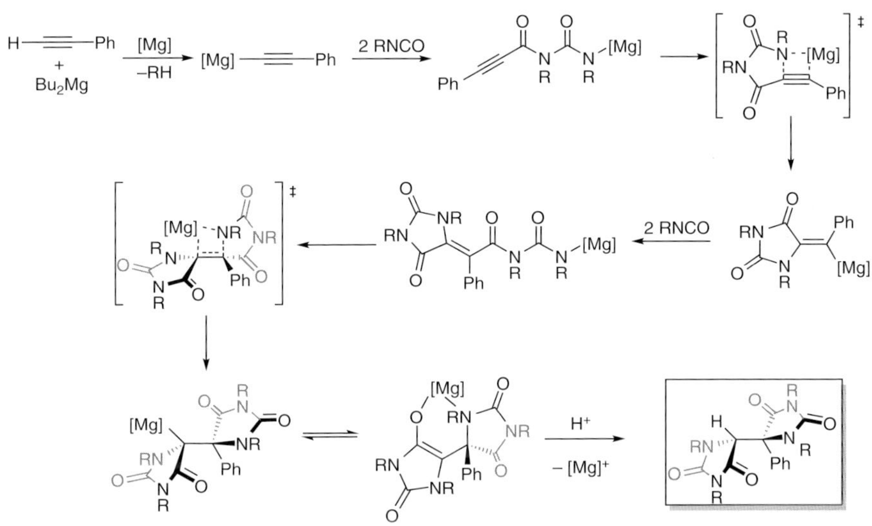

Scheme 13.13 Cascade of reactions for the Mg-mediated 100% atom-efficient assembly of bis(imidazolin-2,4-dione) molecules from terminal alkynes and isocyanates.

In a subsequent advance, use of the bis-hexamethyldisilazides $[M\{N(SiMe_3)_2\}_2(THF)_2]$ (M = Mg, Ca, Sr) enabled the elaboration of this chemistry to a generalized one-pot catalytic regime for the atom-efficient synthesis of imidazolidine and thiazolidine derivatives [68, 69]. The reactions were carried out via the initial catalytic formation of a propargylamidine as shown in Scheme 13.11. Subsequent addition of a stoichiometric equivalent of isocyanate or isothiocyanate was found to initiate a further cascade of reactions involving isocyanate insertion into the catalytically active group 2 amidinate, intramolecular Ae-N insertion of the alkynyl residue, and protonolysis of the cyclized alkaline earth vinyl intermediate by the remaining propargylamidine. This latter step regenerates the group 2 propargylamidinate and provides the imidazolidin-2-one derivative (Scheme 13.14). Support for this mechanism was provided by the isolation of the magnesium 2-(propargylamidino)imidate complex (**11**).

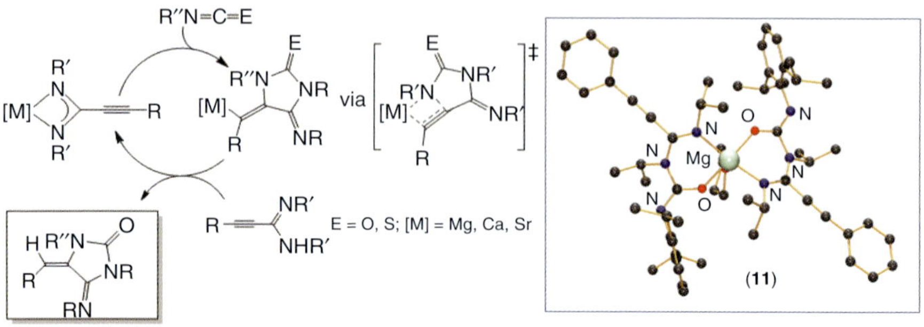

Scheme 13.14 Mechanism for the alkaline earth-catalyzed formation of imidazolidin-2-ones from propargylamidines and isocayanates and the X-ray structure of the isolated intermediate (**11**).

13.3.5 Activation and Deoxygenation of C—O Multiple Bonds

Although other chapters of this monograph are dedicated specifically to the s-block-catalyzed hydrosilylation and hydroboration of polar multiple bonds, it is worth highlighting that the complete reductive cleavage of C—O multiple bonds may also be achieved under appropriate conditions. Earlier research had demonstrated that the catalytic deoxygenative hydrosilylation of CO_2 to methane with Et_3SiH may be achieved through the action of a catalytic quantity of the decamethylscandocenium hydrido-tris(pentafluorophenyl)borate species, $[Cp^*_2Sc\{HB(C_6F_5)_3\}]$ (**12**) [70]. An important consideration in utilization of CO_2 is the ultimate carbon oxidation state, which may be achieved. In contrast to the complete deoxygenation achieved with **12**, the β-diketiminato Mg and Ca hydrido-tris(pentafluorophenyl)borate derivatives, **13** and **14**, have been found to react with CO_2 to provide formate complexes prior to the highly selective reductive hydroboration of CO_2 with pinacolborane (HBpin) to provide the methanol equivalent, CH_3OBpin (Scheme 13.15) [71, 72]. In a similar manner, and emphasizing the importance of catalyst selection, alkali metal hydridotriphenylborates $M[HBPh_3]$ (M = Li, **15**; Na, **16**; K, **17**), characterized as tris{2-(dimethylamino)ethyl}amine (L) complexes $[(L)M][HBPh_3]$, have been shown to catalyze the hydroborative reduction of CO_2 to give formoxyborane HCO_2Bpin without any overreduction [73, 74]. In contrast, the terminal magnesium hydride compound, $[\kappa^3\text{-TismPr}^{i\text{-Benz}}]$ MgH, (**18**, $\text{TismPr}^{i\text{-Benz}}$ = tris[(1-isopropylbenzimidazol-2-yl)-dimethylsilyl]methyl), in combination with $B(C_6F_5)_3$ generates catalytic systems for the hydrosilylation of CO_2 by R_3SiH (R_3SiH = $PhSiH_3$, Et_3SiH, and Ph_3SiH) to afford sequentially the bis(silyl)acetal, $H_2C(OSiR_3)_2$, and CH_4 [75].

Although the precise nature of the C—O activation and cleavage steps associated with the reduction of CO_2 by **13**–**18** have yet to be fully elucidated, some insight has recently been provided by a study of the deoxygenative

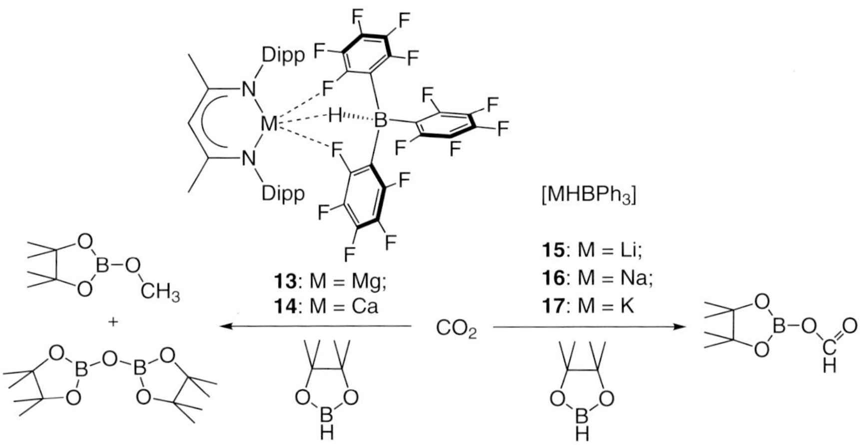

Scheme 13.15 Hydroboration of CO_2 with pinacolborane mediated by s-block catalysis.

reduction of isoelectronic organic isocyanates, RNCO. In this case, reduction of a range of isocyanate small molecules to methyl amine products was achieved through their hydroboration with HBpin in the presence of a β-diketiminato organomagnesium catalyst (**19**, Scheme 13.16) [76]. Although borylated amide and *N*,*O*-bis(boryl)hemiaminal species were identified as intermediates during the reductive catalysis, the overall reduction and C—O activation was shown to be metal-mediated and proposed to occur through a sequence of magnesium formamidato, formamidatoborate, and magnesium boryloxide derivatives. Supported by a density functional theory (DFT) study, the stability of the borate species was suggested to imply that the onward progress of the deoxygenation reaction is crucially dependent on the further activation provided by the Lewis acidic HBpin substrate.

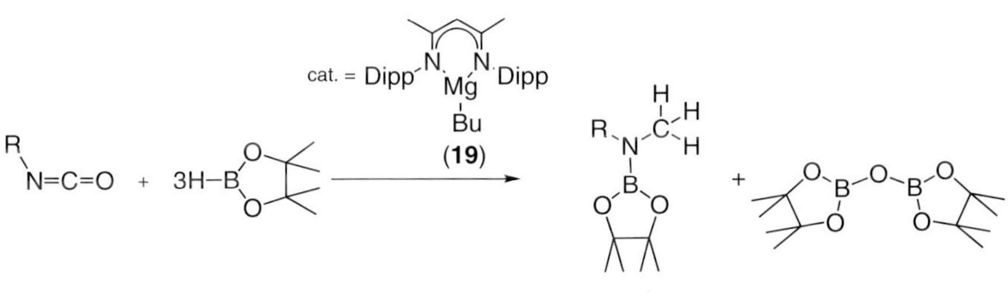

Scheme 13.16 Deoxygenative hydroboration of organoisocyanates catalyzed by compound 19.

Reactions of β-diketiminato magnesium and calcium hydrides (**20** and **21**) with 1 atmosphere of CO have been shown to result in a reductive coupling process to produce the corresponding derivatives of the *cis*-ethenediolate dianion (**22** and **23**, Scheme 13.17a) [77, 78]. Although the C—C coupling process was not catalytic, reactions of CO with $PhSiH_3$ in the presence of **20** and **21** resulted in catalytic reduction to methylsilane and methylene silylether products, respectively. Based on a DFT analysis of the magnesium-centered catalysis, these reactions

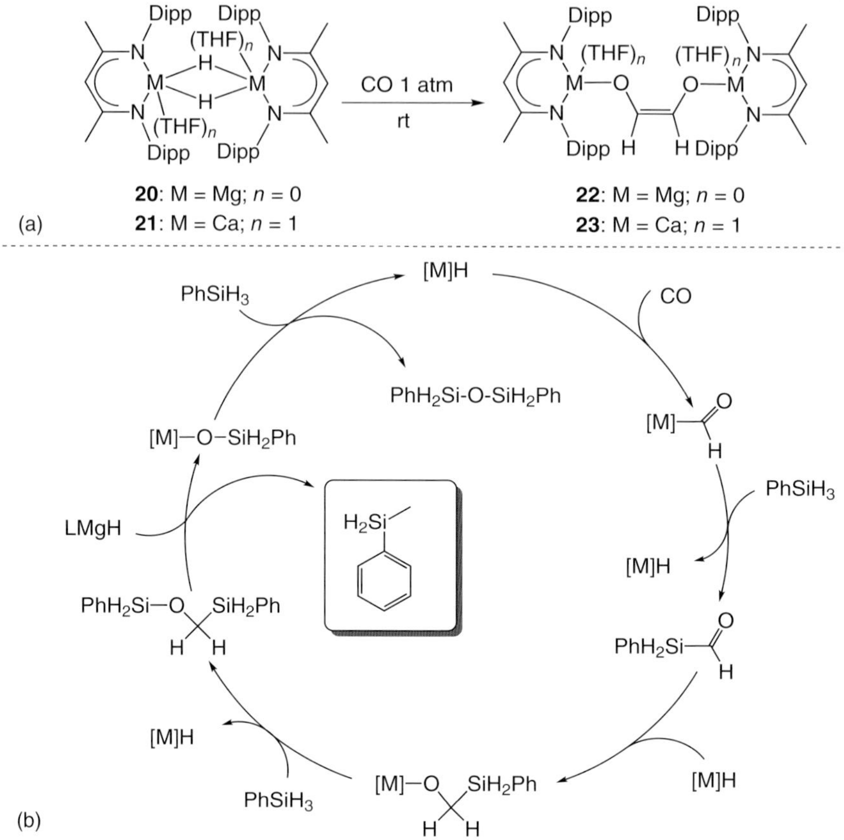

Scheme 13.17 (a) Reductive coupling of CO by **20** and **21**; (b) catalytic hydrosilylation/C—O activation catalyzed by **20**.

were proposed to ensue via the interception of initially formed group 2 formyl intermediates (Scheme 13.17b).

The carbonylation reactivity has also been extended to a variety of magnesium and calcium amides. With primary amido complexes, CO insertion ensues with nitrogen-to-carbon migration of hydrogen to form formamidate species (Scheme 13.18) [79]. Stoichiometric reduction of the resultant magnesium and calcium formamidates with pinacolborane resulted in the synthesis of the

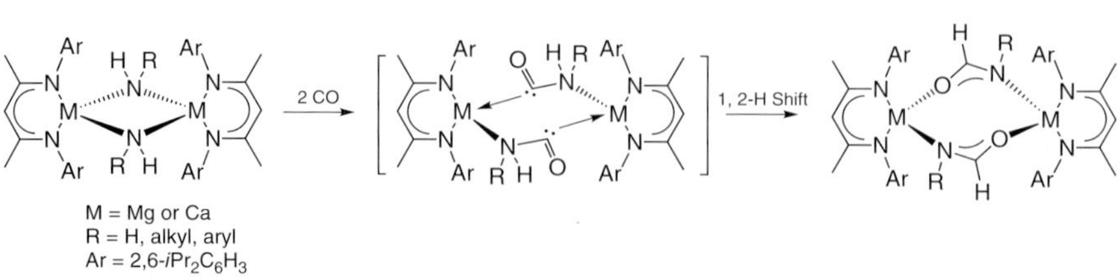

Scheme 13.18 Proposed reaction pathway for the formation of magnesium and calcium amidates through the direct carbonylation of the alkaline earth primary amides.

corresponding N-borylated methylamines, a process that was proposed to take place via a cascade of reactivity that is analogous to that inferred during the deoxygenative hydroboration of organic isocyanates.

13.4 Single-Electron Transfer Steps in s-Block-Centered Catalysis

As outlined by the introductory comments of this chapter, a majority of s-block-mediated reactions, which may be implicated in catalysis, are two-electron processes in which SET is precluded by the unambiguous d^0 electron configurations associated with the various cations in their conventional M^+ and M^{2+} oxidation states. Notably, however, there is a significant body of research in which reagents derived from group 1 and group 2 elements are implicated in stoichiometric SET and radical-based reactivity. Although these reactions are not catalytic in the group 1 reagent, sodium and potassium *tert*-butoxides, for example, have attracted significant recent attention as reagents to effect a wide range of transition metal-free inter- and intramolecular aryl and heteroaromatic cross-coupling reactions (Scheme 13.19) [80–97]. Although the radical initiation step continues to be the subject of some conjecture [92, 98–101], such "metal-free" reactivity, on the basis of electron paramagnetic resonance (EPR) evidence [102] and radical trapping experiments, is generally proposed to occur through the generation of radical anions.

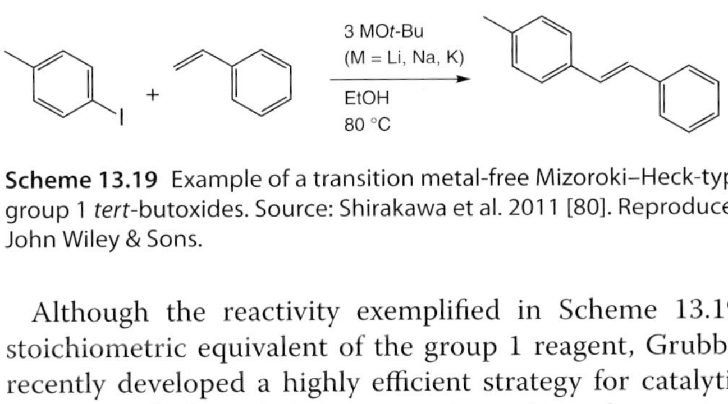

Scheme 13.19 Example of a transition metal-free Mizoroki–Heck-type reaction mediated by group 1 *tert*-butoxides. Source: Shirakawa et al. 2011 [80]. Reproduced with permission of John Wiley & Sons.

Although the reactivity exemplified in Scheme 13.19 requires at least a stoichiometric equivalent of the group 1 reagent, Grubbs and coworkers have recently developed a highly efficient strategy for catalytic C—Si bond formation through the KO*t*Bu-promoted reaction of arenes and heteroarenes with hydrosilanes (Scheme 13.20) [32, 103–106].

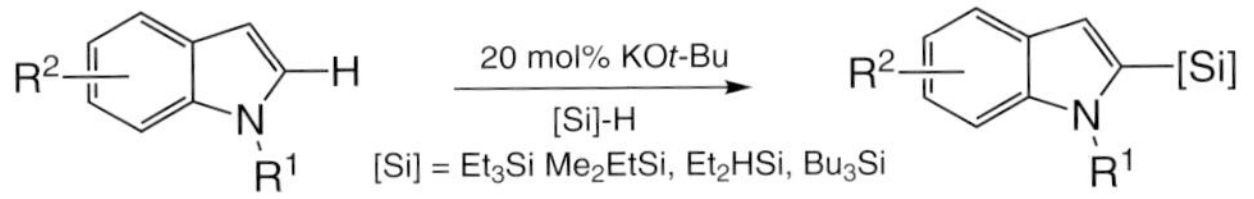

Scheme 13.20 The KO*t*-Bu-catalyzed Grubbs–Stoltz system for the silylation of aromatic heterocycles.

The cross-dehydrogenative process proceeds in the absence of hydrogen acceptors, ligands, or additives and is scalable to greater than 100 g under optionally solvent-free conditions for the silylation of a wide variety of N-, O-, and S-containing heteroarenes. In addition, the *ortho*-silylation of anisole and the directing group-free C(sp^3)–H silylation of toluene were achieved. Although extensive mechanistic studies have been performed, a consensus has yet to emerge and a number of plausible mechanisms, both ionic and heterolytic, have been proposed [107, 108]. The identification of ionic intermediates by electrospray ionization-mass spectrometry (ESI-MS) has led to the proposal of a mechanism dependent on the assembly of a key pentacoordinate silane reagent, the deprotonated aromatic substrate, and the *tert*-butoxide catalyst. Subsequent heterolysis of the Si—H bond, deprotonation of the heteroarene, addition of the heteroarene carbanion to the silyl ether, and dissociation of *tert*-butoxide from silicon are then proposed to lead to the silylated heteroarene product [107]. On the face of it, this may be viewed as a "conventional" metathesis-based dehydrocoupling analogous to that depicted in Scheme 13.2a. Further experiments, however, performed both in the presence of the radical inhibitor 2,2,6,6-tetramethylpiperidin-1-yl-oxyl (TEMPO) and radical clock experiments in conjunction with DFT analysis have implicated the generation of silyl radical intermediates and support C—Si bond formation through silyl radical addition and subsequent β–H scission to regenerate the silyl radical (Scheme 13.21) [108, 109].

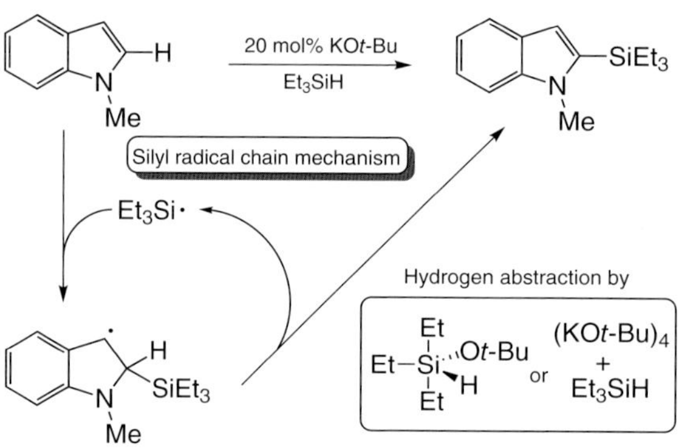

Scheme 13.21 Postulated silyl radical mechanism for the KO*t*-Bu-catalyzed silylation of aromatic heterocycles.

Stoichiometric processes have also dominated attempts to mediate the chemistry of group 2 reagents through SET. Although the reactivity of Jones' famous magnesium(I) reagents, such as the β-diketiminate derivative (**24**), most likely involve single-electron transfer, this stoichiometric chemistry will not be discussed and the interested reader is directed to more focused reviews [110, 111]. Grignard reagents, RMgX, however, have been found to react with 9-diazo-10-anthrone by SET [112], which occurs in competition with nucleophilic attack at a polar C=O bond. The group of Studer has conducted extensive research into the use of stoichiometric quantities of the nitroxyl radical TEMPO to effect oxidative C—C coupling of the organic residues of a variety of aryl, alkenyl, and aryl Grignard reagents (Scheme 13.22) [113–117], while Severin and

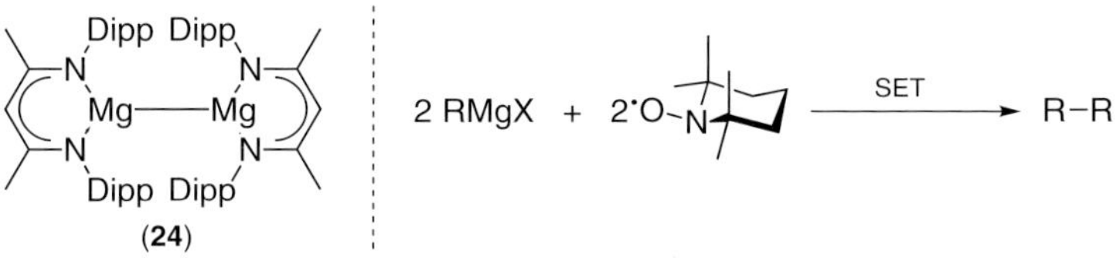

Scheme 13.22 The structure of compound **24** and the TEMPO-mediated oxidative C—C coupling of Grignard reagents.

coworkers has demonstrated that similar homocoupling may also be achieved through the use of nitrous oxide as the stoichiometric oxidant [118, 119].

Although in these latter cases it is likely that the resultant magnesium species are derivatives of reduced TEMPO anions, none of these compounds have been isolated. Mulvey and coworkers reported the first group 1 and 2 complexes of TEMPO, noting not only its coordination as a neutral two-electron donor but also its "chameleonic" ability to adopt an anionic electron configuration when ligating group 2 cations in complexes of the form $[\{(Me_3Si)N\}Mg\{\mu\text{-TEMPO}\}]_2$ [120]. A subsequent report from this group of SET from elemental group 1 metals to TEMPO yielded a structurally diverse series of [M{TEMPO}] compounds (M = Li, Na, K, Rb, and Cs) [121, 122]. The more widespread exploitation of redox-active ligands would, thus, appear a viable means with which to access multiple oxidation states for molecular systems derived from otherwise redox inactive s-block metals [123, 124]. Despite several reports of redox activity in di-imine-based s-block complexes [125–141], the only utilization of a catalytic quantity of a group 2 reagent in a redox-based system also employs TEMPO in conjunction with the β-diketiminato magnesium *n*-butyl and hydride reagents (**19** and **20**, respectively) [142]. Although both compounds reacted with an equimolar quantity of TEMPO to provide the same species, (**25**, Scheme 13.23a), no onward reaction with $PhSiH_3$ could be observed. Addition of a 10 mol% quantity of $[Mg\{N(SiMe_3)_2\}_2(THF)_2]$, however, to a mixture of silane and TEMPO produced H_2 and $TEMPOSi(H)_2Ph$, formed via a catalytic manifold reliant upon both SET and σ-bond metathesis (Scheme 13.23b)

13.5 "Beyond" Hydrofunctionalization and Dehydrocoupling

In common with the vast majority of the chemistry described in the earlier chapters of this monograph, all of the aforementioned reactivity is derived from the s-block-centered insertion of polarized multiple bonds or the activation of prepolarized σ-bonded Y—H substrates in which a p-block heteroatom (Y) is bonded to hydrogen. Although this "restriction" has given rise to a plethora of new processes over the past decade and a half, even greater catalytic diversity may be envisaged if the pathways depicted in Scheme 13.2 are further generalized to encompass the activation of nonhydrogenous heteroatom–heteroatom bonds. Although this aspect of s-block-based catalysis is very much in its nascent phase, examples are beginning to emerge [143].

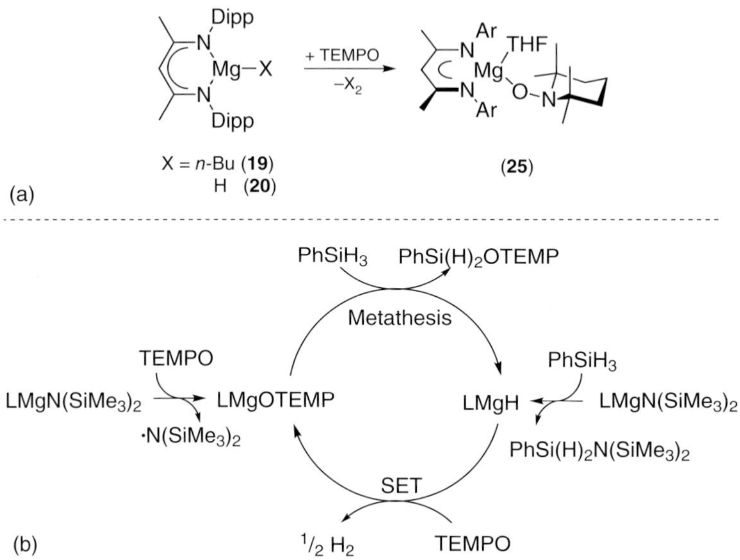

Scheme 13.23 (a) The synthesis of compound **25**; (b) Magnesium-catalyzed dehydrogenative silane oxidation by TEMPO.

In an advance that took the catalytic cross metathesis of heteroatom–heteroatom-bonded substrates "beyond dehydrocoupling," Hill and coworkers have reported that alkaline earth reagents catalyze boron–nitrogen bond formation through the "desilacoupling" of amines, RR′NH (R = alkyl, aryl; R′ = H, alkyl, aryl), and the commercially available silaborane, pinBSiMe$_2$Ph (**26**, Scheme 13.24a) [144]. The viability of a magnesium-centered activation of the silaborane substrate was established through its stoichiometric reaction with the organomagnesium derivative **19** (Scheme 13.24b). Although this process was again envisaged to proceed as a polarized Mg—C/Si—B σ-bond metathesis, the resultant magnesium silyl derivative (**27**) was found to be

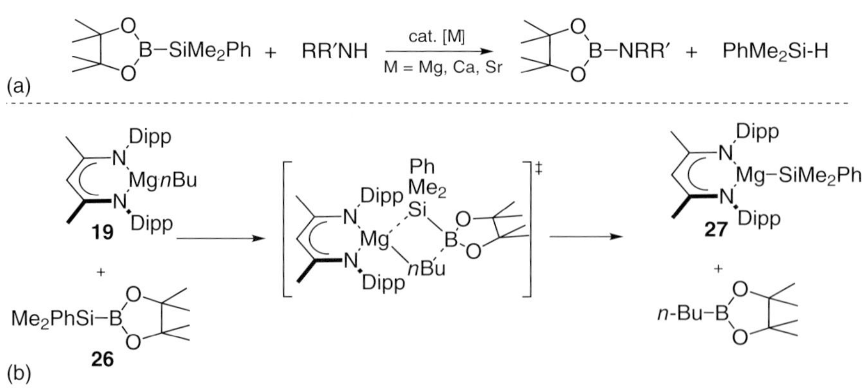

Scheme 13.24 (a) The alkaline earth-catalyzed "desilacoupling" of silaboranes and amines; (b) stoichiometric synthesis of the magnesium silyl, **27**.

unreactive toward N—H-bonded substrates. Application of the less sterically encumbered bis(trimethylsilyl)amides, $[M\{N(SiMe_3)_2\}_2(THF)_2]$ [Me = Mg, Ca, Sr], however, provided smooth conversion to the aminoborane products for a range of aliphatic and aromatic amines. Although some variation in rate was observed (Mg < Ca < Sr) across the three catalytic systems, in all instances, catalysis also yielded a stoichiometric quantity of Me_2PhSiH to provide the first examples of catalytic main group element–element coupling that are not dependent on the concurrent elimination of H_2.

In a similar vein, the first examples of the application of s-block reagents for the catalytic addition of the nonhydrogeneous Si—C-bonded substrate Me_3SiCN to polarized C=O multiple bonds have been very recently described. Sen and coworkers have reported that the amidinatocalcium iodide derivative, compound **28**, which had previously been shown to effect the hydroboration of aldehydes and ketones, also enables the "cyanosilylation" of a similar range of substrates (Scheme 13.25) [145]. The catalysis was found to be very effective for a wide range of carbonyl compounds under ambient conditions. As neither the substrate nor the precatalyst contains a reactive Y—H or M—X bond, the authors suggested that the mechanism differs from the metathesis/insertion sequence (Scheme 13.2b) proposed for many calcium-catalyzed hereofunctionalization processes. Rather, experimental and computational studies led the authors to conclude that carbonyl attack on the silicon center is facilitated by polarization of the Si—CN bond at the Lewis acidic calcium center. Although this viewpoint is likely to hold some validity, it is notable that the identical transformation may also be catalyzed by the magnesium(I) species, **24**, with loadings as low as 0.1 mol% [146]. In this latter very recent report, however, no mechanistic evidence was presented despite the catalysis being claimed to provide the first example of a "truly catalytic application" of a Mg(I) complex.

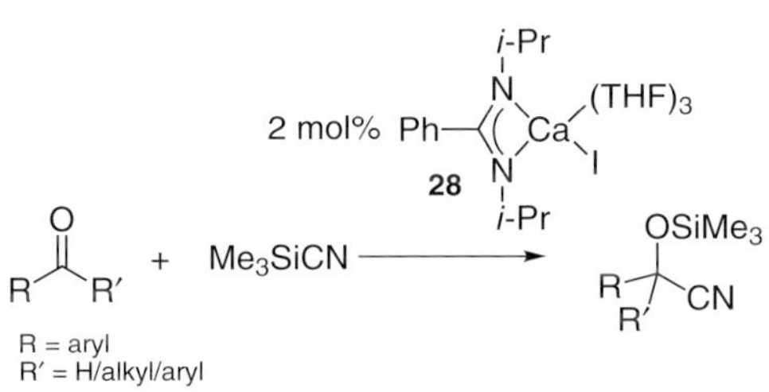

Scheme 13.25 Carbonyl cyanosilylation catalyzed by the amidinatocalcium complex, **28**.

13.6 Conclusions and Conjecture

Although the chemistry outlined in this chapter is necessarily disparate, some attempt has been made to place the reactivity described into the broad categories framed by Schemes 13.1 and 13.2 and the introductory comments. Although it can be realistically claimed that s-block-derived catalysis has now emerged as a field of study in its own right, the advances of the last decade do not threaten the current pre-eminence of transition metals. The range, specificity, and potency of the catalytic reactivity described, not just in this chapter but elsewhere in this monograph, remain comparably narrow. From this perspective, the challenges presented by some of "holy grails" of homogeneous molecular catalysis (e.g. the selective oxidation or transformation of C(sp^3)—H bonds and the productive reduction of dinitrogen) might appear as insuperable as ever. If the relative maturity of both areas of study is also taken into account, however, the portents are good for continued rapid expansion of the variety of reaction types, which are amenable to transformation under group 1- or 2-mediated catalysis. This chapter has emphasized that the range of applicable substrates, whether singly or multiply bonded, generally possess a prepolarized disposition. Although this is true for all cases of catalysis aside from the hydrogenation chemistry cited in the introduction, recent advances have demonstrated that the superficially similar heterolytic activation of the nonpolar B—B bonds of diborane(4) substrates may be achieved with stoichiometric quantities of potassium- or magnesium-based reagents such as **19** (Scheme 13.26) [147, 148]. Similar, albeit limited, precedent also exists for the calcium-centered activation of thermodynamically robust and kinetically inert C—F and C(sp^2)—H bonds by calcium n-alkyls such as **29** [14, 149], selective

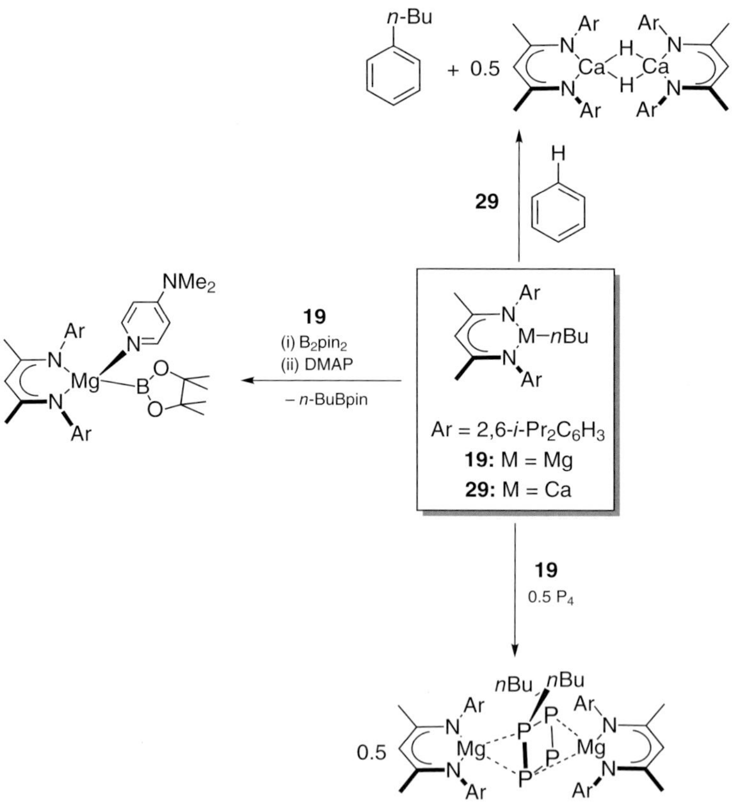

Scheme 13.26 Catalytically unexploited group 2-mediated reactivity.

cleavage of P—P bonds within elemental white phosphorus by compound **19** [150], and the reversible insertion of C=C bonds into the Mg—Mg bond of compound **24** [151]. None of this reactivity has yet been realized in a catalytic system but amply highlights the future potential of even the currently existing reagent types. Although the intent of this chapter was to provide coverage of "one off" demonstrations of s-block catalysis, which may have escaped attention elsewhere, the author hopes that it also serves to highlight the abundant opportunities that exist. The elaboration of more sophisticated or stereoinfluential ligand systems, the development of further redox-based protocols, and the more widespread catalytic application of mixed metal species are obvious areas for growth and which will naturally diversify the scope of the applicable reactivity. From this standpoint, it is self-evident that, although some groundwork has been performed, the real challenges remain to be met.

References

1 Harder, S. (2010). *Chem. Rev.* 110: 3852–3876.
2 Barrett, A.G.M., Crimmin, M.R., Hill, M.S., and Procopiou, P.A. (2010). *Proc. R. Soc. A Math. Phys. Eng. Sci.* 466: 927–963.
3 Crimmin, M.R. and Hill, M.S. (2013). Homogeneous catalysis with organometallic complexes of group 2. In: *Alkaline-Earth Metal Compounds: Oddities and Applications*, vol. 45 (ed. S. Harder), 191–241. Heidelberg: Springer.
4 Westerhausen, M. (2009). *Z. Anorg. Allg. Chem.* 635: 13–32.
5 Krieck, S. and Westerhausen, M. (2017). *Inorganics* 5: 17.
6 Sarazin, Y. and Carpentier, J.F. (2016). *Chem. Rec.* 16: 2482–2505.
7 Alexander, J.S. and Ruhlandt-Senge, K. (2002). *Eur. J. Inorg. Chem.* 2002: 2761–2774.
8 Alexander, J.S. and Ruhlandt-Senge, K. (2004). *Chem. Eur. J.* 10: 1274–1280.
9 Buchanan, W.D., Allis, D.G., and Ruhlandt-Senge, K. (2010). *Chem. Commun.* 46: 4449–4465.
10 Rochat, R., Lopez, M.J., Tsurugi, H., and Mashima, K. (2016). *ChemCatChem* 8: 10–20.
11 Waterman, R. (2013). *Organometallics* 32: 7249–7263.
12 Barrett, A.G.M., Brinkmann, C., Crimmin, M.R. et al. (2009). *J. Am. Chem. Soc.* 131: 12906–12907.
13 Schuhknecht, D., Lhotzky, C., Spaniol, T.P. et al. (2017). *Angew. Chem. Int. Ed.* 56: 12367–12371.
14 Wilson, A.S.S., Hill, M.S., Mahon, M.F. et al. (2017). *Science* 358: 1168–1171.
15 Weeber, A., Harder, S., Brintzinger, H.H., and Knoll, K. (2000). *Organometallics* 19: 1325–1332.
16 Harder, S., Feil, F., and Knoll, K. (2001). *Angew. Chem. Int. Ed.* 40: 4261–4264.
17 Harder, S., Feil, F., and Weeber, A. (2001). *Organometallics* 20: 1044–1046.
18 Harder, S. and Feil, F. (2002). *Organometallics* 21: 2268–2274.

19 Piesik, D.F.J., Habe, K., and Harder, S. (2007). *Eur. J. Inorg. Chem.* 2007: 5652–5661.
20 Spielmann, J., Buch, F., and Harder, S. (2008). *Angew. Chem. Int. Ed.* 47: 9434–9438.
21 Crimmin, M.R., Barrett, A.G.M., Hill, M.S. et al. (2007). *Organometallics* 26: 2953–2956.
22 Crimmin, M.R., Arrowsmith, M., Barrett, A.G.M. et al. (2009). *J. Am. Chem. Soc.* 131: 9670–9685.
23 Dunne, J.F., Fulton, D.B., Ellern, A., and Sadow, A.D. (2010). *J. Am. Chem. Soc.* 132: 17680–17683.
24 Melen, R.L. (2016). *Chem. Soc. Rev.* 45: 775–788.
25 Crimmin, M.R., Casely, I.J., and Hill, M.S. (2005). *J. Am. Chem. Soc.* 127: 2042–2043.
26 Harder, S. and Spielmann, J. (2012). *J. Organomet. Chem.* 698: 7–14.
27 Liptrot, D.J., Hill, M.S., Mahon, M.F., and Wilson, A.S.S. (2015). *Angew. Chem. Int. Ed.* 54: 13362–13365.
28 Bellini, C., Carpentier, J.F., Tobisch, S., and Sarazin, Y. (2015). *Angew. Chem. Int. Ed.* 54: 7679–7683.
29 Leich, V., Spaniol, T.P., and Okuda, J. (2015). *Chem. Commun.* 51: 14772–14774.
30 Bauer, H., Alonso, M., Farber, C. et al. (2018). *Nat. Catal.* 1: 40–47.
31 Zeng, G.X. and Li, S.H. (2010). *Inorg. Chem.* 49: 3361–3369.
32 Toutov, A.A., Liu, W.-B., Betz, K.N. et al. (2015). *Nature* 518: 80–84.
33 Werner, T. and Koch, J. (2010). *Eur. J. Org. Chem.* 2010: 6904–6907.
34 Guo, Z.Q., Wei, X.H., Tong, H.B. et al. (2015). *J. Organomet. Chem.* 783: 73–76.
35 Mojtahedi, M.M., Akbarzadeh, E., Sharifi, R., and Abaee, M.S. (2007). *Org. Lett.* 9: 2791–2793.
36 Rajesh, K. and Berke, H. (2013). *Adv. Synth. Catal.* 355: 901–906.
37 Waddell, D.C. and Mack, J. (2009). *Green Chem.* 11: 79–82.
38 Crimmin, M.R., Barrett, A.G.M., Hill, M.S., and Procopiou, P.A. (2007). *Org. Lett.* 9: 331–333.
39 Barman, M.K., Baishya, A., and Nembenna, S. (2015). *J. Organomet. Chem.* 785: 52–60.
40 Day, B.M., Mansfield, N.E., Coles, M.P., and Hitchcock, P.B. (2011). *Chem. Commun.* 47: 4995–4997.
41 Day, B.M., Knowelden, W., and Coles, M.P. (2012). *Dalton Trans.* 41: 10930–10933.
42 Cronin, L., Manoni, F., O'Connor, C.J., and Connon, S.J. (2010). *Angew. Chem. Int. Ed.* 49: 3045–3048.
43 Berberich, H. and Roesky, P.W. (1998). *Angew. Chem. Int. Ed.* 37: 1569–1571.
44 Deacon, G.B., Gitlits, A., Roesky, P.W. et al. (2001). *Chem. Eur. J.* 7: 127–138.
45 Burgstein, M.R., Berberich, H., and Roesky, P.W. (2001). *Chem. Eur. J.* 7: 3078–3085.
46 Zuyls, A., Roesky, P.W., Deacon, G.B. et al. (2008). *Eur. J. Org. Chem.* 2008: 693–697.
47 Shi, J.C., Guo, Z.Q., Wei, X.H. et al. (2011). *Synlett*: 1937–1939.

48 Liu, Q., Guo, Z.Q., Han, H.F. et al. (2015). *Polyhedron* 85: 15–19.
49 Guo, Z.Q., Wang, S., Tong, H.B. et al. (2013). *Inorg. Chem. Commun.* 33: 68–72.
50 Guo, Z.Q., Wang, Y.K., Yang, J.H., and Wei, X.H. (2017). *Polyhedron* 130: 13–17.
51 Hernan-Gomez, A., Bradley, T.D., Kennedy, A.R. et al. (2013). *Chem. Commun.* 49: 8659–8661.
52 Orzechowski, L., Jansen, G., and Harder, S. (2006). *J. Am. Chem. Soc.* 128: 14676–14684.
53 Orzechowski, L. and Harder, S. (2007). *Organometallics* 26: 2144–2148.
54 Brinkmann, C., Barrett, A.G.M., Hill, M.S. et al. (2012). *Organometallics* 31: 7287–7297.
55 Li, D.Y., Shi, K.J., Mao, X.F. et al. (2014). *Tetrahedron* 70: 7022–7031.
56 Barrett, A.G.M., Crimmin, M.R., Hill, M.S. et al. (2009). *Chem. Commun.* (17): 2299–2301.
57 Arrowsmith, M., Crimmin, M.R., Hill, M.S. et al. (2013). *Organometallics* 32: 4961–4972.
58 Avent, A., Crimmin, M., Hill, M., and Hitchcock, P. (2005). *Organometallics* 24: 1184–1188.
59 Lee, L., Berg, D.J., and Bushnell, G.W. (1995). *Organometallics* 14: 5021–5023.
60 Heeres, H.J., Nijhoff, J., Teuben, J.H., and Rogers, R.D. (1993). *Organometallics* 12: 2609–2617.
61 Rochat, R., Yamamoto, K., Lopez, M.J. et al. (2015). *Chem. Eur. J.* 21: 8112–8120.
62 Ong, T.-G., O'Brien, J.S., Korobkov, I., and Richeson, D.S. (2006). *Organometallics* 25: 4728–4730.
63 Barrett, A.G.M., Crimmin, M.R., Hill, M.S. et al. (2008). *Organometallics* 27: 6300–6306.
64 Arrowsmith, M., Crimmin, M.R., Hill, M.S. et al. (2014). *Dalton Trans.* 43: 14249–14256.
65 Schwamm, R.J. and Coles, M.P. (2013). *Organometallics* 32: 5277–5280.
66 Barrett, A.G.M., Boorman, T.C., Crimmin, M.R. et al. (2008). *Chem. Commun.* (41): 5206–5208.
67 Hill, M.S., Liptrot, D.J., and Mahon, M.F. (2013). *Angew. Chem. Int. Ed.* 52: 5364–5367.
68 Arrowsmith, M., Shepherd, W.M.S., Hill, M.S., and Kociok-Kohn, G. (2014). *Chem. Commun.* 50: 12676–12679.
69 Arrowsmith, M., Hill, M.S., and Kociok-Kohn, G. (2015). *Chem. Eur. J.* 21: 10548–10557.
70 Berkefeld, A., Piers, W.E., Parvez, M. et al. (2013). *Chem. Sci.* 4: 2152–2162.
71 Anker, M.D., Arrowsmith, M., Bellham, P. et al. (2014). *Chem. Sci.* 5: 2826–2830.
72 Anker, M.D., Arrowsmith, M., Arrowsmith, R.L. et al. (2017). *Inorg. Chem.* 56: 5976–5983.
73 Mukherjee, D., Osseili, H., Spaniol, T.P., and Okuda, J. (2016). *J. Am. Chem. Soc.* 138: 10790–10793.

74 Mukherjee, D., Shirase, S., Spaniol, T.P. et al. (2016). *Chem. Commun.* 52: 13155–13158.
75 Rauch, M. and Parkin, G. (2017). *J. Am. Chem. Soc.* 139: 18162–18165.
76 Yang, Y., Anker, M.D., Fang, J. et al. (2017). *Chem. Sci.* 8: 3529–3537.
77 Lalrempuia, R., Kefalidis, C.E., Bonyhady, S.J. et al. (2015). *J. Am. Chem. Soc.* 137: 8944–8947.
78 Anker, M.D., Hill, M.S., Lowe, J.P., and Mahon, M.F. (2015). *Angew. Chem. Int. Ed.* 54: 10009–10011.
79 Anker, M.D., Kefalidis, C.E., Yang, Y. et al. (2017). *J. Am. Chem. Soc.* 139: 10036–10054.
80 Shirakawa, E., Zhang, X.J., and Hayashi, T. (2011). *Angew. Chem. Int. Ed.* 50: 4671–4674.
81 Sun, C.L. and Shi, Z.J. (2014). *Chem. Rev.* 114: 9219–9280.
82 Studer, A. and Curran, D.P. (2014). *Nat. Chem.* 6: 765–773.
83 Shang, X.J. and Liu, Z.Q. (2016). *Org. Biomol. Chem.* 14: 7829–7831.
84 Yanagisawa, S., Ueda, K., Taniguchi, T., and Itami, K. (2008). *Org. Lett.* 10: 4673–4676.
85 Guastavino, J.F., Buden, M.E., and Rossi, R.A. (2014). *J. Org. Chem.* 79: 9104–9111.
86 Buden, M.E., Guastavino, J.F., and Rossi, R.A. (2013). *Org. Lett.* 15: 1174–1177.
87 Roman, D.S., Takahashi, Y., and Charette, A.B. (2011). *Org. Lett.* 13: 3242–3245.
88 Bhakuni, B.S., Kumar, A., Balkrishna, S.J. et al. (2012). *Org. Lett.* 14: 2838–2841.
89 De, S., Mishra, S., Kakde, B.N. et al. (2013). *J. Org. Chem.* 78: 7823–7844.
90 Cuthbertson, J., Gray, V.J., and Wilden, J.D. (2014). *Chem. Commun.* 50: 2575–2578.
91 Liu, W., Cao, H., Zhang, H. et al. (2010). *J. Am. Chem. Soc.* 132: 16737–16740.
92 Zhou, S.Z., Anderson, G.M., Mondal, B. et al. (2014). *Chem. Sci.* 5: 476–482.
93 Zhou, S.Z., Doni, E., Anderson, G.M. et al. (2014). *J. Am. Chem. Soc.* 136: 17818–17826.
94 Chan, T.L., Wu, Y.N., Choy, P.Y., and Kwong, F.Y. (2013). *Chem. Eur. J.* 19: 15802–15814.
95 Beyer, A., Buendia, J., and Bolm, C. (2012). *Org. Lett.* 14: 3948–3951.
96 Doni, E., Zhou, S.Z., and Murphy, J.A. (2015). *Molecules* 20: 1755–1774.
97 Liu, X., Zhao, X., Liang, F.S., and Ren, B.Y. (2018). *Org. Biomol. Chem.* 16: 886–890.
98 Stephens, D.E., Lakey-Beitia, J., Burch, J.E. et al. (2016). *Chem. Commun.* 52: 9945–9948.
99 Barham, J.P., Coulthard, G., Emery, K.J. et al. (2016). *J. Am. Chem. Soc.* 138: 7402–7410.
100 Zhang, L., Yang, H., and Jiao, L. (2016). *J. Am. Chem. Soc.* 138: 7151–7160.
101 Patil, M. (2016). *J. Org. Chem.* 81: 632–639.
102 Yi, H., Jutand, A., and Lei, A.W. (2015). *Chem. Commun.* 51: 545–548.
103 Toutov, A.A., Salata, M., Fedorov, A. et al. (2017). *Nat. Energy* 2: 17008.

104 Toutov, A.A., Betz, K.N., Schuman, D.P. et al. (2017). *J. Am. Chem. Soc.* 139: 1668–1674.

105 Toutov, A.A., Betz, K.N., Haibach, M.C. et al. (2016). *Org. Lett.* 18: 5776–5779.

106 Toutov, A.A., Liu, W.B., Betz, K.N. et al. (2015). *Nat. Protoc.* 10: 1897–1903.

107 Banerjee, S., Yang, Y.F., Jenkins, I.D. et al. (2017). *J. Am. Chem. Soc.* 139: 6880–6887.

108 Liu, W.B., Schuman, D.P., Yang, Y.F. et al. (2017). *J. Am. Chem. Soc.* 139: 6867–6879.

109 Smith, A.J., Young, A., Rohrbach, S. et al. (2017). *Angew. Chem. Int. Ed.* 56: 13747–13751.

110 Stasch, A. and Jones, C. (2011). *Dalton Trans.* 40: 5659–5672.

111 Jones, C. (2017). *Nat. Rev. Chem.* 1: 0059.

112 Bruni, P., Carloni, P., Conti, C. et al. (1996). *Tetrahedron* 52: 6795–6802.

113 Murarka, S. and Studer, A. (2011). *Adv. Synth. Catal.* 353: 2708–2714.

114 Maji, M.S., Murarka, S., and Studer, A. (2010). *Org. Lett.* 12: 3878–3881.

115 Maji, M.S., Pfeifer, T., and Studer, A. (2010). *Chem. Eur. J.* 16: 5872–5875.

116 Maji, M.S. and Studer, A. (2009). *Synth. Stuttgart* 2009: 2467–2470.

117 Maji, M.S., Pfeifer, T., and Studer, A. (2008). *Angew. Chem. Int. Ed.* 47: 9547–9550.

118 Tskhovrebov, A.G., Solari, E., Scopelliti, R., and Severin, K. (2014). *Organometallics* 33: 2405–2408.

119 Kiefer, G., Jeanbourquin, L., and Severin, K. (2013). *Angew. Chem. Int. Ed.* 52: 6302–6305.

120 Forbes, G.C., Kennedy, A.R., Mulvey, R.E., and Rodger, P.J.A. (2001). *Chem. Commun.* (15): 1400–1401.

121 Armstrong, D.R., Balloch, L., Crawford, J.J. et al. (2012). *Chem. Commun.* 48: 1541–1543.

122 Balloch, L., Drummond, A.M., Garcia-Alvarez, P. et al. (2009). *Inorg. Chem.* 48: 6934–6944.

123 Munha, R.F., Zarkesh, R.A., and Heyduk, A.F. (2013). *Dalton Trans.* 42: 3751–3766.

124 Chirik, P.J. and Wieghardt, K. (2010). *Science* 327: 794–795.

125 Fedushkin, I.L., Khvoinova, N.M., Skatova, A.A., and Fukin, G.K. (2003). *Angew. Chem. Int. Ed.* 42: 5223–5226.

126 Fedushkin, I.L., Skatova, A.A., Cherkasov, V.K. et al. (2003). *Chem. Eur. J.* 9: 5778–5783.

127 Fedushkin, I.L., Skatova, A.A., Chudakova, V.A. et al. (2003). *Eur. J. Inorg. Chem.* 2003: 3336–3346.

128 Fedushkin, I.L., Chudakova, V.A., Fukin, G.K. et al. (2004). *Russ. Chem. Bull.* 53: 2744–2750.

129 Fedushkin, I.L., Chudakova, V.A., Skatova, A.A. et al. (2004). *Z. Anorg. Allg. Chem.* 630: 501–507.

130 Fedushkin, I.L., Skatova, A.A., Lukoyanova, A.N. et al. (2004). *Russ. Chem. Bull.* 53: 2751–2762.

131 Fedushkin, I.L., Chudakova, V.A., Skatova, A.A., and Fukin, G.K. (2005). *Heteroat. Chem.* 16: 663–670.

132 Fedushkin, I.L., Morozov, A.G., Rassadin, O.V., and Fukin, G.K. (2005). *Chem. Eur. J.* 11: 5749–5757.
133 Fedushkin, I.L., Skatova, A.A., Fukin, G.K. et al. (2005). *Eur. J. Inorg. Chem.* 2005: 2332–2338.
134 Fedushkin, I.L., Skatova, A.A., Hummert, M., and Schumann, H. (2005). *Eur. J. Inorg. Chem.* 2005: 1601–1608.
135 Fedushkin, I.L., Chudakova, V.A., Skatova, A.A., and Fukin, G.K. (2006). *Heteroat. Chem* 17: 618–618.
136 Fedushkin, I.L., Makarov, V.M., Rosenthal, E.C.E., and Fukin, G.K. (2006). *Eur. J. Inorg. Chem.* 2006: 827–832.
137 Fedushkin, I.L., Chudakova, V.A., Hummert, M., and Schumann, H. (2008). *Z. Naturforsch., B: J. Chem. Sci.* 63: 161–168.
138 Fedushkin, I.L., Morozov, A.G., Chudakova, V.A. et al. (2009). *Eur. J. Inorg. Chem.* 2009: 4995–5003.
139 Fedushkin, I.L., Nikipelov, A.S., Skatova, A.A. et al. (2009). *Eur. J. Inorg. Chem.* 2009: 3742–3749.
140 Razborov, D.A., Lukoyanov, A.N., Baranov, E.V., and Fedushkin, I.L. (2015). *Dalton Trans.* 44: 20532–20541.
141 Fedushkin, I.L., Skatova, A.A., Bazyakina, N.L. et al. (2013). *Russ. Chem. Bull.* 62: 1815–1828.
142 Liptrot, D.J., Hill, P.M.S., and Mahon, M.F. (2014). *Angew. Chem. Int. Ed.* 53: 6224–6227.
143 Hill, M.S., Mahon, M.F., and Robinson, T.P. (2010). *Chem. Commun.* 46: 2498–2500.
144 Liptrot, D.J., Arrowsmith, M., Colebatch, A.L. et al. (2015). *Angew. Chem. Int. Ed.* 54: 15280–15283.
145 Yadav, S., Dixit, R., Vanka, K., and Sen, S.S. (2018). *Chem. Eur. J.* 24: 1269–1273.
146 Wang, W.F., Luo, M., Li, J. et al. (2018). *Chem. Commun.* 54: 3042–3044.
147 Neeve, E.C., Geier, S.J., Mkhalid, I.A.I. et al. (2016). *Chem. Rev.* 116: 9091–9161.
148 Pecharman, A.F., Colebatch, A.L., Hill, M.S. et al. (2017). *Nat. Commun.* 8: 15022.
149 Barrett, A.G.M., Crimmin, M.R., Hill, M.S. et al. (2007). *Angew. Chem. Int. Ed.* 46: 6339–6342.
150 Arrowsmith, M., Hill, M.S., Johnson, A.L. et al. (2015). *Angew. Chem. Int. Ed.* 54: 7882–7885.
151 Boutland, A.J., Carroll, A., Lamsfus, C.A. et al. (2017). *J. Am. Chem. Soc.* 139: 18190–18193.

Index

Early Main Group Metal Catalysis: Concepts and Reactions,
First Edition. Edited by Sjoerd Harder.

h